Student & Parent

One-Stop Interne

Log on to

www.mathmatters1.com

Online Study Tools

- Extra Examples
- Self-Check Quizzes
- Chapter Assessment Practice
- Standardized Test Practice

Online Resources

- Math*Works* Careers
- Links to Chapter Themes
- Math Behind the Market
- Multilingual Glossary

GLENCOE
MATHEMATICS

MathMatters 1

An Integrated Program

mathmatters1.com

Lynch

Olmstead

De Forest-Davis

Glencoe

New York, New York
Columbus, Ohio
Chicago, Illinois
Peoria, Illinois
Woodland Hills, California

Glencoe

The *McGraw·Hill* Companies

Send all inquiries to:
Glencoe/McGraw-Hill
8787 Orion Place
Columbus, OH 43240-4027

ISBN: 0-07-868174-X   *MathMatters 1 Student Edition*

2   3   4   5   6   7   8   9   10   058/111   13   12   11   10   09   08   07   06   05

Contents in Brief

Chicha Lynch currently teaches Honors Advanced Algebra II at Marin Catholic High School in Kentfield, California. She is a graduate of the University of Florida. She was a state finalist in 1988 for the Presidential Award for Excellence in mathematics teaching. Currently, Ms. Lynch is a participating member of the National Council of Teachers of Mathematics as well as a long-time member of California Math Council North.

Eugene Olmstead is a mathematics teacher at Elmira Free Academy in Elmira, New York. He earned his B.S. in Mathematics at State University College at Geneseo in New York. In addition to teaching high school, Mr. Olmstead is an instructor for T^3, Teachers Teaching with Technology, and has participated in writing several of the T^3 Institutes. In 1991 and 1992, Mr. Olmstead was selected as a state finalist for the Presidential Award for Excellence in mathematics teaching.

Kenneth De Forest-Davis is the mathematics department chairperson at Beloit Memorial High School in Beloit, Wisconsin. Mr. De Forest-Davis earned his B.A. in Mathematics and Computer Science from Beloit College in Wisconsin. He later completed his M.A.T from Beloit College, where he received the Von-Eschen-Steele Excellence in Teaching Award.

Reviewers and Consultants

These educators reviewed every chapter and gave suggestions for improving the effectiveness of the mathematics instruction.

Tamara L. Amundsen
Teacher
Windsor Forest High School
Savannah, Georgia

Kyle A. Anderson
Mathematics Teacher
Waiakea High School
Milo, Hawaii

Murney Bell
Mathematics and Science
    Teacher
Anchor Bay High School
New Baltimore, Michigan

Fay Bonacorsi
High School Math Teacher
Lafayette High School
Brooklyn, New York

Boon C. Boonyapat
Mathematics Department
    Chairman
Henry W. Grady High School
Atlanta, Georgia

Peggy A. Bosworth
Retired Math Teacher
Plymouth-Canton High School
Canton, Michigan

Sandra C. Burke
Mathematics Teacher
Page High School
Page, Arizona

Jill Conrad
Math Teacher
Crete Public Schools
Crete, Nebraska

Nancy S. Cross
Math Educator
Merritt High School
Merritt Island, Florida

Mary G. Evangelista
Chairperson, Mathematics
    Department
Grove High School
Garden City, Georgia

Timothy J. Farrell
Teacher of Mathematics and
    Physical Science
Perth Amboy Adult School
Perth Amboy, New Jersey

Greg A. Faulhaber
Mathematics and Computer
    Science Teacher
Winton Woods High School
Cincinnati, Ohio

Leisa Findley
Math Teacher
Carson High School
Carson City, Nevada

Linda K. Fiscus
Mathematics Teacher
New Oxford High School
New Oxford, Pennsylvania

Louise M. Foster
Teacher and Mathematics
    Department Chairperson
Frederick Douglass High School
Altanta, Georgia

Darleen L. Gearhart
Mathematics Curriculum
    Specialist
Newark Public Schools
Newark, New Jersey

Faye Gunn
Teacher
Douglass High School
Atlanta, Georgia

Dave Harris
Math Department Head
Cedar Falls High School
Cedar Falls, Iowa

Barbara Heinrich
Teacher
Wauconda High School
Wauconda, Illinois

Margie Hill
District Coordinating Teacher
    Mathematics, K-12
Blue Valley School District
    USD229
Overland Park, Kansas

Suzanne E. Hills
Mathematics Teacher
Halifax Area High School
Halifax, Pennsylvania

Robert J. Holman
Mathematics Department
St. John's Jesuit High School
Toledo, Ohio

Eric Howe
Applied Math Graduate Student
Air Force Institute of Technology
Dayton, Ohio

Daniel R. Hudson
Mathematics Teacher
Northwest Local School District
Cincinnati, Ohio

Susan Hunt
Math Teacher
Del Norte High School
Albuquerque, New Mexico

continued

Todd J. Jorgenson
Secondary Mathematics
   Instructor
Brookings High School
Brookings, South Dakota

Susan H. Kohnowich
Math Teacher
Hartford High School
White River Junction, Vermont

Mercedes Kriese
Chairperson, Mathematics
   Department
Neenah High School
Neenah, Wisconsin

Kathrine Lauer
Mathematics Teacher
Decatur High School
Federal Way, Washington

Laurene Lee
Mathematics Instructor
Hood River Valley High School
Hood River, Oregon

Randall P. Lieberman
Math Teacher
Lafayette High School
Brooklyn, New York

Scott Louis
Mathematics Teacher
Elder High School
Cincinnati, Ohio

Dan Lufkin
Mathematics Instructor
Foothill High School
Pleasanton, California

Gary W. Lundquist
Teacher
Macomb Community College
Warren, Michigan

Evelyn A. McDaniel
Mathematics Teacher
Natrona County High School
Casper, Wyoming

Lin McMullin
Educational Consultant
Ballston Spa, New York

Margaret H. Morris
Mathematics Instructor
Saratoga Springs Senior High
   School
Saratoga Springs, New York

Tom Muchlinski
Mathematics Resource Teacher
Wayzata Public Schools
Plymouth, Minnesota

Andy Murr
Mathematics
Wasilla High School
Wasilla, Alaska

Janice R. Oliva
Mathematics Teacher
Maury High School
Norfolk, VA

Fernando Rendon
Mathematics Teacher
Tucson High Magnet
   School/Tucson Unified
   School District #1
Tucson, Arizona

Candace Resmini
Mathematics Teacher
Belfast Area High School
Belfast, Maine

Kathleen A. Rooney
Chairperson, Mathematics
   Department
Yorktown High School
Arlington, Virginia

Mark D. Rubio
Mathematics Teacher
Hoover High School—GUSD
Glendale, California

Tony Santilli
Chairperson, Mathematics
   Department
Godwin Heights High School
Wyoming, Michigan

Michael Schlomer
Mathematics Department Chair
Elder High School
Cincinnati, Ohio

Jane E. Swanson
Math Teacher
Warren Township High School
Gurnee, Illinois

Martha Taylor
Teacher
Jesuit College Preparatory School
Dallas, Texas

Cheryl A. Turner
Chairperson, Mathematics
   Department
LaQuinta High School
LaQuinta, California

Linda Wadman
Instructor, Mathematics
Cut Bank High School
Cut Bank, Montana

George K. Wells
Coordinator of Mathematics
Mt. Mansfield Union
   High School
Jericho, Vermont

TABLE OF CONTENTS

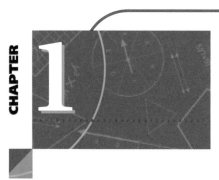

CHAPTER 1

Theme: Health and Fitness

Math*Works* Careers
Nutritionist 15
Recreation Worker 33

Applications
advertising 36
community service 13, 21
fitness 7, 18, 39
health 21
manufacturing 7, 27
market research 36
marketing 35
nutrition 7, 26
retail 9, 12, 17, 31
safety 9
sports 11, 13, 27, 34

Standardized Test Practice
Multiple Choice 46
Short Response/Grid In 47
Extended Response 47

Data and Graphs

Page 2

Measurement

Theme: Travel and Vacation

Math*Works* Careers
Concierge 61
Cartographer 79

Applications
agriculture 73
architecture 65, 73, 83
construction 69
entertainment 59, 76, 87
fashion 54, 57
fitness 58
hobbies 59, 69
horticulture 75
interior design 64, 93
landscaping 66, 72, 82
manufacturing 93
recreation 92
retail 77
sports 67, 68, 69, 82, 83, 86
travel 54, 59, 65

Standardized Test Practice
Multiple Choice 98
Short Response/Grid In 99
Extended Response 99

Page 61

Real Numbers and Variable Expressions

Theme: **Sports**

Math*Works* Careers
Sports Statistician 113
Fitness Trainer 131

Applications
astronomy 135, 138
entertainment 129, 135
finance 107, 126, 129
fitness 116
food service 120
gardening 111
hobbies 106, 122
retail 117
safety 145
sports 107, 110, 117, 121, 126,
128, 129, 134, 135, 138, 139,
144, 145
travel 115
weather 107, 109, 111, 126

Standardized Test Practice
Multiple Choice 150
Short Response/Grid In 151
Extended Response 151

Page 100

Two- and Three-Dimensional Geometry

Theme: Retail Marketing

Math*Works* Careers
Packaging Designers 165
Billboard Assembler 183

Applications

Standardized Test Practice

Page 153

Equations and Inequalities

Theme: Recycling

Math*Works* Careers
Solid Waste Disposal Staff 217
Financial Analyst 237

Applications
business 234
environment 243
finance 210, 248
fitness 221, 234
food service 231
hobbies 214, 249
industry 221
market research 231
part-time job 219, 230
photography 224
recycling 211, 214, 224, 235,
   239, 249
retail 213, 214, 234, 241, 248
safety 230
sports 209, 214, 225, 242
travel 210, 220, 235
transportation 247
weather 233

Standardized Test Practice
Multiple Choice 254
Short Response/Grid In 255
Extended Response 255

Page 204

Equations and Percents

Theme: Living on Your Own

Math*Works* Careers
Financial Counselor 269
Political Scientist 289

Applications
business 283
entertainment 265
finance 267, 283
fitness 291
food service 295
living on your own 263, 267,
   273, 277, 283, 287, 293, 295
part-time job 276
real estate 262, 287, 292
retail 261, 266, 273, 277, 286,
   292
sports 266
spreadsheets 277
travel 263, 272, 283, 292
weather 262

Standardized Test Practice
Multiple Choice 300
Short Response/Grid In 301
Extended Response 301

Page 271

7

Functions and Graphs

Theme: Velocity

Math*Works* Careers
Truck Driver 313
Ship Captain 333

Applications
architecture 336
business 321
carpentry 327
finance 317
fitness 320, 330
food service 320
health 321
manufacturing 317, 331
navigation 308
part-time job 316, 340
retail 341
safety 337
shipping 331
sports 331, 336
travel 311, 314, 326, 336
velocity 311, 316, 321, 327, 340

Standardized Test Practice
Multiple Choice 346
Short Response/Grid In 347
Extended Response 347

Page 339

Relationships in Geometry

Theme: Animation

Math*Works* Careers
Animator 361
Video Editor 379

Applications
animation 354, 359, 365, 366,
373, 377, 383
civil engineering 354
construction 355
health 373
interior design 371
landscaping 377
machinery 359
safety 363
sports 357, 377

Standardized Test Practice
Multiple Choice 388
Short Response/Grid In 389
Extended Response 389

Page 349

Polynomials

Theme: Natural Disasters

Math*Works* Careers
Meteorologist 403
Disaster Relief Coordinator
403

Applications
agriculture 410
architecture 407
astronomy 410
construction 425
earthquakes 397
engineering 407
finance 397
interior design 425
landscaping 400, 424
manufacturing 401
money 396
photography 409
retail 411
volcanoes 400
weather 397, 417

Standardized Test Practice
Multiple Choice 430
Short Response/Grid In 431
Extended Response 431

Page 390

10

Probability

Theme: Genetics

Math*Works* Careers
Genetic Counselor 445
Research Geneticist 463

Applications
construction 457, 466
engineering 443
entertainment 441, 461
finance 467
genetics 438, 447, 453, 459,
   461
industry 443, 460
literature 448, 459
political science 460, 461
recreation 449
retail 442, 450, 452
safety 459
sports 442, 453, 458, 466
travel 452
weather 439

Standardized Test Practice
Multiple Choice 472
Short Response/Grid In 473
Extended Response 473

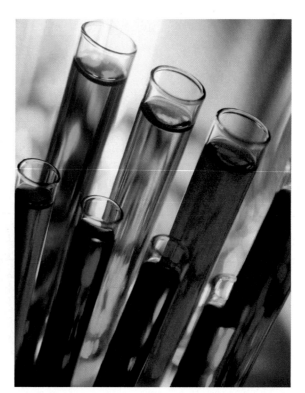

Page 432

Theme: Mysteries

Math*Works* Careers
Investigator 487
Cryptographer 505

Applications
advertising 491
architecture 479
earth science 488
entertainment 494, 500
fashion 493
landscaping 508
literature 491
mystery 485, 490, 500, 509
political science 495
recreation 484, 494
sports 492, 495, 501
transportation 495

Standardized Test Practice
Multiple Choice 514
Short Response/Grid In 515
Extended Response 515

Reasoning

Page 475

Student Handbook

How to Use Your MathMatters Book

Welcome to *MathMatters*! This textbook is different from other mathematics books you have used because *MathMatters* combines mathematics topics and themes into an integrated program. The following are recurring features you will find in your textbook.

Chapter Opener This introduction relates the content of the chapter and a theme. It also presents a question that will be answered as part of the ongoing chapter investigation.

Are You Ready? The topics presented on these two pages are skills that you will need to understand in order to be successful in the chapter.

Build Understanding The section presents the key points of the lesson through examples and completed solutions.

Try These Exercises Completing these exercises in class are an excellent way for you to determine if you understood the key points in the lesson.

Practice Exercises These exercises provide an excellent way to practice and apply the concepts and skills you learned in the lesson.

Extended Practice Exercises Critical thinking, advanced connections, and chapter investigations highlight this section.

Mixed Review Exercises Practicing what you have learned in previous lessons helps you prepare for tests at the end of the year.

MathWorks This feature connects a career to the theme of the chapter.

Problem Solving Skills Each chapter focuses on one problem-solving skill to help you become a better problem solver.

Look for these icons that identify special types of exercises.

 CHAPTER INVESTIGATION Alerts you to the on-going search to answer the investigation question in the chapter opener.

 WRITING MATH Identifies where you need to explain, describe, and summarize your thinking in writing.

 MANIPULATIVES Shows places where the use of a manipulative can help you complete the exercise.

 TECHNOLOGY Notifies you that the use of a scientific or graphing calculator or spreadsheet software is needed to complete the exercise.

 ERROR ANALYSIS Allows you to review the work of others or your own work to check for possible errors.

Data and Graphs

THEME: Health and Fitness

Collecting, interpreting, and organizing data are essential for living and working in the real world. One area where data is constantly being collected, interpreted, and displayed is health and fitness. You can find important nutrition and health data on food labels, exercise equipment, and in newspapers, magazines, and books.

- **Nutritionists** (page 15) use data about food, health, and exercise to plan nutrition programs and diets. They also manage cafeterias and food services for institutions such as schools and hospitals.

- **Recreation workers** (page 33) plan, organize, and manage recreation activities offered at places such as playgrounds, parks, clubs, camps, and tourist attractions.

Math Online

mathmatters1.com/chapter_theme

Fitness Trail Map

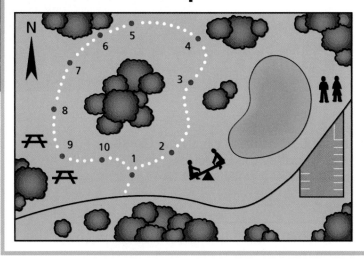

Data Activity: Fitness Trail

Fitness trails are designed to provide an individual with a complete workout while walking or hiking in nature. These paths are often scenic, with workout stations that exercise each of the major muscle groups. The exercises can include sit-ups, jumping jacks, pull-ups, and leg lifts.

Use the fitness trail map for Questions 1–5.

1. How many stations are on this fitness trail?

2. How many miles of fitness trail are between Stations 1 and 5?

3. Which two stations appear to have the shortest distance between them?

4. Which two stations appear to have the greatest distance between them?

5. To proceed from Station 8 to Station 9, which geographic direction will you travel?

CHAPTER INVESTIGATION

Many teenagers exercise regularly. Some teenagers participate in athletics because they enjoy competition, teamwork, and mastering physical skills. Exercise also maintains one's health, fitness, and well-being.

Working Together

Conduct a survey to find out more about teenagers and fitness. Organize your data and display it visually. Use the Chapter Investigation icons to check your progress.

Are You Ready?

Refresh Your Math Skills for Chapter 1

The skills on these two pages are ones you have already learned. Use the examples to refresh your memory and complete the Exercises. For additional practice on these and more prerequisite skills, see pages 536–544.

WORKING WITH WHOLE NUMBERS

You have used whole numbers since you first started to count. These problems will help you get ready for a new year of math.

Solve.

1. $\begin{array}{r} 4{,}657 \\ +12{,}987 \end{array}$	**2.** $\begin{array}{r} 15{,}121 \\ +\ 2{,}608 \end{array}$	**3.** $\begin{array}{r} 9498 \\ -1285 \end{array}$	**4.** $\begin{array}{r} 254{,}316 \\ -\ 19{,}467 \end{array}$
5. $\begin{array}{r} 284 \\ \times\ 83 \end{array}$	**6.** $\begin{array}{r} 2{,}615 \\ \times\ 308 \end{array}$	**7.** $14\overline{)2{,}348}$	**8.** $37\overline{)24{,}970}$
9. $\begin{array}{r} 9056 \\ -8427 \end{array}$	**10.** $\begin{array}{r} 84{,}000 \\ -17{,}258 \end{array}$	**11.** $\begin{array}{r} 257{,}611 \\ +579{,}834 \end{array}$	**12.** $\begin{array}{r} 7{,}003 \\ +73{,}037 \end{array}$
13. $7\overline{)59{,}800}$	**14.** $32\overline{)205{,}821}$	**15.** $\begin{array}{r} 26{,}804 \\ \times\ 580 \end{array}$	**16.** $\begin{array}{r} 515{,}697 \\ \times\ 1{,}205 \end{array}$

ROUNDING NUMBERS

When working with large numbers or with decimals, it's often helpful to be able to round an exact number to something that is easier to work with mentally.

Locate the digit in the place you are rounding to. If the digit to its right is 5-9, increase the digit you are rounding to by 1. If the digit to the right is 0-4, do not change the digit you are rounding to. In both cases, all digits after the place value you are rounding to are replaced with zeros.

Examples Round **257,329** to the nearest thousand.
The digit following 7 tells you how to round.
257,329 rounds to 257,000.

Round **7.058** to the nearest tenth.
The digit following 0 tells you how to round.
7.058 rounds to 7.1.

Round each number to the nearest hundred, thousand, and ten thousand.

17. 256,125   **18.** 124,120   **19.** 29,866   **20.** 570,648

Round each number to the nearest unit, tenth, and hundredth.

21. 128.069   **22.** 264.429   **23.** 56.987   **24.** 20.862

WORKING WITH DECIMALS

Decimals follow rules similar to those for whole numbers, but there are a few differences. For addition and subtraction, you line up the decimal points. For multiplication and division, you do not.

Solve.

25. 14.78
       + 5.798

26. 26.579
       +338.6

27. 349.2
       − 58.65

28. 437.09
       −294.8

29. 8.95
       × 0.7

30. 0.159
       × 1.08

31. 22)42.57

32. 0.5)128

PROBLEM SOLVING FIVE-STEP PLAN

Mathematics is about solving problems. You will solve many problems this year, and it may be helpful to follow a specific plan.

Here's a basic plan you can use when solving problems.

Read Read the problem carefully. Make sure you know what is being asked.

Plan Make a plan with one or more steps. Is there enough information? Do you need to get information from another source? What operations do you need to use?

Solve Use your plan to solve the problem.

Answer Reread the question. Make sure your solution answers the question.

Check Rework the problem to check your answer. Use a different method if possible.

Use the problem solving plan to solve each problem.

33. Michele wants to run 20 mi by the end of the month. She has run 12 mi so far. She knows she will be running another 5 mi this week. She has one more week left. How many miles will Michele need to run during the last week?

34. A warehouse has received shipments of a small part in the following quantities—17,500; 10,850; 24,600; and 14,400. They have orders pending for 75,000 of these parts. Do they have enough to ship all the orders? If not, how many more of these parts do they need?

35. A class at Pattonville High School wants to go on a field trip to Thunder Park. Admission is $17.75 per student and $25.25 per adult. There are 35 students and 9 adults. How much does the class need to raise to pay admission for all attending?

Collect and Interpret Data

Goals
- Choose a procedure to sample a population.
- Interpret data from tables, charts, and survey results.

Applications   Fitness, Manufacturing, Nutrition, Retail, Music

In a group, discuss methods you might use to find types of movies that are popular among teenagers.

1. Make a list of the methods.

2. Which method would be the most time-consuming?

3. Which method would be the most expensive?

4. Which method would you prefer? Why?

◥ BUILD UNDERSTANDING

Data are pieces of information that can be gathered through interviews, records of events or questionnaires. Collecting information from an entire group, or *population*, is called a census. To save time and money, many people conduct a survey, or poll.

Surveys and polls provide data about a part, or **sample**, of the population. Choosing the best method to gather data includes carefully wording the question. Effective surveys ask questions that require a specific answer, such as yes, no, red or 26.

To make decisions about a population based on a sample, choose an appropriate sampling method.

Reading Math

The word *data* is the plural form of the Latin word *datum*. The word *data* must be followed with the plural form of a verb.

The data are correct.

These data were collected.

Random Sampling Each member of the population has an equal chance of being selected. The members are chosen independently of one another. An example is putting 100 names in a hat, then drawing 20 of the names.

Convenience Sampling The sample is chosen only because it is easily available. An example is polling ten students who happen to be sitting near you in the cafeteria.

Systematic Sampling After a population is ordered in some way, the sample is chosen according to a pattern. For example, select every tenth item from a long list.

Cluster Sampling Members of the population are chosen at random from a particular part of the population and then polled in clusters. For example, choose areas of a city at random. Visit the pet shops in these areas, asking every pet owner to name a favorite brand of pet food.

Example 1

FITNESS Which reflects the random sampling method that a health club owner might use to identify the most popular exercise machine in the club?

    A. Ask the first 20 members who enter the club one morning.

    B. Ask the members whose phone numbers end with the digit 7.

    C. Ask members who live on the six busiest streets in town.

Solution

Choice B. Since phone numbers are assigned at random, those members whose phone numbers end in 7 represent a random sample.

Example 2

MANUFACTURING In an assembly line, every twentieth rowing machine out of 500 is quality tested. Two are defective.

a. What kind of sampling does this situation represent?

b. What might be an advantage of this kind of sampling?

c. What might be a disadvantage of this kind of sampling?

Solution

a. This example represents systematic sampling.

b. An advantage is that the sample comes from the whole population.

c. Other defective machines may not be in the sample group.

Example 3

Which automobile has faster acceleration?

Acceleration	Car A (seconds)	Car B (seconds)
Zero to 30 mi/h	1.8	3.2
Zero to 40 mi/h	2.8	4.7
Zero to 50 mi/h	3.9	6.4
Zero to 60 mi/h	5.1	8.8
Zero to 70 mi/h	6.8	11.6
Zero to 80 mi/h	8.6	15.0

Solution

Car A accelerates from zero to each measured speed in fewer seconds than Car B. Car A has better acceleration.

◤ TRY THESE EXERCISES

NUTRITION In a survey of 50 people who work at home, nine out of ten eat breakfast.

1. What kind of sampling does this situation represent?

2. What might be an advantage of this kind of sampling?

3. What might be a disadvantage of this kind of sampling?

DATA FILE For Exercises 4–6, use the travel data on page 534.

4. What reason did most of the people polled give as the main purpose of their travel?

5. What reason was the least popular among the people polled?

6. What kind of sampling do you think was used to gather this information? Explain.

7. Give an example of random sampling.

8. Give an example of cluster sampling.

9. Give an example of convenience sampling.

10. Give an example of systematic sampling.

PRACTICE EXERCISES • For Extra Practice, see page 545.

11. Which of these reflects the cluster sampling method that a committee of students might use to find how many adults would attend a school fair?

   a. Survey the adults who leave a supermarket on Tuesday between 5:00 P.M. and 6:00 P.M.

   b. Survey the members of the men's and women's softball teams.

   c. Survey the adults who live in houses chosen at random on the streets surrounding the school.

Identify the method used to find the most popular video arcade game. List advantages and disadvantages of the method.

12. Poll every twelfth person who enters the local video arcade.

13. Poll a randomly chosen sample of 12 students in your school.

14. Poll the first 12 people you meet in a video-rental store.

15. Randomly select people to call from the telephone book.

16. **WRITING MATH** Could the day of the week a poll is taken affect the results of a survey? Explain.

For Exercises 17–19, use the table at the right.

17. In a keyboarding class, seven students achieved these words/minute scores on a test. Who is the fastest typist?

18. From this data, can you tell which typist is the most accurate?

19. Which typist is exactly twice as fast as Jasper?

Typist	Words/min
Tyrone	31
Benjiro	58
Carmen	39
Latanya	44
Jasper	27
Rashida	47
Spenser	61

MUSIC You need to determine teenagers' favorite musical instrument. Identify the sampling method, and list its advantages and disadvantages.

20. Poll your friends.

21. Poll one person in each homeroom.

22. Poll the football team.

23. Poll every fifth person in the lunch line.

RETAIL A shopping mall needs to fill two vacant restaurants in the food court. To find which foods shoppers like best, the first 100 people at the food court during lunch hour are surveyed.

24. What kind of sampling is represented by this situation?

25. How could this method be changed to produce a systematic sample?

26. How could the survey be conducted to produce a random sample?

27. Which sample would be most useful to the mall owners? Explain.

For Exercises 28–30, use the table at the right.

A sample of 500 people was asked if they preferred orange, grapefruit, or tomato juice.

Juice	Replies
Orange	183
Grapefruit	48
Tomato	123
No Opinion	146

28. How many people had an opinion?

29. Based on the survey, which juice is liked the best?

30. Did more than half of the people polled vote for orange juice?

31. DATA FILE Refer to the graph on the contents of garbage cans shown on page 529. How do you think the data were collected?

SAFETY To find out how many teenagers wear seatbelts, every tenth student in the cafeteria line is polled.

32. What kind of sampling is represented by this situation?

33. Give an example of how the student group could be polled using each of the other types of sampling.

34. Which method would you use? Give reasons for your choice.

■ EXTENDED PRACTICE EXERCISES

Data are gathered using the methods described. Give reasons why the conclusions are biased.

35. *Survey* People living near railroad tracks are asked if they want the tracks moved. *Conclusion* Most people want the tracks moved.

36. *Survey* People in the ticket line for *Threat of the Martians* are asked about their favorite type of movie. *Conclusion* Most people like science fiction movies best.

37. CHAPTER INVESTIGATION Write a survey question about your topic. Decide on a method of sampling and the number of people in your sample.

■ MIXED REVIEW EXERCISES

Add or subtract. (Basic math skills)

38. $4685 + 1709$ **39.** $17,505 - 8,936$ **40.** $30,710 - 19,666$ **41.** $23,006 + 10,084$

Find the fractional part. (Basic math skills)

42. $\frac{1}{3}$ of 360   **43.** $\frac{1}{4}$ of 360   **44.** $\frac{1}{9}$ of 360   **45.** $\frac{3}{4}$ of 360

1-2 Measures of Central Tendency and Range

Goals
- Use the measures of central tendency.
- Find the range of a set of data.

Applications Business, Community service, Sports, Retail

Work in groups of two or three students.

1. Count the number of books you brought to school today.

2. Record the number of books for each student in the class.

3. Is there a number that occurs more frequently than others?

4. Arrange the data in numerical order. What is the middle number? If the number of data is even, there are two middle numbers.

5. What is the difference between the greatest and the least number of books?

6. Guess the average number of books for the students in your class.

◢ BUILD UNDERSTANDING

Four statistical measures that help you describe a set of data are the mean, median, mode and range.

The **mean**, or arithmetic average, is the sum of the values divided by the number of items of data.

The **median** is the middle value when the data are arranged in numerical order. If the number of items of data is odd, the median is the middle value. If the number of items of data is even, add the two middle values and divide by 2.

The **mode** is the value that occurs most frequently in the set of data. Some sets of data have no mode. Some have more than one mode.

The **range** is the difference between the greatest and the least values in a set of data.

Mean, median, mode, and range enable you to analyze data and determine how the data are grouped. Because the mean, median, and mode locate centers of a set of data, these terms are called **measures of central tendency**.

To analyze a set of data, it usually helps to begin by arranging the data in order from least to greatest.

Check Understanding

Find the median of the following set of data.

14 37 31 25 19 28

Example 1

Seven students had the following number of books.

$$4 \quad 3 \quad 6 \quad 5 \quad 5 \quad 7 \quad 5$$

Use data given to find each of the following.

a. mean   **b.** median   **c.** mode   **d.** range

Solution

a. Find the sum of the data. Then divide by the number of values.

$4 + 3 + 6 + 5 + 5 + 7 + 5 = 35 \quad \dfrac{35}{7} = 5$

The mean is 5.

b. When the data are arranged in order, the middle number is 5.

3,   4,   5,   5,   5,   6,   7

The median is 5.

c. The number that occurs most frequently is 5.

3   4   5 $\boxed{5}$ 5   6   7

The mode is 5.

d. Find the difference between the greatest and least values.

$7 - 3 = 4$

The range is 4.

Example 2

Which measure of central tendency best represents the data?

a. advertised March temperature in a resort brochure

b. most popular politician after a debate

Solution

a. The mean of daily temperatures during March gives you an average temperature to use in the brochure.

b. The mode indicates how many people preferred the politician.

> ### Math: Who, Where, When
>
> As a child, the German mathematician Karl Friedrich Gauss (1777–1855) astounded his teacher with his almost instantaneous response to his teacher's challenging problem. Karl was able to find the sum of the numbers from 1 to 100 mentally.

◥ TRY THESE EXERCISES

SPORTS During a gymnastics competition, the judges award the following scores.

$$7.0 \quad 7.0 \quad 9.6 \quad 9.4 \quad 10.0 \quad 8.8 \quad 9.2 \quad 9.8 \quad 9.0 \quad 8.5$$

1. Find the mean to the nearest tenth.   **2.** Find the median.

3. Find any modes.   **4.** Find the range.

5. WRITING MATH Which measures of central tendency best describe the average score from the data above? Explain.

Which measures of central tendency best represent each data set?

6. number of people with red hair in a sample of 100 people

7. target blood pressure for a 16-year-old

8. favorite color among your classmates

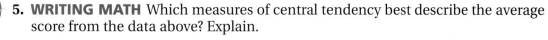

SPORTS A cyclist's competition times in seconds for one season are shown.

9.1 8.7 9.2 9.0 8.7 8.9 15.0 9.2

9. Find the mean to the nearest tenth.   10. Find the median.

11. Find any modes.   12. Find the range.

13. Explain which measure of central tendency best indicates the cyclist's ability.

14. If the cyclist competed in a ninth race and achieved a time equal to the mean, which measure(s) of central tendency would not change?

DATA FILE For Exercises 15–18, use the data on the maximum speeds of animals on page 518.

15. Find the mean of the speeds.   16. Find the median of the speeds.

17. Find any modes of the speeds.   18. Find the range of the speeds.

Which measure of central tendency best represents each data set?

19. average mass of an adult polar bear

20. most requested song at a radio station

21. middle value of nurses' salaries

22. average height of a female student in your grade

BUSINESS The table shows the salaries for the executives of an exercise equipment company.

Position	Salary
President	$125,000
Vice President	$ 86,000
Plant Manager	$ 71,500
Accounting Manager	$ 53,000
Personnel Manager	$ 41,250
Research Manager	$ 40,975

23. Find the mean.   24. Find the median.

25. Find any modes.   26. Find the range.

27. Which measure of central tendency best describes the average executive salary? Explain.

28. **CALCULATOR** Which measures of central tendency are best found using your calculator, and which are found easily without a calculator?

RETAIL The stores at a shopping center recorded the number of customers they had each day of the week.

Store	Mon.	Tues.	Wed.	Thurs.	Fri.	Sat.
K Electronics	206	184	212	253	267	184
Computer Store	212	241	208	279	296	197
Jeans City	198	147	164	196	281	162
Record Shop	146	162	180	234	294	184
A & M Store	553	607	692	980	714	833

29. Find the mean number of customers at Jeans City each day.

30. Find the median number of customers at Jeans City each day.

31. How many customers should Jeans City expect in one 30-day month?

32. On which three days of the week should Jeans City employ extra help?

33. Find the mean number of customers for the five stores on Thursday.

34. Which store might need to hire part-time employees to work on Thursdays?

35. Find the range of the number of customers for Record Shop in one week?

COMMUNITY SERVICE For Exercises 36–42, use the table, which shows the amount of time that a sample of students spent each week volunteering.

Name	Time (hours)
Ravi	$1\frac{1}{2}$
Shawna	$2\frac{1}{3}$
Jordan	$1\frac{1}{2}$
Yesenia	$1\frac{3}{4}$
Cathy	$1\frac{1}{4}$
DeJuan	$1\frac{1}{4}$
Lisbeth	$2\frac{3}{4}$
Phil	$1\frac{1}{2}$

36. Find the mean to the nearest tenth.

37. Find the median.

38. Find any modes.

39. Find the range.

40. How would your answers to Exercises 36–39 change if each person had spent 30 min more volunteering?

41. If Shawna volunteered for 5 h, how would each measure of central tendency change?

42. Use your solution to Exercise 41 to find which measure changes the most and which changes the least.

DATA FILE For Exercises 43 and 44, use the data on calories spent in activities on page 530.

43. For a 150-lb person, find the mean and median of the calories spent in all of the activities.

44. Explain the difference between the mean result and the median result.

EXTENDED PRACTICE EXERCISES

45. Calculate the mean and median of five consecutive whole numbers. What conclusion can you draw?

46. **CRITICAL THINKING** Does your conclusion from Exercise 45 hold true for six consecutive whole numbers? Explain.

47. **WRITING MATH** In 1997, the minimum wage was raised from \$4.75/h to \$5.15/h. How do you think this may affect the median income? Explain.

Create sample data to illustrate your answers to Exercises 48–50.

48. If the median of a set of data is 15, must one of the data items be 15?

49. If the mean of a set of data is 15, must one of the data items be 15?

50. If the mode of a set of data is 15, must one of the data items be 15?

51. **CHAPTER INVESTIGATION** Find the mean, median, range, and any modes of the data you collected in your survey.

MIXED REVIEW EXERCISES

Round each number to the nearest hundred, thousand, and ten thousand. (Basic math skills)

52. 85,461   **53.** 1,034,510   **54.** 151,151   **55.** 379,897

Solve. (Basic math skills)

56. A large box holds twice as many nails as a medium box, which holds twice as many nails as a small box. If 3150 nails are used to fill one box of each size, how many nails are in each box?

Review and Practice Your Skills

Identify the sampling method in each situation.

1. The first name on each page of a telephone book is used for a sample.

2. The names of all the people in a club are put in a box. Ten names are drawn.

3. People in six cities are interviewed after they leave a new movie.

4. You ask five friends to name their favorite sport.

Number of Customers at Two Stores		
Month	**Store A**	**Store B**
Jan	231	305
Feb	345	346
Mar	341	298
Apr	467	275
May	510	265
Jun	498	302

Use the table for Exercises 5-7.

5. Describe what is happening to the number of customers for Store A.

6. Describe what is happening to the number of customers for Store B.

7. From this information, which store will have more customers in the second half of the year?

Find the mean for each data set.

8. 9, 12, 15, 15, 18, 24, 28, 30, 32

9. 0.1, 0.5, 0.7, 0.7, 0.9, 1.3, 1.4

10. 135, 120, 142, 110, 155, 115, 133

11. 9, 3, 11, 4, 12, 5, 7, 9, 12

12. 23, 45, 54, 61, 28, 93, 42, 46, 58, 31, 24, 43, 46, 37, 25, 39, 41, 64, 33, 35

Find the median for each data set.

13. 20, 21, 24, 45, 28

14. 120, 135, 110, 105, 115, 120, 135

15. 8, 2, 3, 4, 2, 6, 4, 5, 1, 1, 3

16. 2.1, 2.5, 2.4, 2.8, 1.9, 3.2

Find any modes for each data set.

17. 14, 15, 18, 13, 18, 14, 13, 13, 10, 17

18. 3, 1, 2, 4, 3, 4, 2, 5, 4, 3, 2, 2, 3

19. 150, 175, 125, 150, 150, 125, 225, 200

20. 5.5, 6.5, 5.5, 6.5, 7.5, 7.5, 5.5, 5.5

21. 10, 12, 14, 18, 12, 15, 14, 18, 17, 15, 14, 17, 14, 12, 13, 16, 17, 12, 14

Find the range for each data set.

22. 0.6, 0.5, 0.9, 1.2, 0.3, 1.1, 0.2, 0.8

23. 23, 27, 28, 29, 19, 32, 25, 26, 22

24. 240, 231, 358, 226, 239, 258, 214, 254

25. 8, 9, 6, 4, 3, 9, 7, 4, 5, 9, 6

26. 4.5, 4.6, 4.9, 3.2, 4.5, 4.7, 3.1, 4.2, 3.6, 4.1, 3.7, 3.2, 3.9, 4.3, 4.5, 3.7

PRACTICE ◤ LESSON 1-1–LESSON 1-2

Identify the sampling method in each situation. (Lesson 1-1)

27. For each name on a list, a coin is tossed to determine if the person is included in the sample.

28. A nation-wide book store chain introduces a new product in four cities to find out if people will buy it.

29. A list of names is put in alphabetical order. Then, every tenth person is chosen for the sample.

Use the table for Exercises 30 and 31. (Lesson 1-1)

30. Which club has the highest total membership?

31. From this information, what can you conclude about club membership at the school?

Club	Grade 9	Grade 10
Chess	12	18
Yearbook	8	23
Computer	16	19
Science	5	14

Find the mean, median, range, and any modes for each data set. (Lesson 1-2)

32. 140, 120, 120, 110, 110, 130, 150

33. 8, 3, 5, 3, 3, 5, 4, 9, 6, 2, 5, 7

34. 2.5, 2.2, 2.4, 2.5, 2.6, 2.5, 2.7, 2.3

35. 40, 45, 40, 30, 35, 25, 55, 20

MathWorks — Career – Nutritionist
Workplace Knowhow

Nutritionists promote healthy eating to prevent and treat disease and physical disorders. They manage food service systems for schools, hospitals and other institutions. They work with individuals to achieve health goals such as reducing cholesterol, controlling blood pressure or weight, or increasing muscle mass.

Use the nutrition label for Exercises 1–5.

1. How many ounces are in this package?

2. How many servings would provide 50% of the Recommended Daily Allowance of niacin?

3. How many more grams of carbohydrates are in this product than fat?

4. How many calories are in $\frac{1}{2}$ a serving?

5. How many calories are in 2 servings?

Nutrition Facts
Serving Size 15 crackers (28g/1 oz)
Servings Per Container 20

Amount Per Serving	
Calories 110 Calories from Fat 15	
	% Daily Value*
Total Fat 2g	3%
Saturated Fat 0g	0%
Cholesterol 0mg	0%
Sodium 95mg	4%
Total Carbohydrate 23g	8%
Dietary Fiber Less than 1g	2%
Sugars 7g	
Protein 2g	

Vitamin A 0% • Vitamin C 0%	
Calcium 0% • Iron 8%	
Thiamine 15% • Riboflavin 10%	
Niacin 10%	

1-3 Stem-and-Leaf Plots

Goals ■ Read and create stem-and-leaf plots.

Applications Fitness, Political science, History, Sports

Work with a partner.

1. Find your pulse at the side of your neck or at your wrist. Count the number of beats in your pulse for exactly 1 min.

2. Record the pulse rates for the class.

3. Which pulse rate was the fastest?

4. Which pulse rate occurred most frequently?

5. Suggest two ways to record the data.

▼ BUILD UNDERSTANDING

A **stem-and-leaf plot** organizes and displays data. The last digits of the data values are the **leaves**. The digits in front of the leaves are the **stems**.

Data in a stem-and-leaf plot can be analyzed by looking for the greatest and least values, outliers, clusters and gaps. Data values that are much greater than or much less than most of the other values can be called **outliers**. **Clusters** are isolated groups of values. **Gaps** are large spaces between values.

Example 1

The stem-and-leaf plot shows the heights in inches of 22 people.

Find the following.

a. least and greatest values

b. outliers

c. clusters

d. gaps

```
4 | 7 8 9 9
5 | 1 5 7 7 8 9 9
6 | 2 2 2 2 3 3 5
7 | 1 1 2
8 | 4
```

5|1 represents 51 inches.

Solution

a. The shortest student is 47 in. tall and the tallest is 84 in. tall.

b. The value 84 is an outlier. It is far from the other values.

c. There are several clusters of values: those in the high 40s, those in the high 50s, and those in the low 60s.

d. The greatest gaps are between 65 and 71 and between 72 and 84.

When creating a stem-and-leaf plot, use the data values to decide what the stems will be. Then plot the leaves of each data value to the right of their stems. The vertical line separates the stems from the leaves.

Example 2

Organize these pulse rates into a stem-and-leaf plot.

| 51 | 73 | 79 | 90 | 66 | 71 | 86 | 64 | 67 | 81 | 67 | 53 | 59 |
| 91 | 73 | 87 | 86 | 61 | 66 | 68 | 58 | 93 | 76 | 73 | 82 | 76 |

Solution

Step 1 Identify the least and greatest values of the data.

Step 2 Write the stems (5, 6, 7, 8, 9) in a column.

Step 3 Draw a vertical line to the right of the stems.

Step 4 Write the leaves to the right of their stems.

Step 5 Rewrite the data in order from least to greatest on a second plot.

Step 6 Write a key for the data.

Stems	Leaves

7|3 represents a pulse rate of 73.

```
Stems    Leaves
  5 | 1  3  8  9            Least pulse rate is 51.
  6 | 1  4  6  6  7  7  8
  7 | 1  3  3  3  6  6  9
  8 | 1  2  6  6  7
  9 | 0  1  3               Greatest pulse rate is 93.
```

◄ TRY THESE EXERCISES

1. Organize these data into a stem-and-leaf plot.

61	70	36	52	49	77	62
37	55	32	44	82	96	73
55	61	43	38	37	56	59
62	32	46	58	63	72	84

2. Find the greatest value.

3. Find the least value.

SPORTS The stem-and-leaf plot shows the number of points scored by a high school basketball team during a recent season. Use the stem-and-leaf plot to identify the following.

4. number of games in the season

5. greatest number of points

6. least number of points

7. possible outliers

8. clusters

9. gaps

10. mode(s)

```
3 | 4
4 | 0  2
5 | 1  3  7  8  9
6 | 0  0  3  5  5  6  6  6  7  7  8
7 | 0  3  4  4  5  5  6  7  8
8 | 7  7  8
9 | 2  2  3  9
```

5|1 represents 51 points.

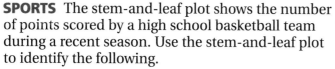

The stem-and-leaf plot shows the test scores for a math class.

11. How many students took this test?

12. How many scores are above 90?

13. How many scores are below 65?

14. How many scored 85?

15. Find the mean, median, mode and range for the class.

9	2 5 8
8	0 2 5 5 7
7	1 3 5 5 5 7 8 9
6	0 3 5 5 8
5	0 5 8

7|1 represents 71 points.

FITNESS Use the following data that represents the number of sit-ups by students in 1 min.

27	10	41	25	48	16	50	32	12
30	39	28	12	27	8	42	18	37
13	25	43	36	20	35	19	51	24

16. Create a stem-and-leaf plot of the data.

17. How many students are represented?

18. What is the greatest number of sit-ups?

19. What is the least number of sit-ups?

20. How many students did 14 sit-ups?

21. How many did 21 sit-ups?

22. Identify the clusters.

23. Identify the gaps.

24. Which measure of central tendency best represents the average number of sit-ups?

25. Make a stem-and-leaf plot for the data in the table at the right. The ones digit is the stem. The tenths digit is the leaves.

Walking on the Job

Job	mi/day
Hospital Nurse	5.3
Security Officer	4.2
Retail Salesperson	3.5
Waiter/Waitress	3.3
Doctor	2.5
Architect	2.3
Secretary	2.2
Accountant	1.8
Teacher	1.7
Lawyer	1.5
Housewife	1.3
Magazine Editor	1.3
Radio Announcer	1.1
Dentist	0.8

DATA FILE For Exercises 26 and 27, use the data on the number of calories burned during various kinds of exercise by a person weighing 110 lb on page 530.

26. Make a stem-and-leaf plot.

27. About how many calories are burned by a 110-lb person if he or she walks 3.5 mi/h for an hour, bikes at 13 mi/h for a half hour, then plays tennis for an hour?

POLITICAL SCIENCE These data represent the number of voters in hundreds at 40 polling stations.

30	58	39	52	46	29	27	44	20	29
23	32	24	62	35	40	42	60	34	62
67	38	29	29	69	41	30	41	28	53
53	52	35	49	41	50	23	15	29	15

28. Make a stem-and-leaf plot for the data in which 2 | 3 represents 2300 voters.

29. **WRITING MATH** Write two questions based on the data. Then answer them. Ask two other students your questions. Answer their questions.

EXTENDED PRACTICE EXERCISES

HISTORY A *back-to-back stem-and-leaf plot* organizes two sets of data so that they can be compared. The stems form the center column. The back-to-back stem-and-leaf plot shows age data for the first three U.S. presidents. Use the table and back-to-back stem-and-leaf plot for Exercises 36–44.

30. Copy and complete the back-to-back stem-and-leaf plot to include the data for all 42 presidents.

Age at Inauguration		Age at Death
	9	0
	8	3
	7	
1	6	7
7 7	5	
	4	

1 | 6 | 7 represents 61 years old and 67 years old.

31. How many presidents were inaugurated when they were 50–59 years old?

32. Who were the youngest and oldest presidents to be inaugurated, and how old were they?

33. Which president died the youngest? How old was he?

34. Why are there dashes in the bottom of the Age-at-death column?

35. Why is there a dash in the Age-at-death column of the table for the 24th president?

Identify the following.

36. possible outliers

37. clusters

38. gaps

39. YOU MAKE THE CALL The stem-and-leaf plot at the right shows the data from the fourth, fifth, and sixth presidents. Is it correct? If not, make the correct stem-and-leaf plot.

Age at Inauguration		Age at Death
	9	
	8	5 0
	7	3
	6	
7 7 8	5	

Ages of U.S. Presidents at Inauguration and at Death

	Name	Term	Age at inaug.	Age at death
1.	Washington	1789-1797	57	67
2.	J. Adams	1797-1801	61	90
3.	Jefferson	1801-1809	57	83
4.	Madison	1809-1817	57	85
5.	Monroe	1817-1825	58	73
6.	J.Q. Adams	1825-1829	57	80
7.	Jackson	1829-1837	61	78
8.	Van Buren	1837-1841	54	79
9.	W.H. Harrison	1841	68	68
10.	Tyler	1841-1845	51	71
11.	Polk	1845-1849	49	53
12.	Taylor	1849-1850	64	65
13.	Filmore	1850-1853	50	74
14.	Pierce	1853-1857	48	64
15.	Buchanan	1857-1861	65	77
16.	Lincoln	1861-1865	52	56
17.	A. Johnson	1865-1869	56	66
18.	Grant	1869-1877	46	63
19.	Hayes	1877-1881	54	70
20.	Garfield	1881	49	49
21.	Arthur	1881-1885	50	56
22.	Cleveland	1885-1889	47	71
23.	B. Harrison	1889-1893	55	67
24.	Cleveland	1893-1897	-	-
25.	McKinley	1897-1901	54	58
26.	T. Roosevelt	1901-1909	42	60
27.	Taft	1909-1913	51	72
28.	Wilson	1913-1921	56	67
29.	Harding	1921-1923	55	57
30.	Coolidge	1923-1929	51	60
31.	Hoover	1929-1933	54	90
32.	F.D. Roosevelt	1933-1945	51	63
33.	Truman	1945-1953	60	88
34.	Eisenhower	1953-1961	62	78
35.	Kennedy	1961-1963	43	46
36.	L.B. Johnson	1963-1969	55	64
37.	Nixon	1969-1974	56	81
38.	Ford	1974-1977	61	-
39.	Carter	1977-1981	52	-
40.	Reagan	1981-1989	69	93
41.	G.H.W. Bush	1989-1993	64	-
42.	Clinton	1993-2001	46	-
43.	G.W. Bush	2001-	54	-

MIXED REVIEW EXERCISES

Use the table for Exercises 40–42. (Lesson 1-2)

40. Find the mean, median, and mode of the ages of the Presidents at inauguration. Round to the nearest whole number.

41. Which measure of central tendency is the best indication of the average age of a President at inauguration?

42. Find the mean, median, and mode of the ages of the Presidents at death.

1-4 Problem Solving Skills: Circle Graphs

In a **circle graph**, or *pie chart*, a visual shows how data is divided into categories that do not overlap. Each **sector**, or slice, is a percentage of the total number of data. The entire circle represents 100% of the data. A **key**, or *legend*, describes the data in each sector. To solve a circle graph problem, it helps to break it into small parts. This strategy is often called **solve a simpler problem**.

Problem Solving Strategies

Guess and check

Look for a pattern

✔ Solve a simpler problem

Make a table, chart or list

Use a picture, diagram or model

Act it out

Work backwards

Eliminate possibilities

Use an equation or formula

Problem

Use the circle graph to find the number of teenagers who participate in each sport.

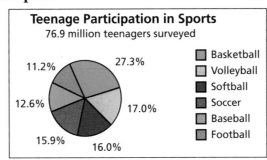

Teenage Participation in Sports
76.9 million teenagers surveyed

27.3%
11.2%
12.6%
17.0%
15.9%
16.0%

■ Basketball
■ Volleyball
■ Softball
■ Soccer
■ Baseball
■ Football

Solve the Problem

To find the number of teenagers for each sport, multiply the percent by the number of teenagers surveyed.

Baseball	12.6% of 76.9 M = $0.126 \cdot 76.9 \approx 9.69$	About 9.7 M teens play baseball.
Basketball	27.3% of 76.9 M = $0.273 \cdot 76.9 \approx 20.99$	About 21 M teens play basketball.
Football	11.2% of 76.9 M = $0.112 \cdot 76.9 \approx 8.61$	About 8.6 M teens play football.
Soccer	15.9% of 76.9 M = $0.159 \cdot 76.9 \approx 12.23$	About 12.2 M teens play soccer.
Softball	16.0% of 76.9 M = $0.16 \cdot 76.9 \approx 12.30$	About 12.3 M teens play softball.
Volleyball	17.0% of 76.9 M = $0.170 \cdot 76.9 \approx 13.07$	About 13.1 M teenagers play volleyball.

◥ TRY THESE EXERCISES

A survey was taken of 80 customers at a sporting goods store concerning their favorite form of exercise. Use the circle graph for Exercises 1–4.

1. How many customers prefer bicycling?

2. How many customers prefer running/walking?

3. How many customers prefer aerobics or weight training?

4. What is the total percent of the sectors in the circle graph?

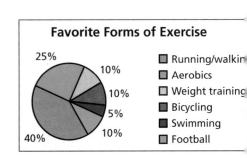

Favorite Forms of Exercise

25%
10%
10%
5%
10%
40%

■ Running/walkin
■ Aerobics
□ Weight training
■ Bicycling
■ Swimming
□ Football

◣ PRACTICE EXERCISES

Five-step Plan

1 Read
2 Plan
3 Solve
4 Answer
5 Check

COMMUNITY SERVICE A youth group raised a total of $1983 for a local homeless shelter. The group leader made a circle graph of the money earned but forgot to make a legend. Identify which fund-raiser should appear in each sector.

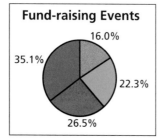

Fund-raising Events

16.0%
35.1%
22.3%
26.5%

5. The spaghetti dinner earned $525.

6. The bake sale earned $318.

7. The magazine sale earned $697.

8. The car wash earned $443.

For Exercises 9 and 10, use the circle graphs.

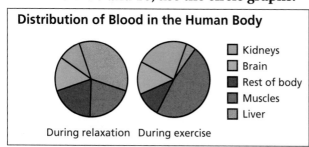

Distribution of Blood in the Human Body

☐ Kidneys
☐ Brain
■ Rest of body
■ Muscles
☐ Liver

During relaxation   During exercise

9. Estimate how many times as much blood is in the muscles during exercise as during relaxation.

10. Estimate how many times as much blood is in the liver during relaxation as during exercise.

DATA FILE For Exercises 11–12, use the data on the main elements of the Earth's crust on page 526.

11. What percent of the Earth's crust is iron?

12. What percent of the Earth's crust is silicon, aluminum, or calcium?

HEALTH Use the table that shows the average number of hours Americans sleep each night.

Hours of sleep	%
Less than six	13
Six	18
Seven	31
Eight	28
More than eight	10

13. How many hours does the average American sleep?

14. What percent of people surveyed sleep more than 6 h/night?

15. Draw a circle graph of the data. Estimate the size of the sectors.

16. **SPREADSHEET** Use spreadsheet software to construct a circle graph using the data on sleep. Then compare your results to the circle graph you drew in Exercise 15.

17. **ERROR ALERT** In the graph at the right, the sectors are labeled incorrectly. Arrange the labels correctly.

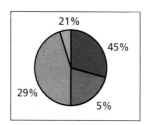

21%
45%
29%
5%

◣ MIXED REVIEW EXERCISES

Round each number to the nearest unit, tenth, and hundredth. (Basic math skills)

18. 85.039   19. 5.198   20. 19.567   21. 57.983

22. The manufacturers of Lucky Stars cereal decided to place a coupon for a free box of cereal inside every 100th box. Does this distribution use random, cluster, convenience, or systematic sampling? (Lesson 1-1)

Review and Practice Your Skills

PRACTICE ◾ LESSON 1-3

Use the stem-and-leaf plot.

1. age of the oldest person

2. age of the youngest person

3. range of the ages

4. mode or modes

5. median age

Ages of People in a Club

1	2 4 4 4 6
2	0 1 3 7
3	0 2 4 5 5 6 8 8 9
4	3 6 6
5	2 3

5|2 represents 52 years old.

Make a stem-and-leaf plot for each data set.

6. ages: 23, 45, 54, 41, 28, 23, 42, 46, 58, 31, 24, 43, 46, 37, 25, 39, 41, 24, 33

7. test scores: 65, 72, 66, 84, 78, 79, 86, 64, 60, 76, 74, 70, 62, 76, 81, 66, 74, 80

8. heights (inches): 48, 52, 56, 61, 73, 49, 58, 57, 63, 64, 70, 62, 58, 56, 61, 45, 63

9. game scores (1000s): 12, 13, 16, 24, 31, 25, 18, 16, 14, 25, 23, 17, 26, 18, 22

PRACTICE ◾ LESSON 1-4

Use the circle graph shown. Identify which budget item should appear in each sector.

10. Rent: $800

11. Transportation: $700

12. Food: $550

13. Other: $450

Monthly Budget

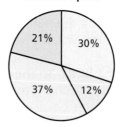

Use the circle graph shown. Identify which sport should appear in each sector.

14. Baseball: 1776 people

15. Football: 1440 people

16. Hockey: 576 people

17. Basketball: 1008 people

Favorite Sport

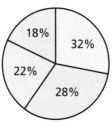

Use the circle graph shown. Identify which person should appear in each sector.

18. Kelly: 22,272 votes

19. Garcia: 118,784 votes

20. Wong: 33,408 votes

21. Burns: 11,136 votes

Election Results

Identify the sampling method in each situation. (Lesson 1-1)

22. You ask every person sitting in your row in a theater if they like the movie.

23. A list of people is identified and ordered by their birth dates. Then, every fifth person is chosen for the sample.

Find the mean, median, range and any modes for each data set. (Lesson 1-2)

24. 61, 28, 93, 42, 46, 58, 31, 24, 43, 46, 37

25. 245, 250, 250, 245, 235, 225, 245

26. 10.5, 10.2, 10.3, 10.3, 10.7, 10.4

27. 1, 3, 2, 3, 3, 5, 2, 3, 2, 3, 1, 2, 3

Make a stem-and-leaf plot for each data set. (Lesson 1-3)

28. test scores: 65, 68, 92, 75, 58, 81, 68, 70, 75, 65, 83, 65, 58, 63, 75

29. ages: 17, 22, 40, 18, 14, 31, 30, 25, 20, 16, 18, 38, 14, 20, 41, 16, 14

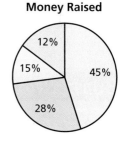

Money Raised

Use the circle graph shown. Identify which grade should appear in each sector. (Lesson 1-4)

30. Grade 6: $336

31. Grade 8: $540

32. Grade 7: $144

33. Grade 9: $180

Mid-Chapter Quiz

1. Define the following in your own words: data, poll, population, sample, random sampling, convenience sampling, cluster sampling and systematic sampling. (Lesson 1-1)

2. Give an example to illustrate the meaning of each word in Question 1. (Lesson 1-1)

3. Would you use mean, median, mode, or range to find the age most students graduate from high school? (Lesson 1-2)

4. Find the mean, median, mode and range for this data: 18, 25, 19, 22, 27, 20. (Lesson 1-2)

5. The median for a set of data is 50. Does 50 need to be one of the pieces of the data? Give an example to illustrate your answer. (Lesson 1-2)

6. Find the mode for this set of data: 0, 1, 3, 4, 6, 7, 135, 135. Why is the mode not a useful piece of information for the data? (Lesson 1-2)

7. Estimate the percents for a circle graph for the data given. A: 153, B: 218, C: 175 (Lesson 1-4)

Use the following scores for Question 8.

| 61 | 70 | 36 | 52 | 49 | 77 | 62 | 37 | 55 | 32 | 44 | 82 | 96 | 73 |

| 55 | 61 | 43 | 38 | 37 | 56 | 59 | 62 | 32 | 46 | 58 | 63 | 72 | 84 |

8. Construct a stem-and-leaf plot showing the scores. (Lesson 1-3)

1-5 Frequency Tables and Pictographs

Goals ■ Interpret frequency tables and pictographs.
■ Create frequency tables and pictographs.

Applications Nutrition, Art, Manufacturing, Sports, Music

Brad interviewed people about their favorite television programs. He used this code to record the following data.

S Sports		**Q** Quiz shows	
M Mysteries		**V** Music videos	
O Soap operas		**A** Adventure	
N News		**C** Comedies	

1. How many people were interviewed according to the data shown below?

V C O N Q A M S M O A O Q C C

N A C N M Q S O A S Q A C S N

Q A Q S A A M S O V N C A C S

2. What is the most popular type of television program?

BUILD UNDERSTANDING

A **frequency table** shows how often an item appears in a set of data. A tally mark is used to record each response. The total number of marks for a given response is the *frequency* of that response.

Example 1

Use the frequency table of Brad's data.

a. How many more chose sports programs than news?

b. Which two programs together have the same frequency as adventures?

Types of TV Programs

Program	Tally	Frequency
Sports	ЖΗΤ ΙΙ	7
Mysteries	ΙΙΙΙ	4
Soap operas	ЖΗΤ	5
News	ЖΗΤ	5
Quiz shows	ЖΗΤ Ι	6
Music videos	ΙΙ	2
Adventure	ЖΗΤ ΙΙΙΙ	9
Comedies	ЖΗΤ ΙΙ	7

Solution

a. Seven people chose sports. Five people chose news. $7 - 5 = 2$, so two more people chose sports than news.

b. As many people chose adventures as the following pairs of programs.
sports and music videos   mysteries and soap operas
mysteries and news   comedies and music videos

A picture graph, or **pictograph**, displays data with graphic symbols. The key identifies the number of data items represented by each symbol. The symbols often represent rounded amounts.

To read a pictograph, start by interpreting the key. Then multiply the value of one symbol by the number of symbols in a row to find the value for the row.

Example 2

Use the pictograph at the right.

a. In which country is the most fish eaten? About how many pounds of fish does the average person in that country eat in a year?

b. How much fish does the average person in the U.S. eat in a year?

c. The average Norwegian eats about 110 lb of fish/yr. How many fish symbols are needed to represent this information on the pictograph?

Fish Consumption around the World
(pounds, per person, per year)

Key: 🐟 = 10 lb

Solution

a. The average person in Japan eats about 150 lb of fish/yr.

b. The average person in the U.S. eats about 45 lb of fish/yr.

c. Eleven fish symbols are needed to represent the amount of fish eaten by Norwegians.

Example 3

DATA FILE Use the data on the 2000 Summer Olympics on page 533 to construct a pictograph for the number of gold, silver and bronze medals won by Great Britain.

**Great Britain Medal Standings
2000 Summer Olympics**

Key: 🏅 = 2 medals

Solution

To construct a pictograph, choose a symbol for the key. Then determine a value for the symbol. Let one symbol represent two medals. Draw the symbols to represent the data. Label the pictograph with a title and key.

◥ TRY THESE EXERCISES

Use the table below to answer Exercises 1–3.

1. Make a pictograph of the data shown, except for the U.S. and Russia.

2. How many countries have sent one person into space?

3. Why weren't the U.S. and Russia included in the pictograph?

Number of Individuals Who Have Flown in Space

Country	Individuals	Country	Individuals
United States	251	India	1
Russia/CIS	98	Italy	3
Austria	1	Japan	5
Canada	6	Mexico	1
Cuba	1	Saudi Arabia	1
France	8	United Kingdom	1
Germany	10	Vietnam	1

NUTRITION Use the pictograph for Exercises 4–7.

Amount of Sodium in Some Foods

Pizza (1 slice)	🧂🧂🧂🧂
Tomato juice (1 cup)	🧂🧂🧂🧂🧂🧂🧂🧂
Pita bread (1 pocket)	🧂🧂
Potato salad (1 cup)	🧂🧂🧂🧂🧂🧂🧂🧂🧂🧂🧂🧂🧂
Carrot cake (1 slice)	🧂🧂🧂🧂

Key: 🧂 = 100 mg

4. Which food listed contains the most sodium? Approximately how much sodium does it contain?

5. Which food listed contains the least sodium? Approximately how much sodium does it contain?

6. Approximately how much sodium is in two slices of pizza?

7. There are 1075 mg of sodium in 1 cup of soup. How many symbols would represent this information on the pictograph?

Use the table on caffeine in beverages.

Amount of Caffeine in 12 oz of Various Beverages

Beverages	Milligrams of caffeine
Cola	45
Caffeinated water	45
Iced tea	70
Regular coffee	280
Hot chocolate	10

8. Make a pictograph of the data on caffeine in beverages.

9. What symbol did you choose to represent the data?

10. How many times more caffeine is there in 12 oz of regular coffee than in 12 oz of hot chocolate?

PRACTICE EXERCISES • For Extra Practice, see page 546.

ART The prices in dollars of paintings sold at an art auction are shown.

1800 750 600 600 1800 1350 300 1200 750 600 750 2700

600 750 300 750 600 450 2700 1200 600 450 450 300

11. Make a frequency table of the data.

12. What price was paid most often for the artwork?

13. What is the average price paid for artwork at this auction?

14. How many artworks sold for at least $600 and no more than $1200?

15. **WRITING MATH** How are frequency tables and pictographs similar? How are they different?

MUSIC Use the pictograph for Exercises 16–20.

Student Preferences–Kinds of Music

Rock	♪ ♪ ♪ ♪ ♪ ♪
Jazz	♪ ♪ ♪
Country/western	♪ ♪ ♪
Classical	♪ ♪
Blues	♪ ♪ ♪ ♪
Other	♪ ♪ ♪

Key: ♪ = 50 students

16. Which type of music is most popular? Least popular?

17. About how many students prefer country/western music?

18. About how many more students prefer blues than classical music?

19. How many more students prefer rock than prefer country/western music?

20. Give examples of the kinds of music "other" could represent. How many students might prefer each of the "others" you listed?

Weight (pounds)	Frequency
19.6	1
19.7	6
19.8	7
19.9	4
20.0	9
20.1	13
20.2	8
20.3	1
20.4	1

MANUFACTURING Dog food packaged by a machine in 20-lb bags varies slightly in actual weight. Use the table for Exercises 21–24.

21. Make a pictograph of the data.

22. If a 0.2-lb difference from the target weight is acceptable, how many bags were defective?

23. How many bags weighed more than 20 lb?

24. Does the equipment usually overfill or underfill the bags?

25. WRITING MATH In a frequency table, is the survey response with the highest frequency always the mode? Is it always the median? Explain.

SPORTS Roberto has scored the greatest number of points on the basketball team. Here are his scores for a recent season's games.

Game	1	2	3	4	5	6	7	8
Points	12	20	16	12	20	4	16	24

26. Construct a pictograph of the data. For the key, use a basketball symbol to represent four points.

27. In which game did he score the greatest number of points?

28. In which game did he score the least number of points?

EXTENDED PRACTICE EXERCISES

Use the data on points scored by hockey players in a season.

Player	Kelly	Green	Tookey	Currie	Smith	Charron
Points	60	74	45	36	40	48

29. Construct a pictograph to show the points scored by the players.

30. What is an appropriate scale for your pictograph?

31. Find the mean.   **32.** Find the median.

33. Find any modes.   **34.** Find the range.

35. CRITICAL THINKING Explain how each measure of central tendency can be used to describe the data.

MIXED REVIEW EXERCISES

Multiply or divide. (Basic math skills)

36. $578 \cdot 27$ **37.** $5924 \div 8$

38. $28,962 \cdot 408$ **39.** $57,075 \div 15$

40. Record the number of times each digit is used in Exercises 36–39. Include the item numbers but not the answers. Make a circle graph to display the data. (Lesson 1-4)

1-6 Bar Graphs and Line Graphs

Goals
- Interpret and construct bar graphs and line graphs.
- Make predictions from bar graphs and line graphs.

Applications   Biology, Retail, Health, Earth science

Answer each question using Sarah's growth chart.

1. What was Sarah's height at age 4?

2. How much taller was Sarah at age 10 than at age 5?

3. Between which two consecutive years did Sarah grow the most?

4. How tall do you think Sarah will be when she is 11 years old?

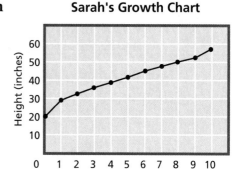

BUILD UNDERSTANDING

In a **bar graph**, horizontal or vertical bars display data. A scale is used to show intervals. To read a bar graph, look at the top edge of each bar. Match that edge with the number on the scale to find the value of that bar.

Example 1

Use the bar graph to answer the following questions.

a. Which bar represents the longest river? About how long is that river?

b. About how much longer is the Nile than the Yangtze?

c. Which rivers are over 5400 km long?

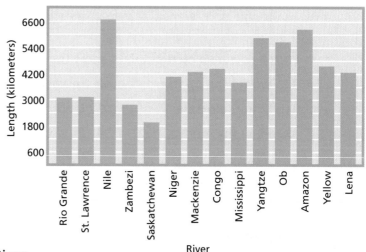

Solution

a. The longest bar represents the Nile River. The top of the bar is very close to the 6600-km mark on the scale. The Nile is about 6600 km long.

b. The Nile is about 800 km longer than the Yangtze.

c. The Nile, the Ob, the Yangtze and the Amazon are each over 5400 km long.

Example 2

Make a bar graph of the data.

Solution

To construct a bar graph, choose and label an appropriate scale. Draw and label the bars. Write a title for the graph.

Children's Injuries Caused by Toys

Toy	Number of injuries
Marbles	1845
Children's wagons	7935
Blocks and pull toys	2799
Nonwheeled riding toys	4022
Wheeled riding toys	6553
Balloons	1913
Toy guns	2581
Flying toys	4263

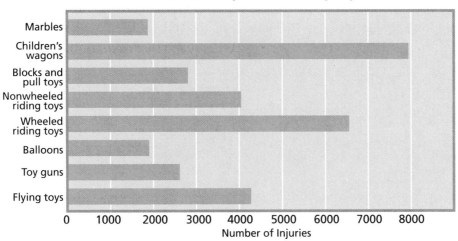

On a **line graph**, points representing data are plotted, then connected with line segments. Because the points are connected in sequence, a line graph shows trends, or changes, in data over a period of time.

To read a line graph, locate the first data point. Relate it to the corresponding labeled points on the vertical and horizontal scales. Do the same for each of the other data points.

Example 3

Construct a line graph for the data.

Solution

To construct a line graph, draw and label the horizontal and vertical axes. Locate and mark the data points.

Draw a straight line to connect the points. Write a title for the graph.

Average Monthly Rainfall in Tokyo, Japan

Month	Jan.	Feb.	Mar.	Apr.	May	June
Rainfall (millimeters)	49.9	71.5	106.4	129.2	144.0	176.0

Month	July	Aug.	Sept.	Oct.	Nov.	Dec.
Rainfall (millimeters)	135.0	148.5	216.4	194.1	96.5	54.4

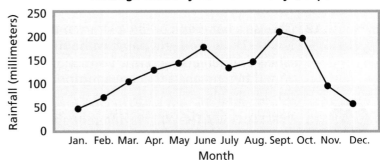

Use the bar graph for Exercises 1–3.

Blue Jeans Sales for One Week

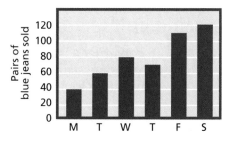

1. Which day of the week had the lowest number of sales? Approximately how many pairs were sold?

2. Which day of the week had the most sales? Approximately how many pairs were sold?

3. Which days had over 100 pairs of jeans sold?

Number of Bacteria

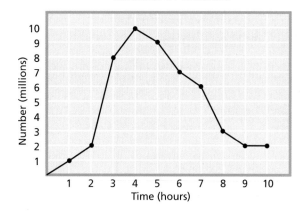

BIOLOGY Use the line graph showing the number of bacteria in a culture at different times.

4. When were the least number of bacteria present?

5. How many fewer bacteria were present at 7 h than at 4 h?

6. What can you state about the longevity of this particular bacterium from the line graph?

Use the table on automobiles in the U.S.

7. Make a bar graph for the data on automobiles in the U.S.

8. Make a line graph for the data.

Automobiles in the U.S.

Year	Cars registered
1960	61,671,390
1970	89,243,557
1980	121,600,843
1990	133,700,497
2000	133,621,420

PRACTICE EXERCISES • For Extra Practice, see page 547.

Use the bar graph on smallpox for Exercises 9–11.

9. How many times as many smallpox cases were there in 1951 compared to the number of cases in 1971?

10. Estimate in which years the number of smallpox cases was about twice the number that occurred in 1971.

11. Estimate in which year the number of smallpox cases was about one-third the number of cases that occurred in 1947.

Cases of Smallpox Worldwide

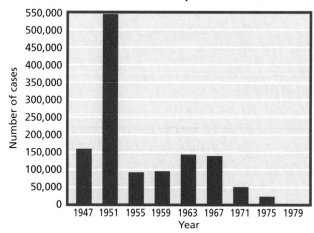

12. Choose a bar graph or line plot from this lesson. Draw it on grid paper without labels or intervals. Describe a situation with different labels that are appropriate for this graph.

13. **WRITING MATH** What is different about the line graph than the other graphs studied so far? When is a line graph more appropriate than one of the other graphs?

Use the table on large bodies of water for Exercises 14–16.

14. Construct a horizontal bar graph for the data.

15. The average depth of the Caribbean Sea is 2575 m. Which bodies of water listed here are shallower?

16. The average depth of the Red Sea is 540 m. Which bodies of water listed here are deeper?

Depths of Large Bodies of Water

Body of water	Average depth (meters)
Hudson Bay	90
Mediterranean Sea	1500
Arctic Ocean	1330
Atlantic Ocean	3740
Pacific Ocean	4190
Black Sea	1190

RETAIL The line graph shows changes in the price of a handheld calculator from 1965 to 2000.

17. What was the price of the calculator in 1965?

18. In which 5-year period did the price of a calculator drop the most?

19. In which year was the price of the calculator the least?

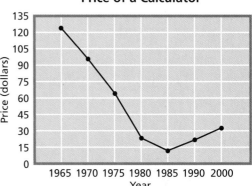

Price of a Calculator

DATA FILE For Exercises 20–23, use the data on life expectancy on page 531.

20. What was the projected life expectancy of a baby boy born in 1960?

21. What is the projected life expectancy of a baby girl born in 2001?

22. Draw a line graph for the data.

23. Predict what the life expectancies will be for boys and girls born in 2010. Give reasons for your predictions.

■ EXTENDED PRACTICE EXERCISES

DATA FILE For Exercises 24–27, use the multiple-line graph on high school sports participation on page 532.

24. What year did boys' and girls' participation begin to drop?

25. Which line shows the most frequent decrease?

26. About how many boys and girls participated in sports in 2001–2002?

27. **CHAPTER INVESTIGATION** Create a bar graph of your data.

■ MIXED REVIEW EXERCISES

Add. (Basic math skills)

28. $3.5 + 13 + 27.56$

29. $54.1 + 28.75 + 1.952$

30. $258 + 327.15 + 15.6$

31. $25.3 + 64.01 + 39.9$

32. If every one-hundredth box of cereal contains a prize, what is the greatest number of prizes that could be in 15,000 boxes of cereal? (Basic math skills)

Review and Practice Your Skills

PRACTICE ◣ LESSON 1-5

Refer to these test scores.

85	95	75	85	90	80	90	85	100	95
90	80	85	80	85	85	75	95	85	90

1. Make a frequency table.
2. Which score occurs most often?
3. What is the average score?
4. How many test scores were lower than 85?

Refer to these ages.

14	16	16	15	17	16	13	17	16	14	13	17	17	15	16	16
16	15	13	16	14	15	14	15	16	15	15	16	14	16	14	17

5. Make a frequency table.
6. Which age occurs most often?
7. What is the average age?
8. How many people are older than 15 years?

Refer to the pictograph.

9. Which category brought in the most money?
10. How much higher were the sales of clothing than the sales of automotive products?
11. Find the total dollar sales shown by the graph.

Sales Last Month

Sporting goods	$ $ $ $ $
Clothing	$ $ $ $ $ $
Electronics	$ $ $ $
Automotive	$ $ $ $
Other	$ $

Key: $ = $1000

PRACTICE ◣ LESSON 1-6

Refer to the bar graph.

12. On which day did the person spend twice as much time working out as on Thursday?

13. Find the total number of hours shown.

14. Find the average number of minutes spent working out per day for this week.

Refer to the line graph.

15. Describe what happened to the amount of taxes in the time period shown on the graph.

16. What is the difference between taxes paid in 1982 and taxes paid in 1988?

17. Between which two years did the amount of taxes increase the most?

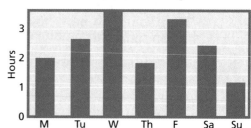

Time Spent Working Out

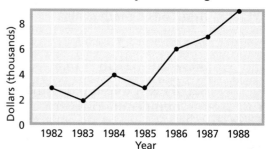

Income Taxes Paid by the Average American

PRACTICE ◣ LESSON 1-1–LESSON 1-6

On the circle graph, which color should appear in each sector? (Lesson 1-4)

18. red: 192 people   **19.** tan: 228 people   **20.** blue: 180 people

21. Four students from each grade are chosen at random to represent the students in an entire school. Identify the sampling method. (Lesson 1-1)

Favorite Color

30%
38%
32%

Refer to these test scores.

| 80 | 90 | 65 | 70 | 75 | 90 | 90 | 70 | 80 | 85 | 90 |
| 85 | 70 | 75 | 80 | 90 | 65 | 80 | 80 | 70 | 80 | 75 |

22. Find the mean, median, range and any modes. (Lesson 1-2)

23. Make a stem-and-leaf plot. (Lesson 1-3)

24. Make a frequency table. (Lesson 1-5)

Refer to the line graph. (Lesson 1-6)

25. The bike ride started at 9:30 A.M. The group stopped for one hour for lunch. At what time did lunch end?

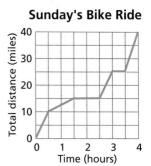

Sunday's Bike Ride

Math*Works*
Workplace Knowhow
Career – Recreation Worker

Recreation workers plan, organize, instruct and manage recreation programs. They plan leisure activities and athletic programs at playgrounds, camps and health clubs. Recreation workers teach the correct use of equipment and facilities. They research new activities and hobbies, in order to offer unique, interesting, and appropriate opportunities for people. These jobs can be part-time, seasonal and volunteer.

Use the frequency table for Questions 1–5.

1. How do you think this data was collected?

2. Make a pictograph of the data.

3. Which activity is the most popular?

4. Which activity is the least popular?

5. Which two activities are closest to the median?

Activity	Frequency
Aerobics	21
Bowling	12
Choir	11
Cooking	17
Crafts	12
Hiking	7
Quilting	10
Softball	25
Swimming	12
Tennis	15

Scatter Plots and Lines of Best Fit

Goals
- Read and create scatter plots.
- Use lines of best fit to identify trends.

Applications Market research, Sports, Advertising, Retail

Measure and record in inches the heights of your classmates and the length of their feet.

1. Display these data in a graph using one axis for foot size and the other axis for height. Make a point on the graph to represent each student.

2. What pattern do you see in the data displayed in your graph?

3. Are there any points that do not fit the pattern as well as other points on your graph? If so, what could be the reason for this?

4. How is the length of your foot related to your height?

▼ BUILD UNDERSTANDING

A **scatter plot** displays two sets of related data on the same set of axes. Points represent the data in a scatter plot, but they are not connected. There can be more than one point for any number on either axis.

Example 1

SPORTS The scatter plot displays the number of home runs by baseball team members.

a. How many players are represented?

b. How many home runs did the player with 10 years of experience hit?

c. What is the range of the number of home runs?

d. What was the "average" number of home runs?

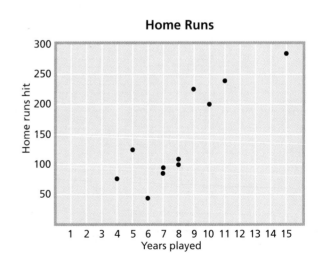

Home Runs

Solution

a. Eleven players were surveyed.

b. The point corresponding to 10 yr lies at 200 home runs.

c. The range is 283 − 46 or 237.

d. Most of the points fall near 100 home runs.

On some scatter plots, a **line of best fit**, or **trend line**, can be drawn near most of the points. A line of best fit that slopes up and to the right indicates a positive correlation among the data. A **positive correlation** means that as the horizontal axis values increase, the vertical axis values tend to increase.

A line of best fit that slopes down and to the right indicates a negative correlation. In a **negative correlation**, as the horizontal axis values increase, the vertical axis values tend to decrease. It is possible that there is no correlation.

Use a straightedge to estimate the line of best fit. To draw a line of best fit accurately requires a graphing utility or special computer program.

Example 2

Use the data randomly collected from newspaper advertisements to answer the following questions.

a. Draw a scatter plot and estimate a line of best fit.

b. Is there a positive or negative correlation between a car's age and price?

c. About how much would a 10-year-old car cost from this data?

d. About how old would an $8000 car be from this data?

Midsize Car Ages and Prices

Age (years)	Price ($100s)
2	150
5	105
8	45
4	95
6	90
3	90
9	25
2	105
5	75
8	60
6	65
4	85
3	100
5	60
4	110
7	80
3	130
6	80
7	55
2	130

Solution

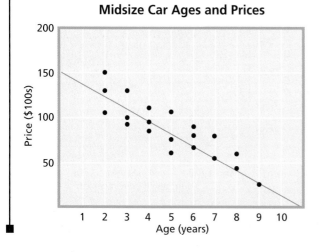

Midsize Car Ages and Prices

a. Your line of best fit may differ because of estimation.

b. The correlation is negative because the price decreases as the age increases. The line of best fit slopes down and to the right.

c. A 10-year-old car might cost about $1000.

d. An $8000 car would be about 5 years old.

TRY THESE EXERCISES

MARKETING A store manager reviews last year's sales of skates to plan next year's marketing.

1. What were the sales in March?

2. Which month had sales of about $4500?

3. Is there a positive or negative correlation?

4. What is the range of sales?

5. When were sales of skates the greatest?

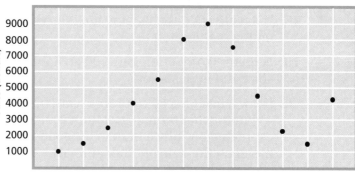

In-line Skate Sales

MARKET RESEARCH A survey asked students the number of hours they watched television daily, then compared the data to the students' test scores.

Hours Spent Watching Television and Test Score Averages

Hours	0.5	1	1.5	1.5	2	2.5	2.5	3	3	3.5	4	4.5	5	5.5	6
Test Scores	100	85	88	82	75	85	80	68	70	65	65	60	55	60	50

6. Draw a scatter plot and line of best fit for the data.

7. Use the line of best fit to predict the score of a student who watched no television.

8. Is there a positive or negative correlation between the data?

9. Which data point lies farthest from the line of best fit? What could account for this piece of data?

10. **GRAPHING** Use a graphing calculator or computer to create the scatter plot and line of best fit for the data above. Does it match the one you created by hand?

PRACTICE EXERCISES • For Extra Practice, see page 548.

ADVERTISING A health food store recorded the number of advertisements and the amount of sales over several weeks.

11. Find the sales resulting from 7 advertisements.

12. Predict the number of advertisements needed to result in $4000 in sales.

13. Is there a positive or negative correlation between the number of advertisements and the amount of sales?

14. Which data point lies farthest from the line of best fit?

15. What could account for this piece of data?

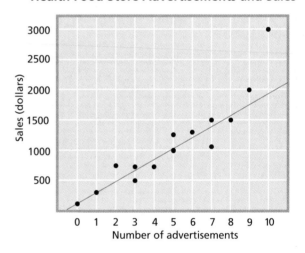

Health Food Store Advertisements and Sales

DATA FILE For Exercises 16–21, use the data on gestation, longevity, and incubation of animals on page 519.

16. Choose 15 animals. Make a scatter plot. Make the horizontal axis life expectancy in years and the vertical axis gestation in days.

17. What intervals did you choose for each axis?

18. What effect does changing the intervals have on the scatter plot?

19. Is there a correlation between an animal's life span and its gestation?

20. Draw a line of best fit for this scatter plot. Assuming that human gestation is nine months, estimate how long a human life span would be based on the data of other animals.

21. Explain why humans live longer than estimated from the data.

Use the data table.

22. Make a scatter plot and line of best fit on a single set of axes using different colors for men and women.

23. Is there a positive or a negative correlation?

24. Do any of the data points lie far from the line of best fit?

25. Are men's or women's times changing at a faster rate? Explain why you believe this to be true.

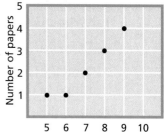

World Record Times for the 800-m Run

Year	Women	Men
1925	144	111.9
1935	135.6	109.7
1945	132	106.6
1955	125	105.6
1965	118	104.3
1975	117.5	104.1
1985	113.3	101.7
1995	113.3	101.1

26. **YOU MAKE THE CALL** Samuru made the scatter plot to show the number of papers he has written for school. His parents will buy him a computer if he will be assigned more than four papers next year. Is Samuru correct when he tells his parents that the line of best fit for this data proves that he needs a computer?

27. **WRITING MATH** How does the line of best fit of a scatter plot differ from a line graph? When is one more appropriate than the other?

EXTENDED PRACTICE EXERCISES

28. **CRITICAL THINKING** Does a correlation imply that one event causes the other event to occur? Give an example to support your answer.

29. Calculate the simple interest of 5% on $100 for 8 years. Then plot the ending balances for years 1–8.

30. What is special about the scatter plot from Exercise 29? In what other situations would this type of scatter plot occur?

MIXED REVIEW EXERCISES

Use the data table at the top of the page to answer Exercises 31–35. (Lesson 1-6)

31. Make a double line graph to display the data in the table.

32. Make a double bar graph to display the data in the table.

Name the graph—line graph, bar graph or scatter plot—that is most appropriate for each situation. (Lesson 1-6 and 1-7)

33. to compare changes over time

34. to compare women's and men's times for a particular year

35. to see if two sets of data follow a similar pattern

DATA FILE For Exercises 36–39, use the data on dinosaur height on page 518. (Lesson 1-6)

36. Make a bar graph of the heights of the dinosaurs shown in the table.

37. Which dinosaur was the tallest?   38. Which dinosaur was the shortest?

39. How many dinosaurs were over 25 ft tall?

40. Add: $\frac{1}{2} + 3\frac{1}{2} + 5\frac{1}{4} + 2\frac{1}{4} + 10\frac{1}{2}$ (Basic Math Skills)

Box-and-Whisker Plots

Goals ■ Read and create box-and-whisker plots.

Applications Business, Sports, Fitness

In 1974 Hank Aaron broke Babe Ruth's career home-run record. The number of home runs Aaron hit each year is shown.

1. Write the number of home runs in numerical order.

2. Draw a box around the middle half of the data.

3. Find and compare the range of the entire set of data and the range of the middle half of the data.

4. Find the mean number of home runs Aaron hit.

Hank Aaron's Career Home Runs

Year	Home runs
1954	13
1955	27
1956	26
1957	44
1958	30
1959	39
1960	40
1961	34
1962	45
1963	44
1964	24
1965	32
1966	44
1967	39
1968	29
1969	44
1970	38
1971	47
1972	34
1973	40
1974	20
1975	12
1976	10

◤ BUILD UNDERSTANDING

Box-and-whisker plots identify trends and summarize information. The distribution of data is divided into four equal parts. The three numbers that separate a set of data into these four parts are called **quartiles**. The box represents the middle 50% of the data.

Example 1

Make a box-and-whisker plot for Hank Aaron's home run data.

Solution

Arrange the data in numerical order and find the median. The median of the data set is the second quartile. Identify the median of the lower half of the data, which is the first quartile. The median of the upper half of the data is the third quartile.

10 12 13 20 24 26 27 29 30 32 34 34 38 39 39 40 40 44 44 44 44 45 47

Least value   First quartile   Median   Third quartile   Greatest value

Above a number line, graph the median, the first and third quartiles, the least value and the greatest value. Draw a box from the first quartile to the third quartile. Draw a vertical line at the median. Draw horizontal lines to connect the least value to the first quartile and the greatest value to the third quartile. These lines are called *whiskers*. Label the box-and-whisker plot.

Hank Aaron's Career Home Runs

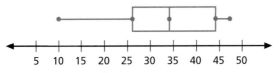

5 10 15 20 25 30 35 40 45 50

Think Back

The median is the middle value when the data are arranged in numerical order. To find the median in an even number of data, add the two middle values and divide by 2.

Example 2

Use the box-and-whisker plot to answer the following questions.

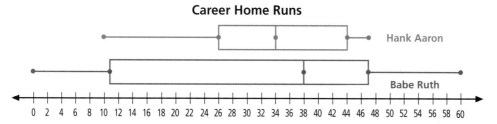

Career Home Runs

Hank Aaron

Babe Ruth

0 2 4 6 8 10 12 14 16 18 20 22 24 26 28 30 32 34 36 38 40 42 44 46 48 50 52 54 56 58 60

a. Who hit the most home runs in a single season? Who hit the least?

b. Who has the highest median?

c. Who hit in the middle 50% most frequently? What does this mean?

d. Which player had more years with a higher number of home runs than his personal average?

Solution

a. Ruth has the lowest and highest number of home runs in one season, 0 and 60 respectively.

b. Ruth has the highest median of 38.

c. The range of the middle 50% for Aaron is 18, and for Ruth it is 36. This indicates that Aaron hit home runs more consistently near the median.

d. Aaron's greatest value is 7 more than the third quartile, and Ruth's greatest value is 13 more than the third quartile. This indicates that Ruth had more high years.

◥ TRY THESE EXERCISES

1. What percent of the data does the box of a box-and-whisker plot represent?

2. What does the vertical bar in the box of a box-and-whisker plot represent?

3. When is the median of a box-and-whisker plot not in the center of the box?

FITNESS Two gyms gather data about the number of customers that visit each location. For each gym's data, find the following.

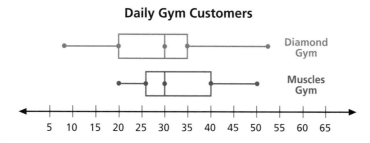

Daily Gym Customers

Diamond Gym

Muscles Gym

5 10 15 20 25 30 35 40 45 50 55 60 65

4. median   5. first quartile   6. third quartile   7. range

8. least number of customers   9. greatest number of customers

10. For each gym, find the range of the middle 50%.

Refer to the box-and-whisker plot.

Treadmill Prices

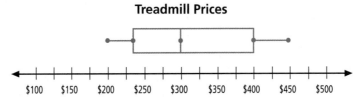

11. What are the highest and lowest prices?

12. What is the range of the prices?

13. What is the median price?

14. What is the range of the middle 50%?

15. In which interval are the prices most closely clustered? (What is the typical price range of a treadmill?)

16. **WRITING MATH** How does a box-and-whisker plot differ from other graphs, such as bar graphs, line graphs and scatter plots?

17. **DATA FILE** Refer to the data on the number of cars and bikes in different countries on page 535. Outliers are values that are far from the rest of the data set. They are marked with bullets in a box-and-whisker plot. Are there any outliers in the data? If so, what are they?

18. Make a box-and-whisker plot for the percent of carbohydrates in brands of cereal. Data shown represents percent listed on nutrition labels.

56	3	48	22	32	44	42	42
40	21	37	36	8	32	30	40
26	55	5	46	44	38	41	33

The box-and-whisker plot below displays the data in the two tables.

19. What is the range for each set of data? Which range is greater?

20. What is the median for each set of data? Which median is greater?

21. Identify any outliers.

22. Which countries produce energy at an amount that falls in the middle 50%?

23. Which countries use energy at an amount that is in the middle 50%?

Countries that Produce the Most Energy

Country	Btus (quadrillion)
United States	72.58
Russia	39.68
China	37.41
Saudi Arabia	20.39
Canada	17.29
Great Britain	11.49
Iran	9.60
India	9.33
Norway	9.28
Venezuela	8.84

Countries that Use the Most Energy

Country	Btus (quadrillion)
United States	93.36
China	37.04
Russia	25.98
Japan	21.37
Germany	14.44
Canada	12.20
India	11.55
Great Britain	10.05
France	9.87
Italy	7.63

Countries that Produce and Use the Most Energy (quadrillion Btus)

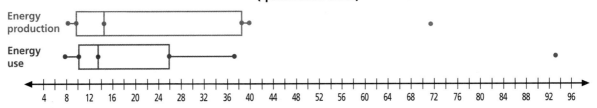

DATA FILE Refer to the data on the ten most popular boys' and girls' athletic programs on page 532.

24. Draw a box-and-whisker plot of the number of participants using different colors for boys and girls. Round to the nearest hundred thousand if you are not using a calculator.

25. Does the plot for boys or for girls have the greatest median?

26. Do boys or girls have the greatest range in the middle 50%?

27. Which of the following can be determined by looking at a box-and-whisker plot: mean, median, mode, range?

28. **STATISTICS** Use a graphing calculator to graph a box-and-whisker plot of the data on boys from Exercise 24. Enter the data using the LIST feature and set the Xmin, Xmax and Xscl for your viewing window. Ignore the y-value settings of the viewing window. Experiment with the x-value settings to find an appropriate viewing window so that you can interpret the plot.

29. Enter the data on girls in a second list in your graphing calculator. Graph both box-and-whisker plots at the same time. Adjust the x-value settings of your viewing window.

30. **CHECK YOUR WORK** Use the LIST and STATISTIC features of your graphing calculator to verify answers to Exercises 28–29.

EXTENDED PRACTICE EXERCISES

Refer to the data in the table.

31. Draw a box-and-whisker plot for the data.

32. What is the range of the middle 50% of the data?

33. Why is the median closer to the left side of the box?

34. Compare the median price to the mean price, and explain any differences.

35. **CRITICAL THINKING** Describe a set of data in which there is only one whisker in its box-and-whisker plot.

36. **CHAPTER INVESTIGATION** Make a box-and-whisker plot of the data from your survey.

MIXED REVIEW EXERCISES

Use the data in the table above for Exercises 37–41. (Lessons 1-3 and 1-6)

37. Round data to the nearest whole dollar. Display data in a stem-and-leaf plot.

38. How many activities cost from $10 to $19?

39. In which range of $10 ($0–$9, $10–$19, and so on) do most activities fall?

40. Make a bar graph to show the data in the stem-and-leaf plot. Make one bar for each range of $10.

41. Which display is more appropriate for showing a comparison of the number of activities at each price range?

42. Which display is more appropriate for showing a range of data?

Cost of Activities

Activity	Price (dollars)
Zoo	10.60
Museum	6.89
Go carts	3.18
School sporting event	2.00
Bowling	2.39
Golfing	47.70
Movie theater	7.75
Miniature golf	3.27
Indoor rock climbing	13.25
Roller skating	5.30
Amusement park	37.09
Movie rental	3.95
Indoor ice skating	7.42
Batting cages	5.48
Snow skiing	53.00
White water rafting	45.58
Racquetball	9.54
Tennis	7.95

Chapter 1 Review

VOCABULARY ◣

Choose the word from the list that best completes each statement.

1. The middle value when data are arranged in numerical order is called the __?__ .

2. A __?__ shows the distribution of data by dividing it into four equal parts.

3. __?__ are pieces of information that can be gathered through interviews, records of events or questionnaires.

4. A tally mark is used in a __?__ to show how often an item appears in a set of data.

5. Trends, or changes, in data over a period of time can be shown in a __?__ .

6. The value that occurs most frequently in a set of data is the __?__ .

7. In __?__ , each member of the population has an equal chance of being selected.

8. Data values that are much greater than or much less than most of the other data are called __?__ .

9. A __?__ demonstrates how the data are divided into categories that do not overlap.

10. In some scatter plots, a __?__ can be drawn that is near most of the data points.

a. box-and-whisker plot

b. circle graph

c. cluster sampling

d. data

e. frequency table

f. line graph

g. line of best fit

h. median

i. mode

j. outliers

k. random sampling

l. range

LESSON 1-1 ◣ Collect and Interpret Data, p. 6

▶ Surveys and polls provide data about a part, or **sample**, of the population. Some ways of sampling are *random, convenience, systematic* and *cluster*.

A population of homeowners is surveyed to determine the benefits of various home-heating systems. Name the kind of sampling that is represented.

11. Every homeowner on streets in four neighborhoods is surveyed.

12. A telephone survey is conducted of names randomly chosen from a phone book.

13. Which of the samples in Exercises 11 and 12 would be the most valuable? Explain.

In a supermarket, various stations are set up for customers to sample and compare products offered in the store.

14. Which type of sampling does this represent?

15. Even though this may not be representative of the entire community, why is this a good method of sampling for this situation?

LESSON 1-2 ◣ Measures of Central Tendency and Range, p. 10

▶ Three statistical measures that help you describe a set of data are *mean*, *median* and *mode*. Because these measures locate centers of a set of data, these terms are called **measures of central tendency**.

▶ The **range** is the difference between the greatest and the least values in a set of data.

One class had the exam scores shown.

53	68	82
61	91	69
72	96	72
67	66	55
79	87	67
90	84	62
98	67	55
81	75	79

16. Find the mean.

17. Find the median.

18. Find the mode.

19. Find the range.

20. What does the value of the range tell about the exam scores?

21. Which measure of central tendency is least like the other two?

22. If you wanted to show how well the class did on their exams, which measure of central tendency would you choose? Explain your reasoning.

LESSON 1-3 ◣ Stem-and-Leaf Plots, p. 16

▶ A **stem-and-leaf plot** organizes and displays data. The last digits of the data values are the *leaves*. The digits in front of the leaves are the *stems*.

Organize the exam grades from Exercises 16–22 into a stem-and-leaf plot.

23. What numbers did you use for the stems in your plot?

24. Which stem has the greatest number of leaves?

25. Which stem has the least number of leaves?

26. Identify any outliers in the data.

27. Where were the gaps of 4 or more in the data?

28. Does this data have any clusters? If so, name them.

LESSONS 1-4, 1-5 and 1-6 ◣ Graphs, p. 20

▶ Each *sector* in a **circle graph** is a percentage of the total number of data. The entire circle represents 100% of the data.

▶ A **pictograph** displays data with graphic symbols.

▶ In a **bar graph**, horizontal or vertical bars display data.

A survey was taken of 60 customers at a grocery store concerning their favorite flavor of ice cream.

29. How many customers prefer chocolate?

30. How many customers prefer rocky road?

31. How many customers prefer vanilla or strawberry?

32. Make a pictograph of the data. Let one symbol represent 3 customers.

33. Make a bar graph for the data.

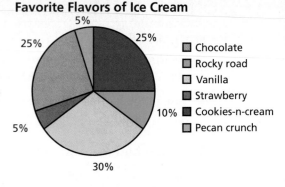

Favorite Flavors of Ice Cream

- ☐ Chocolate
- ☐ Rocky road
- ☐ Vanilla
- ☐ Strawberry
- ☐ Cookies-n-cream
- ☐ Pecan crunch

34. Make a line graph for the data in the table.

Time (hours)	0	0.5	1	1.5	2
Distance (miles)	0	25	50	75	100

LESSON 1-7 ◤ Scatter Plots and Lines of Best Fit, p. 34

▶ A **scatter plot** displays two sets of related data on the same set of axes. A *line of best fit* can be drawn near most of the points.

Use the table on heart rates.

35. Make a scatter plot for the data.

36. Draw a line of best fit for the scatter plot.

37. Is there a positive or negative correlation?

38. Describe what the correlation means.

39. What is the range of the heart rates given?

40. Would you expect the pattern shown by the line of best fit to continue if you ran for 30 minutes? Explain.

Running (min)	Heart rate (beats/min)
0	65
1	68
2	70
3	70
4	75
5	86
6	92
7	100
8	120
9	123
10	140

LESSON 1-8 ◤ Box-and-Whisker Plots, p. 38

▶ A **box-and-whisker** plot identifies trends and summarizes information. The distribution of data is shown by dividing it into four equal parts called *quartiles*.

Use the list of exam grades from Exercises 16–22.

41. What are the least and greatest values?　　**42.** What are the first and third quartiles?

43. Make a box-and-whisker plot.　　**44.** What is the range of the middle 50%?

45. Where are the scores most closely clustered?

46. If 68 were the lowest passing score on this test, what does the box-and-whisker plot show about the percent of the class who passed the test?

CHAPTER INVESTIGATION

EXTENSION Write an article for the school or community newspaper in which you describe your poll. Describe why you decided on the topic and the question, why you chose the particular population to poll, and the data that resulted. Display your data in a table, graph or chart.

Chapter 1 Assessment

Answer each question.

1. To test the quality of a shipment of light bulbs, a sample of 500 was chosen randomly from different lots. In the sample, 14 bulbs were defective. What kind of sampling does this represent?

2. What is an advantage of the type of sampling in Question 1?

Use the circle graph of 780 students for Questions 3–6.

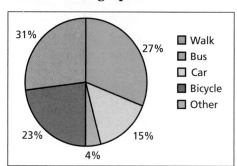

3. Which method of travel is used by the greatest number of students?

4. How many students walk to school?

5. Which method of travel is used by 120 students?

6. Make a pictograph of the data. Let one symbol represent 30 students.

Linda took a survey of people's ages and recorded the data in the stem-and-leaf plot. Use her plot for Questions 7–13.

How many people were in each age group?

7. 0–9   8. 10–19   9. 30–39

How many people were there of each age?

10. 66   11. 15   12. 57

13. Can a scatter plot be drawn from the data? Explain.

```
0 | 7 9 9
1 | 1 4 5 5 6 6 6 8 9
2 | 2 2 3 4 5 5 6 7 7 8
3 | 1 1 1 2 3 3 5 8
4 | 1 2 2 5 5 6 7 9 9
5 | 3 5 6 6 7
6 | 0 2 4 6 6 6
7 | 4 5 8
8 | 6
```

4 | 1 **represents 41 years.**

Find the median of each set of data.

14. 6, 12, 8, 2   15. 115, 115, 130, 110

The bar graph gives the number of tapes sold in one store in a week. On which day were more tapes sold?

16. Tuesday or Wednesday

17. Monday or Thursday

18. Friday or Saturday

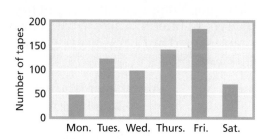

Use the number of television repairs needed during five years for Questions 19–20.

3	2	0	0	4	5	7	3	2	1
1	3	6	4	8	3	10	4	0	2

19. Find the mean, median, mode and range of the data.

20. Draw a box-and-whisker plot of the data.

Standardized Test Practice

Part 1 Multiple Choice

Record your answers on the answer sheet provided by your teacher or on a sheet of paper.

1. A radio newscaster asked every fifth caller to answer the question "Do you think the President is doing a good job?" Which term describes the sample population? (Lesson 1-1)

   Ⓐ random
   Ⓑ clustered
   Ⓒ systematic
   Ⓓ convenience

Use the data for Questions 2 and 3.

Daily Cafeteria Sandwich Sales					
98	124	83	116	89	93
102	126	101	133	105	100
114	132	94	94	122	108
135	112	126	137	136	111
86	130	104	121	138	107

2. The number 55 is the ___?___ of the data. (Lesson 1-2)

   Ⓐ mean
   Ⓑ median
   Ⓒ mode
   Ⓓ range

3. What number would you use for the stems in a stem-and-leaf plot of this data? (Lesson 1-3)

   Ⓐ 1–9
   Ⓑ 8–13
   Ⓒ 4–14
   Ⓓ 10–14

4. The graph represents the ages of people in a survey of Internet users. What percent are represented by the 21–30 age group? (Lesson 1-4)

Age of Internet Users

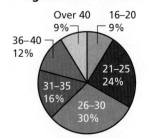

   Ⓐ 24%
   Ⓑ 30%
   Ⓒ 54%
   Ⓓ 70%

5. What are the *most common* and the *least common* shoe sizes in the group sampled? (Lesson 1-5)

Female Students' Shoe Sizes

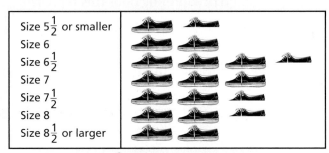

Key: represents 40 students.

   Ⓐ $6\frac{1}{2}$ and $5\frac{1}{2}$ or smaller
   Ⓑ 6 and $8\frac{1}{2}$ or larger
   Ⓒ 6 and $6\frac{1}{2}$
   Ⓓ $5\frac{1}{2}$ or smaller and $6\frac{1}{2}$

The bar graph shows the number of animals in five zoos in the U.S.

Animals in Selected Zoos

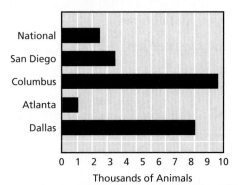

6. Which zoo has about three times the number of animals as the National Zoo? (Lesson 1-6)

   Ⓐ Atlanta
   Ⓑ Columbus
   Ⓒ Dallas
   Ⓓ San Diego

7. The zoo with about 3500 animals is— (Lesson 1-6)

   Ⓐ San Diego
   Ⓑ National
   Ⓒ Dallas
   Ⓓ Atlanta

Part 2 Short Response/Grid In

Record your answers on the answer sheet provided by your teacher or on a sheet of paper.

8. To discuss leadership issues shared by all United States Senators, the President asks 4 of his closest colleagues in the Senate to meet with him. What kind of sampling does this represent? (Lesson 1-1)

The number of boats going through a strait was recorded during 10 days. (Lesson 1-2)

36, 41, 58, 45, 36, 39, 52, 40, 43, 47

9. Find the mean, median, and mode for the data.

10. Suppose the 47 should be 94. Which measure of central tendency would change? Which measure would then best represent the data?

The stem-and-leaf plot shows the number of hamsters sold during one week for various pet stores in the community. Use this plot for Questions 10-13. (Lessons 1-3 and 1-8)

```
0 | 8 9
1 | 5 6 8 8
2 | 2 2 4 5 5 5
3 | 4 4 6 6 9 9 9
4 | 1 1 2 3 3 5 5 6 7 7
5 | 2 3 3 4
6 | 0 0 1
7 | 1
```

6 | 0 represents 60 hamsters.

11. How many stores sold 43 hamsters?
12. Name the mode(s) of the data.
13. Make a box-and-whisker plot of the data.
14. Which part of the graph contains the most closely clustered data?

Test-Taking Tip

(A) (B) (C) (D)

Question 9
Carefully reread your response to make sure that you thoroughly answered each part of the question.

Math Online mathmatters1.com/standardized_test

15. Draw a circle graph representing the data in the table. (Lesson 1-4)

How many free music files have you downloaded from the Internet?	
100 files or less	76%
100–500 files	16%
501–1000 files	5%
More than 1000 files	3%

16. If 1500 people were surveyed, how many downloaded 100 files or less? (Lesson 1-4)

Part 3 Extended Response

Record your answers on a sheet of paper. Show your work.

17. Students were asked how far they travel, in miles, from home to school each morning. (Lesson 1-3)

1.7	4.3	5.6	0.5	2.5	3.1	3.7
1.9	5.0	6.8	3.4	3.7	2.1	4.3
6.2	0.8	2.4	3.7	2.5	0.9	1.9
5.6	7.4	4.3	3.6	13.6	8.1	5.0

a. Make a stem-and-leaf plot of the data.
b. How many students were surveyed?
c. How many students travel 10 mi or more?
d. How many travel less than 2.6 mi?

18. The chart shows the attendance at an amusement park during a peak week in the summer.

Sun	45,643
Mon	16,347
Tue	19,346
Wed	26,477
Thu	36,545
Fri	42,463
Sat	48,658

a. What is the mean? (Lesson 1-2)
b. What type of graph would you use to display these data? Explain your choice. (Lessons 1-4, 1-5, 1-6)

Measurement

THEME: Travel and Vacation

Units of measure, perimeter, area, and ratios can help you understand and enjoy travel, sightseeing and vacation. Traveling uses mathematical skills associated with measurement. Long before you begin to plan a vacation, people in many types of careers perform tasks that ensure your trip is a success.

- Hotels and resorts employ **concierges** (page 61) to assist travelers by providing services. Concierges use measurements of time, temperature, cost, and distance to help make a person's stay more enjoyable.

- **Cartographers** (page 79) use precision and accuracy combined with proportions and scale drawings to create maps used by travel agents, transportation specialists, families, and others.

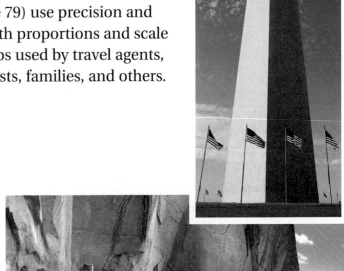

Math Online

mathmatters1.com/chapter_theme

National Parks and Monuments

Park or Monument	Location	Year Dedicated	Estimated Number of Annual Visitors	Height	Total Acreage
Aztec Ruins National Monument	Aztec, NM	1934	63,000	NA	320
Booker T. Washington National Monument	Hardy, VA	1956	25,000	NA	224
George Washington Carver National Monument	Diamond, MO	1953	49,000	NA	210
Mount Rushmore National Monument	Keystone, SD	1939	2,598,600	60 ft*	NA
Navajo National Monument	Tonalea, AZ	1909	100,000	NA	360
Pu'uhonua o Honaunau National Historical Park	Honaunau, HI	1961	478,000	NA	182
Statue of Liberty	New York, NY	1886	over 4,000,000	305 ft	NA
Washington Monument	Washington, DC	1885	841,000	555 ft	NA

* faces from chin to crown

Data Activity: National Parks and Monuments

1. Which national park or monument has about twice as many visitors annually as the George Washington Carver National Monument?

2. Suppose the Navajo National Monument had 100,000 visitors each year since it was dedicated, and the Pu'uhonua o Honaunau National Historical Park had 478,000 visitors each year since it was dedicated. Which monument has had the most visitors to date?

3. What is the difference in square feet between the largest park and the smallest park? (Hint: 1 acre = 43,560 ft^2)

CHAPTER INVESTIGATION

The Statue of Liberty is a monument to an international friendship between France and the United States. The 225-ton statue was transported to New York in 350 pieces that were packed in 214 crates. The monument was reassembled in the New York Harbor in 1886. The Statue of Liberty was completely restored in time for her centennial on July 4, 1986.

Working Together

Use a photograph and the dimensions given in the Data File to make a scale drawing of the Statue of Liberty. Your scale drawing will be a simple representation of the statue. Use the Chapter Investigation icons to guide your group to a finished drawing.

Are You Ready?

Refresh Your Math Skills for Chapter 2

The skills on these two pages are ones you have already learned. Use the examples to refresh your memory and complete the exercises. For additional practice on these and more prerequisite skills, see pages 536–544.

MEASUREMENT

You can use English or metric units of measure to find mass, distance and capacity. You use these measurements every day.

Write *true* or *false* for each statement.

1. You use grams to measure distances.

2. The customary unit of measure for distances smaller than 1 ft is the inch.

3. A ruler can be used to measure the mass of an object.

4. "Milli" is a prefix meaning 100.

5. Kilometers are larger than meters.

6. The metric system is based on powers of ten.

SIMPLIFYING FRACTIONS

English measurements are often given in fractions. You may need to simplify fractions to write a measurement as a fraction.

Examples

$$3 \text{ ft } 10 \text{ in.} = \underline{\ ?\ } \text{ ft} \qquad 126 \text{ in.} = \underline{\ ?\ } \text{ ft}$$

$$3 \text{ ft } 10 \text{ in.} = 3\frac{10}{12} \text{ ft} \qquad 126 \text{ in.} = \frac{126}{12} \text{ ft}$$

$$\frac{(10 \div 2)}{(12 \div 2)} = \frac{5}{6} \qquad \frac{126}{12} = 10\frac{6}{12} = 10\frac{1}{2}$$

$$3\frac{10}{12} \text{ ft} = 3\frac{5}{6} \text{ ft} \qquad 126 \text{ in.} = 10\frac{1}{2} \text{ ft}$$

Write as a mixed number.

7. 14 ft 8 in.

8. 9 lb 6 oz

9. 12 gal 2 qt

10. 22 lb 12 oz

11. Write 46 in. in feet.

12. Write 49 ft in yards.

13. Write 12,320 ft in miles. There are 5280 ft in one mile.

14. Write 258 c in gallons. There are 16 c in one gallon.

FRACTION OPERATIONS

English measurements often require the use of fractions. Knowing how to compute fractions will make working with measurements easier.

Examples

Addition	Subtraction	Multiplication	Division

$$6\frac{1}{8} = 6\frac{1}{8}$$
$$+ 4\frac{1}{2} = 4\frac{4}{8}$$
$$10\frac{5}{8}$$

$$10\frac{1}{3} = 10\frac{8}{24} = 9\frac{32}{24}$$
$$- 6\frac{5}{8} = 6\frac{15}{24} = 6\frac{15}{24}$$
$$3\frac{17}{24}$$

$$\frac{3}{8} \cdot \frac{4}{9} = \frac{12}{72}$$
$$\frac{1\cancel{3}}{\cancel{2}8} \cdot \frac{1\cancel{4}}{\cancel{3}9} = \frac{1}{6}$$

$$1\frac{3}{5} \div \frac{2}{3} =$$
$$\frac{8}{5} \div \frac{2}{3} =$$
$$\frac{4\cancel{8}}{5} \cdot \frac{3}{\cancel{2}} = \frac{12}{5} = 2\frac{2}{5}$$

Perform each operation. Write answers in simplest form.

15. $2\frac{3}{8} + \frac{13}{4}$

16. $5\frac{1}{4} + 2\frac{5}{6}$

17. $6\frac{2}{9} - 2\frac{2}{3}$

18. $12\frac{7}{16} - 5\frac{3}{8}$

19. $\frac{7}{9} \cdot \frac{1}{5}$

20. $2\frac{1}{3} \cdot 6\frac{1}{2}$

21. $\frac{3}{5} \div \frac{9}{10}$

22. $4\frac{2}{3} \cdot \frac{1}{8}$

QUADRILATERALS

Quadrilaterals are figures with four sides. Parallelograms, rectangles, and squares are three kinds of quadrilaterals. Being able to identify quadrilaterals is helpful when applying formulas for area.

Examples

Parallelogram   Rectangle   Square

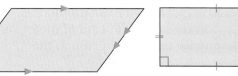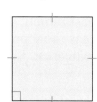

Identify each shape. Write all the names that apply for each shape.

23.

24.

25.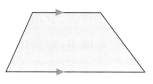

26.

27.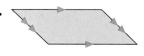

28.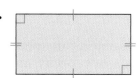

2-1 Units of Measure

Goals ■ Choose appropriate units of measure.
■ Estimate measures.

Applications Travel, Construction, Cooking, Fashion

Work in small groups. Discuss measuring tools you used in the last month.

1. Set up a table like the one shown. Complete six more rows of the table.

Object measured	Tool used	Units labeled on tool (✔)	Exact measure (✔)
Distance between two cities on a map	Fingernail		
Milk	Liquid measuring cup	✔	✔

2. Name four situations when exact measures are not necessary.

▼ BUILD UNDERSTANDING

Measurement is the process used to find sizes, quantities, or amounts. A measure that is nearly exact is said to be **accurate**. Using the most appropriate tool for a measurement will result in a more accurate measure. For example, using a tape measure around a cylinder is more accurate than using a wooden ruler.

A measurement is always an approximation—it is never exact. **Precision** of a measurement is the exactness to which a measurement is made. A measure is more precise when you use a smaller unit of measure. Measuring to the nearest $\frac{1}{16}$ in. is more precise than measuring to the nearest $\frac{1}{2}$ in.

Example 1

Which unit of measure gives a more precise measurement?

a. meter, centimeter   **b.** pint, quart

Solution

a. A centimeter is more precise since it is a smaller unit of measure.

b. A pint is more precise since it is a smaller unit of measure.

Linear objects are measured using tools such as a ruler, yardstick or meter stick. Liquids, or capacity, are measured with cups, measuring spoons or beakers. Scales measure mass or weight. There are many other measuring tools such as a compass, a protractor, an eye dropper, a caliper and a stop watch.

Estimation provides an approximate measure when an exact measure is not necessary or available. Three estimation strategies are shown.

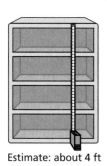

Estimate: about 4 ft

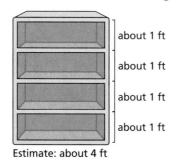

about 1 ft

about 1 ft

about 1 ft

about 1 ft

Estimate: about 4 ft

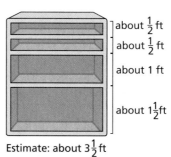

about $\frac{1}{2}$ ft

about $\frac{1}{2}$ ft

about 1 ft

about $1\frac{1}{2}$ ft

Estimate: about $3\frac{1}{2}$ ft

Example 2

Choose the appropriate unit to estimate the measure.

a. width of this textbook: inch, foot, or yard **b.** fruit juice in a bottle: milliliter, liter, or kiloliter

Solution

a. inch   **b.** liter

Estimates include a unit of measure in order to be understood. To say "the estimated length of a desk is 5" is unclear. However, 5 ft is a clear estimation.

Example 3

Choose the best estimate for each measure.

a. length of a classroom: 30 cm, 30 ft, or 30 yd

b. diameter of a basketball: 24 ft, 24 mm, or 24 cm

Solution

a. 30 ft   **b.** 24 cm

TRY THESE EXERCISES

Which unit of measure gives a more precise measurement?

1. meter, decimeter   **2.** inch, centimeter   **3.** yard, foot

Name the tool or tools that would best measure each object.

4. width of a door   **5.** a person's waist   **6.** cooking oil for a recipe

Choose the appropriate unit to estimate the measure.

7. flour in a bread recipe: cup, teaspoon, or liter

8. diameter of a quarter: meter, decimeter, or millimeter

Choose the best estimate for each measure.

9. length of a newborn baby: 47.5 cm, 47.5 mm, or 47.5 ft

10. juice in a can: 355 oz, 355 pt, or 355 mL

11. TRAVEL Michael measures a distance on a map by using his fingernail as a unit of measure. He estimates the length of his fingernail to represent 30 mi. Is this an accurate way of measuring the distance of a trip? Explain.

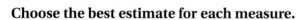

▼ PRACTICE EXERCISES • For Extra Practice, see page 549.

Which unit of measure gives a more precise measurement?

12. milliliter, liter

13. ounce, pound

14. mile, yard

15. decimeter, centimeter

16. gram, kilogram

17. teaspoon, tablespoon

Name the tool or tools that would best measure each object.

18. human foot

19. distance on a map

20. the fastest runner

21. body temperature

22. width of a football field

23. picture frame

Choose the appropriate unit to estimate the measure.

24. mass of a hamster:
gram, kilogram, or metric ton

25. fabric for a blouse:
yard, millimeter, or mile

26. box of cereal:
milligram, ounce, or pound

27. adult dose of cough syrup:
teaspoon, liter, or cup

28. height of a tree:
centimeter, mile, or foot

Choose the best estimate for each measure.

29. water in a fishbowl: 1 pt, 1 c, or 1 gal

30. height of a skyscraper: 300 m, 300 mi, or 300 in.

31. length of an eyelash: 6.3 cm, 6.3 in., or 6.3 mm

32. mass of a paper clip: 1 g, 1 lb, or 1 cm

33. water in a swimming pool: 2000 g, 2000 gal, or 2000 c

34. WRITING MATH A painter is preparing to paint a client's kitchen. She has to estimate the amount of paint needed to complete the job. How accurate must her estimate be? Does her estimate have to be precise?

35. FASHION How accurate or precise must a tailor be when fitting a client for clothing? Should a tailor estimate measurements? Explain.

Measure the length of a dollar bill to the given unit of measure.

36. nearest inch

37. nearest $\frac{1}{4}$ in.

38. nearest $\frac{1}{16}$ in.

For Exercises 39–42, you need a ruler and three objects to measure.

39. Sketch these three objects, clearly marking how you intend to measure each one.

40. Estimate each measurement and write the method used.

41. Measure each item accurately. Record the measurements and tools used.

42. Did you overestimate or underestimate? How did you determine this?

43. DATA FILE Refer to the data on the water supply of the world on page 528. Why is the volume of large bodies of water measured in cubic miles, not gallons?

Create a table like the one shown. Complete the Estimate column first.

44. length of a pencil

45. length of a stamp

46. length of a key

47. width of your hand

48. width of a calculator

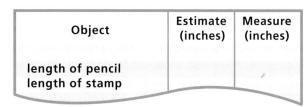

Object	Estimate (inches)	Measure (inches)
length of pencil		
length of stamp		

■ EXTENDED PRACTICE EXERCISES

GRAPHING If an estimate is the same as the actual measurement then the point (measurement, estimate) lies on line *p*. For example, if an object measures 5 in., and you estimated it to be 5 in., then the graph of its point lies on line *p*.

Use your completed table from Exercises 44–48 and the LIST feature of your graphing calculator to find the following.

49. Plot the data from the table where the coordinates of the points are (measurement, estimate).

50. Identify the objects overestimated.

51. Identify the objects underestimated.

52. By looking at the graph, how can overestimates and underestimates be determined?

■ MIXED REVIEW EXERCISES

Use the following data. (Lesson 1-3)

   36 48 22 34 47 41 36 25 29 32 37 44 26 29 22 31 38 39 44 43 41 20

53. Display the data in a stem-and-leaf plot.

54. What are the outliers?

55. Where does the data cluster?

56. Display the data in a box-and-whisker plot. (Lesson 1-8)

57. Give the difference between the value of the mean and the value of the median of the data at the right. (Lesson 1-2)

58. Explain how you could use a pictograph to show the relative values of U.S. coins to one another. (Lesson 1-5)

7	9	2
3	8	9
5	1	9
4	3	6
8	0	1

2-2

Work with Measurements

Goals
- ■ Convert units of measure.
- ■ Perform basic operations using units of measure.

Applications   Fitness, Hobbies, Entertainment, Travel

DATA FILE   Refer to the data on the longest bridge spans in the world on page 520.

1. Explain why you think the lengths are given in feet rather than miles and in meters rather than kilometers.

2. If you change the bridge spans to miles, will the number of units be greater or fewer than the number of units in feet? Explain.

3. If you change the bridge spans to kilometers, will the number of units be greater or fewer than the number of units in meters? Explain.

▼ BUILD UNDERSTANDING

The metric system is a **decimal system** in which you can convert measurements by multiplying or dividing by powers of ten. The chart shows the relationships of the prefixes used to name metric units.

thousands	hundreds	tens	ones	tenths	hundredths	thousandths
kilo	hecto	deka	no prefix	deci	centi	milli
1 km = 1000 m	1 hm = 100 m	1 dam = 10 m	1 m	1 dm = 0.1 m	1 cm = 0.01 m	1 mm = 0.001 m
1 kg = 1000 g	1 hg = 100 g	1 dag = 10 g	1 g	1 dg = 0.1 g	1 cg = 0.01 g	1 mg = 0.001 g
1 kL = 1000 L	1 hL = 100 L	1 daL = 10 L	1 L	1 dL = 0.1 L	1 cL = 0.01 L	1 mL = 0.001 L

To convert from a larger unit to a smaller unit, multiply (move decimal point to the right) by the appropriate power of ten. To convert from a smaller unit to a larger unit, use division (move decimal point to the left) by the appropriate power of ten.

Example 1

Complete.

a. $10 \text{ m} = \underline{\ ?\ } \text{ cm}$ **b.** $2 \text{ m} = \underline{\ ?\ } \text{ mm}$

c. $40 \text{ cm} = \underline{\ ?\ } \text{ m}$ **d.** $3500 \text{ mm} = \underline{\ ?\ } \text{ m}$

Solution

a. $10 \text{ m} = 10 \cdot 100 \text{ cm} = 1000 \text{ cm}$ **b.** $2 \text{ m} = 2 \cdot 1000 \text{ mm} = 2000 \text{ mm}$

c. $40 \text{ cm} = 40 \div 100 \text{ m} = 0.4 \text{ m}$ **d.** $3500 \text{ mm} = 3500 \div 1000 \text{ m} = 3.5 \text{ m}$

In the English system, multiply by equivalent factors to convert from a larger unit to a smaller unit. To convert from a smaller unit to a larger unit, divide by equivalent factors.

Example 2

Perform the following conversions.

a. Change $\frac{3}{4}$ ft to inches.   **b.** Change 29 qt to gallons.

Problem Solving Tip

The answer to Example 2, part a, is a fraction of a foot. It must be less than 12 in.

Solution

a. Multiply to change from a larger unit to a smaller unit.

$$\frac{3}{4} \cdot 12 = \frac{3 \cdot 12}{4} = \frac{36}{4} = 9 \text{ in.} \qquad \text{1 ft = 12 in.}$$

b. Divide to change from a smaller unit to a larger unit.

$29 \div 4 = 7$ with remainder 1   1 gal = 4 qt

29 qt = 7 gal 1 qt   The remainder has the same unit as the given measurement.

Example 3

FASHION Ramona has a dress pattern that calls for 7 yd of lace. She already has 4 ft of lace. How much lace does she need to buy?

Solution

$7 \cdot 3 \text{ ft} = 21 \text{ ft}$ Change 7 yd to the equivalent number of feet. 1 yd = 3 ft

Find the difference between the amount she needs and the amount she already has.

21 ft − 4 ft = 17 ft

$17 \div 3 = 5$ with remainder 2

Ramona needs to buy 5 yd 2 ft of lace.

Example 4

Calculate. Write each answer in simplest form.

a. 3 lb 9 oz
    + 1 lb 9 oz

b. 8 gal 1 qt
    − 6 gal 2 qt

c. 3(4 ft 9 in.)

d. 5L $\div$ 8 = __?__ mL

Solution

a. 3 lb 9 oz
    +1 lb 9 oz

    4 lb 18 oz   18 oz is more than 1 lb.

    4 lb 18 oz = 4 lb + (1 lb + 2 oz)
        = (4 lb + 1 lb) + 2 oz
        = 5 lb 2 oz

c. 3(4 ft 9 in.) = 12 ft 27 in.

Rewrite 27 in. using feet and inches.

12 ft 27 in. = 12 ft + (2 ft + 3 in.)

    = 14 ft 3 in.

b. Rewrite 8 gal 1 qt, since 2 qt can't be subtracted from 1 qt.

8 gal 1 qt = (7 gal + 4 qt) + 1 qt = 7 gal 5 qt

    7 gal 5 qt
  − 6 gal 2 qt

    1 gal 3 qt

d. Since 1 L = 1000 mL, 5 L = 5000 mL.

5 L $\div$ 8 = 5000 mL $\div$ 8 = 625 mL

Complete.

1. 5 ft = __?__ in.
2. 1 gal = __?__ c
3. $3\frac{1}{4}$ lb = __?__ oz
4. 1760 yd = __?__ mi
5. 2 km = __?__ m
6. 4000 g = __?__ kg

7. **FITNESS** Levona jogged 520 yd. How many more yards must she jog before she has jogged 1 mi?

Complete. Write each answer in simplest form.

8.   4 c 6 fl oz
     + 1 c 4 fl oz

9.   1 yd 2 ft
     ×   5

10.   7 lb 2 oz
     −   10 oz

11. 5$\overline{)6\text{ ft 3 in.}}$

12. 200 mL + 750 mL = __?__ L
13. 3 kg – 145 g = __?__ g
14. 4.15 cm · 10 = __?__ mm
15. 7.2 m ÷ 6 = __?__ cm

Complete.

16. 48 fl oz = __?__ pt
17. 30 in. = __?__ ft
18. 8 L = __?__ mL
19. 4 yd = __?__ ft
20. $3\frac{1}{2}$ T = __?__ lb
21. 112 in. = __?__ ft __?__ in.
22. 32 oz = __?__ lb
23. 56 in. = __?__ ft __?__ in.
24. 5 qt = __?__ gal __?__ qt
25. 112 oz = __?__ lb
26. 12.75 g = __?__ mg
27. 46 ft = __?__ yd __?__ ft
28. 7 m = __?__ cm
29. 2 L = __?__ mL
30. 46 oz = __?__ lb __?__ oz
31. 1254 mg = __?__ g
32. 5.5 km = __?__ m
33. 3500 g = __?__ kg

Complete. Write each answer in simplest form.

34.   6 lb 12 oz
     + 4 lb 9 oz

35.   9 ft 2 in.
     − 8 ft 8 in.

36.   8 yd
     − 2 yd 1 ft

37.   5 gal 2 qt
     +1 gal 3 qt

38.   2 c 3 fl oz
     ×   4

39.   1 lb 8 oz
     ×   6

40. 5 yd 1 ft ÷ 2

41. 1 ft 8 in. ÷ 4

42. 3 qt 1 pt = __?__ pt
43. 4 · 65 mm = __?__ cm
44. 1.6 L ÷ 5 = __?__ mL
45. 1.25 cm · 14 = __?__ mm
46. 120 mL + 860 mL = __?__ L
47. 1.9 kg – 1.25 kg = __?__ g
48. 250 mm ÷ 4 = __?__ cm
49. 2.4 kg + 600 g = __?__ kg

50. **HOBBIES** Maria bought 1224 cm of ribbon for her art picture at $0.75/m. How much did she pay for ribbon, to the nearest cent?

51. **WRITING MATH** Don is making loaves of raisin bread, and the recipe calls for 2 lb of raisins. He has three 12-oz boxes of raisins. Explain the math that Don must use to find out if he has enough raisins for the bread.

52. Chun knows that his size $10\frac{1}{2}$ shoe is about 12 in. long. By walking slowly along the length of a room with one foot touching the other, heel to toe, Chun can estimate the length of the room. Chun's son, Kim, wears a shoe that is half as long as Chun's. If Kim uses his father's method of measuring the length of a 24-ft long room, how many steps will Kim take?

53. **ENTERTAINMENT** The running time on the videotape of a movie that Dina rented was 108 min. Dina began watching the movie at 7:45 P.M. At what time did the movie end?

54. **DATA FILE** Refer to the data on annual emissions and fuel consumption for an average passenger car on page 528. How many milligrams per mile of nitrogen oxides are emitted into the air?

■ EXTENDED PRACTICE EXERCISES

Write each answer in simplest form.

55. $\begin{array}{r} 2 \text{ gal } 2 \text{ qt } 1 \text{ pt} \\ +1 \text{ gal } 3 \text{ qt } 1 \text{ pt} \\ \hline \end{array}$

56. $\begin{array}{r} 4 \text{ yd } 1 \text{ ft } 3 \text{ in.} \\ -1 \text{ yd } 2 \text{ ft } 9 \text{ in.} \\ \hline \end{array}$

57. $\begin{array}{r} 4 \text{ gal} \\ -1 \text{ gal } 3 \text{ qt } 1 \text{ pt} \\ \hline \end{array}$

Replace ■ with the measurement that makes each statement true.

58. 7 lb 6 oz + ■ = 18 lb 2 oz

59. ■ ÷ 6 = 1 ft 2 in.

60. ■ − 3 ft 5 in. = 1 ft 7 in.

61. 5 · ■ = 2 c 4 fl oz

62. **CHAPTER INVESTIGATION** Refer to the data on the Statue of Liberty on page 521. Leave out the size of the fingernail for this activity. Round each English dimension to the nearest multiple of 2 ft. Round up each metric dimension to the nearest 0.5 m. Your group must choose which system to use in the scale drawing. Make a list of the rounded dimensions.

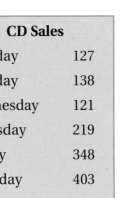

■ MIXED REVIEW EXERCISES

Use the data in the frequency table.
(Lessons 1-5 and 1-6)

63. Display the data in a bar graph.

64. Display the data in a line graph.

65. Which graph makes it easier to compare sales throughout the week?

66. Which graph makes it easier to see how sales rise and fall throughout the week?

CD Sales	
Monday	127
Tuesday	138
Wednesday	121
Thursday	219
Friday	348
Saturday	403

Review and Practice Your Skills

PRACTICE ▰ LESSON **2-1**

Which unit of measure gives a more precise measurement?

1. yard, foot

2. kilogram, milligram

3. pint, gallon

4. liter, milliliter

5. minute, second

6. millimeter, centimeter

Name the tool or tools that would best measure each object.

7. a winning race car

8. depth of a swimming pool

9. indoor temperature

10. growth of a tomato plant

11. cheese for a recipe

12. large dog

Choose the appropriate unit to estimate each measure.

13. weight of calculator: gram, kilogram, ton

14. water in a bathtub: ounce, milliliter, gallon

15. height of a back pack: foot, meter, inch

16. car trip: centimeter, kilometer, yard

Choose the best estimate for each measure.

17. pitcher of lemonade: 1 cup, 1 L, 1 mL

18. picture frame width: 2.5 cm, 2.5 dm, 2.5 m

19. fabric for curtains: 4 yd, 4 mi, 4 in.

20. small paperback book: 150 kg, 150 mg, 150 g

21. head of a nail: 2.3 cm, 2.3 mm, 2.3 m

22. a tricycle ride: 8 cm, 8 yd, 8 km

PRACTICE ▰ LESSON **2-2**

Complete.

23. 90 in. = __?__ ft

24. 4 lb = __?__ oz

25. 850 m = __?__ km

26. 2.3 km = __?__ m

27. 400 mL = __?__ L

28. 5 cups = __?__ fl oz

29. 100 mg = __?__ g

30. 20 ft = __?__ yd __?__ ft

31. 16 kg = __?__ g

32. 14 pt = __?__ qt

33. 25 cm = __?__ mm

34. 3.1 m = __?__ cm

35. $3\frac{1}{2}$ yd = __?__ in.

36. 50 qt = __?__ gal

37. $5\frac{1}{4}$ T = __?__ lb

Complete. Write each answer in simplest form.

38.   2 gal 3 qt
      + 4 gal 1 qt
---

39.   5 lb 2 oz
      − 1 lb 10 oz
---

40.   6 ft 3 in.
      − 1 ft 9 in.
---

41.   8 yd 2 ft
      + 2 yd 2 ft
---

42. 2 ft 3 in.
     ×   5
---

43. 6 c 6 fl oz
     ×   10
---

44. 3 gal 1 qt ÷ 2

45. 4 yd 11 in. ÷ 5

46. 7.5 kg ÷ 3 = __?__ g

47. 2.1 L · 6 = __?__ mL

48. 3 km − 700 m = __?__ km

49. 4.5 mm + 5 cm = __?__ cm

50. 8 m − 600 cm = __?__ cm

51. 40 mL + 1.5 L = __?__ mL

52. 41 g · 20 = __?__ kg

53. 95 km ÷ 50 = __?__ m

54. 4 kg · 2 = __?__ g

PRACTICE ◣ LESSON 2-1–LESSON 2-2

Choose the appropriate unit to estimate each measure. (Lesson 2-1)

55. dimensions of a garden: yard, mile, inch

56. water in a lake: gallon, ounce, kiloliter

57. salt in can of soup: kilogram, gram, milligram

58. length of a car: centimeter, foot, kilometer

Choose the best estimate for each measure. (Lesson 2-1)

59. length of playground: 30 m, 30 cm, 30 km

60. height of a room: 10 yd, 10 ft, 10 in.

61. bag of dog food: 4.75 mg, 4.75 kg, 4.75 g

62. dose of medicine: 10 mL, 10 L, 10 kL

Complete. (Lesson 2-2)

63. 3.5 L = __?__ mL

64. 30 in. = __?__ ft

65. 850 mm = __?__ cm

66. 56 oz = __?__ lb

67. 4.5 g = __?__ mg

68. 10 qt = __?__ gal __?__ qt

Complete. Write each answer in simplest form. (Lesson 2-2)

69. $\begin{array}{r} 6\text{ ft } 2\text{ in.} \\ -\ 2\text{ ft }10\text{ in.} \\ \hline \end{array}$

70. $\begin{array}{r} 4\text{ h }20\text{ min} \\ +\ 3\text{ h }45\text{ min} \\ \hline \end{array}$

71. $\begin{array}{r} 6\text{ qt }1\text{ pt} \\ -\ 1\text{ qt }3\text{ pt} \\ \hline \end{array}$

72. $\begin{array}{r} 4\text{ gal }1\text{ qt} \\ \times\qquad\quad 4 \\ \hline \end{array}$

73. 3 m − 175 cm = __?__ cm

74. 2.3 kg · 3 = __?__ g

75. 2.1 L ÷ 3 = __?__ mL

Math*Works* — Career – Concierge
Workplace Knowhow

Hotels in cities and resorts often offer concierge (kōn-syerzh) services. The concierge provides guests with information about the city or resort. Typical responsibilities include making reservations, securing tickets and providing directions. A concierge estimates distance, cost, temperature and time.

1. An international guest with ballet tickets needs to know the distance to the theater. The distance is 2.5 mi. Estimate the distance in kilometers.

2. The standard taxi fare is $1.44/mi. Estimate the taxi fare from the hotel to the restaurant then to the sports arena. The restaurant is 12 mi from the hotel, and the arena is 7 mi from the restaurant.

Tables in the hotel banquet room are 2 ft by 5 ft. The trays of food from the caterer are 2 ft by 2.5 ft. Each tray of food serves 8 people.

3. How many trays of food can be placed on each table?

4. How many tables does the concierge need to have in the room to serve 80 people?

Perimeters of Polygons

Goals
- Find the perimeter of polygons.
- Solve problems involving perimeter.

Applications Architecture, Design, Landscaping

Use a paper clip as a unit of measure to find the distance around the front cover of a book.

1. How many units long is the cover?

2. How many units wide is the cover?

3. What is the fastest way to find the distance around the cover? Did you discover a shortcut?

4. If another book's cover measures the same distance around but has a different length and width, what are possible dimensions? Explain how you can use the same shortcut described in Question 3.

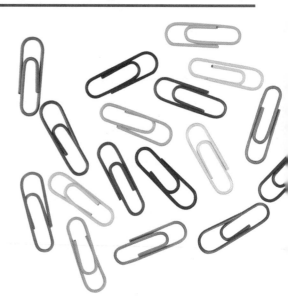

◣ BUILD UNDERSTANDING

The distance around a plane figure is called the **perimeter**. You can find the perimeter by adding the lengths of its sides. Since a rectangle has two pairs of equal sides, you can find the perimeter by doubling the given length and doubling the given width, then adding the results.

Perimeter of a Rectangle	$P = 2l + 2w$ where l is the length and w is the width.

Example 1

Name the figure and find its perimeter.

a.
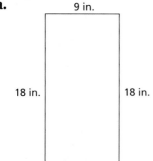

9 in.

18 in.   18 in.

9 in.

b.
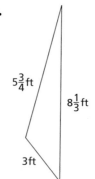

$5\frac{3}{4}$ ft $8\frac{1}{3}$ ft

3 ft

c.
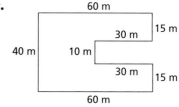

60 m

40 m   10 m   30 m   15 m

30 m   15 m

60 m

Solution

a. Rectangle

$P = 2l + 2w$

$P = 2(18) + 2(9)$

$P = 36 + 18$

$P = 54$ The perimeter is 54 in.

b. Triangle

$P = 3 + 5\frac{3}{4} + 8\frac{1}{3}$ Add the lengths of the sides.

$P = 3 + 5\frac{9}{12} + 8\frac{4}{12}$ Write equivalent fractions with common denominators.

$P = 16\frac{13}{12}$

$P = 16 + 1\frac{1}{12}$

$P = 17\frac{1}{12}$ The perimeter is $17\frac{1}{12}$ ft.

c. Polygon

$P = 40 + 60 + 15 + 30 + 10 + 30 + 15 + 60$

$P = 260$ The perimeter is 260 m.

Example 2

Use a metric ruler to measure the sides of the rectangle in centimeters. Then find the perimeter.

Solution

The length of the rectangle is 5.5 cm. The width is 2 cm.

$P = 2(5.5) + 2(2)$ $P = 2l + 2w$

$P = 11 + 4$

$P = 15$ The perimeter is 15 cm.

A square is a rectangle with four equal sides. The perimeter of a square can be found by multiplying the length of one side by four.

Perimeter of a Square	$P = 4s$ where s is the length of one side.

Example 3

Estimate the perimeter of a square rug with $s = 31$ cm.

Solution

Round 31 cm to 30 cm. Then use the perimeter formula of a square.

$P = 4s$

$P = 4(30)$

$P = 120$ The perimeter of the rug is approximately 120 cm.

Check Understanding

Explain the similarities and differences in the formula for the perimeter of a rectangle and the formula for the perimeter of a square.

Find the perimeter of each figure with its given dimensions.

1.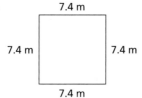

2.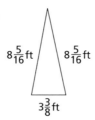

3.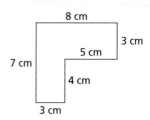

4. square: 2.7 ft

5. rectangle: 7 in., 9 in.

6. triangle: 15 cm, 20 cm, 22 cm

7. Use a ruler to measure the sides of the triangle in inches. Find the perimeter.

Estimate the perimeter of the following objects.

8. picture frame: 5.1 in. by 3.8 in.

9. mirror: 77 cm by 93 cm

10. INTERIOR DESIGN An interior designer needs to order wallpaper border. The room dimensions are 10 ft by 12 ft. How many feet of border should he order?

11. WRITING MATH If the wallpaper border in Exercise 10 only comes by the yard, how would he decide how many yards to order?

For Exercises 12–20, find the perimeter of each figure drawn on the grid. Each unit of the grid represents 1 ft.

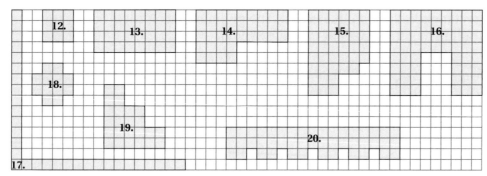

Find the perimeter of each figure with its given dimensions.

21. square: 89 mm

22. rectangle: 4.9 yd, 3.5 yd

23. triangle: 43 cm, 24 cm, 37 cm

24. rectangle: 10 cm, 5 cm

Estimate the perimeter of each object.

25. square rug: 5.8 m

26. calendar: 13.6 in. by 12.8 in.

27. computer monitor: 47 cm by 33 cm

28. wall: 2.8 m by 1.3 m

29. TRAVEL A road travels all the way around Lake Tahoe. The road is approximately 72 mi long. Is the perimeter of the lake greater than or less than 72 mi? Explain.

Use a ruler to measure the sides of each figure in centimeters. Then find each perimeter.

30.

31.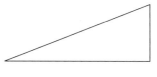

32. Measure the figures in Exercises 30 and 31 to the nearest $\frac{1}{8}$ in. Find the perimeter.

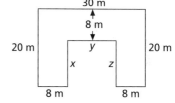
Lake Tahoe, Nevada

ARCHITECTURE Find the perimeter in meters of each rectangular structure in the table.

33. Izumo Shrine

34. Step Pyramid of Zosar

35. Parthenon

36. Palace of the Governors

Noted Rectangular Structures

Structure	Country	Length (meters)	Width (meters)
Izumo Shrine	Japan	10.9	10.9
Palace of the Governors	Mexico	96.0	11.0
Parthenon	Greece	69.5	30.9
Step Pyramid of Zosar	Egypt	125.0	109.0

◼ EXTENDED PRACTICE EXERCISES

Find each unknown dimension in the figure.

37. x

38. y

39. z

40. Find the perimeter.

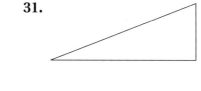

 41. WRITING MATH Study the dimensions of the figure used in Exercises 37-40. Write a description of what such a figure could represent. Why might someone need to know the figure's perimeter?

42. CRITICAL THINKING Find the cost of enclosing a rectangular courtyard that is 12.4 m long and 5.6 m wide with fencing that costs $7.29/m.

◼ MIXED REVIEW EXERCISES

DATA FILE For Exercises 43–46, refer to the data about popular U.S. rollercoasters on page 521. (Lessons 1-1 and 1-2)

43. Find the mean, median and mode of the length of the rollercoasters.

44. Which measure of central tendency is the best indication of the center of the data? Explain.

45. How many rollercoasters are less than 5000 ft long?

46. Is this data a random sample of all U.S. rollercoasters? How can you tell?

Areas of Parallelograms and Triangles

Goals ■ Solve problems using area formulas.

Applications Sports, Landscaping, Construction

Draw a rectangle and a parallelogram on grid paper as shown.

1. Estimate the area of each figure by counting the square units. How do the areas compare?

2. Cut out the parallelogram. Then cut it on the dotted line. Rearrange the two pieces to form a rectangle. How is this new rectangle related to the original?

3. Could you cut the parallelogram another way to form the rectangle?

4. What does this tell you about the areas of parallelograms and rectangles?

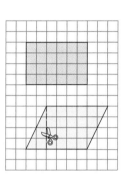

◤ BUILD UNDERSTANDING

Area is the amount of surface enclosed by a geometric figure. Area is measured in square units. Formulas provide a method of calculating the area of particular shapes.

Area of a Rectangle	$A = l \cdot w$ where l is the length and w is the width.

A square is a rectangle with all four sides equal. The formula for its area can be shortened to $A = s^2$, where s is the length of each side.

Example 1

LANDSCAPING How many square yards of sod are needed to cover a professional football field?

Solution

Use the formula for the area of a rectangle.

$A = l \cdot w$

$A = 120 \text{ yd} \cdot 53\frac{1}{3} \text{ yd}$

$A = 6400 \text{ yd}^2$

A professional football field needs 6400 yd² of sod.

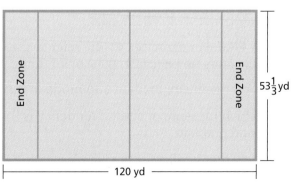

Example 2

SPORTS In major league baseball, a pitcher must throw the ball over home plate and between the batter's chest and knees. The shaded region *ABCD* is called the strike zone. Find the area of the strike zone for the batter shown.

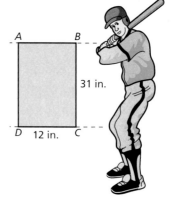

Solution

$A = l \cdot w$

$A = 31 \cdot 12 = 372$ The area of the strike zone is 372 in.²

The area of a parallelogram is related to the area of a rectangle. Because of this relationship, the area formula for a parallelogram is similar to the area formula for a rectangle.

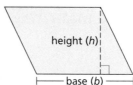

Area of a Parallelogram	$A = b \cdot h$ where *b* is the length of the base and *h* is the height.

Example 3

Find the area of the parallelogram.

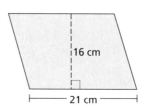

Solution

$A = b \cdot h$

$A = 21 \cdot 16 = 336$

The area of the parallelogram is 336 cm².

The area of a triangle is one-half the area of a parallelogram with the same base and height. The base of a triangle can be any one of its sides. The height of a triangle is the perpendicular distance from the opposite vertex to the base.

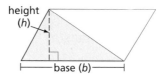

Area of a Triangle	$A = \frac{1}{2} \cdot b \cdot h$ where *b* is the length of the base and *h* is the height.

Example 4

Find the area of the triangle shown.

Solution

$A = \frac{1}{2} \cdot b \cdot h$

$A = \frac{1}{2} \cdot 2.6 \cdot 3.1$

$A = 4.03$

The area of the triangle is 4.03 m².

Find the area of each figure given the dimensions.

1. rectangle: 8 in. by 17 in.

2. square: 10-ft sides

3. parallelogram: $b = 19$ m, $h = 16$ m

4. triangle: $b = 10$ cm, $h = 8$ cm

Find the areas of each strike zone. The width of home plate is 12 in.

5. 39 in.

6. 30 in.

7. 27 in.

8. 35 in.

9. WRITING MATH Why is area given in square units and perimeter given in linear units?

Find the area of each figure given the dimensions.

10. rectangle: $\frac{7}{10}$ cm by $\frac{3}{10}$ cm

11. parallelogram: $b = 27$ m, $h = 70$ m

12. square: 10.01-yd sides

13. triangle: $h = 14.8$ km, $b = 10.6$ km

14. parallelogram: $b = 16$ in., $h = 8$ in.

15. square: 56-mm sides

16. triangle: $h = 308$ mi, $b = 119$ mi

17. rectangle: 30.75 in. by 18.61 in.

SPORTS Find the areas of each former player's strike zone. The dimensions of the strike zone for each player are listed in the table.

18. Jose Canseco

19. Carlton Fisk

20. Pete Incaviglia

21. Rickey Henderson

22. Lance Parrish

Player	Strike zone dimensions (inches)	
	Length	Width
Jose Canseco	27	12
Carlton Fisk	32	12
Pete Incaviglia	37	12
Rickey Henderson	34	12
Lance Parrish	28	12

Find the area of each figure.

23. 24 cm, 24 cm

24. 30 in., 42 in.

25. 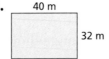 40 m, 32 m

26.  0.8 m, 1.2 m

27. The Great Pyramid's base is a square with sides measuring 746 ft. How many acres (to the nearest whole acre) does it cover? (Hint: 43,560 ft^2 = 1 acre)

28. ERROR ALERT Bruce is a contractor. His client called with the dimensions of the floor he is going to carpet. The dimensions he gave Bruce are 12 ft by 4.5 m. What must Bruce first do to find the area?

29. CONSTRUCTION A contractor needs to install fencing around a ranch resort. The area of the resort is 200 mi². Can she find the perimeter with this information? Why or why not? Draw examples to prove your answer.

30. GEOMETRY SOFTWARE Use geometry software to draw a triangle. Find the area of the triangle. Rotate the triangle, and find the area again. Did the area change? Why or why not?

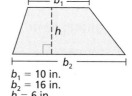

SPORTS Use the dimensions in the table to find the area of each court in square feet and square meters.

31. tennis, singles

32. tennis, doubles

33. badminton, singles

34. badminton, doubles

35. table tennis, table

36. table tennis, floor area

Sport		Court dimensions	
		(feet)	(meters)
Tennis	singles	27 × 78	8.23 × 23.77
	doubles	36 × 78	10.97 × 23.77
Badminton	singles	17 × 44	5.18 × 13.41
	doubles	20 × 44	6.10 × 13.41
Table tennis	table	5 × 9	1.52 × 2.74
	floor area	19 ft 6 in. × 39 ft 4 in.	5.94 × 11.99

▉ EXTENDED PRACTICE EXERCISES

Identify which figure has the greater area.

37. triangle A, $b = 0.704$ m, $h = 0.4$ m

triangle B, $b = 0.75$ m, $h = 0.75$ m

38. square C, $s = \dfrac{154}{22}$ ft

square D, $s = 7.018$ ft

39. A trapezoid has one pair of parallel sides labeled b_1 and b_2. Find the area of the trapezoid shown using the formula $A = \frac{1}{2} h \cdot (b_1 + b_2)$.

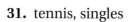

$b_1 = 10$ in.
$b_2 = 16$ in.
$h = 6$ in.

40. CRITICAL THINKING Draw a rectangle, a square, and a triangle that have the same area.

41. WRITING MATH The area of a rectangle is 64 ft². How can this area be changed to square yards?

▉ MIXED REVIEW EXERCISES

Complete. (Lesson 2-2)

42. 32 oz = __?__ lb

43. 5 yd = __?__ ft

44. 25 ft = __?__ yd __?__ ft

45. $2\frac{3}{4}$ T = __?__ lb

46. 5 qt = __?__ pt

47. 56 fl oz = __?__ pt __?__ c

48. HOBBIES Michele needs the following lengths of wood to build a frame— 5 in., 5 in., 6 in., and 6 in. She has a board that is $1\frac{2}{3}$ ft long. Does she have enough wood?

Review and Practice Your Skills

PRACTICE ◼ LESSON 2-3

Find the perimeter of each figure. Each unit of the grid represents 1 ft.

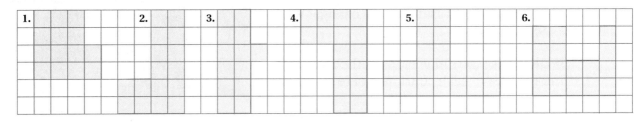

Find the perimeter of each figure.

7. triangle: 13 in., 10 in., 20 in.
8. pentagon: 4 cm, 3 cm, 3 cm, 5 cm, 5 cm
9. rectangle: 4 m by 1.5 m
10. square: 1.5 km
11. square: 300 ft
12. rectangle: 5 yd by 12 yd

Estimate the perimeter of each rectangular object.

13. square table top: 28 in.
14. window: 98 in. by 32 in.
15. small rug: 42 in. by 29 in.
16. square quilt: 1.85 m
17. garden: 28 ft by 19 ft
18. picture frame: 6.8 cm by 4.2 cm

PRACTICE ◼ LESSON 2-4

Find the area of each figure.

19. rectangle: 0.5 m by 2.4 m
20. parallelogram: $b = 10$ mm, $h = 14$ mm
21. square: 12 in. sides
22. square: 0.6 m sides
23. triangle: $h = 20$ cm, $b = 15$ cm
24. triangle: $h = 8$ ft, $b = 11$ ft
25. parallelogram: $b = 3\frac{1}{2}$ yd, $h = 6$ yd
26. rectangle: 40 in. by 25 in.

Find the area of each figure.

27.
28.
29.
30.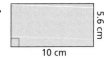

31.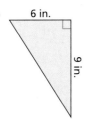
32.
33.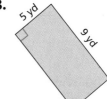
34.

Choose the best estimate for each measure. (Lesson 2-1)

35. 4-hour drive: 200 ft, 200 m, 200 mi

36. heavy letter: 75 kg, 75 g, 75 mg

37. water in small bucket: 1.3 mL, 1.3 L, 1.3 kL

38. front yard of house: 22 yd, 22 mi, 22 in.

Complete. Write each answer in simplest form. (Lesson 2-2)

39. 5.1 kg = __?__ g

40. 100 oz = __?__ lb __?__ oz

41. 40 cm = __?__ mm

42. 85 g + 1.5 kg = __?__ g

43. 1.5 L ÷ 5 = __?__ mL

44. 40 mm · 8 = __?__ cm

Find the perimeter of each figure. (Lesson 2-3)

45. rectangle: 10 in. by 8 in.

46. square: 4 yd

47. triangle: 8 cm, 10 cm, 15 cm

48. pentagon: 5 ft, 2 ft, 2 ft, 4 ft, 4 ft

Find the area of each figure. (Lesson 2-4)

49. triangle: $h = 12$ ft, $b = 7$ ft

50. rectangle: 4.3 cm by 10 cm

51. square: 1.2 cm sides

52. parallelogram: $b = 6$ in., $h = 9$ in.

Mid-Chapter Quiz

Complete. (Lesson 2-2)

1. 18 in. = __?__ ft __?__ in.

2. 16 L = __?__ mL

3. 3.4 kg + 700 g = __?__ kg

4. $\frac{1}{4}$ mi = __?__ ft

5. 345 cm = __?__ m

6. 2.30 cm · 12 = __?__ mm

7. **CHEMISTRY** In her laboratory, Mala has two beakers, each with a capacity of 500 mL, and another with a capacity of 1.25 L. Can the three beakers accommodate 4.2 L of liquid? (Lesson 2-2)

8. If molding costs $1.25/ft, how much does it cost to frame a painting that measures 4 ft by 3 ft? (Lesson 2-3)

9. How many 6 in.-by-9 in. tiles are needed to cover a wall that measures 72 in. by 63 in.? (Lesson 2-4)

10. Which has the greatest area? How much greater is it than the next greatest area? (Lesson 2-4)

rectangle: $l = 5$ m, $w = 7$ m

parallelogram: $b = 4.5$ m, $h = 8$ m

triangle: $b = 9$ m, $h = 4$ m

Problem Solving Skills: Quantity and Cost

Formulas are mathematical tools that can guide you to a solution. The strategy **use an equation or formula** is appropriate when you have data that can be substituted into a formula or equation. To use formulas in solving problems, first choose a formula. For example, to find the amount of wallpaper needed to cover the walls of a room, use a formula for area. To find the amount of fencing needed to enclose a field, use a formula for perimeter.

Problem Solving Strategies

Guess and check

Look for a pattern

Solve a simpler problem

Make a table, chart or list

Use a picture, diagram or model

Act it out

Work backwards

Eliminate possibilities

✔ Use an equation or formula

Problem

LANDSCAPING The rectangular park needs to be enclosed by a fence and covered with sod.

a. How many yards of fencing need to be ordered?

b. The sod costs $1.10/yd^2 and is ordered by the square yard. Find the total cost of sod to cover the entire park.

Solve the Problem

a. To find the amount of fencing, find the park's perimeter.

$P = 2l + 2w$ Use the perimeter formula for a rectangle.

$P = 2(88.5) + 2(72.5)$ $l = 88.5, w = 72.5$

$P = 177 + 145 = 322$

A total of 322 yd of fencing is needed.

b. To find the cost of sod, first find the area of the park.

$A = l \cdot w$ Use the area formula for a rectangle.

$A = 88.5 \cdot 72.5 = 6416.25$

Since sod is ordered by the square yard, order 6417 yd^2.

Then, multiply 6417 by the cost of sod per yard. $6417 \cdot 1.10 = 7058.7$

The total cost of sod to cover the entire park is $7058.70.

88.5 yd

72.5 yd

◥ TRY THESE EXERCISES

Write _P_ or _A_ to tell if you would use a perimeter or an area formula to find the following.

1. amount of fencing

2. amount of wallpaper border

3. amount of carpet

4. amount of fabric in a quilt

5. A garden is 7.9 m by 14 m. How many meters of fence will enclose the garden?

Reading Math

The symbol ' stands for feet. The symbol " stands for inches.

Read 12' × 20'6" as "twelve feet by twenty feet, six inches."

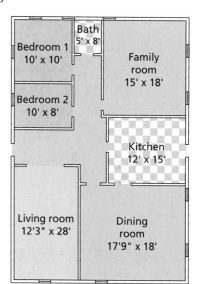

ARCHITECTURE Architects design many types of buildings. They draw plans for houses, such as the plan shown. Use the plan for Exercises 6–27.

An architect wants to install decorative molding around the ceilings in some of the rooms. The decorative molding costs $1.79/ft. Use a formula to find how much molding will be needed for each room.

6. family room

7. kitchen

8. dining room

9. bedroom 1

10. bedroom 2

11. living room

12–17. Find the molding cost for each room in Exercises 6–11.

18. Find the total cost of the molding for all six rooms.

Carpet will be installed in several of the rooms and is ordered by the square yard. Find the area to determine how much carpeting needs to be ordered for each room. (Hint: Convert all dimensions to yards before you use the area formula.)

19. family room

20. living room

21. bedroom 1

22. bedroom 2

23–26. The carpet costs $32/yd^2. Find the cost of carpeting each room in Exercises 19–22.

27. What is the total cost of carpeting all four rooms?

28. WRITING MATH Why do you suppose carpet companies have so many carpet remnants?

AGRICULTURE Use the diagram showing the layout of the farm.

29. What is the area of land used to grow hay?

30. It costs $0.63/ft^2 to fertilize the vegetable garden. What is the total cost?

31. A fence is to enclose the property around the house. The dimensions of the house are 35 ft by 35 ft. At least how many feet of fencing are needed?

32. Each apple tree requires 25 ft^2 of space. How many apple trees can there be in the orchard?

33. How many square feet of the farm are used neither for the house property nor for crops?

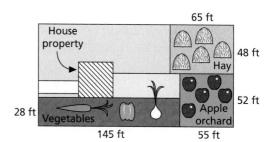

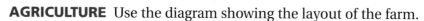

Name the kind of sampling represented by each situation. (Lesson 1-1)

34. A toy store polls every customer about their favorite toy.

35. A student polls every tenth student in the cafeteria line about how they will vote in the Student Council elections.

36. A forest ranger walks 500 yards and tests the water level of all the trees within 5 ft of where she stopped. She continues doing this, walking around the forest perimeter.

2-6 Equivalent Ratios

Goals
- Read and write ratios in lowest terms.
- Find an equivalent ratio.

Applications Horticulture, Cooking, Entertainment, Retail

An **analogy** is a conclusion that if two or more things agree with one another in some respects, then they probably will agree in others.

1. Work with a partner. Supply the missing word in this analogy.

    Hot is to cold as new is to ___?___ .

2. With a partner, make up four new analogies. Exchange papers with another group and find the missing words in the analogies.

3. How does the word relationship in each analogy show comparison?

4. The word *analogy* is from the Greek word *analogos* that means "in due ratio." How is the relationship between two numbers in a ratio similar to the relationship between two words in an analogy?

▼ BUILD UNDERSTANDING

A **ratio** compares one number to another. A ratio can be written five ways.

$$1 \text{ to } 4 \qquad 1:4 \qquad \frac{1}{4} \qquad 0.25 \qquad 25\%$$

When a ratio is written as a fraction, write it in lowest terms. The order of numbers in a ratio always matches the order of their labels. For example, the ratio of 5 hats for 12 people is 5 : 12, not 12 : 5.

Football statisticians keep track of the percent of successful field goals made by an individual kicker using ratios.

$$\text{kicking ratio} = \frac{\text{successful field goal kicks}}{\text{attempted field goals}}$$

If a player has 15 successful kicks out of 20 attempts, that player's kicking ratio is $\frac{15}{20}$, or 0.75. When ratios are expressed as decimals, the comparison is always to 1.

<aside>
Check Understanding

Which way of writing a ratio best shows that it is comparing two numbers by division?
</aside>

Example 1

Write each ratio as a fraction in lowest terms.

a. 12 c to 96 c   **b.** 65 min : 3 h   **c.** 5 m to 2.5 cm

Solution

Convert one of the quantities if necessary so that both terms of the ratio are in the same unit of measure.

a. $\dfrac{12}{96} = \dfrac{1}{8}$ **b.** $\dfrac{65}{180} = \dfrac{13}{36}$ **c.** $\dfrac{500}{2.5} = \dfrac{200}{1}$

If two ratios represent the same comparison, then they are **equivalent ratios**. Equivalent ratios are found by multiplying or dividing each term by the same nonzero number.

Example 2

Find an equivalent ratio. Write the ratio as a fraction.

a. 2 to 3   **b.** $12 : 72$ **c.** $\dfrac{3}{4}$

Solution

Write the ratio as a fraction. The solution will depend on the number by which you choose to multiply or divide the numerator and denominator.

a. $\dfrac{2}{3} = \dfrac{2 \cdot 2}{3 \cdot 2} = \dfrac{4}{6}$ **b.** $\dfrac{12}{72} = \dfrac{12 \div 12}{72 \div 12} = \dfrac{1}{6}$ **c.** $\dfrac{3}{4} = \dfrac{3 \cdot 2}{4 \cdot 2} = \dfrac{6}{8}$

Example 3

HORTICULTURE Carlos feeds his houseplants with a mixture of liquid plant food and water. This mixture calls for 3 parts plant food to 8 parts water. What is the ratio of plant food to water? How much plant food should he mix with 40 fl oz of water?

Solution

The ratio of plant food to water is $\dfrac{3}{8}$. For each additional 8 fl oz of water, Carlos needs an additional 3 fl oz of plant food. Use this pattern to make a table. Extend the table to 40 fl oz of water. Carlos should mix 15 fl oz of plant food with 40 fl oz of water.

Fluid ounces of plant food	3	6	9	12	15
Fluid ounces of water	8	16	24	32	40

◤ TRY THESE EXERCISES

Write each ratio two other ways.

1. 1 to 2   **2.** $\dfrac{4}{8}$ **3.** 10 to 9   **4.** $9 : 13$

Write each ratio as a fraction in lowest terms.

5. 8 cm to 3 cm   **6.** 5 kg to 2 kg   **7.** 25 qt to 5 qt

Find three equivalent ratios.

8. $\dfrac{1}{4}$ **9.** $4 : 6$ **10.** 2 out of 7   **11.** $\dfrac{30}{50}$

Are the ratios equivalent? Write *yes* or *no*.

12. 1 to 3, 25 to 75   **13.** $3 : 2, 15 : 5$ **14.** $\dfrac{7}{3}, \dfrac{42}{18}$

15. EDUCATION Julio answered 51 out of 60 questions correctly on a history test. He answered 17 out of 20 questions correctly on a math quiz. What is the ratio of correct answers to questions for the test and the quiz?

Write each ratio two other ways.

16. 6 to 9   **17.** 10 : 3   **18.** 5 : 4   **19.** $\frac{1}{3}$

20. $\frac{6}{7}$ **21.** 7 : 10   **22.** $\frac{8}{9}$ **23.** 4 to 1

Write each ratio as a fraction in lowest terms.

24. 5 cm : 2 cm   **25.** 4 min to 20 min   **26.** 1.5 m to 30 cm

27. 16 gal to 12 qt   **28.** 2 h to 6 h   **29.** $1.40 : $0.55

Write three equivalent ratios for each given ratio.

30. $\frac{6}{7}$ **31.** 15 to 18   **32.** 3 : 4   **33.** 16 : 20

34. 5 to 3   **35.** $\frac{7}{8}$ **36.** $\frac{8}{9}$ **37.** 3 : 10

Are the ratios equivalent? Write _yes_ or _no_.

38. 4 to 5, 12 to 15   **39.** 8 : 2, 16 : 9   **40.** $\frac{14}{20}, \frac{170}{100}$ **41.** $\frac{8}{14}, \frac{24}{44}$

42. $\frac{6}{7}, \frac{36}{42}$ **43.** 2 : 5, $\frac{100}{250}$ **44.** 30 : 98, 10 : 15   **45.** $\frac{4}{5}, \frac{12}{18}$

Find the ratio of height to base in each rectangle. Express the ratio as a fraction in lowest terms.

46. 24 cm   **47.**   **48.**

1 m

 50 cm
3 m

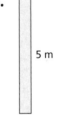
5 m
65 cm

49. DATA FILE Refer to the data about calories spent in activities on page 530. Write ratios for calories spent studying to each weight in the table. Which weight has the greatest ratio? What conclusions can you draw from the different ratios?

COOKING A recipe calls for 4 parts cornstarch to 7 parts water.

50. What is the ratio of cornstarch to water?

51. How much cornstarch should be mixed with 28 parts water?

ENTERTAINMENT The Weiskers are at an amusement park from 10:00 A.M. to 9:00 P.M. It rains a total of 3 h that day.

52. Write a ratio for the total number of hours the Weiskers spend at the park to the number of rainy hours.

53. Write a ratio for the number of rainy hours to dry hours.

54. YOU MAKE THE CALL In Tanesha's class, 12 out of 16 students received a grade of B or better. Tanesha said that 1 out of 4 received less than a B. Is she correct? Explain.

55. RETAIL A car dealership is running a clearance sale. It sells 6 out of 9 midsize cars and 24 out of 45 economy cars. Write the ratio for midsize cars sold and the ratio for economy cars sold. Which ratio is greater?

Write a ratio equivalent to the following.

56. $a : b$, if $a = 5$ and $b = 4$

57. x to y, if $x = 45$ and $y = 50$

58. $\frac{d}{e}$, if $d = 17$ and $e = 25$

◤ EXTENDED PRACTICE EXERCISES

RETAIL Gold jewelry is measured in karats (k), with pure gold being 24 k. In a 10-k gold ring, the ratio of the mass of pure gold to the mass of the ring is 10 : 24.

59. Write the gold content of an 18-k gold ring as a ratio in lowest terms.

60. Write the gold content of a 14-k gold earring as a ratio in lowest terms.

NUMBER SENSE Ratios can be written with three numbers. For example, 30 : 85 : 60 means 30 compared to 85 compared to 60.

61. Write the ratio 30 : 85 : 60 in lowest terms.

62. Write two equivalent 3-number ratios for 30 : 85 : 60.

63. Copy and complete the ratio table shown. The ratio of A to B is 3 : 4. The ratio of B to C is 1 : 3.

part A	?	6	?	12	?
part B	4	?	12	?	20
part C	?	?	?	48	?

64. CHAPTER INVESTIGATION Use your list of rounded dimensions to determine a scale factor. The maximum size poster board is 36 in. by 24 in. For the English system, use 1 in. = ■ ft. For the metric system, use 1 cm = ■ m. The scale factor does not have to be a whole number, but choose a value that provides lengths easily read on a ruler. For example, choosing 1 in. = 10 ft means that the hand length is 1.6 in., which is not easily located on a ruler. In your group, eliminate factors that are not practical. Write your group's scale factor as a fraction.

◤ MIXED REVIEW EXERCISES

Complete. (Lesson 2-2)

65. 5 m = __?__ km

66. 4.6 cm = __?__ mm

67. 500 mL = __?__ L

68. 1.5 kg = __?__ g

69. 6 mm = __?__ cm

70. 423 g = __?__ kg

Convert the following measurements. (Lesson 2-2)

71. 5 qt to cups

72. 1500 m to kilometers

73. 3 kg to grams

74. An eyedropper holds 5 mL of medicine. How many doses of medicine does a 1-L bottle hold? (Lesson 2-2)

Review and Practice Your Skills

PRACTICE ◣ LESSON 2-5

Write *P* or *A* to tell if you would use a perimeter or area formula.

1. cost of cleaning a rug
2. seed to plant a lawn
3. wood for a picture frame
4. trim to edge a tablecloth
5. stones to surround a flower bed
6. size of a living room

Solve.

7. How much flooring is needed for a kitchen that is 15 ft long and 12 ft wide?
8. How much fence is needed around a lot that is 60 ft square?
9. How much carpet do you need for a room that is 18 ft long and 15 ft wide?

PRACTICE ◣ LESSON 2-6

Write each ratio two other ways.

10. $\frac{2}{3}$
11. $1:8$
12. 4 to 9
13. $\frac{10}{3}$

14. 7 to 3
15. 12 to 7
16. $\frac{5}{4}$
17. $5:2$

Write each ratio as a fraction in lowest terms.

18. 25 cm to 10 cm
19. 15 kg : 3 kg
20. 24 oz to 3 lb
21. 2 ft : 6 in.
22. 4 gal to 2 qt
23. 45 sec : 3 min

Write three equivalent ratios for each given ratio.

24. $4:18$
25. $\frac{10}{8}$
26. $6:30$
27. 8 to 20

28. $\frac{3}{15}$
29. 20 to 25
30. $15:10$
31. $\frac{4}{3}$

Are the ratios equivalent? Write *yes* or *no*.

32. 3 to 2, 12 to 10
33. $\frac{2}{3}, \frac{12}{20}$
34. $1:5, 3:15$
35. 2 to 9, $\frac{4}{18}$

36. $12:18, \frac{2}{9}$
37. $6:8,$ 3 to 4
38. $\frac{14}{30}, \frac{7}{15}$
39. $\frac{4}{6}, 3:2$

Find the ratio of width to length in each rectangle. Express the ratio as a fraction in lowest terms.

40.

12 cm

1 m

41.

10 in.

2 ft

Choose the best estimate for each measure. (Lesson 2-1)

42. capacity of pitcher: 60 gal, 60 fl oz, 60 pt   **43.** length of pencil: 8.2 cm, 8.2 mm, 8.2 km

Complete. Write each answer in simplest form. (Lesson 2-2)

44. 2500 lb = __?__ T   **45.** 40 ft = __?__ yd   **46.** 5 m − 125 cm = __?__ cm

Find the perimeter of each figure with the given dimensions. (Lesson 2-3)

47. rectangle: 3.5 m by 2 m   **48.** triangle: 5 ft, 11 ft, 12 ft

Find the area of each figure with the given dimensions. (Lesson 2-4)

49. parallelogram: $b = 2$ yd, $h = 1\frac{1}{2}$ yd   **50.** triangle: $h = 3.5$ cm, $b = 4$ cm

Write *P* or *A* to tell if you would use a perimeter or area formula. (Lesson 2-5)

51. paint to cover a wall   **52.** fence to enclose a playground

Are the ratios equivalent? Write *yes* or *no*. (Lesson 2-6)

53. $\frac{8}{15}, \frac{24}{45}$ **54.** $\frac{2}{3}, 18:12$ **55.** $20:15, 4:3$ **56.** $\frac{10}{18}, 5$ to 6

Math*Works* Workplace Knowhow — Career – Cartographer

A cartographer creates maps. The maps represent areas as small as your neighborhood and as large as the Earth. To represent a specific area on a map, a cartographer creates a scale drawing. Distances on the map are proportionately drawn to the actual distances. The scale of the map varies with the size of the actual area and the size of the map. A cartographer uses ratios, proportions, measurements, angle measurements, measurement tools, and measurement conversions.

Use a ruler to measure each map distance in inches. Then use the scale 1 in. : 15 mi to find the actual distance.

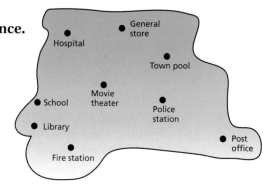

1. school to library

2. town pool to post office

3. fire station to general store

4. Which two locations are farthest apart?

5. Which two locations are closest together?

2-7 Circumference and Area of a Circle

Goals ■ Find the circumference and area of circles.

Applications Architecture, Sports, Landscaping

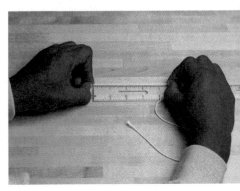

Complete Question 1. Then work with a partner to complete Questions 2–7.

1. Use a compass to draw three circles, each one on a separate paper. Adjust your compass so that each circle has a different radius.

2. Choose one circle. Cut a piece of string the length of the circle's diameter.

3. Carefully fit string around the circle in Question 2. Cut the string so that it represents the distance around the circle.

4. Measure the length of each string in centimeters. Record lengths in a table with columns named *diameter* and *distance around*.

5. Repeat Questions 2–4 for the other circles.

6. Add a column to your table named $\frac{\text{distance around}}{\text{diameter}}$. Calculate this ratio for each row. Round to the nearest hundredth.

7. Describe the ratios calculated in Question 6. What significance does this have to circles? Name the symbol for this ratio.

Math: Who, Where, When

Srinivasa Ramanujan (1887–1920) was a mathematician from India. He is credited with over 6000 theorems, including a formula for calculating a decimal approximation for π. His formula was similar to one used by mathematicians in 1987 when they determined the value of π to more than 100 million decimal places.

▼ BUILD UNDERSTANDING

A circle is a set of all points in a plane that are a fixed distance from a given point in the plane.

The point is the center of the circle, and the fixed distance is the **radius**. The **diameter** of a circle is a segment that passes through the center and is twice the radius. The distance around a circle is its **circumference**.

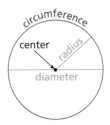

For a circle of any size, the ratio of the circumference (*C*) to the diameter (*d*) is always equal to the same number, $\frac{C}{d} = \pi$. This number is represented by the Greek letter π (pi).

Circumference of a Circle	$C = \pi d$ or $C = 2\pi r$ where *d* is the diameter and *r* is the radius of the circle.

Pi (π) is an irrational number. An irrational number cannot be expressed as a fraction. To make it easier to work with π, use the following approximations. The symbol ≈ means *is approximately equal to*. However, your calculator has a ⎡2nd⎤ [π] key that allows you to make more precise calculations.

$$\pi \approx 3.14 \quad \text{or} \quad \pi \approx \frac{22}{7}$$

Example 1

Find the circumference.

a. Use 3.14 for π.

5 in.

b. Use $\frac{22}{7}$ for π.

7 mm

Solution

a. $C = \pi d$

$C \approx 3.14 \cdot 5$

$C \approx 15.7$

The circumference is approximately 15.7 in.

b. $C = 2\pi r$

$C \approx 2 \cdot \frac{22}{7} \cdot 7$

$C \approx 44$

The circumference is approximately 44 mm.

> **Check Understanding**
>
> In the solutions of Example 1, why is the $\approx$ symbol used?

A circle can be cut into equal sections. These sections can then form a figure that approximates a parallelogram. The length of the base of this parallelogram is one-half the circumference of the circle. The height of the parallelogram equals the radius of the circle.

A formula for the area of a circle can be derived from the formula for the area of a parallelogram.

$A = b \cdot h$

$A = \frac{1}{2}C \cdot r \qquad\qquad b = \frac{1}{2}C, h = r$

$A = \frac{1}{2} \cdot (2 \pi r) \cdot r$

$A = \pi r \cdot r$

$A = \pi r^2$

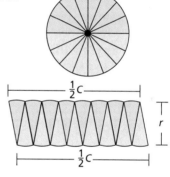

Area of a Circle	$A = \pi r^2$ where r is the radius.

Example 2

Find the area of circle P to the nearest square centimeter. Use 3.14 for π.

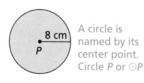

8 cm
P

A circle is named by its center point. Circle P or $\odot P$

Solution

$A = \pi r^2$

$A \approx 3.14 \cdot 8^2$ Substitute 8 for r.

$A \approx 3.14 \cdot 64$

$A \approx 200.96$

$A \approx 201$ Round to the nearest whole number.

The area of circle P is approximately 201 cm^2.

> **Technology Note**
>
> If your calculator has a $\boxed{x^2}$ key, use 8 and the $\boxed{x^2}$ instead of 8 × 8.

Example 3

SPORTS Discus throwing is an Olympic field event. The circular discus has a radius of 11 cm. Find the circumference and the area of the discus. Round your answers to the nearest whole number.

Solution

$C \approx 2 \cdot 3.14 \cdot 11 \qquad C = 2\pi r$ $A \approx 3.14 \cdot (11)^2 \qquad A = \pi r^2$

$C \approx 69.08 \approx 69$ $A \approx 3.14 \cdot 121 \approx 380$

The circumference of the discus is about 69 cm, and the area is about 380 cm².

◣ TRY THESE EXERCISES

Find the area. Use 3.14 or $\frac{22}{7}$ for π, as appropriate. If necessary, round your answer to the nearest tenth.

1. 18 cm

2. 14 cm

3.  13.5 in.

4–6. Find the circumferences of the circles in Exercises 1–3.

Find the area of a circle with the given diameters. If necessary, round your answers to the nearest tenth.

7. 8 ft   **8.** 9.8 m   **9.** 10 cm

10–12. Find the circumferences of the circles described in Exercises 7–9.

13. LANDSCAPING A lawn sprinkler rotates in a circular fashion. The area that is watered has a diameter of 14 m. Find the area of the lawn watered by the sprinkler.

14. WRITING MATH When finding the circumference or area of a circle, when would you choose to use $\frac{22}{7}$ for π instead of 3.14?

◣ PRACTICE EXERCISES • For Extra Practice, see page 551.

Find the area. If necessary, round your answer to the nearest tenth.

15. 42 mm

16. 12.5 yd

17. 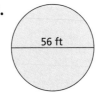 56 ft

18. 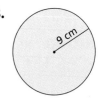 9 cm

19–22. Find the circumferences of the circles in Exercises 15–18.

Find the area of a circle with the given dimensions.

23. $d = 19$ cm   **24.** $r = 5.4$ mm   **25.** $d = 13.5$ ft   **26.** $r = 5\frac{1}{2}$ yd

27–30. Find the circumference of the circles described in Exercises 23–26.

DATA FILE Refer to the data on large telescopes of the world on page 524.

31. Find the circumference of the lens of each refractor telescope.

32. Find the area of the lens of each refractor telescope.

33. Copy and complete the chart.

Radius	Diameter	Circumference	Area
15 cm	■	■	■
■	100 ft	■	■
500 in.	■	■	■
■	0.1 mm	■	■

34. **SPORTS** The diameter of the face-off circle in hockey is 6 m. Find the circumference and area.

35. **ARCHITECTURE** The first Ferris wheel, named after its designer, George Ferris, was built in 1893. The diameter was 76 m. What was its circumference?

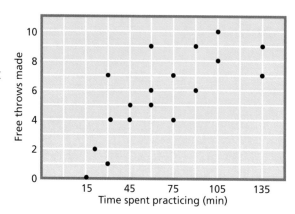

■ EXTENDED PRACTICE EXERCISES

36. **CRITICAL THINKING** Which has the greater distance around: a square with 8-in. sides or a circle with a 9-in. diameter?

37. Which has a greater area: a square with 8-in. sides or a circle with a 9-in. diameter?

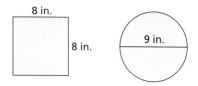

Find the area of each figure to the nearest tenth.

38.

39.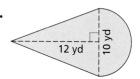

40.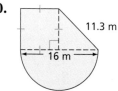

41–43. Find the distance around each figure to the nearest tenth.

■ MIXED REVIEW EXERCISES

Use the scatter plot. (Lesson 1-7)

44. Describe the relationship between the amount of time spent practicing and the number of free throws made.

45 Did all people who spent less than an hour make fewer than 5 free throws? Explain.

46. Suppose you collected data on the number of free throws the team made in games throughout the season. You have one data point for each game. How would you expect the scatter plot to look?

2-8 Proportions and Scale Drawings

Goals
- Solve problems by writing and solving proportions.
- Apply and interpret scale drawings.

Applications Art, History, Sports

Work with a partner.

Each partner should create a simple design on grid paper. (Be sure that your design is drawn only on the grid lines.) Exchange designs with your partner. Copy your partner's design, but make the following change. For every small square that your partner's design covers, your design should cover 4 small squares that are arranged in the shape of a large square. Discuss the results with your group.

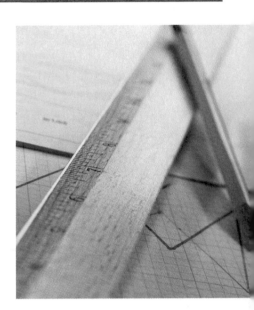

◢ BUILD UNDERSTANDING

A **proportion** is an equation stating that two ratios are equivalent. You can write a proportion two ways.

$$\frac{2}{3} = \frac{6}{9} \qquad \text{or} \qquad 2 : 3 = 6 : 9$$

The numbers 2, 3, 6 and 9 are the **terms** of the proportion. If a statement is a proportion, the **cross-products** of the terms are equal. For $\frac{a}{b} = \frac{c}{d}$, the cross-products are ad and bc.

$$\frac{2}{3} = \frac{6}{9}$$

$$2 \cdot 9 = 18$$
$$3 \cdot 6 = 18$$
cross-products

> **Reading Math**
>
> Read the proportion $\frac{2}{5} = \frac{4}{10}$ as "two is to five as four is to ten."

Example 1

Tell whether each statement is a proportion.

a. $\frac{3}{4} \stackrel{?}{=} \frac{21}{28}$

b. $2 : 7 \stackrel{?}{=} 6 : 10$

Solution

Find the cross-products.

a. $\frac{3}{4} \stackrel{?}{=} \frac{21}{28}$

$3 \cdot 28 = 84$

$4 \cdot 21 = 84$

$84 = 84$, so $\frac{3}{4} = \frac{21}{28}$

Yes, the statement is a proportion.

b. $\frac{2}{7} \stackrel{?}{=} \frac{6}{10}$ Write each ratio as a fraction.

$2 \cdot 10 = 20$

$7 \cdot 6 = 42$

$20 \neq 42$, so $2 : 7 \neq 6 : 10$

No, the statement is not a proportion.

Example 2

Use mental math to complete each proportion.

a. $\dfrac{1}{4} = \dfrac{?}{8}$

b. $12 : ? = 4 : 3$

Solution

a. $1 \cdot 8 = 4 \cdot ?$ Write the cross-products.

      $8 = 4 \cdot ?$ What times 4 equals 8?

      $4 \cdot 2 = 8$

      $\dfrac{1}{4} = \dfrac{2}{8}$

b. $\dfrac{12}{?} = \dfrac{4}{3}$ Write ratios as fractions.

   $12 \cdot 3 = ? \cdot 4$ Write the cross-products.

      $36 = ? \cdot 4$ What times 4 equals 36?

      $9 \cdot 4 = 36$

   $12 : 9 = 4 : 3$

A **scale drawing** is a drawing that represents a real object. All lengths in the drawing are proportional to the object's actual lengths. The ratio of the size of the drawing to the size of the actual object is called the **scale** of the drawing.

To find the actual length of the drawing, set up a proportion.

Example 3

ART The scale of this drawing is 1 in. : 2 ft. Find the actual length of the mural.

3 in.

Solution

Write a proportion and complete by using mental math.

drawing length (inches)→ $\dfrac{1}{2} = \dfrac{3}{?}$
mural length (feet)→

      $2 \cdot 3 = 1 \cdot ?$

        $6 = 1 \cdot ?$ What number times 1 equals 6?

        $6 = 1 \cdot 6$

The actual length of the mural is 6 ft.

Example 4

The scale of a painting is 1 in. : 3 ft. Find the drawing length that would be used to represent an actual length of 12 ft.

Solution

Write a proportion and complete by using mental math.

drawing length (inches)→ $\dfrac{1}{3} = \dfrac{?}{12}$
actual length (feet)→

      $12 = 3 \cdot ?$

      $12 = 3 \cdot 4$

The drawing length is 4 in.

Tell whether each statement is a proportion. Write *yes* or *no*.

1. $\frac{3}{9} \overset{?}{=} \frac{9}{27}$

2. $\frac{2}{5} \overset{?}{=} \frac{12}{11}$

3. $4 : 12 \overset{?}{=} 7 : 36$

4. $7 : 9 \overset{?}{=} 49 : 63$

5. $\frac{5}{9} \overset{?}{=} \frac{25}{45}$

6. $7 : 10 \overset{?}{=} 21 : 40$

Use mental math to solve each proportion.

7. $\frac{3}{2} = \frac{?}{4}$

8. $\frac{?}{6} = \frac{12}{36}$

9. $4 : 3 = ? : 9$

Find the actual or drawing length.

10. Scale, 1 in. : 2 ft
    Drawing length, 8 in.
    Actual length, ?

11. Scale, 1 cm : 7 m
    Drawing length, ?
    Actual length, 14 m

12. Scale, 1 in. : 8 mi
    Drawing length, ?
    Actual length, 32 mi

13. **SPORTS** A baseball pitcher pitched 12 losing games. The ratio of wins to losses is 3 : 2. How many games did the pitcher win?

14. Jane made a scale drawing of her living room. The length of the living room in her drawing is 4 in. The actual length is 20 ft. What scale did Jane use for her drawing?

15. **WRITING MATH** In your own words, describe a proportion.

Tell whether each statement is a proportion. Write *yes* or *no*.

16. $\frac{2}{3} \overset{?}{=} \frac{4}{6}$

17. $\frac{4}{8} \overset{?}{=} \frac{1}{4}$

18. $6 : 8 \overset{?}{=} 3 : 4$

19. $5 : 6 \overset{?}{=} 40 : 48$

20. $\frac{8}{5} \overset{?}{=} \frac{15}{24}$

21. $2 : 7 \overset{?}{=} 6 : 21$

22. $\frac{12}{5} \overset{?}{=} \frac{36}{20}$

23. $11 : 13 \overset{?}{=} 44 : 52$

24. $\frac{7}{22} \overset{?}{=} \frac{21}{66}$

Use mental math to solve each proportion.

25. $\frac{1}{2} = \frac{?}{16}$

26. $\frac{3}{10} = \frac{?}{60}$

27. $? : 12 = 2 : 24$

28. $\frac{?}{15} = \frac{20}{30}$

29. $? : 28 = 7 : 4$

30. $6 : 9 = ? : 27$

31. $\frac{1}{8} = \frac{?}{32}$

32. $3 : 2 = ? : 50$

33. $\frac{5}{12} = \frac{15}{?}$

34. $8 : ? = 2 : 4$

35. $11 : 6 = 55 : ?$

36. $\frac{6}{15} = \frac{?}{45}$

Find the actual or drawing length.

37. Scale, 1 cm : 6 m
    Drawing length, 3 cm
    Actual length, ?

38. Scale, 1 cm : 12 km
    Drawing length, ?
    Actual length, 120 km

39. Scale, 1 in. : 20 ft
    Drawing length, ?
    Actual length, 120 ft

40. Scale, 1 in. : 7 mi
    Drawing length, 9 in.
    Actual length, ?

41. Scale, 1 cm : 5 m
    Drawing length, ?
    Actual length, 55 m

42. Scale, 1 cm : 8 m
    Drawing length, 6 cm
    Actual length, ?

43. ART Devon makes green paint by mixing blue paint and yellow paint in the ratio 1 : 2. How many parts of yellow are mixed with 6 parts of blue to make green?

44. The Crafts Club uses 4 lb of clay to make 20 sculptures. How many sculptures could they make with 12 lb of clay?

45. ENTERTAINMENT A movie company made a large model of a domino. The scale was 1 mm : 5 dm. The width of the model is 75 dm. Find the width of the actual domino.

Mount Rushmore, South Dakota

HISTORY Gutzon Borglum, the sculptor of Mt. Rushmore, created a model of the faces before carving on the mountainside. The scale used to transfer the model to the mountainside was 1 in. : 12 in. The following dimensions are of Washington's head on the mountainside. Find each dimension on the model.

46. forehead to chin: 60 ft

47. width of eye: 11 ft

48. length of nose: 20 ft

49. width of mouth: 18 ft

DATA FILE Refer to the data on dinosaurs on page 518. To make a model of the following dinosaurs with a scale of 1 in. : 5 ft, what would be the approximate height and length of each dinosaur? (Hint: Round the actual height and length to the nearest multiple of 5.)

50. Brachiosaurus   **51.** Tyrannosaurus   **52.** Triceratops

■ EXTENDED PRACTICE EXERCISES

53. CRITICAL THINKING Write a proportion using the numbers 9, 27, 35 and 105.

54. WRITING MATH Write a problem that involves a proportion. Have a classmate solve the problem.

55. Write as many proportions as you can by using the digits 1 through 9. Each digit may be used only once in each proportion. One proportion could be $\frac{1}{3} = \frac{2}{6}$.

56. Make a scale drawing of your classroom. Measure the room to the nearest inch. Choose a scale. Show windows, desks and tables.

57. CHAPTER INVESTIGATION Use your list of rounded dimensions and your scale factor to write and solve a proportion for each dimension in the scale drawing. Organize your data so that the physical features of the Statue are listed in both actual and scale drawing lengths. What is the minimum size poster board that your group could use to make a scale drawing?

■ MIXED REVIEW EXERCISES

Use the table. (Lessons 1-6 and 1-7)

Date April 1 - April 15	1	2	3	4	5	6	7	8	9	10	11	12	13	14	15
Rain accumulation (in.)	0.1	0.1	1.0	1.2	1.2	1.2	3.2	3.2	3.2	4.0	5.5	5.5	5.5	5.5	5.5
Grass height	1.0	1.1	1.2	1.5	1.8	1.9	2.0	2.4	2.8	3.0	3.2	3.5	3.8	4.0	4.1

58. Display the data in a scatter plot.

59. Describe the relationship between rainfall and grass height.

60. Draw a line graph to show the change in grass height.

Review and Practice Your Skills

PRACTICE ◤ LESSON 2-7

Find the area. If necessary, round the answer to the nearest tenth.

1.
14 in.

2.
2 ft

3.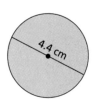
4.4 cm

4.
8 m

5-8. Find the circumference of each circle in Exercises 1-4.

Find the area of a circle with the given dimensions. If necessary, round the answer to the nearest tenth.

9. $r = 20$ cm   **10.** $d = 1.2$ cm   **11.** $r = 100$ in.   **12.** $d = 5$ ft

13. $d = 1$ m   **14.** $d = 5.4$ mm   **15.** $d = 30$ ft   **16.** $r = 3\frac{1}{2}$ yd

17-24. Find the circumference of each circle described in Exercises 9-16.

PRACTICE ◤ LESSON 2-8

Tell whether each statement is a proportion. Write *yes* or *no*.

25. $2 : 11 \stackrel{?}{=} 6 : 66$ **26.** $8 : 3 \stackrel{?}{=} 16 : 9$ **27.** $\frac{8}{18} \stackrel{?}{=} \frac{4}{9}$

28. $\frac{5}{3} \stackrel{?}{=} \frac{20}{15}$ **29.** $4 : 3 \stackrel{?}{=} 24 : 18$ **30.** $20 : 6 \stackrel{?}{=} 12 : 3$

31. $36 : 20 \stackrel{?}{=} 3 : 2$ **32.** $\frac{12}{15} \stackrel{?}{=} \frac{60}{75}$ **33.** $8 : 1 \stackrel{?}{=} 40 : 5$

Use mental math to solve each proportion.

34. $\frac{5}{2} = \frac{50}{?}$ **35.** $3 : 4 = ? : 16$ **36.** $\frac{6}{5} = \frac{36}{?}$

37. $12 : 9 = 4 : ?$ **38.** $\frac{8}{12} = \frac{2}{?}$ **39.** $3 : 7 = ? : 42$

40. $2 : 15 = ? : 30$ **41.** $8 : 4 = 64 : ?$ **42.** $\frac{1}{4} = \frac{?}{20}$

43. $\frac{3}{16} = \frac{9}{?}$ **44.** $\frac{50}{35} = \frac{?}{7}$ **45.** $5 : 8 = ? : 40$

Find the actual or drawing length.

46. Scale, 1 in. : 150 ft
Drawing length, 5 in.
Actual length, ?

47. Scale, 1 cm : 20 cm
Drawing length, ?
Actual length, 400 cm

48. Scale, 1 in. : 100 mi
Drawing length, ?
Actual length, 650 mi

49. Scale, 1 cm : 100 m
Drawing length, ?
Actual length, 650 m

50. Scale, 1 in. : 10 mi
Drawing length, 5 in.
Actual length, ?

51. Scale, 1 cm : 5 m
Drawing length, 3.2 cm
Actual length, ?

52. Scale, 1 in. : 10 ft
Drawing length, 10 in.
Actual length, ?

53. Scale, 1 cm : 10 km
Drawing length, 0.5 cm
Actual length, ?

54. Scale, 1 in. : 150 mi
Drawing length, ?
Actual length, 600 mi

Choose the best estimate for each measure. (Lesson 2-1)

55. water in a small pool: 5 L, 5 mL, 5 kL

56. weight of a paper clip: 1.3 mg, 1.3 kg, 1.3 g

57. height of building: 27 m, 27 cm, 27 km

58. 3-hour walk-a-thon: 14 ft, 14 yd, 14 mi

Complete. Write each answer in simplest form. (Lesson 2-2)

59. 450 km = __?__ m
60. 350 mL = __?__ L
61. 60 in. = __?__ yd __?__ ft

62. 6.4 m ÷ 8 = __?__ cm
63. 25 g · 400 = __?__ kg
64. 50 mL + 1.3 L = __?__ L

Find the perimeter of each figure with the given dimensions. (Lesson 2-3)

65. rectangle: 12 cm by 30 cm
66. square: 20 ft

67. pentagon: 8 in., 6 in., 6 in., 3 in., 3 in.
68. triangle: 2.5 m, 3 m, 4.5 m

Find the area of each figure with the given dimensions. (Lesson 2-4)

69. rectangle: 6.2 m by 10 m
70. parallelogram: $b = 30$ mm, $h = 14$ mm

71. triangle: $h = 12$ in., $b = 18$ in.
72. square: 16 ft sides

Write *P* or *A* to tell if you would use a perimeter or area formula. (Lesson 2-5)

73. tiles to cover a kitchen floor
74. wall around a garden

75. distance run around the block
76. floor space in a restaurant

Are the ratios equivalent? Write *yes* or *no*. (Lesson 2-6)

77. $15 : 5, \dfrac{3}{1}$
78. $\dfrac{4}{5}, \dfrac{12}{10}$
79. $6 : 5, 25 : 30$

80. $\dfrac{9}{6}, \dfrac{3}{2}$
81. 12 to 3, 4 : 1
82. $\dfrac{15}{20}$, 3 to 5

Find the area of a circle with the given dimensions. (Lesson 2-7)

83. $d = 14$ in.
84. $r = 2.4$ cm
85. $r = 8$ m
86. $d = 2\dfrac{1}{2}$ ft

87-90. Find the circumference of each circle described in Exercises 83–86.

Use mental math to solve each proportion. (Lesson 2-8)

91. $\dfrac{8}{18} = \dfrac{?}{9}$
92. $10 : 35 = 2 : ?$
93. $\dfrac{4}{3} = \dfrac{32}{?}$

94. $2 : 7 = ? : 35$
95. $\dfrac{2}{5} = \dfrac{8}{?}$
96. $18 : 12 = ? : 4$

Find the actual or drawing length. (Lesson 2-8)

97. Scale, 1 cm : 20 km
Drawing length, 8 cm
Actual length, ?

98. Scale, 1 in. : 20 ft
Drawing length, ?
Actual length, 140 ft

99. Scale, 1 cm : 50 m
Drawing length, 1.5 cm
Actual length, ?

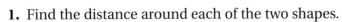

2-9 Areas of Irregular Shapes

Goals ■ Use area formulas to find the area of irregular shapes.
■ Estimate area.

Applications Recreation, Interior design, Manufacturing

Make a loop with a piece of string that is 8 – 10 in. long. Place the
string on grid paper to create a shape. Trace this shape. Create
and trace a different shape on the same grid paper.

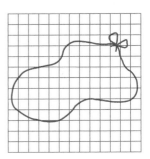

1. Find the distance around each of the two shapes.

2. How can you find the area of each shape? Are your
   measurements accurate or precise, or neither?

3. How many different shapes are possible using this string?

4. The string is used to form each shape, so the perimeters of the shapes
   are equal. Does that mean the areas of the shapes are equal, also? Explain.

◤ BUILD UNDERSTANDING

Finding area may require more than one area formula. One method to find the
area of an **irregular figure** is to separate it into smaller familiar regions. Find the
area of each region. Then add the areas.

Example 1

**Find the area of a field with
the dimensions shown.**

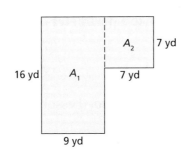

Solution

Total area = $A_1 + A_2$

$A_1 = 16 \cdot 9$ $A_1 = l \cdot w$ $A_2 = 7^2$ $A_2 = s^2$

$A_1 = 144$ $A_2 = 49$

$A_1 + A_2 = 144 + 49 = 193$.   The total area is 193 yd^2.

> **Problem Solving Tip**
>
> State the answer to a
> word problem as a
> sentence, including the
> units of measure.

Another way to find the area is to determine the
dimensions of the field as if it were a complete rectangle A_3.
Find the area of the complete rectangle, and then subtract
the area of the nonexistent shape from the total area A_4.

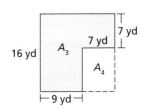

Total area = $A_3 - A_4$

$A_3 = 16 \cdot (9 + 7)$ $A_3 = l \cdot w$ $A_4 = 9 \cdot 7$ $A_4 = l \cdot w$

$A_3 = 256$ $A_4 = 63$

$A_3 - A_4 = 256 - 63 = 193$. The total area is 193 yd^2.

Example 2

New tiles are needed to cover the bottom of the swimming pool shown. The tiles will cost $0.85/ft^2.

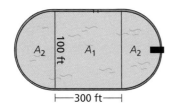

a. How many square feet of tiles are needed?

b. How much will the new tiles cost?

Solution

a. Find the area of the rectangle and two semicircles. Then add the areas. The area of a semicircle is half the area of a circle with the same radius.

Total area = $A_1 + 2A_2$

$A_1 = l \cdot w$ $\qquad\qquad$ $A_2 = \dfrac{1}{2}\pi r^2$

$A_1 = 300 \cdot 100$ $\qquad\qquad$ $A_2 = \dfrac{1}{2}(3.14) \cdot 50^2$

$A_1 = 30,000$ $\qquad\qquad$ $A_2 \approx 3925$

$A_1 + 2A_2 = 30,000 + 2(3925) = 37,850$

The total area is about 37,850 ft^2.

b. Multiply the area by $0.85 to find the total cost of the new tiles.

$0.85 \cdot 37,850 = 32,172.50$ $\quad$ The new tiles will cost $32,172.50.

> **Think Back**
>
> The formula for the area of a circle is $A = \pi r^2$.
>
> The radius of a circle is half the diameter.

Estimating is a useful tool to find the area of irregular shapes that cannot be easily measured, such as a state, an oil spill or your foot. You need to visualize the irregular shape fitting snugly inside one or more polygons or circles. Use the given dimensions of the shape to estimate the dimensions of the polygons or circles. Compute the area of each polygon or circle. The estimated area of the irregular shape is the sum of the areas that fit around it.

Example 3

Estimate the area of the lake shown.

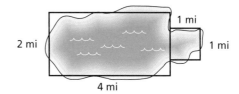

Solution

Enclose the figure in two familiar shapes, such as a rectangle and a square. Estimate the area of each shape, then add the two areas.

Total area = $A_1 + A_2$

$A_1 = 4 \cdot 2$ $\quad A_1 = l \cdot w$ $\qquad\qquad\qquad$ $A_2 = 1^2$ $\quad A_2 = s^2$

$A_1 = 8$ $\qquad\qquad\qquad\qquad\qquad\qquad$ $A_2 = 1$

$A_1 + A_2 = 8 + 1 = 9$ $\quad$ The area of the figure is about 9 mi^2.

Find the area of each figure.

1.
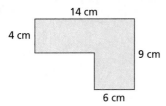
14 cm
4 cm
9 cm
6 cm

2.
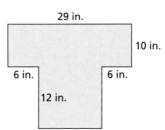
29 in.
10 in.
6 in.   6 in.
12 in.

3.
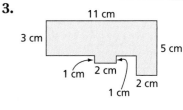
11 cm
3 cm
5 cm
1 cm   2 cm
2 cm
1 cm

4. A rectangular floor measures 8 yd by 10 yd. If carpeting costs $12.75/yd², what is the cost of carpeting the floor?

5. Find the area of a backyard with the dimensions shown.

6. The cost of grass sod is $1.65/yd². Find the cost of sodding the backyard shown.

200 yd
110 yd   120.5 yd
140 yd

7. Outline your hand on grid paper. Estimate its area.

8. WRITING MATH Explain how you estimated the area of your hand in Exercise 7. Why did you choose that method?

Find the area of each figure.

9.
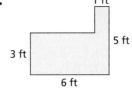
1 ft
5 ft
3 ft
6 ft

10.
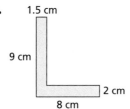
1.5 cm
9 cm
2 cm
8 cm

11.
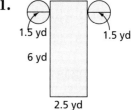
1.5 yd   1.5 yd
6 yd
2.5 yd

12. Egypt's northern border is about 900 km, the southern border is about 1100 km and the height is about 1100 km. Estimate the area of Egypt.

Cairo

Estimate the area of each object using the units of measure given.

13. your footprint in inches   **14.** a pencil in centimeters

15. a house key in centimeters   **16.** the classroom in meters

RECREATION Most sailboats have two sails, the jib and the mainsail. Assume that the sails are triangles. Find the total area of each boat's sails to the nearest tenth.

17.
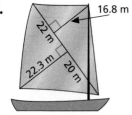
16.8 m
22 m
22.3 m   20 m

18.

23.9 m
19.5 m
10.9 m   8.6 m

19.
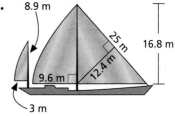
8.9 m
25 m
16.8 m
9.6 m
12.4 m
3 m

20. Find the total area of the figure shown. The triangles have equal areas.

21. INTERIOR DESIGN Yahto is painting a client's rectangular bedroom. One gallon of paint will cover about 350 ft². One wall is 14 ft by 8 ft, and the other wall is 12 ft by 8 ft. How many gallons of paint does he need to cover the four walls?

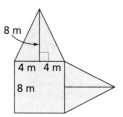

MANUFACTURING A box lid is constructed from the pattern shown.

22. Find the area of the top of the lid.

23. Find the total area of the rectangular sides of the lid.

24. ERROR ALERT Jasmine followed these steps to find the total area of the triangular tabs. Is she correct?

$$\frac{1}{2}(b)(h) = \frac{1}{2}(4)(4) = 8$$

The area of the tabs is 8 cm².

25. Find the total amount of material used to make the lid.

26. Draw a pentagon on grid paper. Estimate its area.

27. Describe the method you used to estimate the area in Exercise 26.

28. WRITING MATH When would you use an estimated area? Explain.

EXTENDED PRACTICE EXERCISES

29. CRITICAL THINKING The figure shown is a circle and a sector, or a portion of a circle. Because the angle of the sector is 90°, it is one-fourth of a circle. Find the total area of the figure.

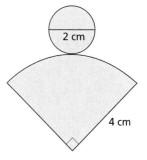

30. Estimate the area of the shaded region at the left.

31. Estimate the area of the unshaded region.

32. Describe the method you used to estimate the areas in Exercises 30 and 31.

33. How would you estimate the area of each region if the grid were missing?

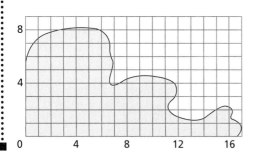

MIXED REVIEW EXERCISES

Find the perimeter and area of each shape.

34.

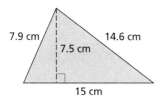

35.

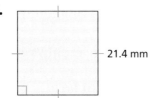

36.

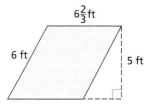

37. The diameter of a circular garden is 6 ft. What is the circumference and area? Round to the nearest tenth.

Chapter 2 Review

VOCABULARY ◣

Choose the word from the list that best completes each statement.

1. __?__ is the amount of surface enclosed by a geometric figure.
2. __?__ is the process used to find the size, quantities or amounts.
3. The __?__ is the distance around a circle.
4. The distance around a plane figure is called the __?__.
5. In the proportion $\frac{2}{5} = \frac{4}{10}$, the __?__ are 2, 5, 4, and 10.
6. In the proportion $\frac{3}{10} = \frac{?}{6}$, $3 \cdot 6 = 10 \cdot ?$ is called the __?__.
7. To find the area of a(n) __?__, separate it into familiar regions.
8. The longest segment of a circle is called the __?__.
9. The closeness of a measurement to its actual measure is called __?__.
10. A(n) __?__ measure is nearly exact.

a.	accurate
b.	area
c.	circumference
d.	cross-product
e.	diameter
f.	irregular figure
g.	measurement
h.	perimeter
i.	precision
j.	proportion
k.	radius
l.	terms

LESSON 2-1 ◣ Units of Measure, p. 52

▶ A measure that is exact is said to be **accurate**. **Precision** is the closeness of a measurement to its exact measure. **Estimation** provides an approximation when an exact measure is not available.

11. Which unit of measurement gives a more precise measurement: gallon or liter?
12. Name the tool or tools that would best measure the height of a flagpole.
13. Choose the best unit to estimate coffee in a mug: cup, teaspoon, or liter?
14. Choose the best estimate for the width of a blade of grass: 4 cm, 4 in., or 4 mm?

LESSON 2-2 ◣ Work With Measurements, p. 56

▶ To convert from a *larger to a smaller unit* in the metric system, multiply (move the decimal right) by the appropriate power of ten. To convert from a *smaller to a larger unit*, use division (move the decimal left) by the appropriate power of ten.

▶ To change units in the English system, multiply or divide by equivalent factors.

Complete. Write each answer in simplest form.

15. 6 ft = __?__ in.
16. 3000 g = __?__ kg
17. 800 m = __?__ mm
18. 8 lb 9 oz + 5 lb 2 oz = __?__ lb __?__ oz
19. 52 ft = __?__ yd __?__ ft
20. 3 c 6 fl oz · 5 = __?__ c __?__ fl oz
21. 360 mm ÷ 9 = __?__ cm

LESSON 2-3 ◼ Perimeter of Polygons, p. 62

▶ The **perimeter** of any plane figure is found by adding the lengths of its sides.

Find the perimeter of each figure.

22.

23.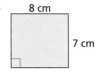

24.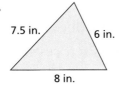

25.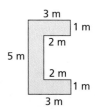

26.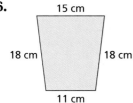

27.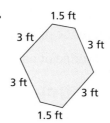

28.

29.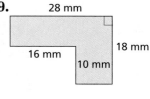

LESSON 2-4 ◼ Areas of Parallelograms and Triangles, p. 66

▶ Area is measured in square units.

Find the area of each figure.

30. triangle: $b = 16$ cm, $h = 5$ cm

31. parallelogram: $b = 18$ m, $h = 7$ m

32.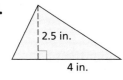

33.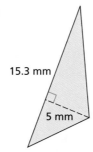

34.

35.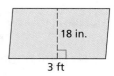

LESSON 2-5 ◼ Problem Solving Skills: Quantity and Cost, p. 72

▶ To use formulas in solving problems, first decide which formula is needed.

Write *P* or *A* to tell if you would use a perimeter or an area formula.

36. amount of wood for a frame

37. amount of paint to paint a wall

38. cost of tiling the floor of a 40 ft by 18 ft room

39. amount of fencing to enclose a yard

40. amount of sealer needed to coat a fence

LESSON 2-6 ◼ Equivalent Ratios, p. 74

▶ A ratio compares one number to another. A ratio can be written three ways.

41. Write 4 to 5 as a ratio two other ways.

42. Write 16 : 24 in lowest terms.

LESSON 2-7 ◤ Circumference and Area of a Circle, p. 80

▶ A **circle** is the set of points in a plane a fixed distance from a given point. The **diameter** passes through the **center** and is twice the **radius**.

Find the area of each circle with the given dimensions. Use 3.14 for π. Round to the nearest tenth.

43. diameter: 8 cm   **44.** radius: 3 yd   **45.** radius: 12.3 in.

46.   **47.**   **48.**   **49.**

50. Find the circumference of each circle described in Exercises 43–45.

51. Find the circumference of each circle in Exercises 46–49.

LESSON 2-8 ◤ Proportions and Scale Drawings, p. 84

▶ A **proportion** is an equation stating that two ratios are equivalent. The **cross-products** in a proportion are equal.

▶ **Scale drawing** lengths are proportional to the actual lengths.

Solve each proportion.

52. $\dfrac{?}{9} = \dfrac{5}{15}$ **53.** $\dfrac{8}{?} = \dfrac{3}{12}$ **54.** $3:8 = ?:40$

55. Drawing length: 3 in. Actual length: 12 ft. What scale is used in the drawing?

LESSON 2-9 ◤ Areas of Irregular Shapes, p. 90

▶ To find the area of irregular shapes, separate the figure into familiar regions.

56. Find the area of the figure in Exercise 25.

Find the area of each figure. Round to the nearest tenth. Use 3.14 for π.

57. 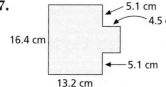   **58.** 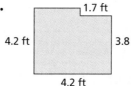   **59.** 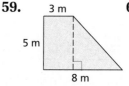   **60.**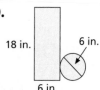

CHAPTER INVESTIGATION

EXTENSION Make a scale drawing of the Statue of Liberty. Use poster board, a photograph, a ruler and your list of scale drawing lengths. Use the photograph for reference. Each group member is accountable for at least one of the duties listed. Display your completed drawing.

▶ Mark the scale lengths on the board as a guide for the drawing.

▶ Draw the outline and features.

▶ Verify that the scale lengths are accurate.

▶ Provide color, background, and key to the drawing.

Chapter 2 Assessment

Solve.

1. Which unit of measure gives a more precise measurement: millimeter, meter?

2. Name the tool or tools that would best measure the milk in a glass.

3. Choose the best estimate for the weight of soup in a can: $10\frac{3}{4}$ oz, $10\frac{3}{4}$ lb, or $10\frac{3}{4}$ kg?

Complete. Write each answer in simplest form.

4. 9 gal = ■ qt

5. 10 lb = ■ oz

6. 2 mi = ■ yd

7. 280 mm = ■ m

8. 5 L = ■ mL

9. 589 mg = ■ g

10. $\quad\begin{array}{r} 6\text{ c }4\text{ fl oz} \\ + 2\text{ c }10\text{ fl oz} \\ \hline \end{array}$

11. $\quad\begin{array}{r} 6\text{ lb} \\ -4\text{ lb }5\text{ oz} \\ \hline \end{array}$

12. $\begin{array}{r} 2\text{ ft }6\text{ in.} \\ \times 3 \\ \hline \end{array}$

13. 13 gal 2 qt ÷ 3

14. 12 m − 750 cm = __?__ m

15. 3.21 cm · 3 = __?__ mm

Find the perimeter and area of each figure.

16.

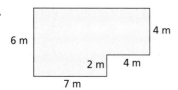

17.

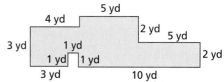

Find the area of each figure.

18. triangle: $b = 14$ in., $h = 7$ in.

19. rectangle: 6 cm by 10 cm

20. parallelogram: $b = 8$ ft, $h = 11$ ft

21. square: 9-yd sides

Solve.

22. Write 5 to 6 as a ratio in two other ways.

23. Write 24:36 in lowest terms.

24. Find three equivalent ratios for 4 out of 9.

25. Is $\frac{5}{3}$ equivalent to $\frac{20}{12}$?

26. Wire fencing costs $6.25 per foot. Find the cost of fencing a right triangular garden whose sides have lengths of 6 ft, 8 ft, and 10 ft.

Find the area and circumference of each circle. Use 3.14 for π.

27. diameter = 49 mm

28. radius = 6.6 ft

29. radius = 17 m

Solve each proportion.

30. $\frac{?}{8} = \frac{5}{20}$

31. $\frac{6}{?} = \frac{9}{3}$

32. $\frac{5}{15} = \frac{?}{21}$

33. $\frac{9}{18} = \frac{3}{?}$

34. A drawing length is 4 cm. The scale is 1 cm = 3 m. Find the actual length.

Standardized Test Practice

Part 1 Multiple Choice

Record your answers on the answer sheet provided by your teacher or on a sheet of paper.

1. The table shows the scores of the first round of golf. What is the mean of the data? (Lesson 1-2)

First Round Golf Scores
78   89   79   73   68   80   78   81   85

  Ⓐ 68
  Ⓑ 78.5
  Ⓒ 79
  Ⓓ 85

2. The scatter plot displays the number of completed receptions by a wide receiver in a football season.

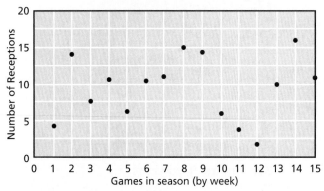

Number of Receptions Caught

Which describes the line of best fit? (Lesson 1-7)

  Ⓐ inverse correlation
  Ⓑ positive correlation
  Ⓒ negative correlation
  Ⓓ no correlation

3. Which is the best choice of a unit to measure a window for mini-blinds? (Lesson 2-1)

  Ⓐ inches
  Ⓑ feet
  Ⓒ meters
  Ⓓ miles

4. A step stool is 2 ft 8 in. tall. If the step stool is expanded by an additional 10 in., what is the total height? (Lesson 2-2)

  Ⓐ 2 ft 2 in.
  Ⓑ 2 ft 10 in.
  Ⓒ 3 ft 4 in.
  Ⓓ 3 ft 6 in.

5. What is the perimeter of a square in which the measure of each side is 64 mm? (Lesson 2-3)

  Ⓐ 32 mm   Ⓑ 128 mm
  Ⓒ 256 mm   Ⓓ 4096 mm

6. The dimensions of a garden are 6.8 ft by 9.3 ft. Which is the area of the garden? (Lesson 2-4)

  Ⓐ 54 ft^2 Ⓑ 60 ft^2
  Ⓒ 63 ft^2 Ⓓ 70 ft^2

7. The area of a triangle is 400 cm^2. If the base is 25 cm, what is the height? (Lesson 2-4)

  Ⓐ 12 cm   Ⓑ 16 cm
  Ⓒ 24 cm   Ⓓ 32 cm

8. The Kratzers are replacing carpeting in the shaded areas. What is the area of carpet they will need? (Lesson 2-9)

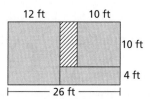

  Ⓐ 260 ft^2
  Ⓑ 276 ft^2
  Ⓒ 324 ft^2
  Ⓓ 364 ft^2

Test-Taking Tip

Ⓐ Ⓑ Ⓒ Ⓓ

Exercises 5, 6, and 7
Draw a diagram if one is not provided in the problem. Label your diagram with the information given.

Preparing for the Standardized Tests
For test-taking strategies and more
practice, see pages 587-604.

Part 2 Short Response/Grid In

Record your answers on the answer sheet provided by your teacher or on a sheet of paper.

9. Amanda is designing a survey to ask her classmates about their favorite foods served in the cafeteria. She plans to survey every tenth student who enters the cafeteria. What type of sampling is she planning to use? (Lesson 1-1)

10. Ms. Worden tested her 5th period class. The scores are 75, 82, 64, 86, 91, 87, 86, 99, 52, 71, 86, 98, 66, 82, and 79. What is the median score? (Lesson 1-2)

11. The pictograph shows the amount of calcium found in fruits.

Calcium Found in Fruit

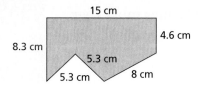

Key: ⌷ = 2 milligrams

How much more calcium is found in 3 raw apricots than 1 raw apple? (Lesson 1-5)

12. What is the perimeter of the figure in centimeters? (Lesson 2-3)

15 cm
8.3 cm
4.6 cm
5.3 cm
5.3 cm
8 cm

13. Which figure has the greatest area? (Lesson 2-4)

   triangle: $b = 15$ in., $h = 7$ in.
   parallelogram: $b = 8$ in., $h = 6$ in.
   rectangle: $l = 13$ in., $w = 4$ in.

14. The scale of a painting and a reproduction is 15 in. to 3 in. If the length of the actual painting is 2 ft, what is the length of the reproduction in inches? (Lesson 2-8)

Part 3 Extended Response

Record your answers on a sheet of paper. Show your work.

15. A survey was taken of 60 consumers about their favorite way to shop. (Lesson 1-4)

Ways Consumers Shop

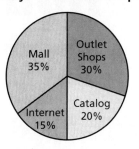

   a. How many consumers prefer to shop on the Internet?

   b. How many consumers prefer to shop at the outlet stores or at the mall?

   c. Which method of shopping is least preferred?

16. Use the bar graph. (Lesson 1-6)

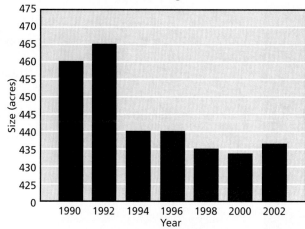

Size of Average U.S. Farm

   a. About how much larger was an average farm in 1992 than in 1998?

   b. How many bars represent farm sizes smaller than 445 acres?

   c. During which year was the average farm size about 440 acres?

Real Numbers and Variable Expressions

THEME: Sports

You use all types of numbers in everyday life. Although you may not realize it, you often use these numbers in variable expressions that you solve mentally. Numbers and variables are essential in sports for both the participants and the spectators. Sports use numbers to record the statistics of a competition. Variables play a role in preparing for the next competition.

- **Sports statisticians** (page 113) keep records of sporting events. They keep running totals of specific information about individual players from every team, as well as the score and penalties assessed.

- **Fitness trainers** (page 131) help athletes maintain and improve their performance. A fitness trainer must keep records of her client's basic health information including physical endurance and ability tests.

Math Online

mathmatters1.com/chapter_theme

Size and Weights of Balls Used in Various Sports

Type	Diameter (centimeters)	Average weight (grams)
Baseball	7.6	145
Basketball	24.0	596
Croquet ball	8.6	340
Field hockey ball	7.6	160
Golf ball	4.3	46
Handball	4.8	65
Soccer ball	22.0	425
Softball, large	13.0	279
Softball, small	9.8	187
Table tennis ball	3.7	2
Tennis ball	6.5	57
Volleyball	21.9	256

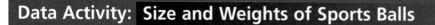

Data Activity: Size and Weights of Sports Balls

Use table to answer Questions 1–4.

1. How many more grams does a handball weigh than a tennis ball?

2. Find the difference in the circumference of a basketball and a soccer ball.

3. Order the objects from least to greatest by diameter. Then order them from least to greatest by average weight. Is there a correlation between the diameter and the average weight of a sports ball? Explain.

4. Use < or > to complete the inequalities according to weight.
   a. volleyball____softball, small   **b.** croquet ball____basketball
   c. golf ball ____handball   **d.** field hockey ball____tennis ball

CHAPTER INVESTIGATION

The equipment used in sports varies, but most sports require some sort of ball. You can probably think of many different types of sports balls. The role that the ball serves in the sport influences the manufacturing decisions about its appearance and measurements.

Working Together

Choose two of the sports balls from the table above that you know bounce. Determine if the size and weight of each sports ball affect its bouncing capabilities. Use the Investigation icons in this chapter to assist you with your group's experiment.

3 Are You Ready?

Refresh Your Math Skills for Chapter 3

The skills on these two pages are ones you have already learned. Use the examples to refresh your memory and complete the exercises. For additional practice on these and more prerequisite skills, see pages 536-544.

MULTIPLICATION AND DIVISION PATTERNS

You can use patterns to help multiply and divide by powers of ten.

Examples Powers of 10 are 1, 10, 100, 1000, 10,000, and so on.

$1.346 \cdot 10 = 13.46$	$230.5 \div 10 = 23.05$
$1.346 \cdot 100 = 134.6$	$230.5 \div 100 = 2.305$
$1.346 \cdot 1000 = 1346$	$230.5 \div 1000 = 0.2305$
$1.346 \cdot 10,000 = 13,460$	$230.5 \div 10,000 = 0.02305$

Move the decimal point the same number of places as the number of zeros in the power of 10. When multiplying, move to the right. When dividing, move to the left.

Find each product or quotient.

1. $2.57 \cdot 10$ **2.** $0.365 \cdot 1000$ **3.** $15.9 \cdot 100$ **4.** $0.035 \cdot 10,000$

5. $16.9 \div 100$ **6.** $145 \div 10$ **7.** $649.22 \div 1000$ **8.** $23.5 \cdot 10,000$

CHOOSING OPERATIONS

Knowing the operation to use is the first step to solving a problem. Sometimes the words used in a problem can give you a clue.

Examples

Addition Key Words

sum
altogether

Subtraction Key Words

difference
remaining

Multiplication Key Words

product
times

Division Key Words

quotient
how many times

Name the operation needed to solve each problem.

9. Senalda drove 150 mi before stopping for lunch. She drove another 227 mi after lunch. How much further did Senalda travel after lunch than before lunch?

10. Nick earns \$8.50/h. How much will he earn in 23 h?

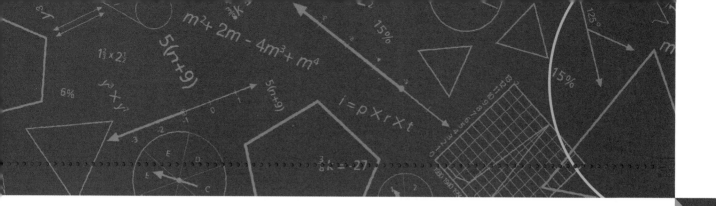

REAL NUMBERS AND THEIR OPPOSITES

Any number that can be written as a decimal is a real number. The opposite value of any number can be found on the number line. A number and its opposite are equidistant from zero, but on opposite sides of zero.

Examples

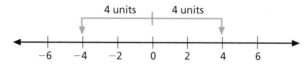

The opposite of 4 is -4. The opposite of -4 is 4.

4 and -4 are the same distance away from zero on opposite sides of zero.

Name the opposite of each number.

11. 16   **12.** -21 **13.** 49   **14.** -62

15. $\dfrac{1}{4}$ **16.** -2.6 **17.** $-\dfrac{2}{3}$ **18.** 3.14

ORDERING REAL NUMBERS

All real numbers can be shown on a number line. A number line can help you compare the values of real numbers.

Examples Graph $\dfrac{1}{3}$, -1.75, 1, $-\dfrac{1}{2}$, and 1.25 on a number line. Write them in order from least to greatest.

In order: -1.75, $-\dfrac{1}{2}$, $\dfrac{1}{3}$, 1, 1.25

Name the value of each letter on the number line as a decimal and as a fraction or mixed number.

19. A **20.** B **21.** C **22.** D

23. E **24.** F **25.** G **26.** H

Write each set of numbers in order from least to greatest.

27. $2.3, -1.9, \dfrac{2}{5}, -1\dfrac{1}{8}, -2$ **28.** $-\dfrac{1}{2}, 1, \dfrac{2}{3}, -0.1, -2$

29. $6\dfrac{1}{8}, -2\dfrac{2}{3}, 1.5, 3.7, -\dfrac{5}{8}$ **30.** $\dfrac{3}{8}, -3.4, -2\dfrac{1}{2}, 2, -1.83$

3-1 Add and Subtract Signed Numbers

Goals
- Add and subtract using a number line or rules.
- Use addition properties to find sums.

Applications Sports, Finance, Weather

In a football game, Team A gained control of the ball on the 50-yd line. Team A's progress in the next eight plays are shown in the table.

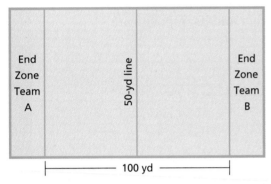

End Zone Team A | 50-yd line | End Zone Team B

|← 100 yd →|

1. On a diagram of a football field, show Team A's position with each play.

2. At the end of the eight plays, how many yards away is Team A from Team B's end zone?

3. On a number line, show how Team A's position changed with the 8 plays. Let 0 represent the 50-yd line.

Yardage – Team A

Play	Gain (yards)	Loss (yards)
1	5	
2	5	
3	7	
4		−6
5	10	
6	12	
7		−3
8	8	

◣ BUILD UNDERSTANDING

Two numbers the same distance from 0 but in opposite directions are **opposites**.

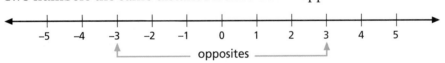

−5	−4	−3	−2	−1	0	1	2	3	4	5

|← opposites →|

Addition Property of Opposites	$a + (-a) = 0$ When opposites are added, the sum is always zero.

A number line can be used to show addition of two numbers. Start at 0. Move to the right to add a positive number. Move to the left to add a negative number.

Example 1

Use a number line to add.

a. $4 + 3$

b. $2 + (-4)$

Solution

a. Begin at 0. Move right 4 places, then 3 places more. The stopping place is +7.
$4 + 3 = 7$

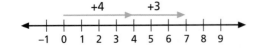

+4   +3

−1	0	1	2	3	4	5	6	7	8	9

b. Begin at 0. Move right 2 places, then left 4 places. The stopping place is −2.
$2 + (-4) = -2$.

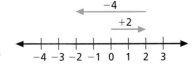

−4

+2

−4	−3	−2	−1	0	1	2	3

The distance a number is from zero on a number line is the **absolute value** of the number. For example, 3 is 3 units from 0, and –3 is also 3 units from 0. So opposite numbers have the same absolute value. The absolute values of 3 and −3 are written as follows.

$$|3| = 3 \quad |-3| = 3$$

You can also add two numbers by using the following rules.

To **add numbers with the same signs**, add the absolute values of the numbers. The sign of the sum is the same sign as the numbers.

To **add numbers with different signs**, subtract the absolute values. The sign of the sum is the sign of the number with the greater absolute value.

The following property shows the rule when one number is zero.

Identity Property of Addition	$a + 0 = a$
	The sum of zero and a number is always that number.

Check Understanding

Find each absolute value.
1. $|5|$ 2. $|-11|$
3. $\left|-\frac{2}{3}\right|$ 4. $|0|$

Example 2

Add.

a. $-7 + (-3)$ **b.** $-8 + 5$ **c.** $-11 + 0$

Solution

a. $|-7| = 7 \quad |-3| = 3$
Since the signs are the same, add the absolute values.
$7 + 3 = 10$

Each number is negative, so the sum is negative.
$-7 + (-3) = -10$

b. $|-8| = 8 \quad |5| = 5$
Since the signs are different, subtract the absolute values.
$8 - 5 = 3$

Since -8 has the greater absolute value, the sum is negative.
$-8 + 5 = -3$

c. Use the identity property of addition.
$-11 + 0 = -11$

Subtraction is much like addition with signed numbers.

Subtraction of Signed numbers	$a - b = a + (-b)$ or $a - (-b) = a + b$
	To subtract a number, add its opposite.

Example 3

Subtract.

a. $-7 - (-6)$ **b.** $-3 - 8$

Solution

a. $-7 - (-6) = -7 + 6$ **b.** $-3 - 8 = -3 + (-8)$
$= -1$ $= -11$

Technology Note

Different calculators require different keystrokes. For Example 3, part b, one calculator may use these keystrokes:
(−) 3 − 8

Another calculator may use:
3 +/− − 8

Experiment with your calculator.

Example 4

HOBBIES A scuba diver dove to an initial depth of 45 ft. He then came back up 17 ft to photograph a school of fish. Find the depth of the school of fish.

Solution

Add. $-45 + 17 = -28$
The school of fish is at a depth of 28 ft.

▼ TRY THESE EXERCISES

Use a number line to add or subtract.

1. $6 + 3$ **2.** $4 + (-3)$ **3.** $-7 - (-2)$ **4.** $4 - 2$

Find each absolute value.

5. $|-8|$ **6.** $|22|$ **7.** $|-3.7|$ **8.** $\left|\dfrac{5}{7}\right|$

Add or subtract using the rules for addition and subtraction.

9. $-6 + 3$ **10.** $7 - 9$ **11.** $-8 + (-5)$

12. $-4 - 4$ **13.** $4 + (-4)$ **14.** $4 + (-12) + (-8)$

15. SPORTS A football team gained 4 yd, lost 10 yd, lost 2 yd, gained 14 yd and then gained 2 yd. What was the net loss or net gain in yards?

▼ PRACTICE EXERCISES • For Extra Practice, see page 553.

Add or subtract. Use a number line or the rules for adding integers.

16. $5 + 5$ **17.** $17 - 21$ **18.** $8 - (-3)$

19. $-\dfrac{1}{2} + \dfrac{3}{4}$ **20.** $-22 - 3$ **21.** $7.3 + (-5.3)$

22. $-5 + (-4)$ **23.** $-6 + (-2)$ **24.** $6 + (-2)$

25. $-9 + (-9)$ **26.** $-16 - (-6)$ **27.** $-\dfrac{3}{5} - \dfrac{7}{10}$

28. $-0.8 + (-0.9)$ **29.** $40 - (-24)$ **30.** $-14 - 8$

31. $\dfrac{4}{5} - \dfrac{5}{7}$ **32.** $-12.9 - (-33.7)$ **33.** $-3 + \dfrac{1}{6}$

Simplify.

34. $12 - (-17) + 33$ **35.** $\dfrac{2}{3} - \dfrac{1}{6} + \dfrac{1}{2}$ **36.** $-4.2 - (-3.7) - 5.9$

37. $-7 + 12 + (-7) + 7$ **38.** $3 + (-9) + (-6) + 12$ **39.** $-4 + (-8) + 9 + (-3)$

Without subtracting, tell whether the answer will be positive or negative.

40. $1.00 - (-2.1)$ **41.** $-16 - (-4)$ **42.** $5 - 18$

43. $12 - (-7)$ **44.** $-72 - 71$ **45.** $-33 + 13$

46. WRITING MATH Will the difference be positive or negative when subtracting a greater number from a lesser number? Explain.

47. SPORTS A hockey arena has a seating capacity of 77,716 people. If there are 52,939 fans in attendance, how many seats are available?

FINANCE On Monday, Novia had $264 in her checking account. On Tuesday, she wrote checks for $69 and $92. On Thursday, she deposited $21.

48. Use paper and pencil to show how to find the balance in the account.

49. CALCULATOR Use a calculator to find the balance in the account.

Replace ■ with <, >, or =.

50. $-8 + 3$ ■ $-3 + (-2)$ **51.** $-3 - (-2)$ ■ $-8 - (-5)$ **52.** $4 + (-8)$ ■ $-6 + 1$

53. $7 + 2$ ■ $-1 + 10$ **54.** $|-6| - |-9|$ ■ $|5| - |-10|$ **55.** $|12| - |-9|$ ■ $|-10| - |-5|$

DATA FILE For Exercises 56 and 57, refer to the data on the surface temperature of the planets on page 525.

56. The temperatures of Venus and Jupiter differ by how many degrees Celsius?

57. The temperatures of Saturn and Pluto differ by how many degrees Fahrenheit?

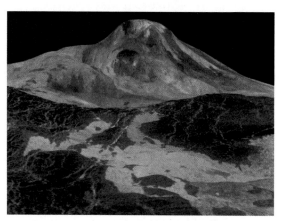

Surface of Venus

◣ EXTENDED PRACTICE EXERCISES

Find the integer represented by ■.

58. $7 - ■ = 2$ **59.** $■ + 2 = 14$

60. $-4 + ■ = 5$ **61.** $■ - (-3) = -8$

WEATHER The highest temperature recorded in Little Rock, Arkansas, is 112°F. The lowest is −5°F. The highest recorded temperature in Concord, New Hampshire, is 102° F. The lowest is −37°F.

62. Find the differences between the high and low temperatures for Little Rock and for Concord.

63. Which city has the greater temperature difference, or range?

◣ MIXED REVIEW EXERCISES

Use the following data. Round to the nearest hundredth. (Lesson 1-2)

145	122	198	121	167	154
135	198	135	175	184	163
125	125	136	154	176	164

64. Find the mean.   **65.** Find the median.

66. Find the mode.   **67.** Find the range.

68. Use estimation to find the approximate sum of the numbers in the data set. (Lesson 2-1)

3-2 Multiply and Divide Signed Numbers

Goals ■ Use rules to multiply and divide signed numbers.

Applications Sports, Weather

Work with a partner. Use Algeblocks and the Quadrant Mat.

Notice the positive and negative signs on a Quadrant Mat. To model multiplication with Algeblocks, use unit blocks along the axes of the Quadrant Mat to represent the factors. To find the product, build a rectangle with unit blocks. The product is the number of blocks in the rectangle with the sign of the quadrant.

Quadrant Mat   **Mat A**   **Mat B**

 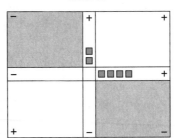 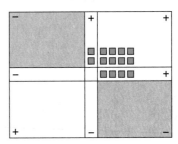

1. What multiplication is modeled by the Algeblocks in Mat A?

2. What is the product of the multiplication in Mat B?

3. Sketch or show 3 · (−3) using Algeblocks and a Quadrant Mat. Name the product.

◥ BUILD UNDERSTANDING

Several techniques can be used to multiply numbers. You can use a model such as Algeblocks, you can look for patterns, or you can follow rules.

| **Product Rules for Signed Numbers** | The product of two numbers with the same sign is positive. |
| | The product of two numbers with different signs is negative. |

Example 1

Multiply.

a. 8 · 4   **b.** −9 · (−6)   **c.** −6 · 5   **d.** 4 · (−2)

Solution

a. 8 · 4 = 32   The product is positive because the signs of the factors are the same.

b. −9 · (−6) = 54   The product is positive because the signs of the factors are the same.

c. −6 · 5 = −30   The product is negative because the signs of the factors are different.

d. 4 · (−2) = −8   The product is negative because the signs of the factors are different.

Division and multiplication are related operations. Therefore, quotient rules are the same as product rules.

| **Quotient Rules for Signed Numbers** | The quotient of two numbers with the same sign is positive. |
| | The quotient of two numbers with different signs is negative. |

Because division and multiplication are inverse operations, you can multiply to check your answer to division.

Example 2

Find each quotient. Then check by multiplying.

a. $56 \div 7$

b. $-42 \div (-7)$

c. $\dfrac{48}{-6}$

d. $\dfrac{-72}{8}$

Math: Who, Where, When

Emmy Noether (1882–1935) was a German-born mathematician and teacher. In 1933 Noether moved to the U.S. She taught mathematics at Bryn Mawr College for two years, until her death. She is considered to be one of the founders of modern algebra.

Solution

a. $56 \div 7 = 8$ The quotient is positive because the signs of the dividend and divisor are the same.

Check: $8 \cdot 7 = 56$

b. $-42 \div (-7) = 6$ The quotient is positive because the signs of the dividend and divisor are the same.

Check: $6 \cdot (-7) = -42$

c. $\dfrac{48}{-6} = -8$ The quotient is negative because the signs of the dividend and divisor are different.

Check: $-8 \cdot (-6) = 48$

d. $\dfrac{-72}{8} = -9$ The quotient is negative because the signs of the dividend and divisor are different.

Check: $-9 \cdot 8 = -72$

Example 3

WEATHER The temperature falls at a constant rate of 2°F/h. Describe the temperature 4 h ago.

Emmy Noether

Solution

The rate at which the temperature falls is represented by a negative number, -2.

The four hours that pass are represented by a negative number, -4.

Find the product.

$$-2 \cdot (-4) = 8$$

Four hours ago, the temperature was 8°F higher.

Example 4

SPORTS A wrestling coach records the weight gains and losses of his 12 team members over a two-month period. Find the average weight gain or loss for the team.

$$-4 \quad 0 \quad 3 \quad -6 \quad 2 \quad -2 \quad -3 \quad -2 \quad -5 \quad 0 \quad -1 \quad -6$$

Solution

Find the sum of the gains and losses.

$$-4 + 0 + 3 - 6 + 2 - 2 - 3 - 2 - 5 + 0 - 1 - 6 = -24$$

Divide the sum by the number of team members.

$$-24 \div 12 = -2$$

The average was a loss of 2 lb for each team member.

◢ TRY THESE EXERCISES

Multiply or divide.

1. $-7 \cdot 8$
2. $8 \div (-4)$
3. $-6 \cdot (-6)$
4. $\dfrac{-15}{3}$
5. $\dfrac{24}{8}$
6. $-4 \cdot 9$
7. $-21 \div (-7)$
8. $2 \cdot (-9)$
9. $-35 \div (-5)$
10. $-4 \cdot (-3)$
11. $-4 \cdot 3$
12. $\dfrac{28}{-4}$

13. **MODELING** Sketch or show $-2 \cdot (-3)$ using Algeblocks. Find the product.

14. **SPORTS** A football team lost 7 yd on each of the last 3 plays. What was the total yardage lost on the plays?

15. The price of a share of stock falls $10 over five days. If the rate of decrease is spread equally over the five days, how much did the price decrease in one day?

16. **WRITING MATH** Use Algeblocks to model the division $-12 \div 3$.

◢ PRACTICE EXERCISES • For Extra Practice, see page 553.

Multiply or divide.

17. $-36 \div 4$
18. $4 \cdot (-8)$
19. $3 \cdot (-7)$
20. $\dfrac{-35}{-5}$
21. $12 \div 3$
22. $-8 \cdot 8$
23. $-9 \cdot (-8)$
24. $40 \div (-8)$
25. $35 \div (-7)$
26. $-5 \cdot 8$
27. $-54 \div 9$
28. $-4 \cdot 5 \cdot (-2)$
29. $\dfrac{18}{-9}$
30. $-4 \cdot (-2)$
31. $\dfrac{-56}{-8}$
32. $-36 \div 6$
33. $14 \cdot 2$
34. $-60 \div (-5)$
35. $-16 \cdot 4$
36. $72 \div (-6)$
37. $-25 \cdot 5$
38. $\dfrac{-36}{12}$
39. $4 \cdot 5 \cdot (-2)$
40. $4 \cdot (-5) \cdot (-2)$

Replace each ■ with <, >, or =.

41. $-4 \cdot (-8)$ ■ $4 \cdot 8$
42. $-3 \div 1$ ■ $-3 \div (-1)$
43. $2 \cdot (-5)$ ■ $-3 \cdot 3$
44. $10 \div (-5)$ ■ $-15 \div (-3)$
45. $-24 \div (-6)$ ■ $\dfrac{-25}{5}$
46. $7 \cdot (-8)$ ■ $-9 \cdot 6$

47. WEATHER The temperature falls at a constant rate of 5°F/h. Describe the temperature 3 h ago.

48. The product of two numbers is −36. The quotient is −4. What are the numbers?

49. A stock portfolio decreased $99 in value. If the decrease was spread equally over 3 days, how did the value of the portfolio change each day?

50. CALCULATOR What keystrokes do you use to solve Exercise 49?

Write the multiplication or division sentence that describes each word phrase. Then find the product or quotient.

51. five times negative eight

52. eight divided by negative four

53. negative two times negative six

54. negative forty divided by five

DATA FILE Refer to the data on record holders from the earth sciences on page 527.

55. About how many times deeper is the Marianas Trench than the Grand Canyon?

56. About how many times older are the oldest slime molds than the oldest birds?

Grand Canyon

■ EXTENDED PRACTICE EXERCISES

Find the number represented by ■.

57. ■ ÷ (−10) = −5   **58.** −24 · ■ = 144

59. −15 · ■ = 75   **60.** ■ ÷ (−15) = 3

61. CRITICAL THINKING You can multiply to check an answer to a division problem. Using this knowledge, explain why zero can be divided by a nonzero number, but no number can be divided by zero.

62. CHAPTER INVESTIGATION One group member holds a yardstick or meter stick against a wall. Another group member holds the ball at the top of the stick and drops the ball so that it will bounce. A third group member records the height of the first three bounces. Repeat with each ball in your experiment.

■ MIXED REVIEW EXERCISES

Estimate. (Lesson 2-1)

63.   4.79
      + 14.2

64.   38.627
      − 17.9

65.   129.5
      × 31.45

66. GARDENING Bill wants to plant marigolds along the border of his garden because they help keep bugs off the plants. Bill's garden is 10 ft wide and 15 ft long. One seed packet provides enough seeds to plant one 7-ft row of marigolds. How many seed packets does Bill need? (Lesson 2-3)

Review and Practice Your Skills

PRACTICE ■ LESSON 3-1

Add or subtract. Use a number line or the rules for adding integers.

1. $-3 + 2$

2. $6 + (-11)$

3. $-6 + (-6)$

4. $-7 + (-15)$

5. $-8.2 + 3.5$

6. $9.1 + (-4.6)$

7. $-8 - 4$

8. $2 - (-6)$

9. $-9 - 14$

10. $4 - 11$

11. $-15 - (-10)$

12. $5 - (-5)$

13. $-0.3 - 6.2$

14. $-5.7 - (-1.7)$

15. $1.8 - (-0.4)$

Simplify.

16. $-14 - 4 + (-5)$

17. $-11 - 3 - (-4)$

18. $2 + (-13) - (6)$

19. $8 - (-2) + (-1.4)$

20. $-4.1 + 5.3 - 6.1$

21. $6.3 - (-0.6) + 2.5$

22. $-3 + (-8) - 10$

23. $7 - (-6) + 15$

24. $-1 - 10 + (-2)$

Without subtracting, tell whether the answer will be positive or negative.

25. $-14 + 20$

26. $-9 - 15$

27. $1 + (-7)$

28. $8 + (-13)$

29. $-12 + (-6)$

30. $-6 - 13$

31. $-1.3 - (-4.8)$

32. $5.2 + (-2.9)$

33. $7.5 - (-1.4)$

34. A mountain climber started at sea level. She then climbed up 450 ft, down 280 ft and up 700 ft. How high above sea level is she?

35. The temperature of molten lava is 1730° C. The surface temperature of Venus is 480°C. Which is hotter? By how much?

PRACTICE ■ LESSON 3-2

Multiply or divide.

36. $-15 \div 5$

37. $\dfrac{-24}{2}$

38. $-11 \cdot (-3)$

39. $56 \div (-7)$

40. $6 \cdot (-4)$

41. $-18 \div 3$

42. $35 \div (-7)$

43. $-4 \cdot (-1)$

44. $-40 \div 10$

45. $-4 \cdot 12$

46. $\dfrac{9}{-9}$

47. $-16 \div (-4)$

48. $-5 \cdot (-8)$

49. $75 \div (-3)$

50. $3 \cdot (-6)$

51. $-8 \cdot (-12)$

52. $\dfrac{-30}{-6}$

53. $-32 \div (-8)$

54. $7 \cdot (-2)$

55. $-6 \cdot 9$

56. $6 \cdot (-10)$

57. $-9 \cdot 20$

58. $-2 \cdot (-15)$

59. $120 \div (-20)$

Replace each ■ with <, >, or =.

60. $-2 \cdot (-3) \; ■ \; 12 \div (-4)$

61. $-30 \div 5 \; ■ \; 30 \div (-5)$

62. $-30 \div 3 \; ■ \; 2 \cdot (-5)$

63. $-12 \cdot (-2) \; ■ \; -60 \div (-5)$

64. $18 \div (-3) \; ■ \; -1 \cdot (-6)$

65. $-4 \cdot 6 \; ■ \; 3 \cdot (-9)$

66. A broken water tank is losing 7 gal of water each minute. Describe the water level in 5 min.

PRACTICE ◼ LESSON 3-1–LESSON 3-2

Add or subtract. (Lesson 3-1)

67. $-6 + (-3)$ **68.** $-12 - (-4)$ **69.** $13 + (-2)$

70. $3.1 - (-0.6)$ **71.** $-1.5 - (-1.5)$ **72.** $-2.9 + 4.1$

Simplify. (Lesson 3-1)

73. $-3 - (-6) + 9$ **74.** $-12 + (-3) - 20$ **75.** $-5 + (-4) + (-11)$

76. $1 - (-9) - (-2)$ **77.** $-0.6 - (-3) + 1.5$ **78.** $2.6 - (1.4) - 8.5$

Multiply or divide. (Lesson 3-2)

79. $-16 \div 2$ **80.** $\dfrac{-18}{-3}$ **81.** $-7 \cdot (-3)$

82. $7 \cdot (-6)$ **83.** $24 \div (-3)$ **84.** $\dfrac{40}{-5}$

Math*Works* Career – Sports Statistician
Workplace Knowhow

Sports statisticians record data from high school, college, and professional sporting events. The data they collect include individual points, team points, assists, serves, information on other teams and more. A sports statistician must have mental math skills. After an event, the data is entered into computer programs to calculate more detailed statistics. Shown are statistics from the first half of a basketball game.

Name	Position	2-points made	2-points attempts	3-points made	3-points attempts	Free throws made	Free throws attemps	Assists	Rebounds
Anderson, J.	forward	||	ЖНГ					|	||
Campbell, P.	guard			)|	ЖНГ|			||	
Kern, W.	center	|||	(||)			|||	ЖНГ		|/| \
McMahon, A.	forward	|	/)						
Smith, K.	forward	||	ЖНГ						||
Stephens, S.	center					|	||		||)
Williams, T.	guard	|)	ЖНГ	)|	|||			(||)	|

Use mental math.

1. What percent of 3-point attempts did Campbell make in the first half?

2. The goal for the centers and forwards is to have a total of 30 rebounds a game. How many rebounds do they need in the second half to meet their goal?

3. The goal for the guards is to have a total of 15 assists a game. How many assists do the guards need in the second half to meet their goal?

4. What percent of free throws did the entire team make in the first half?

Order of Operations

Goals
- Use order of operations to evaluate expressions.
- Use order of operations to solve word problems.

Applications   Retail, Art, Sports, Travel

Think of a simple activity, such as preparing lunch or getting ready for school.

1. List the steps needed to perform the task.

2. Cut your list apart, with each step of the activity on a separate strip of paper.

3. Mix up the strips and place them face down on your desk. Have a partner choose one strip at a time, turning it face up. Place each succeeding strip below the previous one.

4. Read the steps in this new order. Can you still accomplish the task? Explain why the order of the steps is important.

◢ BUILD UNDERSTANDING

Sometimes to achieve a goal, you must follow steps in a specified order. To simplify an expression is to perform as many of the indicated operations as possible. To simplify mathematical expressions, certain steps must be followed in a specific order. These steps are called the **order of operations**.

Order of Operations	**1.** Perform all calculations within parentheses, brackets, and other grouping symbols.
	2. Complete all calculations involving exponents.
	3. Multiply or divide in order from left to right.
	4. Add or subtract in order from left to right.

Sometimes there are no grouping symbols or exponents.

Example 1

Simplify.

a. $7 - 3 \cdot 2 + 9$ **b.** $8 \div 4 + 2 \cdot 3$

Solution

a. $7 - 3 \cdot 2 + 9 = 7 - 6 + 9$ Multiply first.

$\qquad\qquad\qquad = 1 + 9$

$\qquad\qquad\qquad = 10$

b. $8 \div 4 + 2 \cdot 3 = 2 + 2 \cdot 3$ Multiply and divide in order from left to right.

$\qquad\qquad\qquad = 2 + 6$

$\qquad\qquad\qquad = 8$

Example 2

Simplify.

a. $6 - (4 + 8) \div 2$ **b.** $(3 \cdot 4)^2 + 6$

Reading Math

Read the expression in Example 2, part b, as "Three times four, quantity squared, plus six."

Solution

a.
$$6 - (4 + 8) \div 2 = 6 - 12 \div 2$$
$$= 6 - 6$$
$$= 0$$

b.
$$(3 \cdot 4)^2 + 6 = (12)^2 + 6$$
$$= 144 + 6$$
$$= 150$$

The way you enter an expression on your calculator will depend on the type of calculator you have. Most calculators compute according to the order of operations.

Example 3

CALCULATOR Use a calculator to simplify $25 - 2 \cdot 7 + 1$.

Technology Note

Try Example 3 on your calculator. Does your calculator follow order of operations? If not, what keystrokes must you enter?

Solution

Use these keystrokes.

25 $\boxed{-}$ 2 $\boxed{\times}$ 7 $\boxed{+}$ 1 $\boxed{\text{ENTER}}$ 12

Example 4

TRAVEL Anu took 10,000 pesos on her trip to Mexico. She bought 3 pieces of jewelry for 400 pesos each and 2 sombreros for 115 pesos each. Then she exchanged U.S. money for 5000 pesos. How many pesos did she have then?

Solution

$$10,000 - 3 \cdot 400 - 2 \cdot 115 + 5000 = 10,000 - 1200 - 230 + 5000$$
$$= 8570 + 5000$$
$$= 13,570$$

Anu had 13,570 pesos.

▼ TRY THESE EXERCISES

Simplify. Remember to work within parentheses first.

1. $8 - 4 \div 2 \cdot 3$ **2.** $8 \div 4 - 2 + 3$ **3.** $\dfrac{1}{4} + \dfrac{1}{6} \cdot \dfrac{3}{2} - \dfrac{3}{4}$

4. $16 \cdot 4 + 3 \div 3$ **5.** $2.3 \cdot (1.8 + 0.2) \div 2$ **6.** $(12 - 2)4 - 8$

7. $45 \div (3 + 6) + 5$ **8.** $6^2 \div 4 \div 2$ **9.** $6^2 \div (4 \div 2)$

CALCULATOR Simplify using a calculator.

10. $25 \cdot (18 \div 6) - 5$ **11.** $(2 \cdot 16) - 3^3$ **12.** $14 \div 7 + (4 \cdot 2)$

13. FITNESS Colin ran 6 mi last week. He ran twice that distance this week. How far did he run over the two-week period?

14. WRITING MATH Why is it important to have a system for order of operations? Use an example to explain.

PRACTICE EXERCISES • For Extra Practice, see page 554.

Simplify. Use the order of operations.

15. $8 + 3 \cdot 9$

16. $24 - 10 \div 2$

17. $\frac{1}{2} + \frac{3}{4} \cdot \frac{1}{6}$

18. $(-6) \cdot 4 - 4 + 1$

19. $1.4 - 0.4 \cdot 12 + 4.3$

20. $72 \div (3 + 5) \cdot 9$

21. $3 + 2^4 \cdot 4$

22. $49 - (-8) \cdot (42 \div 7)$

23. $(18 + 7) \cdot 4 \div (-2)$

24. $4^3 - (6 \div 3)$

25. $(3 + 2)^4 \cdot 4$

26. $(3 + 3)^2 \cdot \frac{1}{3}$

27. $0.9 \cdot 0.2 + 1.5 \cdot 0.3$

28. $9 \cdot (2 + 5) \cdot 3$

29. $(12 + 15) \div 3 + 20$

30. $7 + 36 \div 6 - 10$

31. $\frac{1}{4}(6 \cdot 2) + 3 \cdot 7$

32. $1.5 \div (-3) + (0.9 \cdot 3.1)$

33. $8 + 6 - 2 \cdot (-2) - 3^2$

34. $4 \cdot 42 \div (56 \div 8 \cdot 3)$

35. $27 \div (7 + 2) \cdot (2 + 3)$

36. $63 \div (7 + 2) - 12 \div 3$

37. $-19 + 2 \cdot (6 - 4)^3$

38. $(65 - 7^2 + 4) \cdot 9$

CALCULATOR Simplify using a calculator.

39. $4^2 + 2 \cdot 6$

40. $9 \cdot 8 - 6 + (-3)$

41. $12 \div 2^2 - 3 \cdot 7$

42. Maya received $10 from her grandmother and three times that amount of money from her parents. She shared the total equally with three sisters. How much money did Maya and her sisters each receive?

43. ART Kirk had $35. He bought 5 tubes of paint for $2.50 each and 3 canvases for $5.50 each. How much money did he have left?

44. YOU MAKE THE CALL John simplified the following expression for his homework. Is his answer correct? Explain. $12 + 6 \div 2 - 4^2 = -7$

Write a numerical expression for each of the following.

45. the product of six and four, minus twenty-five

46. the sum of eight and nine, multiplied by the difference of ten and two

47. five squared, times the difference of thirty and twenty-eight, divided by twenty

Use only one of the symbols +, −, × or ÷ to make each number sentence true.

48. $8 \blacksquare 4 \blacksquare 2 = 1$

49. $(-12) \blacksquare 9 \blacksquare 3 = 0$

50. $6 \blacksquare 2 \blacksquare 4 = 12$

51. $7 \blacksquare 3 \blacksquare 1 = 11$

52. $8 \blacksquare (-3) \blacksquare (-6) = 144$

53. $12 \blacksquare 3 \blacksquare 4 = 1$

Write *true* or *false* for each. For any that are false, insert parentheses to make them true.

54. $4 - 3 \cdot 7 = 7$

55. $\frac{1}{7} \cdot 0 + \frac{1}{8} = \frac{1}{56}$

56. $5 \cdot 24 - 16 = 40$

57. $15 \div 3 - 2 = 3$

58. $16 + 2^2 \div 5 \cdot 3 = 12$

59. $0.4^2 \div 0 + 0.8 = 0.2$

60. WRITING MATH Explain why the key sequences for the expression $14 - 6 \cdot 2 \div 3$ might be different on two different calculators.

◼ EXTENDED PRACTICE EXERCISES

61. SPORTS Mika ran 4 mi farther than the mean distance of the other runners. The others ran 18 mi, 22 mi, 10 mi and 26 mi. How far did she run?

62. Janita opens a savings account with $200. She deposits $22.45, $10.75 and $15.00. Then she withdraws one-half of the total amount. How much does she withdraw from the account?

63. RETAIL Lyn has $60. She wants to purchase a hat for $32.50, two scarves for $15.75 each and one hair ribbon priced at 4 for $1.00. How much more money does she need?

Use parentheses and the symbols +, −, · and ÷ to make each number sentence true.

64. $7 \blacksquare 3 \blacksquare 1 \blacksquare 1 = 9$

65. $8 \blacksquare 3 \blacksquare 6 \blacksquare 5 = 4$

66. $10 \blacksquare 5 \blacksquare 5 \blacksquare 2 = 10$

67. $12 \blacksquare 0 \blacksquare 4 \blacksquare 3 = 1$

68. $30 \blacksquare 5 \blacksquare 5 \blacksquare 23 = 30$

69. $12 \blacksquare 3 \blacksquare 4 \blacksquare 5 = 16$

70. CRITICAL THINKING You can change the value of $3 + 5 \cdot 4 - 1$ by using parentheses. Show four possible ways to do this.

◼ MIXED REVIEW EXERCISES

SPORTS Tara scored the following points for each of her team's basketball games in one season.

5   8   4   10   12   7   8   13   6   11   12   7   12   15

71. Make a box-and-whisker plot of Tara's scores. (Lesson 1-8)

72. Which scores are in the second quartile? (Lesson 1-8)

73. What is the range of Tara's scores? (Lesson 1-2)

74. Find Tara's points per game average for the season. Round to the nearest whole point. (Lesson 1-2)

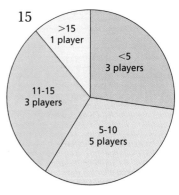

75. ERROR ALERT Tara's coach made a circle graph showing the game average for all the players. Is his graph accurate? Explain. (Lesson 1-4)

3-4 Real Number Properties

Goals
- Categorize numbers according to sets.
- Solve problems involving real number properties.

Applications   Sports, Food service, Accounting

1. Copy the number line shown.

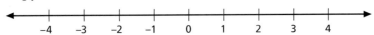

2. Place a dot on each of the following points. $-3, -\frac{1}{2}, 0, 2, 3$ and 3.3

3. Name three numbers that fall between 2 and 3 on the number line.

4. How many numbers fall between 2 and 3 on the number line?

◣ BUILD UNDERSTANDING

In mathematics, numbers are classified by their characteristics. Each of the following groups are **sets** of numbers. The diagram illustrates how the sets relate to each other.

Natural Numbers: $\{1, 2, 3, 4, 5, \ldots\}$

Whole Numbers: $\{0, 1, 2, 3, 4, \ldots\}$

Integers: $\{\ldots, -3, -2, -1, 0, 1, 2, 3, \ldots\}$

Rational Numbers: {any number that can be expressed in the form $\frac{a}{b}$, where a is any integer and b is any integer except 0}

Irrational Numbers: {any number that is a non-terminating, non-repeating decimal, such as $0.12345\ldots$, π or $\sqrt{2}$}

Real Numbers: {all rational and irrational numbers}

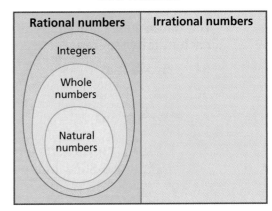

Example 1

To which sets of numbers does each belong?

a. -15 b. $\frac{3}{5}$ c. $0.3571264312\ldots$

Solution

a. -15 is an integer. It is a rational number because it can be expressed as $\frac{-15}{1}$. It is a real number because it is a rational number.

b. $\frac{3}{5}$ is a rational number and, therefore, a real number.

c. $0.3571264312\ldots$ is an irrational number since it is a decimal that does not repeat and does not terminate. It is a real number because it is an irrational number.

Operations on real numbers have some special properties. Commutative and associative properties do not apply to subtraction and division.

Commutative Property	$a + b = b + a$ $\qquad\qquad$ $x \cdot y = y \cdot x$ The order in which numbers are added or multiplied does not affect the sum or product.

For example, $15 + 12 = 12 + 15$, and $3 \cdot 2 = 2 \cdot 3$.

Associative Property	$(a + b) + c = a + (b + c)$ $\qquad$ $(x \cdot y) \cdot z = x \cdot (y \cdot z)$ The grouping of addends or factors does not affect the sum or product.

For example, $(3 + 5) + 6 = 3 + (5 + 6)$, and $(4 \cdot 8) \cdot 2 = 4 \cdot (8 \cdot 2)$.

Distributive Property	$a(b + c) = (a \cdot b) + (a \cdot c)$ $\qquad$ $x(y - z) = (x \cdot y) - (x \cdot z)$ A factor outside parentheses can be used to multiply each term within the parentheses.

For example, $5(3 + 2) = (5 \cdot 3) + (5 \cdot 2)$, and $6(4 - 1) = (6 \cdot 4) - (6 \cdot 1)$.

Example 2

Complete. Tell which property you used.

a. $12 + 7 = \blacksquare + 12$

b. $2(3 + 4) = (2 \cdot 3) + (\blacksquare \cdot 4)$

Solution

a. Use the commutative property. $\quad 12 + 7 = 7 + 12 \qquad$ The unknown number is 7.

b. Use the distributive property. $\quad 2(3 + 4) = (2 \cdot 3) + (2 \cdot 4) \qquad$ The unknown number is 2.

When an operation $(+, -, \cdot, \div)$ on a set of numbers always results in a number that is in the same set, the set of numbers is said to be **closed** under that operation. For example, when you add integers, the sum will always be an integer. Therefore, the integers are closed under addition.

Example 3

State whether each set is closed under the given operation.

a. natural numbers, division

b. rational numbers, subtraction

Solution

a. The natural numbers are not closed under division since $3 \div 4 = \frac{3}{4}$, which is not a natural number.

b. The rational numbers are closed under subtraction since the difference of any two rational numbers is always a rational number.

Example 4

Lina bought a ceramic bowl for $33.25, a tea set for $24.25 and mugs for $46.75. How much did she spend in all?

Solution

$33.25 + 24.25 + 46.75 = 24.25 + (33.25 + 46.75)$ Use the commutative and associative properties.

$$= 24.25 + 80.00$$

$$= 104.25$$

Lina spent $104.25.

▼ TRY THESE EXERCISES

To which sets of numbers does each belong?

1. -4 **2.** $0.983517\ldots$ **3.** 0 **4.** $-\dfrac{3}{4}$

Complete. Name the property you used.

5. $6 \cdot \blacksquare = 7 \cdot 6$ **6.** $5(4 + 2) = (5 \cdot 4) + (\blacksquare \cdot 2)$ **7.** $(3 \cdot \blacksquare) \cdot 9 = 3 \cdot (8 \cdot 9)$

Simplify. Name the properties you used.

8. $8 + 6 + 4$ **9.** $360 \cdot 2$ **10.** $\dfrac{5}{6} \cdot \dfrac{2}{3} \cdot \dfrac{3}{5}$

State whether each set is closed under the given operation.

11. whole numbers, multiplication   **12.** integers, division

13. FOOD SERVICE A server totaled a customer's lunch check. The customer had a salad for $4.75 and tea for $0.59. Tax on the meal was $0.25. Find the total.

▼ PRACTICE EXERCISES • For Extra Practice, see page 554.

To which sets of numbers does each belong?

14. 15.175 **15.** -1 **16.** $-17.56248\ldots$ **17.** $\dfrac{1}{3}$

18. $\dfrac{1}{9}$ **19.** 1400 **20.** $\dfrac{7}{10}$ **21.** 2.75

Complete. Name the property you used.

22. $23 \cdot 48 = 48 \cdot \blacksquare$ **23.** $\blacksquare \cdot (9 + 5) = (4 \cdot 9) + (4 \cdot 5)$ **24.** $9.36 + 2.58 = \blacksquare + 9.36$

25. $6 \cdot (8 \cdot 5) = (\blacksquare \cdot 8) \cdot 5$ **26.** $9(8 - 5) = (9 \cdot \blacksquare) - (9 \cdot 5)$ **27.** $\left(\dfrac{3}{7} + \dfrac{5}{8}\right) + \dfrac{1}{2} = \blacksquare + \left(\dfrac{5}{8} + \dfrac{1}{2}\right)$

Simplify using mental math. Name the properties you used.

28. $0.05 \cdot 1.32 \cdot 200$ **29.** $1.2 + 0.32 + 0.8$ **30.** $9 \cdot 3 \cdot 3$

31. $6 + 9 + 4 + 1$ **32.** $30 \cdot 198$ **33.** $0.25 + 72 + 0.75$

34. $14 + 8 + 206 + 92$ **35.** $5 \cdot 6 \cdot 2$ **36.** $\dfrac{1}{2} + \dfrac{3}{4} + \dfrac{1}{4}$

State whether each set is closed under the given operation.

37. natural numbers, subtraction

38. rational numbers, multiplication

For Exercises 39–45, copy and complete the table. Place a check mark in the column or columns that describe each number.

39. -21

40. 0.25

41. 0

42. $9.246 \ldots$

43. $\dfrac{11}{15}$

44. 100

45. π

	Natural	Whole	Integer	Rational	Irrational	Real
-21						
0.25						
0						
$9.246\ldots$						
$\dfrac{11}{15}$						
100						
π						

46. ERROR ALERT A customer checked her restaurant bill. She had charges of $6.72, $9.01, $2.28, $0.75, $1.25 and $1.98. The written total was $22.00. Is the total correct? If not, what is the correct total?

47. SPORTS For 3 games, James scored the most points for his team. In one game he scored 17 points. In the next game he scored 22 points. In the third game he scored 16 points. Over the 3 games, how many points did he score?

48. DATA FILE Refer to the data on top films on page 522. Find the total gross, in millions, for the five Star Wars films, *Star Wars, The Phantom Menace, Attack of the Clones, Return of the Jedi*, and *The Empire Strikes Back*.

■ EXTENDED PRACTICE EXERCISES

49. WRITING MATH Write a paragraph explaining the relationship among number sets in the real number system.

Decide whether each statement is *true* or *false*. Explain.

50. Any natural number is an integer.

51. The commutative property cannot be used with division.

52. Any irrational number is a whole number.

53. CHAPTER INVESTIGATION Organize your data into a table that records the type of ball, its diameter, its average weight, and the height of the three bounces. Name the type of numbers that you list in each column of the table.

■ MIXED REVIEW EXERCISES

Find each circumference and area to the nearest hundredth. (Lesson 2-7)

54.

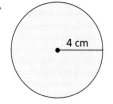

4 cm

55.

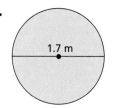

1.7 m

56.

2.5 ft

Review and Practice Your Skills

Simplify. Use the order of operations.

1. $8 + 2 \cdot 11$

2. $(2 + 9) \cdot 6$

3. $8 + 45 \div 5 - 3$

4. $6 - 3 \cdot 4 + 1$

5. $15 + 2 \cdot 6 - 10$

6. $10 \div 2 + 4 \cdot 5$

7. $8 \cdot 10 - 14 \div 7$

8. $(4 + 1)^2 - 10$

9. $(2 - 6) \cdot 100$

10. $40 - 3 \cdot 9$

11. $15 \div 3 + 20 \cdot 2$

12. $16 - 48 \div 6$

13. $25 - 60 \div 6 + 2$

14. $15 + 20 \div 2$

15. $3 \cdot (8 - 9)^2$

16. $(25 + 5) \div (12 - 6)$

17. $3 \cdot 8 + 40 \div 5$

18. $36 \div (4 + 2) + 7$

19. $5^2 + 10 \div 2$

20. $0.5 \cdot (4 + 6) - 1$

21. $5 - 4 \cdot 2 + 10$

22. $10 \cdot 3 - 6 + 20$

23. $2 - 4^2 \cdot (125 - 25)$

24. $16 \div 4 + 3 - 2 \cdot 3$

25. $2.5 \cdot 10 - 4 \cdot 6$

26. $30 + (18 - 16)^3$

27. $7 - 3 \cdot 1 + (4 + 5)$

28. Orlando started a college savings account. He deposited \$200, then added \$25 each week for 8 wk. How much has he deposited at the end of 8 wk?

29. **HOBBIES** A pack of comic collector cards contains 15 cards. Joan buys 4 packs at one store and 2 packs at another. When she gets home she shares the cards with her two brothers, so she and her brothers each have the same number. How many cards does Joan keep?

To which sets of numbers do the following belong?

30. 7

31. $\dfrac{2}{3}$

32. -3

33. $\dfrac{3}{4}$

34. -12

35. 0

36. $0.16116\ldots$

37. 0.45

Complete. Name the property you used.

38. $(-2 + 6) + 3 = -2 + (\blacksquare + 3)$

39. $16 \cdot 29 = \blacksquare \cdot 16$

40. $(-1 \cdot 3) \cdot 8 = -1 \cdot (\blacksquare \cdot 8)$

41. $4(2 - 8) = (4 \cdot \blacksquare) - (4 \cdot 8)$

42. $\blacksquare + 3.6 = 3.6 + 2.1$

43. $-12 + (-15) = -15 + \blacksquare$

44. $\dfrac{2}{3} \cdot \dfrac{1}{5} = \dfrac{1}{5} \cdot \blacksquare$

45. $2(\blacksquare + 3) = (2 \cdot 11) + (2 \cdot 3)$

46. $(8 \cdot 2) \cdot \blacksquare = 8 \cdot (2 \cdot 7)$

Simplify using mental math. Name the properties you used.

47. $1.2 + 3.6 + 4.8$

48. $20 \cdot 16 \cdot 5$

49. $251 \cdot 40$

50. $16 + 83 + 104$

51. $16 \cdot 12 \cdot 0.25$

52. $50 \cdot 498$

53. $\dfrac{5}{6} \cdot \dfrac{4}{9} \cdot \dfrac{6}{5} \cdot \dfrac{1}{4}$

54. $\dfrac{2}{3} + \dfrac{7}{8} + \dfrac{1}{3}$

55. $698 + 214 + 302$

State whether the following sets are closed under the given operation.

56. integers, division

57. whole numbers, subtraction

58. natural numbers, multiplication

59. rational numbers, subtraction

Add, subtract, multiply or divide. (Lessons 3-1 and 3-2)

60. $-11 + (-2)$ **61.** $9 + (-20)$ **62.** $3 - (-11)$ **63.** $-6 - 15$

64. $-8 - (-3)$ **65.** $14 \div (-2)$ **66.** $-5 \cdot 4$ **67.** $-3 \cdot (-10)$

68. $42 \div (-6)$ **69.** $-4 \cdot (-9)$ **70.** $\dfrac{-35}{7}$ **71.** $6 \cdot (-11)$

Simplify. Use the order of operations. (Lesson 3-3)

72. $35 - 2 \cdot 10$ **73.** $6 + 1 \cdot 4^2$ **74.** $(3 - 23) \cdot 100$

75. $20 - (3 + 1)^2 \cdot (-1)$ **76.** $43 + 50 \div 2$ **77.** $5 + 10^2 - 4 \cdot 5$

To which sets of numbers do the following belong? (Lesson 3-4)

78. -14 **79.** $\dfrac{1}{12}$ **80.** 25 **81.** $-2.02002\ldots$

Complete. Name the property you used. (Lesson 3-4)

82. $3.6 \cdot (-0.4) = \blacksquare \cdot 3.6$ **83.** $\left(\dfrac{1}{2} \cdot 6\right) - \left(\dfrac{1}{2} \cdot \blacksquare\right) = \dfrac{1}{2}(6 - 9)$ **84.** $3 + (\blacksquare + 6) = (3 + 9) + 6$

85. $\blacksquare(6 + 2) = (3 \cdot 6) + (3 \cdot 2)$ **86.** $2.34 + \blacksquare = 9.01 + 2.34$ **87.** $(9 \cdot 13) \cdot 7 = 9 \cdot (\blacksquare \cdot 7)$

Mid-Chapter Quiz

Simplify. (Lessons 3-1 and 3-2)

1. $-10 + 4$ **2.** $6 + (-3) + (-2) + 5 - (-8)$ **3.** $456.1 + 3.567 + 21.43$

4. $359.94 \div 0.42$ **5.** $1 - \dfrac{3}{5}$ **6.** $\dfrac{5}{7} \div \dfrac{3}{5}$

7. $6\dfrac{2}{3} \cdot 5\dfrac{1}{4}$ **8.** $0.0032 \cdot 1.25$ **9.** $9 \cdot (-2) \cdot (-1)$

Evaluate. (Lesson 3-3)

10. $(7 + 6)^2 \div 13$ **11.** $(3 + 4) \cdot (9 - 6)^2$ **12.** $8 + 2 - 3 \cdot (9 - 6)^2$

Complete. Name the property you used. (Lesson 3-4)

13. $327 \cdot 33 = \blacksquare \cdot 327$ **14.** $(2 \cdot 4) \cdot 7 \cdot 8 = 2 \cdot (4 \cdot \blacksquare) \cdot 8$ **15.** $8(3 + 6) = 8(3) + \blacksquare(6)$

16. $(10 + 6) + 15 = \blacksquare + (6 + 15)$ **17.** $5 + 21 = 21 + \blacksquare$ **18.** $7 + (5 + 3) = (7 + 5) + \blacksquare$

19. **ART** There are three art exhibits going on simultaneously at the museum. Each exhibit has 5 paintings. Each painting shows 2 men and 3 women. How many people are there in all? (Lesson 3-4)

To which sets of numbers do the following belong? (Lesson 3-4)

20. -137 **21.** $\dfrac{17}{42}$ **22.** $0.12534876\ldots$

23. $\dfrac{31}{39}$ **24.** 0.52 **25.** -77

Variables and Expressions

Goals ■ Write and evaluate variable expressions.

Applications Weather, Finance, Music, Sports

Try the following number trick four times, each time starting with a different number. Suppose you choose 3 for *Step 1* and then add 5 for *Step 2*. This leads to the statement $3 + 5 = 8$.

1. Each time you try the number trick, write a statement for each step.

   Step 1 Choose any number.   *Step 2* Add 5.

   Step 3 Multiply by 2.   *Step 4* Subtract 4.

   Step 5 Divide by 2.   *Step 6* Subtract the original number.

2. Record your answer. What conclusions can you draw?

▼ BUILD UNDERSTANDING

A **variable** is an unknown number usually represented by a letter, such as *a*, *n*, *x*, or *y*. Expressions with at least one variable, such as $n + 5$ and $x \div 9$, are **variable expressions**. You can write variable expressions to describe word phrases.

Example 1

Write a variable expression. Let *n* represent "a number."

a. eight more than a number

b. five less than a number

c. a number divided by three

d. five more than three times a number

Solution

a. $n + 8$ **b.** $n - 5$

c. $n \div 3$ or $\frac{n}{3}$ **d.** $3n + 5$ $\quad$ 3*n* means multiply *n* by 3.

To **evaluate** a variable expression, replace the variable with a number. The number is called a **value** of the variable. Then perform any indicated operations, following the order of operations.

First perform all multiplication and division in order from left to right.

Then complete any addition and subtraction in order from left to right.

Check Understanding

The following Basic Mat represents one of the expressions in Example 1. Is part a, b, c or d represented?

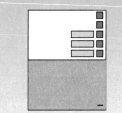

Example 2

Evaluate each expression. Let $x = 2$, $y = 3$, and $z = -2$.

a. $x + 9$ **b.** $18 \div y$

c. $2x + z$ **d.** $5y - 2x$

Solution

a. $x + 9$ **b.** $18 \div y$

$2 + 9 = 11$ $18 \div 3 = 6$

c. $2x + z$ **d.** $5y - 2x$

$2 \cdot 2 + (-2)$ Multiply first. $5 \cdot 3 - 2 \cdot 2$

$4 + (-2) = 2$ Then add or subtract. $15 - 4 = 11$

Example 3

Write an expression for each situation.

a. c pens shared equally among 4 boys

b. d dollars more than $7

c. b baseballs less 7

d. t movie tickets at $8 each

Solution

a. $c \div 4$ or $\dfrac{c}{4}$ **b.** $7 + d$

c. $b - 7$ **d.** $8 \cdot t$ or $8t$

▼ TRY THESE EXERCISES

Write a variable expression. Let n represent "a number."

1. seven greater than a number

2. six less than a number

3. a number divided by nine

4. the sum of negative two and three times a number

Evaluate each expression. Let $a = 3$, $b = 6$, and $c = -2$.

5. $13 - a$ **6.** $b \div 2$ **7.** $5c$ **8.** $2b \div a$

Write an expression to describe each situation.

9. s dollars of savings increased by twenty-seven dollars

10. five more than three times d dollars

11. p dollars of savings equally deposited in five accounts

SPORTS A box holds x golf balls. Three people share them equally.

12. Write an expression to describe the situation.

13. Evaluate the expression. Let $x = 18$.

14. MODELING Sketch or show the expression $4x + 2$ using Algeblocks.

▼ PRACTICE EXERCISES • For Extra Practice, see page 555.

Write a variable expression. Let n represent "a number."

15. half of a number

16. four less than a number

17. six times a number

18. a number increased by nine

19. a number divided by seven

20. the product of ten and a number

21. seventeen greater than four times a number

22. sixteen decreased by a number

Evaluate each expression. Let $a = 2$, $b = 8$, and $c = -4$.

23. $a - 2$

24. $2b$

25. $24 \div c$

26. $21 + a$

27. $b - 3$

28. $7c$

29. $12 - b$

30. $c + 13$

31. $3a$

32. $3a - 2$

33. $3b + 4$

34. $\dfrac{4a}{9}$

35. $a + b$

36. $b - a$

37. $b \div c$

38. $9a - c$

39. $c - 4a$

40. $3b \div a$

41. $7c \cdot 4b$

42. $9a - 2b$

43. $12a \div c$

44. $b \div 2a$

45. $b - 3a$

46. $5c + 6b$

Write an expression to describe each situation.

47. d dollars shared equally by three people

48. p pens at fifty cents each

49. e erasers decreased by two

50. nine more than half of w wallets

MUSIC Jenisa has c compact discs. She gives five away.

51. Write an expression to describe the situation.

52. Evaluate the expression if $c = 38$.

WEATHER The temperature in the morning was 65° F. It rose d degrees before noon.

53. Write an expression to describe the situation.

54. Evaluate the expression if $d = 14$.

FINANCE In May, Chang had d dollars in his bank account. Two months later he had four more than three times that amount.

55. Write an expression to describe the situation.

56. Evaluate the expression if $d = 25$.

Evaluate each expression. Let $a = 2$, $b = 3$, and $c = 6$.

57. $a + c - b$　　　　**58.** $c - 2a + 1$　　　　**59.** $a + \dfrac{12}{b}$

60. $\dfrac{c}{a} + b$　　　　**61.** $3c + 2a + b$　　　　**62.** $12b \div 6 - 1$

DATA FILE For Exercises 63 and 64, refer to the data on the climate of 15 selected U.S. cities on page 535.

63. What temperature do you prefer in the summer? Let x represent this temperature.

64. Find the average daily temperature in Washington, D.C. for the month of July. Write an expression to represent the difference in temperature between your preference, x, and the temperature in Washington, D.C.

65. MODELING Sketch or show the expression $2x + 3y - 5$ using Algeblocks.

Jefferson Memorial, Washington D.C.

EXTENDED PRACTICE EXERCISES

Let x represent an odd number. Write an expression to describe the following.

66. next odd number　　　　**67.** preceding odd number

68. next whole number　　　　**69.** sum of the odd number and the next whole number

Try this number trick.

70. Pick a number. Add 20, multiply by 6, divide by 3, subtract 40, and divide by 2.

71. Write a conclusion about your result.

72. Change the number trick so that the result is twice the original number.

73. WRITING MATH Write a number trick in which the result is your age. Try it out with a classmate.

MIXED REVIEW EXERCISES

Use the pictograph. (Lesson 1-5)

74. How many inches of snow does one snowball represent?

75. How much snow falls, on average, in Denver in December?

76. How much more snow falls in Cleveland in December than in Boston?

77. How much more snow falls in Chicago in December than in Denver?

78. Would this sample data be enough to make a prediction about the amount of snow that falls in any U.S. city in December? Why? (Lesson 1-1)

Average December Snowfall

Baltimore, MD	◯ ◯
Boston, MA	◯ ◯ ◁
Chicago, IL	◯ ◯ ◯ ◯ ◁
Cleveland, OH	◯ ◯ ◯ ◯ ◯ ◯
Denver, CO	◯ ◯ ◯ ◖

Key: ◯ = 2 in.

3-6 Problem Solving Skills: Find a Pattern

Patterns are everywhere. Patterns can be found in the design of a quilt, the interest of a savings account or as a play used in a football game. Recognizing and extending patterns is frequently used as a problem solving strategy. Sometimes a problem is easier to understand and solve if you look for a pattern in the problem.

Problem Solving Strategies

Guess and check

✔ Look for a pattern

Solve a simpler problem

Make a table, chart or list

Use a picture, diagram or model

Act it out

Work backwards

Eliminate possibilities

Use an equation or formula

Problem

SPORTS Renee coaches a volleyball team. She told the players that they are to make 1 successful serve at the first practice, 3 at the second practice, 5 at the third practice, 7 at the fourth practice and so on. If this pattern continues, how many successful serves will each team member complete during the eighth practice?

Solve the Problem

Make a table to organize the information. Then look for a pattern.

Practice	1	2	3	4	5	6	7	8
Successful serves	1	3	5	7	9	11	13	15

Look at the first four entries in the table. Each number of successful serves is increased by two to get the next number. The pattern continues until the eighth practice. If this pattern continues, Renee's team members will each have 15 successful serves at the eighth practice.

TRY THESE EXERCISES

Which rule describes the pattern?

1. 0.08, 0.16, 0.24, 0.32

   A. Add 0.8.

   B. Subtract 0.8.

   C. Add 0.08.

2. 1.5, 1.0, 2.0, 1.5, 2.5

   A. Add 0.5, subtract 1.0.

   B. Subtract 0.5, add 0.5.

   C. Subtract 0.5, add 1.0.

CALCULATOR Use a calculator to find the first four products. Look for a pattern. Predict the fifth product. Use the calculator to check your answer.

3. $10{,}989 \cdot 9 = $ _____
$10{,}989 \cdot 8 = $ _____
$10{,}989 \cdot 7 = $ _____
$10{,}989 \cdot 6 = $ _____

4. $3 \cdot 37{,}037 = $ _____
$6 \cdot 37{,}037 = $ _____
$9 \cdot 37{,}037 = $ _____
$12 \cdot 37{,}037 = $ _____

Five-step Plan

1 Read
2 Plan
3 Solve
4 Answer
5 Check

FINANCE Kenyon plans to put $5 in his savings account in January, $8 in February, $11 in March, $14 in April and so on.

5. How much money will he put into the account in December?

6. How much money will he put into the account for the entire year?

ENTERTAINMENT A movie theater plans to give away 2 tickets on Monday, 8 tickets on Tuesday, 32 tickets on Wednesday and so on for Thursday and Friday.

7. How many tickets will they give away on Friday?

8. How many tickets will they give away for the five days?

9. **SPORTS** Amaya was training for a marathon. She started her training on Monday and ran 2 km. She doubled the amount that she ran each day of the week. On what day of the week did she run 32 km?

10. **WRITING MATH** Find the following five products. Describe the pattern.
$9 \cdot 8 =$ _____, $99 \cdot 8 =$ _____, $999 \cdot 8 =$ _____, $9999 \cdot 8 =$ _____, $99,999 \cdot 8$ _____.

For Exercises 11 and 12, draw the next model.

11.

12.

13. How many dots are in the fourth model for Exercise 11?

14. Predict how many dots will be in the fifth model for Exercise 11.

15. How many dots are in the fifth model for Exercise 12?

16. Predict how many dots will be in the sixth model for Exercise 12.

17. **MODELING** Create another model, like those in Exercises 11 and 12.

18. **NUMBER THEORY** Instances of the Fibonacci sequence of numbers occur frequently in nature. The arrangement of the parts of many plants reflects this sequence. Continue the pattern to find the next three numbers.

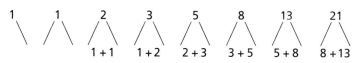

$$1 \quad 1 \quad 2 \quad 3 \quad 5 \quad 8 \quad 13 \quad 21$$
$$1+1 \quad 1+2 \quad 2+3 \quad 3+5 \quad 5+8 \quad 8+13$$

◣ MIXED REVIEW EXERCISES

Find the area of each figure. (Lesson 2-4)

19. rectangle 7 ft long, 4.2 ft wide

20. triangle 7 ft tall, base 4.2 ft wide

21. parallelogram 7 ft long, 4.2 ft wide

22. trapezoid bases of 7 ft and 4 ft, 4.2 ft tall

Review and Practice Your Skills

Write a variable expression. Let n represent "a number."

1. five less than a number
2. twenty divided by a number
3. the product of a number and two
4. a number decreased by six
5. a number increased by three
6. four more than a number
7. two more than three times a number
8. six times a number decreased by one

Evaluate each expression. Let $a = 3$, $b = -1$, and $c = 6$.

9. $2c$
10. $4 - a$
11. $b - 20$
12. $-3 + b$
13. $5a$
14. $8 - c$
15. $c - 9$
16. $-2a$
17. $4a \div (-6)$
18. $10 - b$
19. $c \div (-3)$
20. $6 + a$
21. $a - (-2)$
22. $-10c \div (-5)$
23. $4b - 3$
24. $6 + 2a$
25. $2a - 5c$
26. $12b \div a$
27. $-3b + 4a$
28. $5b \cdot (-a)$
29. $6b - 3c$

Write an expression to describe each situation.

30. h hours at ten dollars per hour
31. two more than half of m miles
32. d dollars shared equally by two people
33. four times b boxes plus ten
34. y years decreased by five
35. five less than m meters

Choose the rule that describes each pattern. Then write the next two numbers.

36. 15, 12.5, 10, 7.5   **a.** Subtract 3.   **b.** Subtract 0.5.   **c.** Subtract 2.5.
37. 4.1, 5.3, 6.5, 7.7   **a.** Add 1.2.   **b.** Add 0.2.   **c.** Add 1.02.
38. 0.0007, 0.07, 7, 700   **a.** Multiply by 100.   **b.** Multiply by 10.   **c.** Divide by 100.
39. 113, 93, 73, 53   **a.** Add 20.   **b.** Subtract 10.   **c.** Subtract 20.
40. 486, 162, 54, 18   **a.** Divide by 30.   **b.** Divide by 3.   **c.** Subtract 36.
41. 0.02, 0.06, 0.18, 0.54   **a.** Add 0.04   **b.** Multiply by 0.3.   **c.** Multiply by 3.
42. 0.08, 0.38, 0.68, 0.98   **a.** Add 0.3.   **b.** Add 0.38.   **c.** Add 0.08.

BIOLOGY During a July study, scientists tagged 15 polar bears. The first animal tagged was numbered PB-10792, the second was numbered PB-10793, and so on.

43. What is the number of the 15th bear tagged?

44. Several months later a polar bear wearing tag number PB-10814 was seen in the area. Is this one of the bears tagged during the July study?

PRACTICE ◣ LESSON 3-1–LESSON 3-6

Add, subtract, multiply, or divide. (Lessons 3-1 and 3-2)

45. $6 + (-7)$
46. $-8 - 13$
47. $-36 \div (-6)$
48. $9 \cdot (-2)$

Simplify. Use the order of operations. (Lesson 3-3)

49. $(6 + 2) \cdot 10 - 3$
50. $5 \cdot 3^2 - 4 \div 2$
51. $4 - (8 - 2)^2 \cdot 10$

Complete. Name the property you used. (Lesson 3-4)

52. $(6 \cdot 2) + (\blacksquare \cdot 8) = 6(2 + 8)$
53. $-0.36 \cdot 5 = \blacksquare \cdot (-0.36)$
54. $\blacksquare + \dfrac{4}{7} = \dfrac{4}{7} + \dfrac{1}{2}$

Evaluate each expression. Let $a = -2$, $b = 5$, and $c = -1$. (Lesson 3-5)

55. $3b - 2$
56. $10 + 3c$
57. $4b \div (-2)$
58. $c - 2a$
59. $4c + 2b$
60. $3a \cdot 2c$

Choose the rule that describes each pattern. Then write the next two numbers. (Lesson 3-6)

61. 287, 28.7, 2.87, 0.287   **a.** Divide by 10.   **b.** Divide by 100.   **c.** Subtract 100.

62. 5.3, 16.3, 27.3, 38.3   **a.** Multiply by 0.3.   **b.** Add 0.3.   **c.** Add 11.

Math*Works* Career – Fitness Trainer
Workplace Knowhow

Fitness trainers develop personalized fitness programs, including aerobics, weight training, and diet. To measure and record a customer's progress, a trainer might calculate the customer's target heart rate. Your target heart rate is the number of beats per minute your heart should pump for you to get the most out of exercise. Fitness trainers use a formula to find a person's target heart rate.

Use the following steps to find your target heart rate.

1. Find your maximum heart rate by subtracting your age from 220.

2. Find your pulse on your wrist or neck. Count your pulse for 10 sec and multiply that number by 6. This is your resting heart rate.

3. Subtract your resting rate from your maximum rate.

4. Multiply that difference by 0.75. Round this number to the nearest whole number.

5. Add the rounded number to your resting heart rate. This number is your target heart rate.

3-7 Exponents and Scientific Notation

Goals ■ Express numbers in exponential form and scientific notation.

Applications Physics, Sports, Astronomy, Biology

Compare the numbers shown. The top number, 10, can be written as the product $1 \cdot 10$. The next number, 100, can be written as the product $10 \cdot 10$.

10
100
1,000
10,000
100,000
1,000,000
10,000,000
100,000,000
1,000,000,000

1. Write each of the other numbers as a product, using only 10 as a factor.

2. Beside each product, write the number of factors.

3. How many factors of 10 are there for the number 100,000,000,000?

4. Think of a rule to find the number of factors in Questions 2 and 3.

◤ BUILD UNDERSTANDING

Large multiples of a number can be easier to use and read by writing them in **exponential form**. A number written in exponential form has a base and an exponent. The **base** tells what factor is being multiplied. The **exponent** tells the number of equal factors.

$$10 \cdot 10 \cdot 10 \cdot 10 = 10^4 \quad \leftarrow \text{exponent}$$
$$\text{base}$$

A number multiplied by itself, $x \cdot x$, is that number **squared**. A number used as a factor three times, $x \cdot x \cdot x$, is that number **cubed**.

Exponents that are one, zero, or negative have special rules.

Any number raised to the first power is that number itself. $\quad 2^1 = 2, 3^1 = 3$

Any nonzero number raised to the zero power is equal to 1. $\quad 5^0 = 1, 10^0 = 1$

Any nonzero number raised to a negative power can be written as a fraction with a numerator of 1 and a denominator that has a positive exponent. $\quad 3^{-2} = \frac{1}{3^2}$

Reading Math

Read 10^2 as "ten squared."

Read 10^3 as "ten cubed."

Read 10^4 as "ten to the fourth power."

Example 1

Write in exponential form.

a. $7 \cdot 7 \cdot 7$ **b.** $0.4 \cdot 0.4 \cdot 0.4 \cdot 0.4 \cdot 0.4$ **c.** $\frac{1}{5 \cdot 5}$

Solution

a. 7^3 **b.** $(0.4)^5$ **c.** $\frac{1}{5^2} = 5^{-2}$

When you write an expression in **standard form**, you calculate the equivalent expression so that the number is written without an exponent.

Example 2

Write in standard form.

a. 3^4 **b.** $(0.2)^0$ **c.** 17^1 **d.** 4^{-2}

Solution

a. $3^4 = 3 \cdot 3 \cdot 3 \cdot 3$ **b.** $(0.2)^0 = 1$ **c.** $17^1 = 17$ **d.** $4^{-2} = \dfrac{1}{4^2}$

$= 81$ $= \dfrac{1}{4 \cdot 4} = \dfrac{1}{16}$

Often, very large or very small numbers are written in **scientific notation**. A number written in scientific notation has two factors. The first factor is a number greater than or equal to 1 and less than 10. The second factor is a power of 10. These numbers are written in scientific notation.

$4 \cdot 10^3$ $9.9 \cdot 10^{-8}$ $6.001 \cdot 10^{20}$

Check Understanding

To write 8000 in scientific notation, is the exponent of 10 positive or negative?

To write 0.026 in scientific notation, is the exponent of 10 positive or negative?

Example 3

Write in scientific notation.

a. 50,000 **b.** 0.016

Solution

a. $50{,}000 = 50{,}000.$

      5.0000 Move the decimal point left so it shows a number between 1 and 10.

      $= 5.0000 \cdot 10^?$ To find the exponent of 10, count the number of decimal places you moved the decimal point to the left.
      $= 5 \cdot 10^4$

b. $0.016 = 01.6 \cdot 10^?$ Move the decimal pont to the right of the first nonzero digit.

      $= 1.6 \cdot 10^{-2}$ Count the number of decimal places you moved the decimal point to the right.

Example 4

Write each number in standard form.

a. $6.47 \cdot 10^8$ **b.** $3.416 \cdot 10^{-4}$

Solution

a. $6.47 \cdot 10^8 = 647000000.$ Move the decimal point to the right 8 places.

      $= 647{,}000{,}000$

b. $3.416 \cdot 10^{-4} = .0003416$ Move the decimal point to the left 4 places.

      $= 0.0003416$

TRY THESE EXERCISES

Write in exponential form.

1. $5 \cdot 5 \cdot 5$

2. $3 \cdot 3 \cdot 3 \cdot 3 \cdot 3$

3. $\dfrac{1}{4 \cdot 4 \cdot 4 \cdot 4}$

Write in standard form.

4. 2^2

5. 4^{-3}

6. -6^1

7. $5.629 \cdot 10^3$

8. $9.04 \cdot 10^5$

9. $4.2 \cdot 10^{-6}$

Write in scientific notation.

10. 7302

11. 0.00009

12. 0.00000002195

13. WRITING MATH Explain why each of the following numbers is not written in scientific notation.

$$0.621 \cdot 10^2 \qquad 1.23 \cdot 5^4 \qquad 1.25 + 10^3$$

14. SPORTS In the U.S., approximately 21 million youths under age 18 participate in basketball, and approximately 13.1 million participate in volleyball. How many more youths participate in basketball than volleyball? Express your answer in scientific notation.

PRACTICE EXERCISES • For Extra Practice, see page 555.

Write in exponential form.

15. $7 \cdot 7 \cdot 7 \cdot 7$

16. $2 \cdot 2 \cdot 2 \cdot 2 \cdot 2 \cdot 2 \cdot 2$

17. $\dfrac{1}{3 \cdot 3 \cdot 3}$

18. $6 \cdot 6$

19. $\dfrac{1}{8 \cdot 8 \cdot 8 \cdot 8 \cdot 8 \cdot 8}$

20. $10 \cdot 10 \cdot 10$

21. $5 \cdot 5 \cdot 5 \cdot 5 \cdot 5$

22. $91 \cdot 91$

23. $\dfrac{1}{4 \cdot 4 \cdot 4 \cdot 4}$

Write in standard form.

24. 3^2

25. -4^2

26. 3^{-5}

27. 9^{-3}

28. 2^1

29. 6^{-4}

30. $(0.5)^2$

31. -7^{-3}

32. 1^{10}

33. 13^0

34. 3^{-4}

35. 6^3

36. $6.39 \cdot 10^4$

37. $6.876 \cdot 10^{-7}$

38. $3 \cdot 10^2$

39. $7.1 \cdot 10^{-3}$

40. $8.62154 \cdot 10^{-8}$

41. $8.7231 \cdot 10^9$

42. $4.124 \cdot 10^3$

43. $9.1302 \cdot 10^{-6}$

Write in scientific notation.

44. 452,968

45. 0.00012

46. 6,000,432

47. 0.0025

48. 0.0000007

49. 892

50. 62,598

51. 0.0000000000196

52. 540,203

53. 1,961,048

54. 0.00087

55. 0.75

56. DATA FILE Refer to the data on the sizes and depths of the oceans on page 526. Express the square miles of each ocean in scientific notation.

57. YOU MAKE THE CALL A student said that 3^2 equals 2^3. Is this student correct? Explain.

ASTRONOMY A light year is the distance that light can travel in one year. One light year equals 9,460,000,000,000 km.

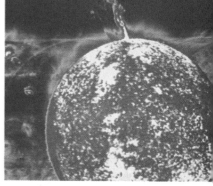

the Sun

58. Express one light year in scientific notation.

59. The mean distance between Earth and the sun is $1.496 \cdot 10^8$ km. Is the distance between Earth and the sun greater than, less than, or equal to one light year?

CALCULATOR Write in standard form. Use a calculator for the computation.

60. 29^5 **61.** 2^{12} **62.** 7^{-5}

Write *equal* if all the expressions in each row are equal. If not, indicate which expression does not belong and explain why.

63. $3 \cdot 3 \cdot 3$ 27^1 3^3 $2.7 \cdot 10^1$

64. $\dfrac{1}{2 \cdot 2 \cdot 2 \cdot 2}$ 0.0625 $\dfrac{1}{16}$ 2^{-4}

◼ EXTENDED PRACTICE EXERCISES

Evaluate. Let $x = 2$, $y = 3$, and $z = 4$.

65. $x^5 - 15$ **66.** $z^{-2} + 2$ **67.** $x^{-2} + y^{-2}$ **68.** $z^4 - x^3$

69. SPORTS A soccer team held a car wash to raise money for new uniforms. They began with $3. Every hour the amount in the fund doubled. The goal was exactly met in 8 h. How long did it take them to meet one half of their goal?

Use mental math to tell which number is greater.

70. 0.0025 or $\dfrac{1}{2^2}$ **71.** 9 or $\dfrac{1}{3^3}$ **72.** $7.987 \cdot 10^3$ or 7987

73. CHAPTER INVESTIGATION The height of the bounce of some balls is a pattern in exponents. Use your table to look for a pattern in your balls' bounces. The bounce can increase or decrease exponentially. Does your table show a pattern in bounces?

◼ MIXED REVIEW EXERCISES

Use the following information for Exercises 74–76. (Lesson 1-1)

ENTERTAINMENT CTA cable company is adding another channel to their offerings. They want to find out what types of programming their subscribers like best. They have asked their customer service employees to poll every subscriber who calls. Subscribers usually call only to report problems or to ask about new features.

74. What kind of sampling is represented by this situation?

75. How could this method be changed to produce a systematic sample?

76. Which sampling method do you think would be of most use to the cable company?

3-8 Laws of Exponents

Goals ■ Apply the laws of exponents.

Applications Astronomy, Sports, Technology

---

1. Find the following products by writing each factor in expanded form. Rewrite the product as a power of ten.

$$10^2 \cdot 10^3 = 10 \cdot 10 \cdot 10 \cdot 10 \cdot 10 = 10^?$$

$$10^4 \cdot 10^5 = \underline{\hspace{2cm}} = 10^? \qquad\qquad 10^1 \cdot 10^2 = \underline{\hspace{2cm}} = 10^?$$

2. How are the exponents in the factors related to the exponent in the product?

3. Find the following quotients by first writing each number in standard form, then dividing. Rewrite the quotient as a power of ten.

$$10^6 \div 10^2 = 1{,}000{,}000 \div 100 = 10{,}000 = 10^?$$

$$10^4 \div 10^3 = \underline{\hspace{2cm}} = 10^? \qquad\qquad 10^5 \div 10^1 = \underline{\hspace{2cm}} = 10^?$$

4. How are the exponents in the dividend and divisor related to the exponent in the quotient?

◤ BUILD UNDERSTANDING

A number can be easier to work with in exponential form. For example, 10^{10} is easier to work with than 10,000,000,000. The **laws of exponents** help when performing operations with numbers written in exponential form.

To find the product of $2^2 \cdot 2^3$, you can write out the factors for each term and then write the product using exponents.

$$2^2 \cdot 2^3 = 2 \cdot 2 \cdot 2 \cdot 2 \cdot 2 = 2^5$$

The product rule is a faster method.

Product Rule	$a^m \cdot a^n = a^{m+n}$ To multiply numbers with the same base, write the base with the sum of the exponents.

Example 1

Use the product rule to multiply.

a. $10^5 \cdot 10^2$ **b.** $5^1 \cdot 5^4$ **c.** $x^2 \cdot x^3$

Solution

a. $10^5 \cdot 10^2 = 10^{5+2}$ **b.** $5^1 \cdot 5^4 = 5^{1+4}$ **c.** $x^2 \cdot x^3 = x^{2+3}$

        $= 10^7$ $= 5^5$ $= x^5$

> **Math: Who, Where, When**
>
> A googol is the number 1 followed by 100 zeros, or 10^{100}. A googolplex is equal to 1 followed by a googol zeros, or 10^{googol}. The 9 year old nephew of an American mathematician, Edward Kasner, invented the name.

To find the quotient $3^4 \div 3^3$, you could write out the factors for each term and then write the quotient using exponents.

$$3^4 \div 3^3 \text{ can be written as } \frac{3^4}{3^3} = \frac{3 \cdot 3 \cdot 3 \cdot 3}{3 \cdot 3 \cdot 3} = 3^1$$

Another method is to use the quotient rule.

Quotient Rule	$a^m \div a^n = a^{m-n}$ To divide numbers with the same base, write the base with the difference of the exponents.

Example 2

Use the quotient rule to divide.

a. $10^3 \div 10^1$ **b.** $2^6 \div 2^3$ **c.** $b^7 \div b^6$

Solution

a. $10^3 \div 10^1 = 10^{3-1}$ **b.** $2^6 \div 2^3 = 2^{6-3}$ **c.** $b^7 \div b^6 = b^{7-6}$

$\quad\quad = 10^2$ $\quad\quad = 2^3$ $\quad\quad = b^1$

Technology Note

Some calculators require different keystrokes when using exponents. For Example 2, part b, one calculator may require these keystrokes:

2 △ 6 ÷ 2 △ 3

Another calculator may require:

2 y^x 6 ÷ 2 y^x 3

Experiment with your calculator.

A number written in exponential form can be raised to a power. You could first write out the factors and then write the product in exponential form.

$$(4^3)^2 = 4^3 \cdot 4^3$$

$$= 4 \cdot 4 \cdot 4 \cdot 4 \cdot 4 \cdot 4$$

$$= 4^6$$

Compare the exponents in the original number to the exponents in the final number to determine their relationship. This relationship is described by the power rule.

Power Rule	$(a^m)^n = a^{mn}$ To raise a power to a power, write the base with the product of the exponents.

Example 3

Use the power rule.

a. $(4^3)^6$ **b.** $(10^2)^{12}$ **c.** $(y^9)^8$ **d.** $(s)^6$

Solution

a. $(4^3)^6 = 4^{3 \cdot 6}$ **b.** $(10^2)^{12} = 10^{2 \cdot 12}$

$\quad\quad = 4^{18}$ $\quad\quad\quad = 10^{24}$

c. $(y^9)^8 = y^{9 \cdot 8}$ **d.** $(s)^6 = s^{1 \cdot 6}$

$\quad\quad = y^{72}$ $\quad\quad\quad = s^6$

Example 4

ASTRONOMY It is believed that 10^4 craters form on the moon every 10^9 years. On average, how many years are there between the formation of one crater and the next?

Solution

Divide the entire span of 10^9 years by the number of craters, 10^4.

$$10^9 \div 10^4 = 10^{9-4} = 10^5 = 100,000$$

A crater forms about every 100,000 years.

◤ TRY THESE EXERCISES

Use the product rule to multiply.

1. $10^4 \cdot 10^2$ **2.** $5^8 \cdot 5^5$ **3.** $m^{21} \cdot m^7$ **4.** $d^6 \cdot d^3$

Use the quotient rule to divide.

5. $9^8 \div 9^2$ **6.** $4^{11} \div 4^0$ **7.** $p^{20} \div p^{10}$ **8.** $a^9 \div a^3$

Use the power rule.

9. $(3^6)^8$ **10.** $(7^2)^{10}$ **11.** $(x^4)^4$ **12.** $(a^8)^1$

13. BIOLOGY The total number of bacteria on a surface was estimated to be 10^{10}. There were 1000 times as many two hours later. How many bacteria are there now?

14. WRITING MATH Is it true that $3^4 \div 3^2 = 3^2 \div 3^4$? Explain.

15. SPORTS The budget for the 1932 winter Olympic games was $1 million. In 1994 the budget was $1 billion. How many times more money was budgeted for the 1994 winter Olympic games than in 1932? Express your answer in exponential form.

◤ PRACTICE EXERCISES • For Extra Practice, see page 556.

Use the product rule to multiply.

16. $10^{21} \cdot 10^8$ **17.** $10^1 \cdot 10^9$ **18.** $5^6 \cdot 5^2$ **19.** $6^2 \cdot 6^3$
20. $n^3 \cdot n^3$ **21.** $d^{11} \cdot d^0$ **22.** $m^9 \cdot m^3$ **23.** $a^6 \cdot a^9$

Use the quotient rule to divide.

24. $10^4 \div 10^2$ **25.** $5^8 \div 5^5$ **26.** $2^8 \div 2^5$ **27.** $6^{16} \div 6^{15}$
28. $b^9 \div b^1$ **29.** $x^{21} \div x^7$ **30.** $y^7 \div y^5$ **31.** $n^6 \div n^3$

Use the power rule.

32. $(45^6)^{10}$ **33.** $(9^8)^8$ **34.** $(5^7)^9$ **35.** $(11^4)^5$
36. $(a^2)^{15}$ **37.** $(x^{20})^5$ **38.** $(n^7)^7$ **39.** $(p^3)^3$

Use the laws of exponents.

40. $(4^{10})^{15}$ **41.** $3^{16} \div 3^8$ **42.** $8^{24} \div 8^3$ **43.** $15^{18} \cdot 15^{14}$

44. $7^{15} \div 7^3$ **45.** $6^{20} \cdot 6^5$ **46.** $(12^5)^5$ **47.** $2^{90} \cdot 2^9$

Find the value of each variable.

48. $6^3 \cdot 6^a = 6^{18}$ **49.** $7^4 \div 7^b = 7^2$ **50.** $(6^c)^3 = 6^{12}$ **51.** $(9^7)^f = 9^{49}$

52. CALCULATOR Use a calculator to write the expression $(5^7)^{10}$ in scientific notation.

53. There are 10^2 cm in a meter and 10^3 m in a kilometer. How many centimeters are in a kilometer?

54. Computers store information in bits. One byte is equal to 2^3 bits. A kilobyte is equal to 2^{10} bytes. How many bits are in a kilobyte?

55. SPORTS The 100-m dash and the 10-km race are track-and-field events. How many times longer is a 10-km race than the 100-m dash? (Remember to convert 10 km to meters.)

56. ASTRONOMY With telescopes we can see about 100 quintillion (10^{20}) stars. It is estimated that about 1 trillionth (10^{-12}) of these stars have planets similar to Earth. Find the estimated number of stars with Earth-like planets.

▰ EXTENDED PRACTICE EXERCISES

Replace ▰ with <, >, or = .

57. $5^8 \div 5^3$ ▰ 5^4 **58.** $3^6 \cdot 3^5$ ▰ 3^{30} **59.** $(7^6)^2$ ▰ $(7^3)^4$

Use the figures shown.

60. Find the number of unit cubes that make up each cube. Write the number using exponents.

61. Why do you think the expression 2^3 is referred to as "two cubed"?

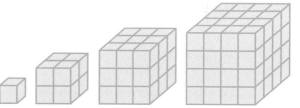

62. WRITING MATH Estimate the number of decimal places in $(1.2)^5$ and $(9.8)^5$. Use a calculator to verify the number of decimal places. Explain your discovery.

▰ MIXED REVIEW EXERCISES

Use the data shown.

Keith's Average Typing Speed

Week	1	2	3	4	5	6
Speed (words per minute)	10	16	19	24	28	36

63. Draw a line graph to show the change in Keith's typing speed during his 6-week class. (Lesson 1-6)

64. What is the ratio of Keith's greatest speed to his speed the first week of class? (Lesson 2-6)

Review and Practice Your Skills

Write in exponential form.

1. $4 \cdot 4 \cdot 4 \cdot 4 \cdot 4 \cdot 4$

2. $\dfrac{1}{3 \cdot 3 \cdot 3}$

3. $5 \cdot 5 \cdot 5 \cdot 5 \cdot 5 \cdot 5 \cdot 5 \cdot 5$

4. $8 \cdot 8 \cdot 8 \cdot 8$

5. $10 \cdot 10 \cdot 10 \cdot 10 \cdot 10$

6. $\dfrac{1}{2 \cdot 2 \cdot 2 \cdot 2 \cdot 2}$

Write in standard form.

7. 5^3

8. 3^{-2}

9. 1^9

10. 2^5

11. 18^1

12. 0.3^2

13. 6^0

14. 10^{-3}

15. $2.4 \cdot 10^4$

16. $1.06 \cdot 10^{-3}$

17. $8.9 \cdot 10^6$

18. $1.245 \cdot 10^{-2}$

Write in scientific notation.

19. 467,800

20. 2,650,000,000

21. 0.091

22. 5,030,000

23. 0.0063

24. 0.0307

25. 34,600,000

26. 0.00034

27. 3,846

28. 0.000082

29. 0.00105

30. 402,600,000,000

31. The thickness of a page in a book is 0.00008 m.

32. The greatest distance from the planet Pluto to the sun is about 7,320,000,000 km.

Use the product rule to multiply.

33. $6^3 \cdot 6^2$

34. $g^1 \cdot g^5$

35. $2^2 \cdot 2^5$

36. $3^3 \cdot 3^4$

37. $c^4 \cdot c^2$

38. $m^3 \cdot m^3$

39. $10^2 \cdot 10^6$

40. $w^5 \cdot w^5$

Use the quotient rule to divide.

41. $3^7 \div 3^2$

42. $d^8 \div d^3$

43. $9^7 \div 9^5$

44. $a^2 \div a^2$

45. $12^5 \div 12^4$

46. $1^9 \div 1^2$

47. $y^{10} \div y^6$

48. $n^6 \div n^4$

Use the power rule.

49. $(13^3)^2$

50. $(4^2)^4$

51. $(2^3)^8$

52. $(8^4)^3$

53. $(x^2)^3$

54. $(h^6)^4$

55. $(q^8)^3$

56. $(m^5)^2$

Use the laws of exponents.

57. $7^6 \div 7^3$

58. $10^3 \cdot 10^6$

59. $2^9 \div 2^2$

60. $(6^2)^8$

61. $4^3 \cdot 4^2$

62. $(3^5)^2$

63. $8^4 \div 8^3$

64. $(5^3)^3$

Find the value of each variable.

65. $2^2 \cdot 2^a = 2^5$

66. $9^8 \div 9^e = 9^3$

67. $6^3 \cdot 6^h = 6^4$

68. $(8^m)^2 = 8^{10}$

69. $(5^3)^d = 5^6$

70. $4^2 \cdot 4^n = 4^8$

71. $(7^x)^4 = 7^{12}$

72. $3^6 \div 3^w = 3^3$

Add, subtract, multiply or divide. (Lessons 3-1 and 3-2)

73. $7 + (-11)$ **74.** $-3 + 12$ **75.** $9 - (-1)$ **76.** $-6 - (-14)$

77. $-18 \div (-2)$ **78.** $4 \cdot (-9)$ **79.** $-2 \cdot (-11)$ **80.** $-54 \div (-6)$

Simplify. Use the order of operations. (Lesson 3-3)

81. $45 - 100 \div 5$ **82.** $56 + 5^2 \cdot 4$ **83.** $(5 + 2) \cdot 10 - 3^2$

Complete. Name the property you used. (Lesson 3-4)

84. $(3 \cdot 7) \cdot 6 = 3 \cdot (7 \cdot \blacksquare)$ **85.** $5(7 - \blacksquare) = (5 \cdot 7) - (5 \cdot 2)$ **86.** $1.8 + 3.95 = 3.95 + \blacksquare$

Evaluate each expression. Let $a = 4$, $b = -2$, and $c = -1$. (Lesson 3-5)

87. $-2a \div (-8)$ **88.** $3b + (-4c)$ **89.** $c \cdot 5b$

90. $-20 - c$ **91.** $2c - 3a$ **92.** $-4b + 5c$

Choose the rule that describes each pattern. Then write the next two numbers. (Lesson 3-6)

93. 6.2, 6.16, 6.12, 6.08   **a.** Add 0.04.   **b.** Subtract 0.4.   **c.** Subtract 0.04.

94. 3, 15, 75, 375   **a.** Add 12.   **b.** Multiply by 5.   **c.** Multiply by 3.

Write in standard form. (Lesson 3-7)

95. 6^{-2} **96.** 7^0 **97.** 10^{-6} **98.** 4^3

99. $3.02 \cdot 10^4$ **100.** $3.8 \cdot 10^{-2}$ **101.** $6.0213 \cdot 10^8$ **102.** $1.005 \cdot 10^{-4}$

Write in scientific notation. (Lesson 3-7)

103. 0.0058   **104.** 38,000   **105.** 0.07   **106.** 1,600,000

107. 67,500,000   **108.** 0.00000912   **109.** 443,000   **110.** 0.000406

111. The average diameter of the virus that causes polio is 0.000025 mm.

112. A microsecond is 0.000001 sec.

Use the laws of exponents. (Lesson 3-8)

113. $8^3 \cdot 8^5$ **114.** $2^6 \cdot 2^2$ **115.** $10^5 \cdot 10^1$ **116.** $3^8 \cdot 3^8$

117. $m^2 \cdot m^0$ **118.** $g^4 \cdot g^2$ **119.** $w^2 \cdot w^6$ **120.** $b^5 \cdot b^1$

121. $2^5 \div 2^4$ **122.** $9^8 \div 9^8$ **123.** $4^6 \div 4^2$ **124.** $15^4 \div 15^1$

125. $n^7 \div n^4$ **126.** $a^2 \div a^1$ **127.** $h^{10} \div h^5$ **128.** $y^6 \div y^5$

129. $(3^4)^2$ **130.** $(8^2)^4$ **131.** $(1^8)^{12}$ **132.** $(5^3)^3$

133. $(q^4)^0$ **134.** $(m^2)^5$ **135.** $(x^8)^3$ **136.** $(c^6)^2$

Find the value of each variable. (Lesson 3-8)

137. $5^3 \cdot 5^h = 5^6$ **138.** $8^5 \div 8^w = 8^4$ **139.** $(12^x)^3 = 12^{12}$ **140.** $(7^2)^d = 7^8$

3-9 Squares and Square Roots

Goals ■ Calculate squares and square roots.

Applications Sports, Safety, Physics

The piece of grid paper illustrates five squares. Square A has an area of one square unit, and the length of each side is one.

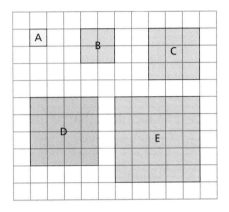

1. Use the diagram to complete the table below.

Square	A	B	C	D	E
Length of a side	1				
Area of square	1				

2. What relationships do you notice between the length of each square and its area?

◤ BUILD UNDERSTANDING

There are two relationships between the area of a square and the length of its side.

The area of a square is equal to the **square** of the length of its side. The square of a number is the product of a number and itself. For example, the square of 8 is 64.

The length of a side of a square is equal to the **square root** of the area. The square root of a number is one of two equal factors of a number. For example, the square root of 64 is 8.

The square root of a number can be the positive factor or the negative factor. The symbol $\sqrt{}$ is called a radical symbol. It is used to indicate the *principal square root*, which is the positive square root of a number. The symbol $-\sqrt{}$ indicates the negative square root, and $\pm\sqrt{}$ indicates both square roots.

Square Root	If $a^2 = b$, then a is a square root of b.

A number a is a square root of another number b if $a^2 = b$.

$$\sqrt{64} = \sqrt{8 \cdot 8} = 8 \qquad \text{8 is the positive square root of 64.}$$

$$-\sqrt{64} = -\sqrt{8 \cdot 8} = -8 \qquad \text{^{-}8 is the negative square root of 64.}$$

$$\pm\sqrt{64} = \pm 8 \qquad \text{Both 8 and ^{-}8 are square roots of 64.}$$

The number 0 has only one square root: $\sqrt{0} = 0$.

A rational number that is a product of two equal factors is a **perfect square**. Perfect squares have square roots that are rational numbers.

Some examples of perfect squares are shown.

$$\sqrt{1} = \sqrt{1 \cdot 1} = 1 \qquad\qquad \sqrt{9} = \sqrt{3 \cdot 3} = 3$$

$$\sqrt{\frac{4}{9}} = \frac{\sqrt{4}}{\sqrt{9}} = \frac{\sqrt{2 \cdot 2}}{\sqrt{3 \cdot 3}} = \frac{2}{3} \qquad \sqrt{0.49} = \sqrt{0.7 \cdot 0.7} = 0.7$$

The number 0 has only one square root: $\sqrt{0} = 0$.

Example 1

Find each square.

a. 5^2 **b.** $(-3)^2$

c. $\left(\frac{1}{4}\right)^2$ **d.** $(0.09)^2$

Solution

a. $5^2 = 5 \cdot 5$ **b.** $(-3)^2 = (-3) \cdot (-3)$

       $= 25$ $= 9$

c. $\left(\frac{1}{4}\right)^2 = \frac{1}{4} \cdot \frac{1}{4}$ **d.** $(0.09)^2 = 0.09 \cdot 0.09$

         $= \frac{1}{16}$ $= 0.0081$

The square root of any positive number that is not a perfect square is an irrational number. This means its value is a non-terminating, non-repeating decimal. You can only approximate its square root.

You can use a calculator to find an approximate square root. Some examples are shown below.

$$\sqrt{5} \approx 2.2361 \qquad\qquad \sqrt{23} \approx 4.7958$$

$$\sqrt{\frac{1}{2}} = 0.7071 \qquad\qquad \sqrt{8.1} \approx 2.8460$$

The values named above are rounded to the nearest thousandth. Your calculator may display more decimal places than four. The more decimal places displayed, the more precise the approximated value is for the square root.

Example 2

Find each square root. Round to the nearest thousandth, if necessary.

a. $\sqrt{23}$ **b.** $-\sqrt{99}$ **c.** $\sqrt{\frac{9}{16}}$

Solution

a. $\sqrt{23} = 4.796$ **b.** $-\sqrt{99} = -9.950$ **c.** $\sqrt{\frac{9}{16}} = \frac{\sqrt{9}}{\sqrt{16}} = \frac{3}{4}$

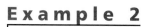

Example 3

SPORTS The area of a square baseball infield is 8100 ft². Find the length of each side of the infield.

Solution

Since the length of a side of a square is equal to the square root of the area, find the square root of 8100.

$$\sqrt{8100} = 90$$

The length of each side of the infield is 90 ft.

◢ TRY THESE EXERCISES

Find each square.

1. 14^2 **2.** $(-4)^2$ **3.** $\left(\dfrac{5}{8}\right)^2$ **4.** $(0.6)^2$

Find each square root.

5. $\sqrt{\dfrac{81}{169}}$ **6.** $\sqrt{0.09}$ **7.** $\sqrt{\dfrac{144}{289}}$ **8.** $-\sqrt{0.25}$

9. $\sqrt{121}$ **10.** $-\sqrt{0.01}$ **11.** $\sqrt{\dfrac{1}{25}}$ **12.** $-\sqrt{64}$

 CALCULATOR Use a calculator to find each square root. Round to the nearest thousandth.

13. $\sqrt{22}$ **14.** $\sqrt{33}$ **15.** $\sqrt{58}$ **16.** $\sqrt{7}$

17. The area of a square stamp is 23 cm². Find the measure of a side. Round your answer to the nearest tenth.

◢ PRACTICE EXERCISES • For Extra Practice, see page 556.

Find each square.

18. 18^2 **19.** 21^2 **20.** 16^2 **21.** 29^2

22. $(-37)^2$ **23.** $(1.7)^2$ **24.** $(-2.1)^2$ **25.** $(0.08)^2$

26. $(0.008)^2$ **27.** $\left(\dfrac{1}{5}\right)^2$ **28.** $\left(\dfrac{3}{7}\right)^2$ **29.** $\left(-\dfrac{8}{9}\right)^2$

30. $\left(\sqrt{8}\right)^2$ **31.** $\left(\sqrt{19}\right)^2$ **32.** $\left(\dfrac{4}{5}\right)^2$ **33.** $\left(-\sqrt{24}\right)^2$

Find each square root.

34. $\sqrt{256}$ **35.** $\sqrt{144}$ **36.** $\sqrt{169}$ **37.** $-\sqrt{100}$

38. $-\sqrt{0.36}$ **39.** $-\sqrt{0.49}$ **40.** $\sqrt{\dfrac{1}{49}}$ **41.** $-\sqrt{\dfrac{4}{81}}$

42. $-\sqrt{\dfrac{16}{25}}$ **43.** $-\sqrt{\dfrac{16}{121}}$ **44.** $-\sqrt{\dfrac{25}{169}}$ **45.** $\sqrt{\dfrac{196}{529}}$

46. $\sqrt{0.81}$ **47.** $-\sqrt{10,000}$ **48.** $\sqrt{\dfrac{1}{4}}$ **49.** $-\sqrt{0.25}$

CALCULATOR Find each square root. Round to the nearest thousandth.

50. $\sqrt{153}$ 51. $\sqrt{527}$ 52. $\sqrt{976}$ 53. $\sqrt{369}$

54. $\sqrt{101}$ 55. $\sqrt{482}$ 56. $\sqrt{815}$ 57. $\sqrt{239}$

58. $\sqrt{12}$ 59. $\sqrt{90}$ 60. $\sqrt{135}$ 61. $\sqrt{712}$

62. **SAFETY** Police officers can measure the skid marks of a car to estimate the speed. The formula $s = \sqrt{24d}$ is used to estimate the speed of a car on a dry concrete road. The speed in miles/hour is represented by s, and d is the distance in feet that the car skidded after the brakes were applied. What was the approximate speed of a car that left skid marks 35 ft long?

63. A ball falls from a 250-ft building. The formula $d = 16t^2$ represents the distance a ball falls in time (t). How long does it take the ball to hit the ground.

64. List three numbers that have square roots between 2 and 3.

 65. **WRITING MATH** Make a chart summarizing the three laws of exponents. Give an example for each law.

EXTENDED PRACTICE EXERCISES

66. **CRITICAL THINKING** Explain why 7 is the best whole number approximation for $\sqrt{47}$.

67. **SPORTS** The area of a square softball infield is 3600 ft^2. Find the perimeter of the infield.

68. The distance you can see on a clear day from a location above Earth can be estimated by the formula $V = 1.22 \cdot \sqrt{A}$. V represents the distance in miles and A is the altitude in feet. The Eiffel Tower is about 1010 ft tall. On a clear day, approximately how many miles can you see from the top of the Eiffel Tower?

PHYSICS The amount of time (t) in seconds it takes an object to fall a distance (d) in meters is expressed in this formula: $t = \sqrt{\dfrac{d}{4.9}}$. Use this formula for Exercises 69 and 70. Round each answer to the nearest tenth.

69. An object fell 40 m. How long did it take the object to hit the ground?

70. A rock falls over a cliff 75 m high. How long will it take the rock to hit the water at the bottom of the cliff?

MIXED REVIEW EXERCISES

Use the data shown.

71. Draw a bar graph for the cable channel data. (Lesson 1-5)

72. Which channel has over twice as many viewers as Law TV? (Lesson 1-2)

73. Which channel represents the mode of this data? (Lesson 1-2)

74. Which channel represents the median? (Lesson 1-2)

Cable channel	Average viewers per hour
All things sports	2300
Golf	870
Movie 24	3620
Chef channel	720
Law TV	1750

Chapter 3 Review

VOCABULARY ◤

Choose the word from the list that best completes each statement.

1. When an operation $(+, -, \cdot, \div)$ on a set of numbers always results in a number in the set, it is ___?___.

2. A number written in ___?___ has a base and an exponent.

3. A ___?___ is an unknown number represented by a letter such as a, n, x, or y.

4. The distance a number is from zero on a number line is the ___?___ of the number.

5. Two numbers are ___?___ if they are the same distance from 0 but in opposite directions.

6. A number written in ___?___ has two factors. The first factor is a number greater than or equal to 1 and less than 10. The second factor is a power of 10.

7. The set of numbers that includes positive and negative whole numbers and zero are ___?___.

8. A rational number that is a product of two equal factors is a ___?___.

9. The ___?___ shows certain steps that must be followed in a specific order to simplify mathematics expressions.

10. Numbers that can be expressed as a fraction are ___?___.

a. absolute value
b. closed
c. exponent
d. exponential form
e. integers
f. opposites
g. order of operations
h. perfect square
i. rational numbers
j. scientific notation
k. variable
l. whole numbers

LESSON 3-1 ◤ Add and Subtract Signed Numbers, p. 104

▶ When opposites are added, the sum is always 0.

▶ The sum of zero and a number is always that number.

▶ To subtract a number, add its **opposite**.

Simplify.

11. $-4 + (-9)$

12. $-0.3 + 1.5$

13. $-\dfrac{1}{2} - \left(-\dfrac{3}{4}\right)$

14. $16 - (-12) + 33$

15. $6 + (-5) + (-6) + 22$

16. $-2 + (-7) + 11 + (-9)$

LESSON 3-2 ◤ Multiply and Divide Signed Numbers, p. 108

▶ The product or quotient of two numbers with the *same sign* is positive.

▶ The product or quotient of two numbers with *different signs* is negative.

Multiply or divide.

17. $-45 \div 9$

18. $\dfrac{-55}{-5}$

19. $-30 \div (-5)$

20. $-8 \cdot (-4)$

21. $\dfrac{21}{-7}$

22. $32 \div (-4)$

LESSON 3-3 ◣ Order of Operations, p. 114

▶ To simplify mathematical expressions, follow the **order of operations**.

Simplify.

23. $7 + 4 \cdot 3$

24. $36 - 15 \div 3$

25. $42 - 6 \cdot (12 \div 4)$

26. $13 + 3 \cdot (8 - 2)^2$

27. $5(4 + 6) - 7 \cdot 7$

28. $18 - 2 \cdot 3(2 + 4)^2$

29. $(21 - 4^2) + (3 + 2)^2$

30. $22 \div 11 \cdot 9 - 3^2$

LESSON 3-4 ◣ Real Number Properties, p. 118

▶ These groups are **sets** of numbers: **Natural, Whole, Integers, Irrational, Rational,** and **Real.**

▶ Real number properties include the commutative, associative, and distributive properties.

To which sets of numbers does each belong?

31. $\dfrac{1}{2}$

32. -5

33. $0.657321\ldots$

34. 0

Complete. Name the property you used.

35. $24 \cdot 16 = 16 \cdot \blacksquare$

36. $\blacksquare (8 + 3) = (4 \cdot 8) + (4 \cdot 3)$

37. $(\blacksquare + 6) + 9 = 5 + (6 + 9)$

LESSON 3-5 ◣ Variables and Expressions, p. 124

▶ Expressions with at least one variable, such as $x + 3$ and $y \div 2$, are **variable expressions**.

Write a variable expression. Let n represent "a number."

38. six more than a number

39. three less than a number divided by eight

40. ten increased by twice a number

Evaluate each expression. Let $a = -1$, $b = 2$ and $c = 22$.

41. $b - 8$

42. $3a + c$

43. $6a \div b$

44. $7a \cdot (-3)b$

45. $abc - 1$

46. $\dfrac{2c}{b}$

47. $a - 5b$

48. $ac - ab$

LESSON 3-6 ◣ Problem Solving Skills: Find a Pattern, p. 128

▶ Sometimes a problem is easier to solve if you can find a pattern.

49. Marty charges $3 for washing 1 window, $5 for 2, $8 for 3, and so on. How much does he charge to wash 4 windows? 10 windows?

50. A store display contains boxes of sneakers. Each row of sneakers has 2 fewer boxes than the row below. The first row has 23 boxes of sneakers. How many boxes will there be in the tenth row?

51. Travis is saving $15 per month to buy a $130 MP3 player. In which month will he have enough saved up if he starts saving in January with $53 in his savings account?

LESSON 3-7 ◣ Exponents and Scientific Notation, p. 132

▶ The **base** of a number in **exponential form** tells the factor being multiplied. The **exponent** tells how many equal factors there are.

▶ A number is in **scientific notation** when it is written as the product of a number greater than or equal to 1 and less than 10 and a power of 10.

52. Write $7 \cdot 7 \cdot 7 \cdot 7 \cdot 7 \cdot 7$ in exponential form.

53. Write 5^{-3} in standard form.

54. Write 630,000 in scientific notation.

55. Write $4.78 \cdot 10^{-5}$ in standard form.

LESSON 3-8 ◣ Laws of Exponents, p. 136

▶ The laws of exponents help when performing operations with numbers written in exponential form.

▶ Product Rule: $a^m \cdot a^n = a^{m+n}$

▶ Quotient Rule: $a^m \div a^n = a^{m-n}$

▶ Power Rule: $(a^m)^n = a^{mn}$

Use the law of exponents.

56. $8^3 \cdot 8^6$ **57.** $x^{15} \div x^9$ **58.** $(4^5)^7$ **59.** $d^9 \cdot d^{11}$

60. $\dfrac{4^6}{4^2}$ **61.** $(6^8)^9$ **62.** $(x^2)^6$ **63.** $\dfrac{f^{10}}{f^7}$

LESSON 3-9 ◣ Squares and Square Roots, p. 142

▶ A number a is a **square root** of another number b if $a^2 = b$.

Find each square.

64. 9^2 **65.** $(-6)^2$ **66.** $\left(\dfrac{3}{4}\right)^2$ **67.** $(0.01)^2$

Find each square root.

68. $\sqrt{121}$ **69.** $-\sqrt{\dfrac{16}{25}}$ **70.** $\sqrt{0.36}$ **71.** $-\sqrt{\dfrac{64}{144}}$

Find each square root to the nearest tenth.

72. $\sqrt{32}$ **73.** $\sqrt{105}$ **74.** $\sqrt{\dfrac{4}{10}}$ **75.** $\sqrt{3844}$

CHAPTER INVESTIGATION

EXTENSION Since two sports balls is not a large enough sample to draw conclusions, compare your findings with those of your class. Make one table that shows all the data collected. Are you able to determine if a sport ball's size and weight affects its bounce? Explain.

Chapter 3 Assessment

Add or subtract.

1. $-5 + (-3) + (-7)$ **2.** $6 + (-6)$ **3.** $-8 + 0 + 8$ **4.** $9 - (-4)$

5. $-23 - (-7)$ **6.** $6 - (-6)$ **7.** $-3 + 2 + 3$ **8.** $8 - (-14) - (-7)$

Multiply or divide.

9. $-4 \cdot (-3) \cdot (-1)$ **10.** $-36 \div 6$ **11.** $-5 \cdot 0 \cdot 5$ **12.** $39 \div (-3)$

13. $\dfrac{-21}{-7}$ **14.** $2 \cdot (-4) \cdot (-7)$ **15.** $-7 \cdot (-2) \cdot (-2)$ **16.** $\dfrac{6}{(-6)}$

Simplify. Use the order of operations.

17. $-\dfrac{1}{2}(3 - 1) + (-2)$ **18.** $-4 \cdot (-5) + (-12) \div 4$ **19.** $3^2 \div (-9) + 8$

20. $(-12 - 2) \cdot 4 - 8$ **21.** $4^2 \div (-4) \div (-2)$ **22.** $-21 \div (-7) - 5 \cdot 3 + (-1)$

To which sets of numbers do the following belong?

23. -9 **24.** $\dfrac{1}{6}$ **25.** 8.24 **26.** $-0.45123\ldots$

Complete. Name the property used.

27. $\blacksquare \cdot (5 + 8) = (7 \cdot 5) + (7 \cdot 8)$ **28.** $18 \cdot 3 = 3 \cdot \blacksquare$

29. $(4 \cdot 8) \cdot \blacksquare = 4 \cdot (8 \cdot 9)$ **30.** $3 \cdot (8 - 6) = (\blacksquare \cdot 8) - (3 \cdot 6)$

State whether the following sets are closed under the given operation.

31. whole numbers, division   **32.** integers, subtraction

Write a variable expression. Let *n* represent "a number."

33. eighteen less than a number

Evaluate each expression. Let *a* = 3, *b* = −2, and *c* = −3.

34. $a - (-5)$ **35.** $2a - (-c)$ **36.** $4b \cdot b + a$ **37.** $8c \div 6b$

Write the number in the indicated form.

38. Write 40,000 and 0.00004 in scientific notation.

39. Write $8 \cdot 10^3$ and $8 \cdot 10^{-3}$ in standard form.

Use the laws of exponents.

40. $4^3 \cdot 4^5$ **41.** $x^{10} \div x^5$ **42.** $(3^4)^2$ **43.** $h^7 \div h^3$

Find each square.

44. 11^2 **45.** $(-3)^2$ **46.** $\left(\dfrac{5}{9}\right)^2$ **47.** $(0.04)^2$

Find each square root. Round to the nearest tenth, if necessary.

48. $\sqrt{9}$ **49.** $-\sqrt{\dfrac{100}{400}}$ **50.** $\sqrt{0.49}$ **51.** $-\sqrt{\dfrac{16}{225}}$

52. $\sqrt{29}$ **53.** $\sqrt{135}$ **54.** $\sqrt{\dfrac{6}{13}}$ **55.** $\sqrt{529}$

Standardized Test Practice

Part 1 Multiple Choice

Record your answers on the answer sheet provided by your teacher or on a sheet of paper.

1. The mean of three scores is 85. Two scores are 78 and 86. What is the third score? (Lesson 1-2)
   - (A) 89
   - (B) 91
   - (C) 92
   - (D) 94

2. How many more Labrador retrievers were registered than German Shepherd Dogs? (Lesson 1-6)

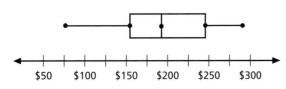

American Kennel Club Registration

   - (A) 25,000
   - (B) 50,000
   - (C) 75,000
   - (D) 100,000

3. What is the range in the prices of the DVD players? (Lesson 1-8)

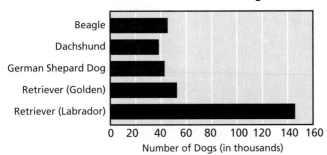

DVD Players

   - (A) $100
   - (B) $150
   - (C) $225
   - (D) $300

4. Find the area of a rectangle whose length is 4 in. and width is 5.4 in. (Lesson 2-4)
   - (A) 5.8 in.^{2}
   - (B) 9.4 in.^{2}
   - (C) 16.4 in.^{2}
   - (D) 21.6 in.^{2}

5. Which ratio is *not* equivalent to 4:3? (Lesson 2-6)
   - (A) 12:16
   - (B) 8:6
   - (C) 20:15
   - (D) 12:9

6. Find the area of the figure. (Lesson 2-9)

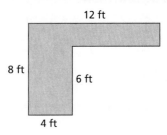

   - (A) 40 ft^2
   - (B) 48 ft^2
   - (C) 56 ft^2
   - (D) 120 ft^2

7. Monique shot the following scores in a four-round golf tournament: $-2, 3, -4, -1$. What was her final score? (Lesson 3-1)
   - (A) -10
   - (B) -4
   - (C) -1
   - (D) 1

8. Your checking account is overdrawn by $50. You write a check for $20. What is the balance now? (Lesson 3-1)
   - (A) $-$70$
   - (B) $-$30$
   - (C) $-$20$
   - (D) $30

9. A technology stock lost 2 points each hour for 6 hours. Find the total points the stock fell. (Lesson 3-2)
   - (A) -12
   - (B) -8
   - (C) 8
   - (D) 12

10. Which expression describes the situation: six less than 5 times d dollars? (Lesson 3-5)
   - (A) $6 - 5d$
   - (B) $6 - (5 - d)$
   - (C) $5(d - 6)$
   - (D) $5d - 6$

11. Find 0.00000506 written in scientific notation. (Lesson 3-7)
   - (A) $5.0 \cdot 10^{-6}$
   - (B) $5.06 \cdot 10^{-6}$
   - (C) $5.0 \cdot 10^{6}$
   - (D) $5.06 \cdot 10^{6}$

Test-Taking Tip

Exercises 2 and 5
Read the question carefully. Look for words like *more, less,* or *not.*

Part 2 — Short Response/Grid In

Record your answers on the answer sheet provided by your teacher or on a sheet of paper.

12. A sample of 1500 students were surveyed on their favorite electronic communication method. Which device did more than half choose as their favorite? (Lesson 1-1)

Favorite Method	Students
e-mail	801
cell phone	324
telephone	25
text message	350

13. During which hour did the greatest amount of snow fall? (Lesson 1-6)

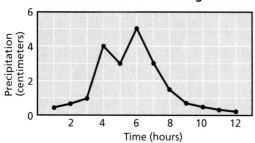

Amount of Snowfall During a Storm

14. Cynthia bought 2100 cm of yarn for a cross-stitch pattern. If the yarn costs $0.55/m, how much did she pay? (Lesson 2-2)

15. The running time for a movie is 134 min. What time will the movie finish if it starts at 1:10 P.M. and has 10 min. of previews? (Lesson 2-2)

16. Find the missing term. (Lesson 2-6)
$$\frac{2}{3} = \frac{n}{13}$$

17. Simpify $2.2(5.6 + 9.1) - 1.7^2$. (Lesson 3-3)

18. Which property is illustrated in the sentence below? (Lesson 3-4)
$$\left(\frac{5}{6} + \frac{1}{3}\right) + \frac{7}{9} = \frac{5}{6} + \left(\frac{1}{3} + \frac{7}{9}\right)$$

19. A ball bounces back 0.8 of its height on every bounce. If a ball is dropped from 150 ft, how high does it bounce on the fourth bounce? (Lesson 3-6)

20. Write $6.1 \cdot 10^4$ in standard form. (Lesson 3-7)

Part 3 — Extended Response

Record your answers on a sheet of paper. Show your work.

21. Javier attached his dog to the corner of his house with a 15-ft leash. The area that the dog can travel is indicated. (Lesson 2-7)

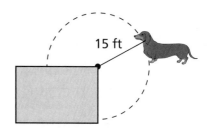

15 ft

a. What is the length of the greatest path the dog can travel?

b. How much area can the dog cover when she is on her leash?

c. Suppose Javier bought a 24 ft-by-24 ft pen for his dog. Does the pen offer more space than being tied to the corner of the house? Explain.

22. The table shows exchange rates for three countries compared to the U.S. dollar on a given day.

Country	Rate
Australia	1.311
Brazil	3.888
South Africa	6.278

a. Write a proportion that could be used to find the cost of an item in U.S. dollars if it costs 24.50 in South African rands.

b. Find the cost of an item in U.S. dollars if it costs 13.67 in Brazilian reais.

4

Two- and Three-Dimensional Geometry

THEME: Marketing

One aspect of geometry is the ability to visualize spatial relationships. One field that uses spatial relationships is marketing. Marketing is the business of promoting products through techniques like advertising and packaging.

A marketing campaign often uses advertisements. Print advertisements require that a three-dimensional object be displayed on a two-dimensional surface. Another aspect of marketing is a product's packaging. A three-dimensional package must be illustrated using two-dimensional geometry, so that it can be designed and built.

- When **packaging designers** (page 183) determine the shape and design of a product's package, they must consider how to display the product's name and marketing slogan on faces of a three-dimensional shape.

- **Billboard assemblers** (page 165) are responsible for transferring an oversized advertisement onto outdoor canvases. Assemblers must measure and accurately place preprinted strips that show the product, logo, and message to passersby.

Math Online

mathmatters1.com/chapter_theme

Advertising Spending (millions of dollars)

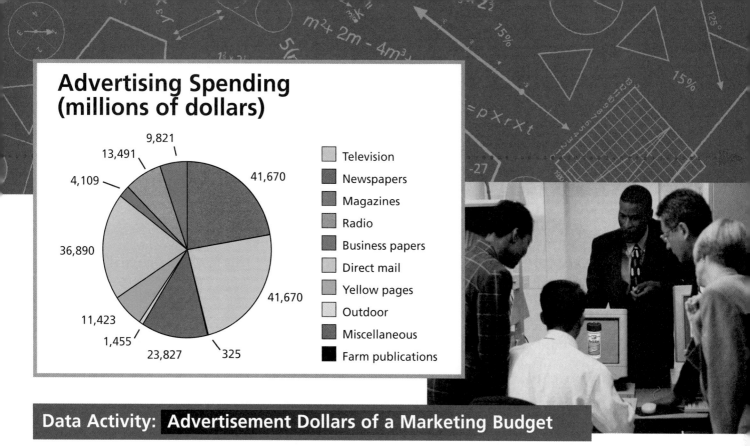

9,821
13,491
4,109
41,670
36,890
41,670
11,423
1,455
23,827
325

- Television
- Newspapers
- Magazines
- Radio
- Business papers
- Direct mail
- Yellow pages
- Outdoor
- Miscellaneous
- Farm publications

Data Activity: Advertisement Dollars of a Marketing Budget

Use the circle graph for Questions 1–5.

1. Which advertising media type is the greatest amount of the total?

2. Which advertising media type is the least amount of the total?

3. What percent of the total advertising dollars is spent on direct mail campaigns?

4. What percent of the advertising dollars is spent on newspaper and magazine advertisements?

5. What media types do you think are included in miscellaneous? Why aren't those media types given their own category?

CHAPTER INVESTIGATION

Packaging can be any size, shape or color. Retail packaging decisions must take into consideration the target audience, how the product will be displayed, the perceived value to the customer and cost of materials. A team of designers and marketers often collaborate to determine a product's package design.

Working Together

Find an unpackaged item at home. The more unusually shaped the item is, the more challenging the investigation will be. Bring it into class and exchange it with a classmate. The Chapter Investigation logo throughout this chapter will help you select a shape and design a package for the item.

The skills on these two pages are ones you have already learned. Use the examples to refresh your memory and complete the exercises. For additional practice on these and more prerequisite skills, see pages 536–544.

BASIC QUADRILATERALS

A quadrilateral is a two-dimensional shape with four sides and four angles. The variations on the length and size of the sides and angles produce several different types of quadrilaterals.

Examples   Here are examples of five different types of quadrilaterals.

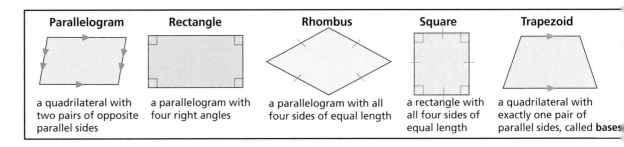

Parallelogram	Rectangle	Rhombus	Square	Trapezoid
a quadrilateral with two pairs of opposite parallel sides	a parallelogram with four right angles	a parallelogram with all four sides of equal length	a rectangle with all four sides of equal length	a quadrilateral with exactly one pair of parallel sides, called **bases**

Classify each quadrilateral with as many names as possible.

1.

2.

3.

4.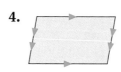

5.

6.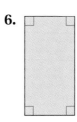

ANGLES

Angles are classified by their size. Triangles are often classified by the size of their angles. Knowing the classifications for angles will help you classify triangles.

Examples

Name of angle	Measure of angle
Acute	between 0° and 90°
Right	90°
Obtuse	between 90° and 180°
Straight	180°

Classify each angle.

7.

8.

9.

10.

AREA

To find the volume of solids, you often need to find the area of the base. Knowing the area formulas for several polygons will make finding volume easier.

Examples   Area of a triangle $= \frac{1}{2}$ base $\cdot$ height $= \frac{1}{2}bh$

Area of a square $= (\text{side})^2 = s^2$

Area of a rectangle $=$ length $\cdot$ width $= lw$

Area of a trapezoid $= \frac{1}{2}(\text{base}_1 + \text{base}_2) \cdot$ height $= \frac{1}{2}(b_1 + b_2)h$

Area of a circle $=$ pi $\cdot (\text{radius})^2 = \pi r^2$

Use the appropriate formula to find the area of each figure. Round to the nearest hundredth as needed.

11.
10.6 in.

12.
0.2 km
0.3 km

13.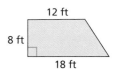
12 ft
8 ft
18 ft

14. A circular garden of radius 25 ft has a stone path around its border. The path is 3 ft wide. Find the area of the garden, and the area of the garden plus the stone path.

15. The area of a second circular garden including its stone path is 1000 ft². The width of the path is 1 ft. What is the area of the garden alone?

4-1 Language of Geometry

Goals
- Identify and classify geometric figures.
- Use a protractor to measure and draw angles.

Applications Navigation, Advertising, Sports

Geometry is all around you. What you know about geometry influences how you understand the world you see.

Work in small groups to answer Questions 1–3.

1. Look at an analog clock. Make a list of what you see on the face of the clock that defines or models geometry.

2. Look at the walls, floor and ceiling of a room. Make a list of what you see that defines or models geometry.

3. Look at a chalkboard (or whiteboard) and tools used to write on it. Name geometric figures modeled by these objects.

◢ BUILD UNDERSTANDING

People understand each other best if they speak the same language. The language of mathematics includes images, words and symbols.

Geometry is built on three terms: point, line and plane. These terms exist without a concrete definition, and are represented by simple figures. The table lists geometric figures and their names, using symbols and descriptions.

Figure	Name	Description
• A	point A	A **point** is a location in space. Although a point has no dimension, it is usually represented by a dot.
B C	$\overleftrightarrow{BC}$ (line BC) or $\overleftrightarrow{CB}$ (line CB)	A **line** is a set of points that extends without end in opposite directions. Two points determine a line. Points on the same line are **collinear points**.
D E	$\overline{DE}$ (line segment DE) or $\overline{ED}$ (line segment ED)	A **line segment** is a part of a line that consists of two endpoints and all points between them.
F G	$\overrightarrow{FG}$ (ray FG)	A **ray** is a part of a line that has one endpoint and extends without end in one direction.
A / B 1 / C	∠ABC (angle ABC), or ∠CBA (angle CBA), or ∠B, or ∠1	An **angle** is formed by two rays with a common endpoint. The endpoint is called the **vertex** of the angle. The rays are called the **sides** of the angle.
X Z Y ℳ	plane XYZ or plane ℳ	A **plane** is a flat surface that extends without end in all directions. It is determined by three noncollinear points. Points on the same plane are **coplanar points**.

Example 1

Write the symbol for each figure.

a. 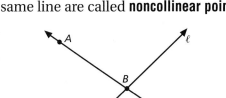 P   Q

b. A —————— B

c. 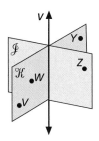 E, D

d. 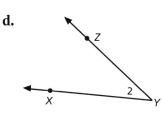 Z, X, 2, Y

Solution

a. $\overleftrightarrow{PQ}$ or $\overleftrightarrow{QP}$ **b.** $\overline{AB}$ or $\overline{BA}$ **c.** $\overrightarrow{DE}$ **d.** $\angle XYZ$, $\angle ZYX$, $\angle Y$ or $\angle 2$

Collinear points are points that lie on the same line. Points that do not lie on the same line are called **noncollinear points**.

Coplanar points are points that lie in the same plane. Points that do not lie in the same plane are called **noncoplanar points**.

Points *A, B* and *C* are collinear.
Points *A, B* and *D* are noncollinear.

Points *V, W,* and *Y* are coplanar.
Points *V, W,* and *Z* are noncoplanar.

Two different lines that intersect have exactly one point in common. In the figure above on the left, the intersection of lines *l* and *m* is point *B*.

Two distinct planes that intersect have exactly one line in common. In the figure above on the right, the intersection of planes *J* and *K* is line *v*.

Example 2

Identify the following.

a. three collinear points in Figure A

b. three noncollinear points in Figure A

c. three coplanar points in Figure B

d. three noncoplanar points in Figure B

Figure A   Figure B

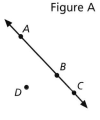 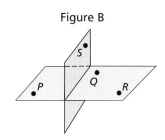

Solution

a. points *A, B,* and *C*   **b.** points *A, B,* and *D*

c. points *P, Q,* and *R*   **d.** points *P, Q,* and *S*

An angle is measured in units called **degrees** (°). On a protractor, one scale shows degree measures from 0° to 180° in a clockwise direction. The other scale shows these degree measures in a counterclockwise direction.

Math: Who, Where, When

Around 300 B.C., Euclid of Alexandria wrote the earliest geometry textbook, *Elements*. It was divided into thirteen books. Six are on elementary plane geometry. Since 1482 at least 1000 editions of *Elements* have been published.

Start reading the scale at 0°. Read the inside scale (counterclockwise) to find the measure of ∠DEF.

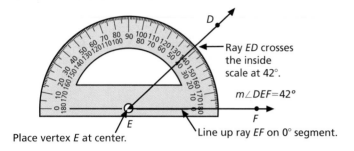

Ray *ED* crosses the inside scale at 42°.

m∠*DEF*=42°

Place vertex *E* at center.

Line up ray *EF* on 0° segment.

Start reading the scale at 0°. Read the outside scale (clockwise) to find the measure of ∠XYZ.

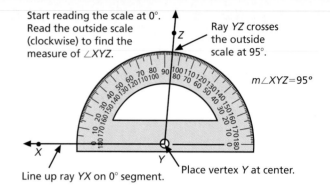

Ray *YZ* crosses the outside scale at 95°.

m∠*XYZ*=95°

Line up ray *YX* on 0° segment.

Place vertex *Y* at center.

Example 3

Use a protractor to draw ∠*JKL* so that m∠*JKL* = 120°.

Solution

Step 1 Draw a ray from point *K* through point *L*.

Step 2 Place the center of the protractor on the vertex, *K*. Place the 0° line of the protractor on $\overrightarrow{KL}$. Locate 120° on the inside scale. Mark point *J* at 120°.

Step 3 Remove the protractor. Draw $\overrightarrow{KJ}$. m∠*JKL* = 120°

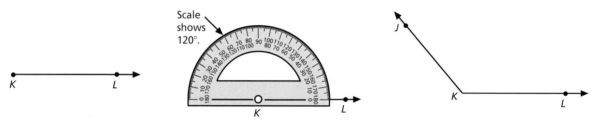

Scale shows 120°.

▼ TRY THESE EXERCISES

Draw each geometric figure. Then write a symbol for each figure.

1. ray *AB*   **2.** line segment *CD*   **3.** angle 3   **4.** angle *EFG*

Identify the following.

5. three collinear points in Figure A

6. three noncollinear points in Figure A

7. three coplanar points in the Figure B

8. three noncoplanar points in Figure B

Figure A

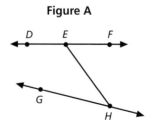

Figure B

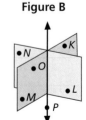

9. Use a protractor to draw an angle with a measure of 40°.

10. Use a protractor to draw an angle with a measure of 93°.

11. Use a protractor to find the measure of ∠*LQM* and ∠*PQN*.

 12. WRITING MATH Write a complete sentence to explain which letter must be in the middle when three letters name an angle.

Draw each geometric figure. Then write a symbol for each figure.

13. point *Z*   **14.** ray *ST*   **15.** angle *EFG*   **16.** line segment *QR*

17. line *LM*   **18.** angle *H*   **19.** plane *WXY*   **20.** line *BC*

Use symbols to complete the following.

21. Name the line four ways.

22. Name two rays with *B* as an endpoint.

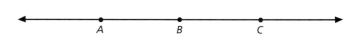

Draw a figure to illustrate each of the following.

23. Points *A*, *B*, and *C* are noncollinear.   **24.** Points *F*, *G*, and *H* are noncoplanar.

25. Line *m* intersects plane *K* at point *N*.   **26.** Planes *R* and *S* intersect at line *t*.

Find the measure of each angle.

27. ∠*NOP*   **28.** ∠*KOP*   **29.** ∠*LOP*   **30.** ∠*MOP*

31. ∠*JOM*   **32.** ∠*JOL*   **33.** ∠*JOK*   **34.** ∠*JON*

Use a protractor to draw an angle of the given measure.

35. 47°   **36.** 110°   **37.** 19°   **38.** 138°

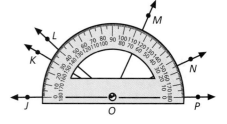

39. NAVIGATION East and west are directions on a compass that are on opposite rays. Identify two other pairs of opposite rays on a compass.

40. WRITING MATH List a real world example of a point, a line and a plane. Explain how your examples differ from the actual geometric figures.

41. MARKETING Describe the geometric elements in the logo.

EXTENDED PRACTICE EXERCISES

Think about your school. Consider that each floor models a plane, each hallway models a line segment, and each door models a point.

42. Name three collinear points.   **43.** Name three noncollinear points.

44. Name three coplanar points.   **45.** Name three noncoplanar points.

46. CRITICAL THINKING Is an angle always a plane figure? Draw an example of an angle that lies on two planes. Explain.

MIXED REVIEW EXERCISES

Add or subtract. (Lesson 3-1)

47. $-25 + (-18)$ **48.** $-23 - (-19)$ **49.** $14 + (-28)$ **50.** $36 - (-72)$

51. SPORTS The high school football team started at the 50-yard line. In the next three plays they gained 25 yards, gained 5 yards, and lost 15 yards. Where are they now?

4-2 Polygons and Polyhedra

Goals
- Identify and classify polygons.
- Identify the faces, edges and vertices of polyhedra.

Applications Transportation

Use a small marshmallow to represent a point and a toothpick to represent a line. You need additional marshmallows and toothpicks to model the following.

1. line segment
2. angle
3. midpoint of a line segment

4. Make a two-dimensional figure that is a closed shape and has a length and width. Model two different figures: a square and a rectangle.

5. Make a two-dimensional figure with the least number of toothpicks and marshmallows possible. Name the figure.

6. Make a three-dimensional figure that is a closed shape and has a length, width, and height. Model two different figures: a cube and a rectangular solid.

7. Make a table of all two-dimensional figures you can model. Include three columns: a sketch of the figure, number of marshmallows used and number of toothpicks used.

8. Repeat Question 7 for all three-dimensional figures you can model.

▼ BUILD UNDERSTANDING

A **polygon** is a two-dimensional closed plane figure formed by joining three or more line segments at only their endpoints. Each line segment is called a **side** of the polygon and joins exactly two others. The point at which two sides meet is a **vertex** (plural: *vertices*). Polygons are classified by their number of sides.

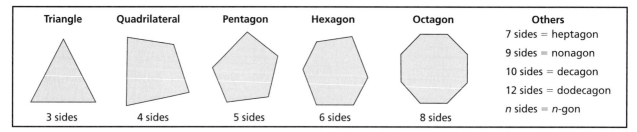

Triangle	Quadrilateral	Pentagon	Hexagon	Octagon	Others
3 sides	4 sides	5 sides	6 sides	8 sides	7 sides = heptagon 9 sides = nonagon 10 sides = decagon 12 sides = dodecagon *n* sides = *n*-gon

A polygon can be named by the letters at its vertices listed in order. An example is triangle *ABD*.

A *quadrilateral* is a polygon with four sides. There are many types of quadrilaterals. The best description of a shape is the most specific.

A polygon is **regular** if all its sides are **congruent** (of equal length) and all its angles are **congruent** (of equal measure). A regular polygon is *equilateral* and *equiangular*.

Problem Solving Tip

Blue marks indicate congruence.

$\overline{BC} \cong \overline{CD}$
$\overline{AE} \cong \overline{ED}$
$\angle CBA \cong \angle DCE$
$\angle BAE \cong \angle CED$

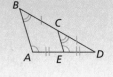

Example 1

Identify each polygon. Explain why it is regular or not regular.

a.

b.
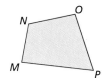

c.

Solution

a. The figure has 5 sides. It is pentagon *ABCDE*. It is regular because the sides are congruent and the angles are congruent.

Technology Tip

Geometry software can be used to draw polygons, measure their sides and measure their angles.

b. The figure is quadrilateral *MNOP*. It is not regular because the sides and the angles are not congruent.

c. The figure has 6 sides. It is hexagon *FGHIJK*. It is not regular because the sides and the angles are not congruent.

A *triangle* is a polygon with three sides. A triangle can be classified by the lengths of its sides. The marks on the figures indicate which sides and angles are congruent.

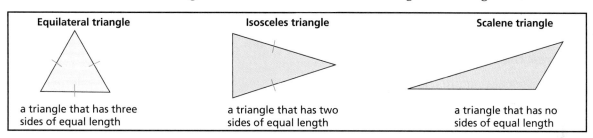

Equilateral triangle	Isosceles triangle	Scalene triangle
a triangle that has three sides of equal length	a triangle that has two sides of equal length	a triangle that has no sides of equal length

A *right triangle* is a triangle having a right angle. An *obtuse triangle* is a triangle having one obtuse angle. An *acute triangle* is a triangle having three acute angles.

Example 2

Classify each triangle.

a.

b.
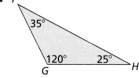

c.
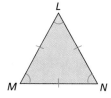

Solution

a. It is an isosceles triangle because two sides have equal length. Angle *E* is a right angle. So △*DEF* is a right triangle.

b. No sides are of equal length so it is a scalene triangle. The $m\angle G > 90°$. So △*GHI* is an obtuse triangle.

c. It is an equilateral triangle because all three sides are of equal length. All of the angles have equal measure. So △*LMN* is an equiangular triangle.

Think Back

Triangles can also be classified by the measures of their angles: acute, right, and obtuse.

Recall that the sum of the angle measures in a triangle equals 180°.

A **polyhedron** (plural: *polyhedra*) is a three-dimensional closed figure formed by joining three or more polygons at their sides.

Each polygon of the polyhedron is called a **face** and joins multiple polygons along their sides. A line segment along which two faces meet is called an **edge**. A point where three or more edges meet is called the **vertex**.

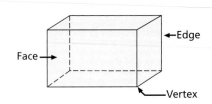

Example 3

Identify the number of faces, vertices, and edges for each figure.

a.

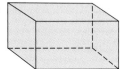

b.

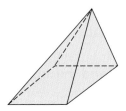

Solution

a. 6 faces, 8 vertices, and 12 edges

b. 5 faces, 5 vertices, and 8 edges

TRY THESE EXERCISES

Draw the following.

1. isosceles acute triangle

2. hexagon that is not regular

3. scalene obtuse triangle

Identify the number of faces, vertices, and edges for each figure.

4.

5.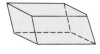

6.

7.

8. WRITING MATH Explain the difference between the word *regular* as it is used in everyday language and its meaning in geometry.

PRACTICE EXERCISES • For Extra Practice, see page 557.

Each figure is a polygon. Identify it and tell if it is regular.

9.

10.

11.

12.

Determine whether each statement is true or false.

13. A rhombus is an equilateral polygon.

14. Every parallelogram is also a rectangle.

15. A polygon can have only three vertices.

16. Equiangular polygons are always equilateral.

Classify each triangle.

17.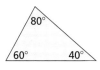

18.

19.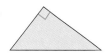

20.

Identify the number of faces, vertices, and edges for each figure.

21.

22.

23.

24.

 25. GEOBOARDS Use geoboards to build three polygons. Build three nonpolygonal figures.

TRANSPORTATION Name the shape of each sign, and state whether it is regular.

26. SPEED LIMIT 50

27. STOP

28.

29. ONE WAY

 30. YOU MAKE THE CALL Danessa said she drew a trapezoid that is a regular polygon. Is this possible? Explain.

EXTENDED PRACTICE EXERCISES

31. The perimeter of a regular octagon is 29.92 cm. What is the length of each side?

32. The perimeter of a regular pentagon is $178\frac{1}{4}$ ft. What is the length of each side?

CRITICAL THINKING Find the unknown angle measure in each triangle.

33.

34.

35.

36.

MIXED REVIEW EXERCISES

Find the perimeter of each figure. (Lesson 2-3)

37.

38.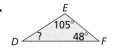

39.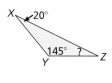

40. A security guard walks the perimeter of the property. She walks 15 yd north, 27 yd east, 6 yd south, 14 yd east, 12 yd south, 31 yd west, 3 yd north, and 10 yd west. Draw a picture of her route. Find the number of feet she walks in one round trip. (Lesson 2-2)

Review and Practice Your Skills

PRACTICE ▰ LESSON 4-1

Draw each geometric figure. Then write a symbol for each figure.

1. line *AB*
2. ray *HK*
3. angle *WXY*
4. plane *DEF*
5. plane *EFG*
6. point *T*
7. line segment *PQ*
8. ray *ST*

Identify the following.

9. three collinear points in Figure A
10. three noncollinear points in Figure A
11. three noncoplanar points in Figure B

Figure A

Figure B

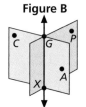

Use a protractor to find the measure of each angle.

12. ∠*APB*
13. ∠*APC*
14. ∠*APD*
15. ∠*BPC*
16. ∠*BPD*
17. ∠*BPE*
18. ∠*CPD*
19. ∠*DPE*

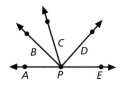

Use a protractor to draw an angle of the given measure.

20. 50°
21. 35°
22. 160°
23. 75°
24. 110°
25. 20°
26. 145°
27. 80°

PRACTICE ▰ LESSON 4-2

Each figure is a polygon. Identify it and tell if it is regular.

28.
29.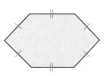
30.
31.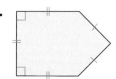

Classify each triangle.

32.
33.
34.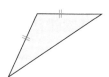
35.

Identify the number of faces, vertices, and edges for each figure.

36.
37.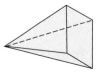
38.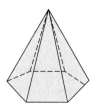
39.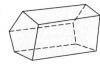

Draw each geometric figure. Then write a symbol for each figure. (Lesson 4-1)

40. line segment *MN*   **41.** angle *DEF*   **42.** plane *FGH*   **43.** ray *PQ*

Use a protractor to draw an angle of the given measure. (Lesson 4-1)

44. 80°   **45.** 135°   **46.** 45°   **47.** 150°

Classify each triangle. (Lesson 4-2)

48. 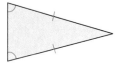   **49.**    **50.**    **51.**

Identify the number of faces, vertices, and edges for each figure.
(Lesson 4-2)

52.    **53.** 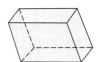   **54.**   **55.**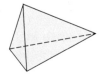

MathWorks — Career – Billboard Assembler
Workplace Knowhow

Outdoor billboards are large, highly visible forms of advertising. One common type of billboard is the poster panel. Its design is printed on a series of paper sheets, which are pre-pasted. Billboard assemblers apply these paper sheets to the face of the poster panel on location. Billboard assemblers must make accurate measurements of the permanent poster panel, the sheets they are going to apply and the amount of water and other materials needed.

For Questions 1–5, use the dimensions of the billboard.

1. What kind of polygon is the billboard?

2. If the sheets are $2\frac{1}{4}$ ft wide and 12 ft high, how many sheets are needed to cover the billboard?

3. A billboard assembler refers to a diagram. The scale of the diagram is 1 in. : 6.25 in. What are the dimensions of a diagram of the billboard shown?

4. A billboard assembler can hand paint features of a billboard. If the cost of hand painting is $3.15/ft^2, how much will it cost to hand paint 25% of the billboard?

5. A billboard assembler needs to install lighting at the base of the billboard. One light should be placed every 3 ft 5 in. How many lights are needed for the billboard?

STOP
and see us
Fred's
Tire Service

12 ft

25 ft

Visualize and Name Solids

Goals
- ■ Identify polyhedra.
- ■ Identify three-dimensional figures with curved surfaces.

Applications Packaging, Government, Manufacturing

Use the figures shown.

1. Which patterns can be folded to form a cube?

2. There are 11 different patterns of 6 squares that can be folded to form a cube. Draw 4 of these patterns.

a.

b.

c.

d.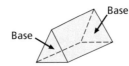

e.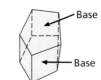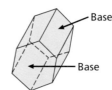

f.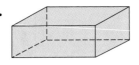

◢ BUILD UNDERSTANDING

Each of these polyhedra is a prism. A **prism** is a three-dimensional figure with two identical, parallel faces, called **bases**. The bases are congruent polygons. The other faces are parallelograms.

A prism is named by the shape of its base. A rectangular prism with edges of equal length is a **cube**.

Base · Base · Base · Base
Base · Base · Base · Base

Rectangular Prism — **Triangular Prism** — **Pentagonal Prism** — **Hexagonal Prism**

Example 1

Identify each polyhedron.

a.   b.   c.   d.

Solution

a. The bases are identical hexagons. The figure is a hexagonal prism.

b. The bases are identical rectangles. The figure is a rectangular prism.

c. The bases are identical triangles. The figure is a triangular prism.

d. The figure is a rectangular prism with all edges of the same length. It is a cube.

A **pyramid** is a polyhedron with only one base. The other faces are triangles. A pyramid is named by the shape of its base.

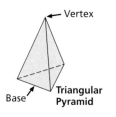
Vertex

Base

Triangular Pyramid

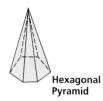
Square Pyramid

Hexagonal Pyramid

Example 2

Identify the pyramid.

a.

b.

c.

Solution

a. The base is a rectangle. The figure is a rectangular pyramid.

b. The base is a pentagon. The figure is a pentagonal pyramid.

c. The base is a triangle. The figure is a triangular pyramid.

Some three-dimensional figures have curved surfaces. A **cylinder** has two identical, parallel, circular bases. A **cone** has one circular base and one vertex. A **sphere** is the set of all points in space that are the same distance from a given point, called the *center* of the sphere.

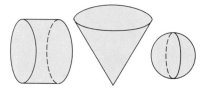

Example 3

Identify the figure.

a.

b.
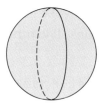

c.
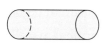

Solution

a. It has a curved surface and one circular base. The figure is a cone.

b. It has a curved surface and no bases, and all points are the same distance from the center. The figure is a sphere.

c. It has a curved surface and two circular bases. The figure is a cylinder.

A **net** is a two-dimensional pattern that when folded forms a three-dimensional figure. Dotted lines indicate folds. To draw a net for a three-dimensional figure, you must know the number and shapes of its base or bases and of its sides.

Example 4

Identify the three-dimensional figure that is formed by the net.

Solution

The figure has two triangular bases and three faces that are rectangles.
It is a triangular prism.

■

◤ TRY THESE EXERCISES

Identify each figure.

1. 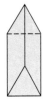   2.   3.   4.   5.

Identify the three-dimensional figure that is formed by each net.

6. 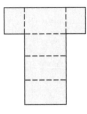   7. 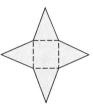   8. 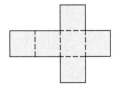   9.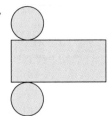

◤ PRACTICE EXERCISES • For Extra Practice, see page 558.

Identify each three-dimensional figure.

10. It has one base that is an octagon. The other faces are triangles.

11. Its two bases are identical, parallel pentagons. The other faces are parallelograms.

12. Its six faces are congruent.

13. It has four faces that are triangles. Its base is a rectangle.

Identify the three-dimensional figure that is formed by each net.

14. 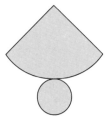   15. 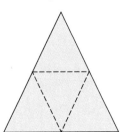   16. 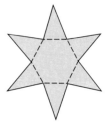   17.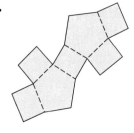

Name three everyday objects that have the given shape.

18. rectangular prism   19. cylinder   20. sphere

Copy and complete the table for Exercises 21–26.

21. triangular prism

22. rectangular prism

23. pentagonal prism

24. triangular pyramid

25. rectangular pyramid

26. pentagonal pyramid

Polyhedron	Number of faces (F)	Number of vertices (V)	Number of edges (E)	F + V − E
triangular prism				
rectangular prism				
pentagonal prism				
triangular pyramid				
rectangular pyramid				
pentagonal pyramid				

27. **WRITING MATH** Study your results from Exercises 21–26. Explain the relationship between the faces, edges and vertices of a polyhedron.

28. **PACKAGING** A graphics artist is designing a cylindrical oatmeal container. Draw the net that she must use.

29. **GOVERNMENT** The offices of the U.S. Department of Defense are located in Arlington, VA, in a building called the Pentagon. Why do you think the building was given this name?

EXTENDED PRACTICE EXERCISES

30. Each base of a prism is a polygon with *n* sides. Write a variable expression that represents the total number of faces of the prism.

31. The base of a pyramid is a polygon with *t* sides. Write a variable expression that represents the total number of edges of the pyramid.

32. **CRITICAL THINKING** Suppose that a cylinder is cut in half using a plane that is perpendicular to the bases. What is the shape of the flat face of each half of the cylinder?

33. **CHAPTER INVESTIGATION** Use heavy paper, tape, and measuring tools to make at least three models of packages for the object. Choose the best package based on safety, durability, and size. Keep all of the models for your final presentation.

MIXED REVIEW EXERCISES

Find the circumference and area of each circle. Round to the nearest tenth. (Lesson 2-7)

34.
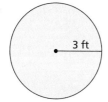
3 ft

35.
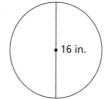
16 in.

36.
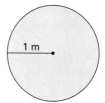
1 m

37. Record the grades you have received in math for the past month. Which measure of central tendency best describes your achievement level? Explain. (Lesson 1-2)

38. **DATA FILE** Refer to the data on the longest bridges in the world on page 520. Find the mean and median length in feet and meters.

Being able to visualize the same object in multiple ways can help you understand the geometry of three-dimensional shapes. When solving problems about three-dimensional shapes, an effective strategy is to **draw a picture, diagram, or model**. You can visually represent a problem either by illustrating it on paper or by modeling it using physical objects.

A net is a two-dimensional figure that when folded, forms the surface of a three-dimensional figure. Nets are used in many real-world applications, such as the design and manufacturing of items that are assembled by the consumer for toys, games and storage boxes.

Problem

MARKETING The Best Breakfast Company manufactures a cereal named Tri-TOPS. As a promotional campaign, each Tri-TOPS box includes two connectable triangles that come in five different colors. Once you have collected enough connectable triangles, you can build a top. You can make a closed polyhedron after the purchase of two boxes. However, if you collect at least two of each color, you can build the Tri-TOP. All collectable triangles are the same size and are equilateral and equiangular. What does a Tri-TOP look like?

Solve the Problem

Assume that you make five purchases. Each time, you receive two triangles of the same color yet a different color from the previous purchase.

The polyhedron that you are trying to build works like a top and spins on one of its vertices. With each purchase, draw or model a net that builds off the previous net. Be certain that you add the two triangles so that one side of the net is a mirror image of the other side. Show a net with 2 triangles, 4 triangles and so on.

First purchase
2 yellow triangles

Second purchase
2 blue triangles

Third purchase
2 red triangles

Fourth purchase
2 green triangles

Fifth purchase
2 purple triangles

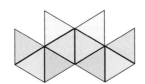

Completed Tri-TOP

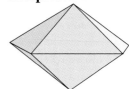

Five-step Plan

1 Read
2 Plan
3 Solve
4 Answer
5 Check

1. You can build closed polyhedra with the number of triangles you have after the third purchase of Tri-TOP cereal. Cut out six equilateral triangles, and use tape to build two different polyhedra.

2. Use your models from Exercise 1. Count the number of faces, edges and vertices in each polyhedra.

Identify the three-dimensional shape formed by each net.

3.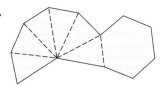

4.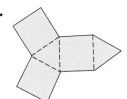

5.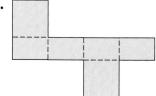

◥ PRACTICE EXERCISES

Draw a net for each three-dimensional figure.

6.

7.

8. Make a scale drawing of the net for a cylinder with height 2 m and circumference 10 m. Use a scale factor of 1 cm : 1 m.

9. **ENTERTAINMENT** James is part of the crew for the school production of *East of the Pyramids*. His task is to design and build two pyramids with square bases. Make a net of a pyramid with a square base. What two-dimensional figures and how many of each will he need to build both pyramids?

10. **PACKAGING** A producer of cheese has an assembly line that makes a specially shaped box for its cracker spread. Use the drawing of the net to sketch the assembled box.

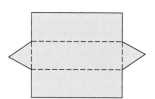

11. The Tri-TOP is a *pentagonal dipyramid*. You can make other polyhedra using ten equilateral triangles. Cut out ten equilateral triangles. Use tape to build as many different polyhedra as you can. Sketch the net of each.

◥ MIXED REVIEW EXERCISES

Find the next three numbers for each sequence. (Lesson 3-6)

12. 14, 31, 48, 65, 82

13. 41, 56, 76, 101, 131

14. −1, 2, 7, 14, 23

15. 1, 2, 11, 12, 21

16. 0.7, 1.2, 1.7, 2.2, 2.7

17. $\frac{1}{4}, \frac{2}{12}, \frac{4}{36}, \frac{8}{108}, \frac{16}{324}$

18. Find the area of an isosceles triangle with a base of 4.2 cm and 3.6 cm height. Round to the nearest hundredth. (Lesson 2-4)

Review and Practice Your Skills

Identify each figure.

1.

2.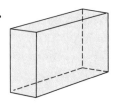

3.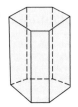

4.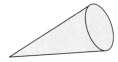

5.

6.

7.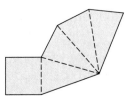

8.

Identify the three-dimensional figure that is formed by each net.

9.

10.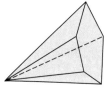

11.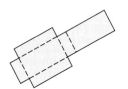

12.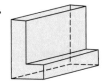

Identify the three-dimensional shape formed by each net.

13.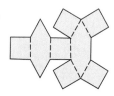

14.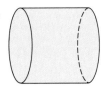

15.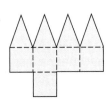

Draw a net for each of the following three-dimensional figures.

16.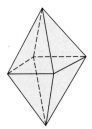

17.

18.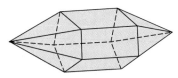

19. Identify three collinear points in the figure. (Lesson 4-1)

Find the measure of each angle. (Lesson 4-1)

20. ∠*BAC* **21.** ∠*CAD* **22.** ∠*CAE* **23.** ∠*DAE*

The sides or angles of a triangle are given. Classify each triangle. (Lesson 4-2)

24. 45°, 45°, 90° **25.** 6 cm, 8 cm, 6 cm **26.** 40°, 60°, 80° **27.** 5 in., 5 in., 5 in.

Identify each figure. (Lesson 4-3)

28.   **29.**   **30.**   **31.**   **32.**

33–35. Identify the number of faces, vertices, and edges for each figure in Exercises 28–32. (Lesson 4-3)

Identify the three-dimensional shape formed by each net. (Lessons 4-3 and 4-4)

36.   **37.**

38. Draw a net for the figure in Exercise 28. (Lesson 4-4)

Mid-Chapter Quiz

Draw a figure to illustrate the following. (Lesson 4-1)

1. Points *D* and *F* are collinear, but points *D*, *E* and *F* are noncollinear.

2. Line *AB* intersects plane 𝒞 at point *G*.

Classify each triangle according to its sides and angles. (Lesson 4-2)

3. a triangle with one right angle and no sides of equal length

4. a triangle with three acute angles and just two sides of equal length

5. a triangle with three acute angles and three sides of equal length

Identify each quadrilateral. (Are You Ready?)

6. It has only one pair of parallel sides. **7.** It has four right angles.

8. It is a parallelogram with four right angles.

Determine the number of faces, vertices and edges for each figure. (Lessons 4-2 and 4-3)

9. hexagonal pyramid **10.** pentagonal prism **11.** cube **12.** square pyramid

4-5 Isometric Drawings

Goals ■ Visualize and represent shapes with isometric drawings.

Applications Carpentry, Construction, Advertising

Draw each three-dimensional object.

1. a cube

2. a rectangular prism

Work with a partner.

3. Compare your drawings. How are they alike? How are they different?

4. How could you make a more realistic drawing?

▼ BUILD UNDERSTANDING

Any two lines in a plane are related in one of two ways.

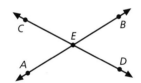

They intersect at a single point.
$\overleftrightarrow{AB}$ and $\overleftrightarrow{CD}$ intersect at point E.

They have no points in common and are **parallel**.
$\overleftrightarrow{FG}$ is parallel to $\overleftrightarrow{HI}$. Write this as $\overleftrightarrow{FG} \parallel \overleftrightarrow{HI}$.

Reading Math

Arrows signify parallel sides. Squares indicate perpendicular sides.

If two lines intersect at right angles, then the lines are **perpendicular**.

$\overleftrightarrow{LM}$ is perpendicular to $\overleftrightarrow{NO}$.
Write this as $\overleftrightarrow{LM} \perp \overleftrightarrow{NO}$.

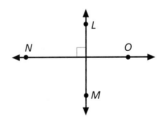

Example 1

Use the figure to name the following.

a. a pair of parallel lines

b. a pair of perpendicular lines

Solution

a. Lines r and s are parallel.

b. Line p is perpendicular to line q.

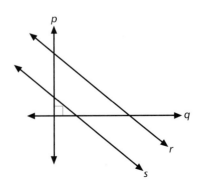

Isometric drawings provide three-dimensional views of an object. By showing three sides of an object, isometric drawings have a "corner" view. The unseen sides can be indicated with dashed lines.

The dimensions of isometric drawings are modeled after the object's actual dimensions. An object's parallel edges are parallel line segments in an isometric drawing.

Example 2

Draw an isometric drawing of the rectangular prism.

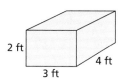

Solution

Use isometric dot paper. Assume that the unit between each pair of dots is 1 ft.

Select a point about halfway across and near the bottom of the paper. From this point, draw a vertical line 2 units long for the prism's height. From the same point, draw a line 4 units long, up and to the right. Draw another line 3 units long, up and to the left.

Draw the top, front and sides of the prism.

Finish by drawing dashed lines to indicate hidden segments. Label the actual measurements of the object. Some isometric drawings don't include dashed lines or measurements.

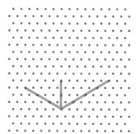 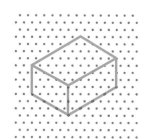 

◥ TRY THESE EXERCISES

Complete the following on one drawing.

1. Draw $\overleftrightarrow{CD}$.

2. Draw $\overleftrightarrow{EF} \perp \overleftrightarrow{CD}$ at point C.

3. Draw $\overleftrightarrow{GH} \perp \overleftrightarrow{CD}$ at point D.

4. What is the relationship between $\overleftrightarrow{EF}$ and $\overleftrightarrow{GH}$?

Make an isometric drawing of the following.

5. a triangular prism

6. a rectangular pyramid

7. a cube

How many cubes are in each isometric drawing?

8.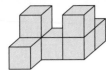

9.

10.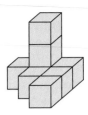

 11. **WRITING MATH** How do parallel and perpendicular lines appear in an isometric drawing?

Use the figure to name the following.

12. all of the parallel lines

13. all of the perpendicular lines

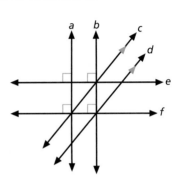

Complete the following on one drawing.

14. Draw $\overleftrightarrow{RS}$.

15. Draw $\overleftrightarrow{RT}$ perpendicular to $\overleftrightarrow{RS}$ at point R.

16. Draw line k parallel to $\overleftrightarrow{RS}$ through point T.

17. What is the relationship between $\overleftrightarrow{RT}$ and line k?

On isometric dot paper, draw the following.

18. cube: $s = 23$ m

19. rectangular prism: $l = 15$ in., $w = 6$ in., $h = 10$ in.

20. cube: $s = 7.5$ ft

21. rectangular prism: 1 unit by 2 units by 3 units

22. How many 1-unit cubes would build the prism in Exercise 21?

 23. **CHECK YOUR WORK** Sketch 1-unit cubes in your drawing from Exercise 21 to check your answer for Exercise 22.

How many cubes are in each isometric drawing?

24.

25.

26.

Make an isometric drawing of each figure.

27.

28.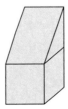

29.

On isometric dot paper, draw the following.

30. Draw the largest cube possible on the paper.

31. Label each vertex with a capital letter of the alphabet.

32. Name each of the edges using the letters assigned to the vertices.

33. How many edges and how many vertices are in this cube?

34. Does every cube have the same number of edges and vertices?

35. **WRITING MATH** How can an isometric drawing of an object help you make a net for the object?

36. **CARPENTRY** Sketch on isometric dot paper a wooden awards stand. The winner will stand at the center of the platform. The second- and third-place winners will be a step lower and to the right and left of the first-place winner.

37. **MODELING** Use heavy paper to construct a three-dimensional object. Make three isometric drawings of the object from different angles.

◼ EXTENDED PRACTICE EXERCISES

38. **ADVERTISING** A graphic artist is designing a billboard. The client's directions follow. Make an isometric drawing of a triangular prism, then surround the prism with realistic forest scenery. Draw an example of a possible first draft of the billboard design.

39. **CONSTRUCTION** You need to replace the steps from the back door to the patio. The house is built on a foundation 2 ft above the ground. The steps should be 4 ft wide. Cement blocks are 1 ft by 0.5 ft by 0.5 ft. Sketch an isometric drawing to show the number of blocks needed and the appearance of the steps.

40. The top, front and side views of a stack of blocks is shown. Make an isometric drawing of the stack.

Top   Front   Side

◼ MIXED REVIEW EXERCISES

Multiply or divide. (Lesson 3-2)

41. -16×3 42. $-44 \div -4$ 43. 12×-6 44. $56 \div -8$

45. **DATA FILE** Refer to the data on the highest and lowest continental altitudes on page 526. What is the range in altitudes between the highest point in Europe and the lowest point in Asia? (Lesson 3-1)

46. Construct a box-and-whisker plot for the following data. (Lesson 1-8)

26 34 49 22 31 52 36 39 25 58 27 44 31 36 47 53 22

Perspective and Orthogonal Drawings

Goals ■ Make perspective and orthogonal drawings.

Applications Construction, Architecture, Engineering, Graphic arts

Work in a group. Refer to the photos shown.

1. Does one elephant appear larger than the others? Explain.

2. Do the railroad tracks appear to meet? Explain.

3. Explain why these photos are realistic even though they don't portray the actual dimensions of the objects.

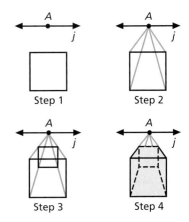

◤ BUILD UNDERSTANDING

A **perspective drawing** is made on a two-dimensional surface in such a way that three-dimensional objects appear true-to-life.

The railroad tracks shown appear to intersect at a point on the horizon. That point is called a **vanishing point**. A **one-point perspective** drawing uses a vanishing point to create depth.

The vanishing point lies on a line called the **horizon line**. The location of vanishing point and horizon line can change, depending on the observer's point of view.

In the drawing shown, point A is the vanishing point. Line l is the horizon line.

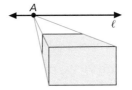

Reading Math

The word *perspective* comes from the Latin word *perspectus*, which means *to see through*.

Example 1

Draw a cube in one-point perspective.

Solution

Step 1 Draw a square to show the front surface of the cube. Draw a horizon line j and a vanishing point A on line j.

Step 2 Lightly draw line segments connecting the vertices of the square to point A.

Step 3 Draw a smaller square whose vertices touch the four line segments. Each side of the smaller square is parallel to the corresponding side of the larger square.

Step 4 Connect the vertices of the two squares. Use dashed segments to indicate the cube edges hidden from view.

Step 1   Step 2

Step 3   Step 4

A perspective drawing shows an object realistically, but it may not provide certain details, such as accurate measurements. An **orthogonal drawing**, or **orthographic projection**, shows the top, front, and side views of an object without distorting the object's dimensions.

These views appear as if your line of sight is perpendicular to the object's top, front, and side. Orthogonal drawings often label the dimensions of the object.

Example 2

Make an orthogonal drawing of the figure.

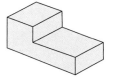

Solution

Imagine looking at the object from the top, front, and side. Draw the object from each view.

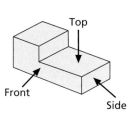

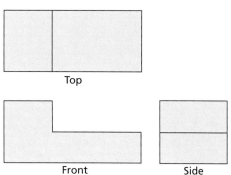

Example 3

CONSTRUCTION Rosario is assembling a swing set. The directions include this orthogonal drawing. How should the finished swing set appear?

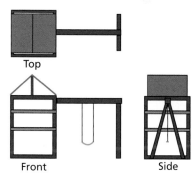

Solution

Visualize how the swing set will appear from each view in the orthogonal drawing.

Trace each figure. Locate and label the vanishing point and horizon line.

1.

2.

3.

4. Make an orthographic drawing of the figure in the isometric drawing.

5. Draw a rectangular prism in one-point perspective.

Match the orthogonal view as though you were looking at the object from the arrow.

6.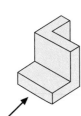

7.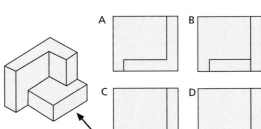

 8. **WRITING MATH** How do perspective and isometric drawings differ?

PRACTICE EXERCISES • For Extra Practice, see page 559.

Trace each figure. Locate and label the vanishing point and horizon line.

9.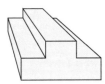

10.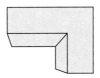

11.

12. Make a one-point perspective drawing of a gift-wrapped box, including ribbon that wraps around the box.

Draw each of the following in one-point perspective.

13. a rectangular pyramid   14. a cylinder   15. a cube

Make an orthogonal drawing of each cube stack, showing top, front and side views.

16.

17.

18.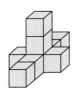

19. ENGINEERING The isometric drawing shows a support for a highway overpass. Make an orthogonal drawing of the overpass.

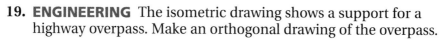

20. WRITING MATH How do parallel lines appear in a one-point perspective drawing?

GRAPHIC ARTS A designer created a model of a box to hold golf clubs.

21. Make an orthogonal drawing of the box.

22. Make a net of the box. Do not include any tabs or flaps for manufacturing.

23. How are the net and the orthographic drawing different? Why would a designer provide both a net and an orthographic drawing to a client?

24. ARCHITECTURE Most skyscrapers are built using steel girders shaped like the capital letter *I*. This shape prevents buckling and can support heavy loads. Make an orthogonal drawing of the I-beam shown.

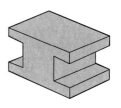

▰ EXTENDED PRACTICE EXERCISES

25. CRITICAL THINKING A perspective drawing of a three-dimensional letter *Z* is shown. Make a one-point perspective drawing of your initials.

26. Perspective drawings can have more than one vanishing point. Draw a two-point perspective of a rectangular prism that is not a cube.

27. CHAPTER INVESTIGATION Make an orthogonal drawing of your package. Include dotted lines for hidden edges, and label the measurements.

Imagine a rectangular box that is 11 in. by 7 in. by 3 in.

28. Make an orthogonal drawing of the box.

29. Find the area of the top of the box.

30. Find the area of the front of the box.

31. Find the area of the side of the box.

▰ MIXED REVIEW EXERCISES

Simplify. (Lesson 3-3)

32. $15 + 28 \div (-4) - (-12)$

33. $4 + 4^2 - 16 + (-10)$

34. $16(28 - 29) + 26 \div (-13)$

35. $[64 + (24 - 2^3)] \div (-8)$

36. $38 - 36 \div (-2) + 7 \times (-4)$

37. $22 + (-45) \div 3^2 + (-10)$

38. Find the area of a trapezoid with bases of 17 cm and 21 cm and height of 12 cm. (Lesson 2-4)

Review and Practice Your Skills

PRACTICE ◼ LESSON 4-5

Complete the following on one drawing.

1. Draw $\overleftrightarrow{AB}$.

2. Draw $\overleftrightarrow{CB}$ perpendicular to $\overleftrightarrow{AB}$ at point B.

3. Draw $\overleftrightarrow{DC}$ parallel to $\overleftrightarrow{AB}$.

4. Draw $\overleftrightarrow{FA}$ parallel to $\overleftrightarrow{CB}$.

5. What is the relationship between $\overleftrightarrow{FA}$ and $\overleftrightarrow{AB}$?

6. Name four pairs of perpendicular lines.

On isometric dot paper, draw the following.

7. cube: $s = 14$ ft

8. rectangular prism: 3 units by 5 units by 2 units

9. cube: $s = 2.3$ cm

10. rectangular prism: $l = 8$ m, $w = 3$ m, $h = 12$ m

11. cube: $s = 11$ yd

12. rectangular prism: $l = 4.5$ in., $w = 7$ in., $h = 10.5$ in.

How many cubes are in each isometric drawing?

13.

14.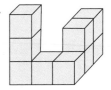

15.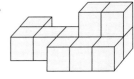

PRACTICE ◼ LESSON 4-6

Trace each figure. Locate and label the vanishing point and horizon line.

16.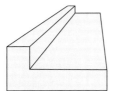

17.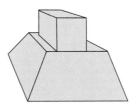

18.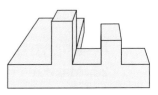

Draw each of the following in one-point perspective.

19. a triangular prism

20. a pentagonal prism

21. a triangular pyramid

22. a rectangular pyramid

Make an orthogonal drawing of each stack of cubes, showing top, front, and side views.

23.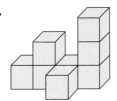

24.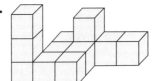

25.

26. Use a protractor to draw an angle measuring 140°. (Lesson 4-1)

The sides or angles of a triangle are given. Classify each triangle. (Lesson 4-2)

27. 45°, 45°, 90°   **28.** 6 cm, 8 cm, 6 cm   **29.** 40°, 60°, 80°   **30.** 5 in., 5 in., 5 in.

31. Identify the number of faces, vertices and edges for Figure A and for Figure C. (Lesson 4-2)

32. Identify Figures A, B and C. (Lesson 4-3)

33. Identify the three-dimensional formed by the net in Figure D. (Lessons 4-3 and 4-4)

34. Draw a net for Figure C. (Lesson 4-4)

35. On isometric dot paper, draw a rectangular prism 5 units by 2 units by 8 units. (Lesson 4-5)

36. Draw a square pyramid in one-point perspective. (Lesson 4-6)

37. How many cubes are in the isometric drawing? (Lesson 4-5)

38. Make an orthogonal drawing of the stack of cubes, showing top, front, and sides views. (Lesson 4-6)

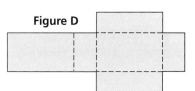

Figure A   Figure B   Figure C

Figure D

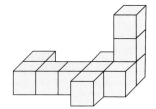

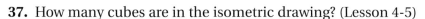

MathWorks Career – Packaging Designer
Workplace Knowhow

Packaging designers use their marketing, manufacturing and computer skills to create packaging that is appropriate for a market. Retail packaging designers must understand the needs of the target audience, then consider the size, shape, weight, durability and function of the product and its package. Other factors may be considered such as the ease of use, child safety standards, display and stacking options, and costs to manufacture. Often, a model or *prototype* of the package is made and then consumers are asked if they would reach for the package when they see it on the shelves in stores.

Answer the following questions.

1. A packaging designer is creating a package in the shape of a rectangular prism to contain the portable stereo. What is its volume?

2. Suppose the stereo needs 4.5 cm of protective filler (styrofoam, bubble wrap, or some other material) to protect it during shipping and handling. What is the volume of the stereo and the filler?

3. Make an orthogonal drawing of a box that would contain the stereo and the filler.

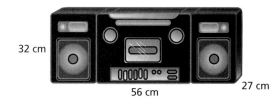

32 cm

56 cm   27 cm

Volumes of Prisms and Cylinders

Goals ■ Find the volumes of prisms and cylinders.

Applications Agriculture, Hobbies, Machinery, Construction

Refer to the cubes for Questions 1–3.

1. How many cubes are in the top layer?

2. How many cubes are in the whole shape?

3. What is the relationship between the dimensions of the shape and the total number of cubes?

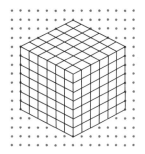

◥ BUILD UNDERSTANDING

Volume is the number of cubic units contained in a three-dimensional figure. Volume is in three dimensions and requires cubic units.

Volume of a Prism	$V = B \cdot h$ where B is the area of the base and h is the height.

You can use this formula to find the volume of any prism.

Triangular prism   Rectangular prism   Cube   Pentagonal prism   Hexagonal prism

Example 1

Find the volume.

a. rectangular prism: 3 ft by 2 ft by 2 ft

b.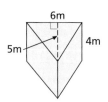

Solution

a. $V = (l \cdot w) \cdot h$ Use $l \cdot w$ to find the area of the base.

$V = 3 \text{ ft} \cdot 2 \text{ ft} \cdot 2 \text{ ft}$ Substitute 3 for l, 2 for w and 2 for h.

$V = 12 \text{ ft}^3$ The volume of the rectangular prism is 12 ft³.

b. $V = \left(\dfrac{1}{2}b \cdot h\right) \cdot h$

$V = \left(\dfrac{1}{2} \cdot 6 \text{ m} \cdot 5 \text{ m}\right) \cdot 4 \text{ m}$

$V = 60 \text{ m}^3$ The volume of the triangular prism is 60 m³.

> **Think Back**
>
> The formula for the area of a triangle is
>
> $A = \dfrac{1}{2}b \cdot h$

To find the volume of a cylinder, multiply the area of the base times the height. The area of the base of a cylinder is the area of a circle, πr^2.

Volume of a Cylinder	$V = \pi r^2 \cdot h$ where r is the radius and h is the height.

Remember from Chapter 3 that 3.14 can be substituted for π. Sometimes it will be more convenient to use $\frac{22}{7}$ for π.

Example 2

Find the volume of the cylinder.

a. Use 3.14 for π.

b. Use $\dfrac{22}{7}$ for π.

c. Use the π key on your calculator.

8 cm

35 cm

Solution

a. $V = \pi r^2 \cdot h$

$V = \pi \cdot 8^2 \cdot 35$

$V \approx 3.14 \cdot 64 \cdot 35$

$V \approx 7033.6 \text{ cm}^3$

b. $V = \pi r^2 \cdot h$

$V = \pi \cdot 8^2 \cdot 35$

$V \approx \dfrac{22}{7} \cdot \dfrac{64}{1} \cdot \dfrac{35^{5}}{1}$

$V \approx 7040 \text{ cm}^3$

c. $V = \pi r^2 \cdot h$

$V = \pi \cdot 8^2 \cdot 35$

$V \approx \pi \cdot 64 \cdot 35$

$V \approx 7037.1675 \text{ cm}^3$

You get different answers because you are using different approximate values for π.

Check Understanding

In Example 2, which answer is most accurate? Explain.

Example 3

CONSTRUCTION A trench has the dimensions shown. Find the amount of earth removed in digging the trench.

10 m

2.5 m

1.2 m

Solution

The trench is a triangular prism. First find the area of one triangular base (B).

$B = \dfrac{1}{2} \cdot b \cdot h$

$B = \dfrac{1}{2} \cdot 2.5 \text{ m} \cdot 1.2 \text{ m}$ Substitute 2.5 for b and 1.2 for h.

$B = 1.5 \text{ m}^2$ The area of the base of the trench is 1.5 m².

Then use the volume formula for prisms.

$V = B \cdot h$

$V = 1.5 \text{ m}^2 \cdot 10 \text{ m}$ Substitute 1.5 for B and 10 for h.

$V = 15 \text{ m}^3$

The volume of earth removed is 15 m³.

Find the volume of the following. Round to the nearest tenth, if necessary.

1. rectangular prism: 12 ft by 6 ft by 4 ft
2. cylinder: $r = 3$ cm, $h = 25$ cm
3. cube: $s = 7.8$ in.
4. cylinder: $d = 1.7$ m, $h = 2.4$ m

5.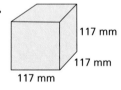
6.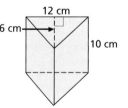
7.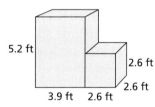

8. Find the volume of a room measuring 3.2 m by 4.8 m by 2.9 m.

9. Find the volume of a cylindrical water drum 4 ft high with a radius of 2.5 ft.

 10. **WRITING MATH** Explain why volume is expressed in cubic units.

▼ PRACTICE EXERCISES • For Extra Practice, see page 559.

Find the volume of each rectangular prism. Round to the nearest tenth, if necessary.

11. $l = 8$ ft, $w = 23$ ft, $h = 17$ ft
12. $l = 5.5$ in., $w = 11$ in., $h = 20.3$ in.
13. $l = 316$ m, $w = 68$ m, $h = 47$ m
14. $l = 0.028$ km, $w = 0.179$ m, $h = 0.263$ m

Find the volume of each cylinder. Round to the nearest tenth, if necessary.

15. $r = \frac{1}{2}$ m, $h = 14$ m
16. $d = 70.5$ m, $h = 31.6$ m
17. $r = 0.2$ cm, $h = 0.5$ cm
18. $h = 15$ yd, $d = 8$ yd

Find the volume. Round to the nearest tenth, if necessary.

19. 20. 21.

Which figure has the greater volume?

22.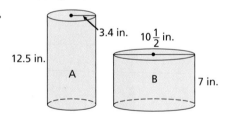
23.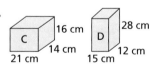

24. **MACHINERY** A car engine has eight cylinders, each with a diameter of 7.2 cm and a height of 8.4 cm. The total volume of the cylinders is called the capacity of the engine. Find the engine capacity.

Copy and complete the table shown for various cylinders.

25. Cylinder A

26. Cylinder B

27. Cylinder C

28. Cylinder D

Cylinder	diameter of base (*d*)	radius of base (*r*)	height (*h*)	Volume (*V*)
A	6 ft	■	8 ft	■
B	18 cm	■	23 cm	■
C	■	6.9 in.	4.2 in.	■
D	■	2.5 m	4.8 m	■

29. **ERROR ALERT** Does the order in which you multiply the base area by the height when finding volume make a difference? Explain.

30. **NETS** Draw a net for a rectangular prism with a volume of 63 in.^{3}

31. **AGRICULTURE** A farmer carries liquid fertilizer in a cylindrical tank that is 10 ft 5 in. long with a 4 ft 6 in. diameter. Find the volume of the tank in cubic feet.

32. **MODELING** Use Algeblocks unit cubes, sugar cubes or other small cubes. If each cube is 1 unit^3, build a rectangular prism that has a volume of 36 unit^3. What is the length, width and height of your prism?

33. **PACKAGING** A designer is creating a cylindrical container to hold 2355 cm^3 of glass beads. To fit in standard shelving, the height of the container must be 30 cm. What is the radius of the container?

EXTENDED PRACTICE EXERCISES

HOBBIES Use this relationship between volume and liquid capacity: 1 L = 1000 cm^3.

34. An aquarium is 40 cm by 28 cm by 43 cm. What is the volume?

35. In liters of water, what is the capacity of the tank?

36. A goldfish requires about 1000 cm^3 of tank space to survive. How many goldfish can this tank support when it is full of water?

For Exercises 37–40, use the figures shown.

37. Find the ratio of the radii of the two cylinders.

38. Find the ratio of the heights of the two cylinders.

39. Find the ratio of the volumes of the two cylinders.

40. **WRITING MATH** Explain how the ratios from Exercises 37–39 compare.

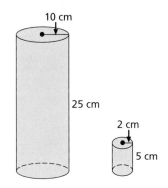

MIXED REVIEW EXERCISES

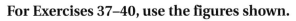

Simplify. (Lessons 3-7 and 3-8)

41. 2^6

42. 3^3

43. 4^6

44. 4^{-2}

45. $5^2 + 5^3$

46. $4^3(4^2)$

47. $2^4 - 2^2$

48. $\dfrac{2^9}{2^6}$

4-8 Volumes of Pyramids and Cones

Goals ■ Find the volumes of pyramids and cones.

Applications Meteorology, Architecture

Use grid paper and a straightedge.

1. Make a net for each of the figures shown.

2. If the cylinder and cone have the same base, which do you think has the greater volume?

3. If the prism and pyramid have the same base, which do you think has the greater volume?

4. How are the figures the same?

5. How are the figures different?

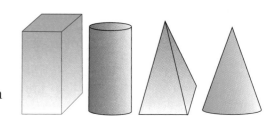

▼ BUILD UNDERSTANDING

The formula for the volume of a pyramid is related to the formula for the volume of a prism.

$$\text{Prism } V = B \cdot h$$

$$\text{Pyramid } V = \frac{1}{3} \cdot B \cdot h$$

The volume of a pyramid with a given base and height is one-third the volume of a prism with the same base and height.

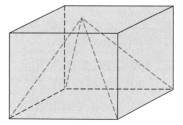

Volume of a Pyramid	$V = \frac{1}{3} \cdot B \cdot h$ where B is the area of the base and h is the height of the pyramid.

Example 1

A rectangular pyramid has a base 6 in. long and 4 in. wide. Its height is 8 in. Find the volume.

Solution

Use the volume formula for a pyramid.

$$V = \frac{1}{3} \cdot B \cdot h$$

$$V = \frac{1}{3} \cdot (6 \cdot 4) \cdot 8$$

$$V = 64$$

The volume of the pyramid is 64 in.^{3}

The formula for the volume of a cone is related to the formula for the volume of a cylinder. The volume of a cone with a given radius and height is one-third the volume of a cylinder with the same radius and height.

Volume of a Cone	$V = \frac{1}{3}\pi r^2 \cdot h$ where r is the radius and h is the height.

Example 2

Find the volume of the cone. Use $\pi \approx 3.14$ and round to the nearest tenth.

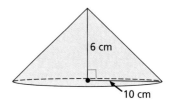

6 cm

10 cm

Solution

Use the volume formula for a cone.

$V = \frac{1}{3}\pi r^2 \cdot h$

$V = \frac{1}{3}\pi (10)^2 \cdot 7$

$V \approx \frac{1}{3} \cdot 3.14 \cdot 100 \cdot 7$

$V \approx 732.7$ The volume of the cone is about 732.7 cm³.

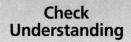

Check Understanding

Find the solution to Example 1 using $\frac{22}{7}$ for π.

Example 3

To the nearest hundredth, find the volume of a cone-shaped water cup with a height of 2.5 in. and a radius of 1.2 in.

Solution

Use the volume formula for a cone.

$V = \frac{1}{3}\pi r^2 \cdot h$

$V = \frac{1}{3}\pi (1.2)^2 \cdot 2.5$

$V \approx \frac{1}{3} \cdot 3.14 \cdot 1.44 \cdot 2.5$

$V \approx 3.77$

The volume of the water cup is about 3.77 in.³

◣ TRY THESE EXERCISES

Find the volume of the following. Round to the nearest tenth if necessary.

1. rectangular pyramid: base is 4 cm by 5 cm, $h = 8$ cm
2. cone: $r = 3.8$ ft, $h = 5.1$ ft

3. square pyramid: $h = 18$ cm, base length is 11 cm
4. cone: $r = 28$ in., $h = 44$ in.

5. The Great Pyramid of Egypt has a square base that is 230 m long. Its height is 147 m. Find the volume.

Find the volume of each figure. If necessary, round to the nearest tenth.

6.
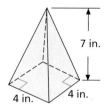
7 in.
4 in.   4 in.

7.
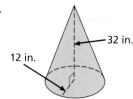
32 in.
12 in.

8.
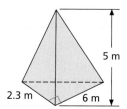
5 m
2.3 m   6 m

9. A pyramid has a rectangular base with length 8 ft and width 3 ft. Its height is 9 ft. Find the volume of the pyramid.

10. Find the volume of a cone 3.8 ft in diameter and 5.1 ft high.

11. Find the volume of a pyramid that is 10.2 m high and has a rectangular base measuring 7 m by 3 m.

▼ PRACTICE EXERCISES • For Extra Practice, see page 560.

Find the volume of each figure. If necessary, round to the nearest tenth.

12.
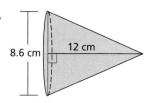
8.6 cm   12 cm

13.
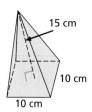
15 cm
10 cm
10 cm

14.
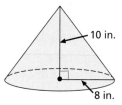
10 in.
8 in.

15.
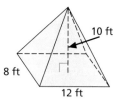
10 ft
8 ft
12 ft

16.
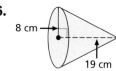
8 cm
19 cm

17.
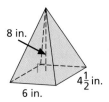
8 in.
$4\frac{1}{2}$ in.
6 in.

18. Find the volume of a pyramid whose height is 10 in. and whose octagonal base has an area of 24.9 in.^{2}

19. Phosphate is stored in a conical pile. Find the volume of the pile if the height is 7.8 m and the radius is 16.2 m.

20. A container for rose fertilizer is in the shape of a square pyramid. Each side of the base is 5 in. and the height is 8 in. What is the capacity of the container?

21. Find the volume of a cone-shaped storage bin with a height of 27 m and a radius of 15 m.

22. METEOROLOGY A tornado is a funnel cloud, meaning it is cone-shaped. If a tornado has a base diameter of about 150 ft and a height of about 1250 ft, find the volume.

DATA FILE Use the data on pyramids on page 520 to find the volume of each square pyramid.

23. Pyramid of the Sun in Mexico; 222.5-m base

24. Step Pyramid of Djoser; 60-m base

Copy and complete each table. Round to the nearest tenth, if necessary.

25. Pyramid A

26. Pyramid B

27. Pyramid C

Solid	Base area	Height	Volume
Pyramid A	180 cm²	20 cm	■
Pyramid B	42 m²	■	140 m³
Pyramid C	■	28 ft	8400 ft³

28. Cone A

29. Cone B

30. Cone C

Solid	Radius	Height	Volume
Cone A	10 m	24 m	■
Cone B	■	12 in.	1808.64 in.³
Cone C	6 cm	■	301.44 cm³

31. YOU MAKE THE CALL Susan says there are 3 ft³ in a cubic yard. Tanisha thinks there are 27 ft³ in a cubic yard. Who is correct?

32. WRITING MATH Explain how the formula for the volume of a pyramid is similar to the formula for the volume of a cone.

■ EXTENDED PRACTICE EXERCISES

33. While camping, Eric and Tyler pitch a tent shaped like a pyramid. The base is a square with sides of 2.5 m, and the volume of the tent is 4.2 m³. Find the height of their tent.

34. The movie theater sells popcorn in the two types of containers shown at the right. Each costs $3.25. Which container is the better deal if the popcorn is filled to the top of both containers? Explain.

35. A souvenir company wants to make snow globes shaped like a pyramid. It decides that the most cost-effective maximum volume of water for the pyramids is 12 in.³. If a pyramid globe measures 4 in. in height, find the area of the base.

36. CHAPTER INVESTIGATION Find the volume of the package that you have designed.

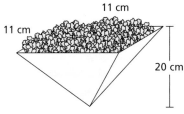

■ MIXED REVIEW EXERCISES

Classify the following triangles as equilateral, iscoceles, or scalene. Then classify them as obtuse, right, or acute. (Lesson 4-2)

37.

38.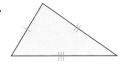

39.

Name each polygon. Describe each polygon as regular or irregular. (Lesson 4-2)

40.

41.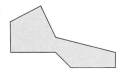

42.

Review and Practice Your Skills

PRACTICE ◤ LESSON 4-7

Find the volume of each rectangular prism. Round to the nearest tenth, if necessary.

1. $l = 5$ in., $w = 12$ in., $h = 3$ in.

2. $l = 6$ yd, $w = 9$ yd, $h = 3$ yd

3. $l = 22$ ft, $w = 11$ ft, $h = 45$ ft

4. $l = 12$ mm, $w = 4$ mm, $h = 8$ mm

5. $l = 1.5$ cm, $w = 6$ cm, $h = 4.5$ cm

6. $l = 3.2$ m, $w = 7.4$ m, $h = 5.1$ m

Find the volume of each cylinder. Round to the nearest tenth, if necessary.

7. $r = 2$ yd, $h = 8$ yd

8. $d = 8$ mm, $h = 15$ mm

9. $r = 1$ cm, $h = 0.25$ cm

10. $d = 20$ in., $h = 35$ in.

11. $r = 5$ ft, $h = 6$ ft

12. $d = 5$ m, $h = 0.2$ m

Find the volume of each figure. Round to the nearest tenth, if necessary.

13.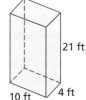

14.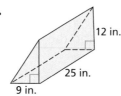

15.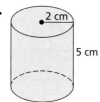

16.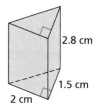

PRACTICE ◤ LESSON 4-8

Find the volume of each figure. Round to the nearest tenth, if necessary.

17. rectangular pyramid: base is 8 ft by 12 ft, $h = 14$ ft

18. cone: $r = 9$ in., $h = 12$ in.

19. square pyramid: edge of base is 6.5 cm, $h = 20$ cm

20. cone: $d = 6$ mm, $h = 8$ mm

21. pentagonal pyramid: area of base is 38.1 m², $h = 14$ m

22. cone: $r = 10$ yd, $h = 20$ yd

Find the volume of each figure. If necessary, round to the nearest tenth.

23.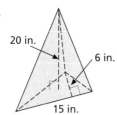

24.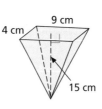

25.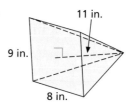

26.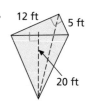

27.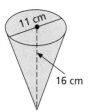

28.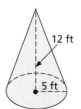

29.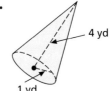

30.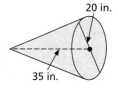

31. Identify three collinear points in the figure. (Lesson 4-1)

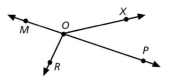

Find the measure of each angle. (Lesson 4-1)

32. *MOR*   **33.** *ROP*   **34.** *ROX*   **35.** *MOX*

The sides or angles of a triangle are given. Classify each triangle. (Lesson 4-2)

36. 12 m, 10 m, 18 m   **37.** 60°, 60°, 60°   **38.** 5 in., 3 in., 5 in.   **39.** 35°, 35°, 190°

Identify each figure. (Lesson 4-3)

40.   **41.**   **42.**   **43.**

Identify the number of faces, vertices, and edges for each figure. (Lesson 4-2)

44. the figure in Exercise 41   **45.** the figure in Exercise 42   **46.** the figure in Exercise 43

Identify the three-dimensional figure that is formed by each net.
(Lessons 4-3 and 4-4)

47. 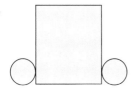   **48.** 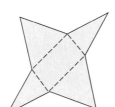   **49.** 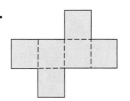   **50.**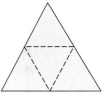

51. On isometric dot paper, draw a cube with *s* = 8 cm. (Lesson 4-5)

How many cubes are in each isometric drawing? (Lesson 4-5)

52. 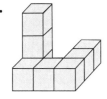   **53.** 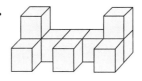   **54.**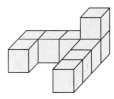

55–57. For each stack of cubes in Exercises 52–54, make an orthogonal drawing
showing top, front, and side views. (Lesson 4-6)

58. Draw a rectangular prism in one-point perspective. (Lesson 4-6)

Find the volume of each figure. (Lessons 4-7 and 4-8)

59. rectangular prism: *l* = 8 cm, *w* = 2.5 cm, *h* = 10 cm   **60.** cube: *s* = 15 in.

61. rectangular pyramid: base is 3 ft by 6 ft, *h* = 14 ft   **62.** cone: *r* = 2 m, *h* = 12 m

4-9 Surface Area of Prisms and Cylinders

Goals ■ Find the surface area of prisms and cylinders.

Applications Design, Health, Recreation, Construction

Each of these nets can be folded to form a polyhedron. Match each net to the polyhedron it would form.

1.   **2.**   **3.**   **4.**

A.   **B.**   **C.**   **D.**

◥ BUILD UNDERSTANDING

The **surface area** of a solid is the amount of material it would take to cover it. Surface area is measured in square units.

Surface Area of a Prism	The surface area of a prism is the sum of the areas of its faces.

Example 1

Find the surface area of the rectangular prism.

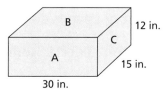

Solution

Area of A

$A = l \cdot w$

$A = 30 \cdot 12$

$A = 360$

Area of B

$A = l \cdot w$

$A = 30 \cdot 15$

$A = 450$

Area of C

$A = l \cdot w$

$A = 15 \cdot 12$

$A = 180$

A rectangular prism has 6 faces. Add the areas of all the faces.

$SA = (2 \cdot 360) + (2 \cdot 450) + (2 \cdot 180)$

$SA = 720 + 900 + 360$

$SA = 1980$

The surface area of the rectangular prism is 1980 in.^{2}

To find the surface area of a cylinder, add the area of the curved surface to the sum of the areas of the two bases.

Surface Area of a Cylinder	$SA = 2\pi rh + 2\pi r^2$ where r is the radius and h is the height of the cylinder.

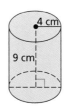

Example 2

Find the surface area of the cylinder. Use $\pi \approx 3.14$. Round to the nearest whole number.

Solution

First, find the area of the curved surface of the cylinder. When "unrolled," this surface is a rectangle with length equal to the circumference of the circle. The width of the rectangle is equal to the height of the cylinder.

$A = 2\pi r \cdot h$

$A \approx 2(3.14)(4) \cdot 9$

$A \approx 226.08$

Find the area of a base. Multiply it by 2.

$A = \pi r^2$ $2A \approx 2 \cdot 50.24$

$A \approx 3.14 \cdot 16$ $2A \approx 100.48$

$A \approx 50.24$

Add the areas to find the surface area.

$SA \approx 226.08 + 100.48 \approx 327$

The surface area of the cylinder is about 327 cm^2.

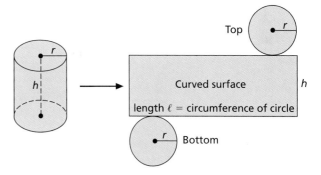

Think Back

A *radius* is a segment whose endpoints are the circle's center and a point on the circle.

The *diameter* contains the circle's center. It has both endpoints on the circle.

Example 3

PACKAGING Find the amount of cardboard on the surface of the cracker box shown.

Solution

You need to find the surface area of the box. There are two square bases and four rectangular sides.

The area of one base is (4.5 in.)^2 = 20.25 in.^{2}

The area of one side is (4.5 in.) $\cdot$ (9.5 in.) = 42.75 in.^{2}

Adding twice the area of a base and four times the area of a side will give the total surface area of the box.

(2 $\cdot$ 20.25) + (4 $\cdot$ 42.75) = 211.5

The amount of cardboard on the surface is about 212 in.^{2}

Find the surface area of each figure. Round to the nearest whole number.

1.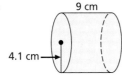
9 cm
4.1 cm→

2.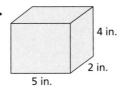
4 in.
2 in.
5 in.

3.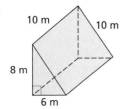
10 m   10 m
8 m
6 m

4.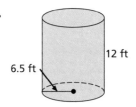
12 ft
6.5 ft

5. cylinder: $r = 22$ in., $h = 49$ in.

6. rectangular prism: 4 m by 8 m by 6 m

7. cube: $s = 2.1$ cm

8. cylinder: $d = 11$ mm, $h = 42$ mm

9. A department store is gift wrapping a box that is 8 in. by 12 in. by 3 in. How much gift wrapping paper is needed to cover the box?

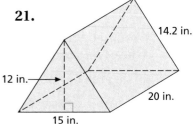

Find each surface area. Round to the nearest whole number.

10. rectangular prism: 4 m by 12 m by 16 m

11. cylinder: $r = 13$ ft, $h = 28$ ft

12. cylinder: $d = 0.8$ in., $h = 1.9$ in.

13. cube: $s = 18.7$ yd

14.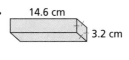
8 ft
3 ft

15.
12 cm
12 cm
12 cm

16.
14.6 cm
3.2 cm

17.
15 cm
8 cm
17 cm
4 cm

18.
12.2 m
8.3 m

19.
12 yd
8 yd
6 yd

20.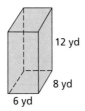
8 dm
6 dm

21.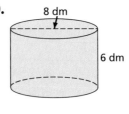
14.2 in.
12 in.
15 in.
20 in.

22. HEALTH A physician's assistant knows that the more surface area a sponge has, the greater number of bacteria can grow on it. Which size sponge would probably have the least amount of bacteria on its surface, 3 in. by 5 in. by 1 in. or 2 in. by 3 in. by 2 in.?

23. A swimming pool has the shape of a rectangular prism 85 ft long, 26 ft wide and 7.5 ft deep. Before the pool is filled with water, the bottom and sides of it must be painted white. If 1 gal of paint will cover about 125 ft², how many gallons of paint are needed?

24. WRITING MATH Describe a shortcut to find the surface area of a cube.

25. Dwayne is scrubbing the walls and ceiling of a rectangular room that is 12 ft by 14 ft by 8 ft. If it takes him 5 min to scrub an area of 24 ft², about how long will it take him to finish the job?

26. RECREATION A tent is 5 ft high. It has a rectangular floor that measures 5 ft by 6.3 ft. How much canvas was needed to make the tent?

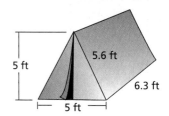

EXTENDED PRACTICE EXERCISES

27. DESIGN A graphic artist is designing a label for a can. The can is 5 in. tall and has a diameter of 2.5 in. What is the area of the paper needed for the label?

CONSTRUCTION A manufacturing plant is in the shape of a rectangular prism with a large cylindrical chimney on top of the roof. The rectangular base of the building is 120 ft by 84 ft, and the height is 35 ft. The chimney, which sits directly on top of the prism, is a cylinder with a diameter of 14 ft and a height of 29 ft.

28. Make a sketch of the building, including the measurements.

29. If a contractor is hired to paint the entire outside of the plant, how many square feet will he need to cover?

30. CHAPTER INVESTIGATION Make a net of your package. Include any tabs or flaps necessary for manufacturing. Find the surface area.

MIXED REVIEW EXERCISES

Name the property modeled by each equation. (Lesson 3-4)

31. $-3 + 3 = 3 + (-3)$

32. $14 \cdot 27 = 14(25) + 14(2)$

33. $(-19 \cdot 18) + (-19 \cdot 12) = -19 \times 30$

34. $47 + (28 + 13) = (28 + 13) + 47$

35. $56 + (-16 + (-12)) = (56 + (-16)) + (-12)$ **36.** $7 \cdot \dfrac{3}{4} = 7\left(\dfrac{1}{8}\right) + 7\left(\dfrac{5}{8}\right)$

Name the opposite of each integer. (Lesson 3-4)

37. -1 **38.** 6 **39.** -10 **40.** 0

Find the volume of each cylinder. (Lesson 4-7)

41. $r = 3$ in., $h = 0.75$ ft **42.** $d = 30$ m, $h = 80$ m

Chapter 4 Review

VOCABULARY ◣

Choose the word from the list that best completes each statement.

1. __?__ is the number of cubic units enclosed by an object.

2. A(n) __?__ is a two-dimensional pattern that, when folded, forms a three-dimensional figure.

3. The __?__ of a solid is the amount of square units of material it would take to cover it.

4. A polygon is __?__ if all its sides are of equal length and all its angles are of equal measure.

5. An angle is measured in __?__.

6. A(n) __?__ shows the top, front, and side views of an object without distorting the object's dimensions.

7. If two lines have no points in common, then they are __?__.

8. A(n) __?__ is a polyhedron with only one base.

9. A three-dimensional figure with two identical and parallel faces is called a(n) __?__.

10. If two lines intersect at right angles, then the lines are __?__.

a.	degrees
b.	net
c.	orthogonal drawing
d.	parallel
e.	perpendicular
f.	perspective drawing
g.	prism
h.	pyramid
i.	regular
j.	sphere
k.	surface area
l.	volume

LESSON 4-1 ◣ Language of Geometry, p. 156

▶ **Collinear points** lie on the same line. **Noncollinear points** do not lie on the same line.

▶ **Coplanar points** lie in the same plane. **Noncoplanar points** do not lie on the same plane.

Draw and label each geometric figure.

11. point *A*   12. ray *PQ*   13. angle *ATK*   14. line segment *GH*

15. Use a protractor to draw an angle whose measure is 125°.

LESSON 4-2 ◣ Polygons and Polyhedra, p. 160

▶ A **polygon** is a two-dimensional, closed figure. Polygons are classified by their number of sides.

▶ A polygon is **regular** if all of its sides are **congruent** and all of its angles are congruent.

▶ A **polyhedron** is a three-dimensional closed figure. Each **face** of the polyhedron is a polygon.

Draw the following.

16. pentagon that is not regular   17. parallelogram that is regular

18. isosceles obtuse triangle

Identify the number of faces, vertices, and edges of each figure.

19.

20.

21.

LESSON 4-3 ◣ Visualize and Name Solids, p. 166

▶ A **prism** is a three-dimensional figure with two identical parallel faces, called **bases**. A rectangular prism with edges of equal length is a **cube**.

▶ A **pyramid** is a polyhedron with only one base. The other faces are triangles.

▶ A **cylinder**, **cone**, and **sphere** are three-dimensional figures with curved surfaces.

Identify each three-dimensional figure.

22. the figures in Exercises 19–21

23. the figure that is formed by the net at the right

24. a figure with a triangle as its base and rectangles as the other faces

25. a figure with three faces as triangles and one triangular base

26. a figure with two circular bases and a curved side

LESSON 4-4 ◣ Problem Solving Skills: Nets, p. 170

▶ When solving problems about three-dimensional shapes draw a picture, diagram, or model.

27. Make a scale drawing of the net of a lid for a box that has a triangular base with lengths of 12 cm. The lid's depth needs to be 4 cm. Use a scale factor of 1 cm : 4 cm.

28. Jaya is to build four cylindrical oil drums out of cardboard for the set of her school play. What types of shapes will she need to build the drums? How many of each shape will she require?

LESSON 4-5 ◣ Isometric Drawings, p. 174

▶ Two lines are **parallel** if they have no points in common.

▶ Two lines are **perpendicular** if they intersect at right angles.

▶ **Isometric drawings** provide three-dimensional views of an object.

Use the figure at the right to name the following.

29. all of the parallel lines

30. all of the perpendicular lines

31. all points of intersection

32. On isometric dot paper, draw a rectangular prism whose dimensions are 1 unit by 2 units by 4 units.

33. How many cubes are in the isometric drawing you drew in Exercise 31?

34. How many cubes are in the isometric drawing at the right?

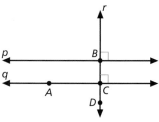

LESSON 4-6 ◼ Perspective and Orthogonal Drawings, p. 178

▶ A **perspective drawing** makes a three-dimensional figure appear true-to-life. A **one-point perspective** drawing uses a vanishing point to create depth.

▶ An **orthogonal drawing** shows undistorted top, front, and side views of an object.

35. Draw a rectangular prism in one-point perspective.

36. Locate and label the vanishing point and horizon line of the figure shown.

37. Make an orthogonal drawing of the figure shown.

38. Make an orthogonal drawing of the figure in Exercise 36.

LESSON 4-7 ◼ Volume of Prisms and Cylinders, p. 184

▶ **Volume of a Prism:** $V = B \cdot h$; where B is the area of the base and h is the height.

▶ **Volume of a Cylinder:** $V = \pi r^2 h$; where r is the radius and h is the height.

Find the volume of the following. Use $\pi \approx 3.14$. Round to nearest tenth, if necessary.

39. rectangular prism: 15 ft by 8 ft by 4 ft

40. cylinder: $r = 5$ cm, $h = 16$ cm

41. cube: $s = 3.4$ yd

42. cylinder: $d = 2.5$ m, $h = 3.6$ m

LESSON 4-8 ◼ Volume of Pyramids and Cones, p. 188

▶ **Volume of a Pyramid**: $V = \frac{1}{3} \cdot B \cdot h$; where B is the area of the base and h is the height.

▶ **Volume of a Cone**: $V = \frac{1}{3}\pi r^2 \cdot h$; where r is the radius and h is the height.

Find the volume of the following. Use $\pi \approx 3.14$. Round to nearest tenth, if necessary.

43. rectangular pyramid: base is 6 in. by 7 in., $h = 5$ in.

44. cone: $r = 2.2$ cm, $h = 4.1$ cm

45. hexagonal pyramid: $B = 9.8$ m^2, $h = 10$ m

46. cone: $d = 3$ ft, $h = 9$ ft

LESSON 4-9 ◼ Surface Area of Prisms and Cylinders, p. 194

▶ The **Surface Area of a Prism** is the sum of the areas of its faces.

▶ **Surface Area of a Cylinder:** $SA = 2\pi rh + 2\pi r^2$; where r is the radius and h is the height.

Find the surface area of each figure. Use $\pi \approx 3.14$. Round to the nearest whole number.

47. cylinder: $r = 20$ cm, $h = 34$ cm

48. cube: $s = 8.1$ m

49. rectangular prism: 2 yd by 7 yd by 5 yd

50. cylinder: $d = 12$ ft, $h = 12$ ft

CHAPTER INVESTIGATION

EXTENSION Present the package you designed to the class. Describe why you selected the design over other packaging options. Explain why your package uses space efficiently, and why it is an effective marketing tool.

Chapter 4 Assessment

Draw and label each geometric figure.

1. ray *AB*

2. line *MN*

3. angle *MJS*

4. line segment *XY*

5. point *H*

6. plane *QRS*

7. planes 𝒜 and ℬ intersect at line *c*

8. coplanar points *R*, *S*, and *T*

9. Use a protractor to draw an angle whose measure is 50°.

Draw the following.

10. scalene triangle

11. regular quadrilateral

12. irregular pentagon

Find the number of faces, vertices, and edges for each figure.

13.

14.

15.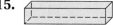

16. Identify the figures in Exercises 13–15.

17. Draw a net for the figure in Exercise 15.

18. How many cubes are in the isometric drawing at the right?

Name one everyday object that has the given shape.

19. cone

20. sphere

21. cylinder

22. Draw a cube in one-point perspective.

23. Locate and label the vanishing point and horizon line of the figure shown.

24. Make an orthogonal drawing of the figure shown.

Find the volume of the following. If necessary round to the nearest tenth.

25. triangular prism: $h = 2$ cm, $B = 36$ cm^2

26. rectangular prism: 3 m by 7 m by 8 m

27. cylinder: $r = 5$ ft, $h = 6$ ft

28. cylinder: $d = 4.6$ cm, $h = 5.2$ cm

29. rectangular pyramid: base is 3 in. by 8 in., $h = 2$ in.

30. cone: $r = 4.2$ cm, $h = 6.6$ cm

Find the surface area of each figure. Use $\pi \approx 3.14$. Round to the nearest whole number.

31.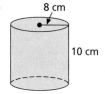
8 cm
10 cm

32.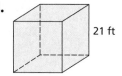
21 ft

33.
11 yd   20 yd   3 yd

Standardized Test Practice

Part 1 Multiple Choice

**Record your answers on the answer sheet
provided by your teacher or on a sheet of paper.**

1. What is the mean weekly salary of someone
   who earns $45,000 per year? (Lesson 1-2)
   - (A) $3750.00
   - (B) $1875.11
   - (C) $937.50
   - (D) $865.38

2. The graph shows the runs scored during
   24 games. In how many games did they
   score 3 or more runs? (Lesson 1-4)

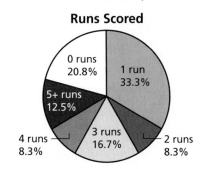

Runs Scored

0 runs 20.8%
1 run 33.3%
5+ runs 12.5%
4 runs 8.3%
3 runs 16.7%
2 runs 8.3%

   - (A) 4
   - (B) 5
   - (C) 9
   - (D) 10

3. Subtract 2 gal 3 qt from 6 gal 2 qt. (Lesson 2-2)
   - (A) 4 gal 1 qt
   - (B) 3 gal 3 qt
   - (C) 3 gal 2 qt
   - (D) 3 gal 1 qt

4. Find the perimeter of the figure. (Lesson 2-3)

15.6 m   21.6 m
11.2 m
8 m   19.6 m
23.7 m

   - (A) 925.0 m
   - (B) 121.3 m
   - (C) 99.7 m
   - (D) 78.1 m

5. Find the area of a circle with a radius of
   2.6 cm. Use $\pi \approx 3.14$. Round to the nearest
   tenth. (Lesson 2-7)
   - (A) 21.2 cm^2
   - (B) 16.3 cm^2
   - (C) 5.3 cm^2
   - (D) 5.2 cm^2

6. Which property describes the statement
   below? (Lesson 3-4)
   $$x(y - z) = xy - xz$$
   - (A) associative property
   - (B) commutative property
   - (C) distributive property
   - (D) identity property

7. What is the best name of a triangle with two
   sides congruent and two angles congruent?
   (Lesson 4-2)
   - (A) scalene, right
   - (B) isosceles
   - (C) right
   - (D) equilateral, equiangular

8. Name the figure. (Lesson 4-3)
   - (A) hexagonal pyramid
   - (B) pentagonal prism
   - (C) triangular prism
   - (D) triangular pyramid

9. Find the volume of a cylinder with a radius of
   1.5 in. and a height of 4 in. Use $\pi \approx 3.14$.
   Round to the nearest tenth. (Lesson 4-7)
   - (A) 7.1 in.^3
   - (B) 11.3 in.^3
   - (C) 18.8 in.^3
   - (D) 28.3 in.^3

10. A box without a top has length 16 in., width
    9 in., and height 13 in. Find its surface area.
    (Lesson 4-9)
    - (A) 730 in.^2
    - (B) 794 in.^2
    - (C) 821 in.^2
    - (D) 938 in.^2

Test-Taking Tip

Exercises 4, 5, 9, and 10
To prepare for a standardized test, review the definitions of
key mathematical terms like *perimeter*, *area*, *volume*, and
surface area.

Preparing for the Standardized Tests
For test-taking strategies and more
practice, see pages 587-604.

Part 2 | Short Response/Grid In

Record your answers on the answer sheet provided by your teacher or on a sheet of paper.

11. The table shows the results of a fund-raiser. How much more money did Room C raise than Room A? (Lesson 1-5)

Room	Amount
A	$121.50
B	$189.23
C	$192.37
D	$188.70

12. The scale of a drawing is 1 in. : 3 ft. A room measures $4\frac{1}{4}$ in. wide. What is the width of the actual room? (Lesson 2-8)

13. Find the area of the figure. Use $\pi \approx 3.14$. Round to the nearest tenth. (Lesson 2-9)

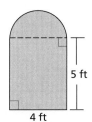

5 ft

4 ft

14. Simplify. Use the order of operations. (Lesson 3-3)

$$\frac{2}{9} + \frac{1}{3} \cdot \frac{5}{6}$$

15. The area of a square is 125 m². What is the length of each side? Round to the nearest tenth. (Lesson 3-9)

16. Identify the number of faces, vertices, and edges of a hexagonal pyramid. (Lesson 4-2)

17. Draw $\overleftrightarrow{FG} \perp \overleftrightarrow{MN}$. (Lesson 4-5)

18. Draw an orthogonal view of the figure below. (Lesson 4-6)

19. The weight of water is 0.029 lb times the volume of water in cubic inches. How many pounds of water would fit into a rectangular child's pool that is 12 in. deep, 3 ft wide, and 4 ft long? (Lesson 4-7)

20. The shape of a fertilizer spreader bin is an upside-down pyramid with a square base measuring 1 ft 4 in. per side. If the bin is 10 in. deep, how many square inches of fertilizer can the spreader hold? Round to the nearest tenth. (Lesson 4-8)

21. A stalactite in Endless Caverns in Virginia is shaped like a cone. It is 4 ft tall and has a diameter at the roof of $1\frac{1}{2}$ ft. Find the volume of the stalactite. Round to the nearest tenth. (Lesson 4-9)

Part 3 | Extended Response

Record your answers on a sheet of paper. Show your work.

22. A 6 in.-by-8 in. rectangle is cut in half, and one half is discarded. The remaining rectangle is again cut in half, and one half is discarded. (Lesson 3-6)

  a. What are the dimensions of the remaining rectangle?

  b. How many more times should the rectangle be cut in half so that the final rectangle has an area less than 1 in.²? What is the area?

23. A flat-screen computer monitor that is 35 cm wide, 10 cm deep, and 40 cm tall needs to be shipped to its customers in a box. (Lesson 4-9)

  a. If 9 cm of foam padding is required to cushion each side of the box, what are the necessary dimensions of the box that will ship the monitor?

  b. How much cardboard is necessary for each box that will ship a monitor?

  c. If cardboard costs $1.68/m², how much will it cost to package 23 monitors?

Equations and Inequalities

THEME: Recycling

Often problems are stated in words. If you can state a problem as an equation or inequality, you can usually find a solution. The problem can be as simple as making the correct change or as complex as engineering a new recyclable plastic.

Recycling reduces the amount of waste produced by humans and decreases the use of natural resources. At the end of the 20th century, the depletion of natural resources became a pressing concern. Statistics on recycling and the environment are compiled by using equations and inequalities.

- **Solid waste disposal staff** (page 217) maintain curbside recycling services by calculating expenses, monthly charges, and the number of households served.

- **Financial analysts** (page 237) manage budgets for many different kinds of businesses. In waste-collection facilities, the financial analyst finds the most cost-effective way to sell recycled materials.

Math Online

mathmatters1.com/chapter_theme

Estimated Waste Generated in the U.S.

Year	1960	1970	1980	1990	1995	2001
Garbage generated (millions of tons)	88	121	143	196	208	229
Resources recycled (percent of total)	6.6	6.6	9.3	17.0	27.0	30.0

Data Activity: Estimated Waste Generated in the U.S.

A formula to calculate the percent of increase or decrease is $p = 100 \frac{(n - q)}{q}$, where p is the percent of increase or decrease, n is the new or most recent number and q is the original number.

Use the formula and the table for Questions 1–3.

1. Calculate the percent of increase for garbage generated for the following time periods.
   a. 1960 to 1970   **b.** 1970 to 1980   **c.** 1980 to 1985
   d. 1985 to 1990   **e.** 1990 to 1995   **f.** 1995 to 2001

2. Calculate the percent of increase for resources recycled for the following time periods.
   a. 1960 to 1970   **b.** 1970 to 1980   **c.** 1980 to 1985
   d. 1985 to 1990   **e.** 1990 to 1995   **f.** 1995 to 2001

3. Does the percent of increase for garbage generated match the percent of increase of resources recycled? Overall, which has a larger percent of increase? How can you explain the difference?

CHAPTER INVESTIGATION

Municipal solid waste (MSW) consists of such items as paper, paperboard, metal, glass, plastic, rubber, leather, textile, wood and more. The Environmental Protection Agency (EPA) uses a standard equation to calculate MSW. The formula is

$$\text{MSW recycling rate (percent)} = 100 \cdot \frac{\text{total MSW recycled}}{\text{total MSW generated}}$$

Working Together

Use the standard equation for MSW and the above table to calculate the MSW recycled. Contact a local recycling center to find the recycling rate in the area you live. Use the Chapter Investigation icons throughout the chapter to guide your group.

5 Are You Ready?

Refresh Your Math Skills for Chapter 5

The skills on these two pages are ones you have already learned. Use the examples to refresh your memory and complete the exercises. For additional practice on these and more prerequisite skills, see pages 536–544.

DISTRIBUTIVE PROPERTY

You can use the distributive property to simplify an expression. Multiply the numbers inside the parentheses individually by the factor outside the parentheses.

Examples

$5x - 6x = x(5 - 6)$
$= -1x$
$= -x$

$9y + 4y = y(9 + 4)$
$= 13y$

Simplify each expression using the distributive property.

1. $14c + 7c$

2. $-16g + 7g$

3. $10t - 24t$

4. $-5m - 8m$

5. $6d + 18d$

6. $2s + 7s - 9s$

OTHER PROPERTIES

Understanding how to use the following properties will help you solve equations.

Examples

Addition Property of Zero	$x + 0 = 0$
Multiplication Property of One	$x \cdot 1 = x$
Commutative Property	$x + 4 = 4 + x$ $4 \cdot x = x \cdot 4$
Associative Property	$(x + 7) + 4 = x + (7 + 4) = x + 11$
	$3 \cdot (6 \cdot x) = (3 \cdot 6) \cdot x = 18 \cdot x = 18x$

Use the properties to simplify each expression. Then name the property used.

7. $418 + (160 - 160)$

8. $245 \cdot (25 \div (5 \cdot 5))$

9. $t \cdot \dfrac{5}{5}$

10. $(m + (-10)) + 14$

11. $9 \cdot (x \cdot (-4))$

12. $-14 + w + 10$

GREATER THAN AND LESS THAN

Inequalities are similar to equations. Instead of an equal sign, inequalities use greater than or less than signs. Knowing how to use these symbols will help you solve inequalities.

Compare. Write >, < or =.

13. $15 \blacksquare -15$

14. $8.4 \blacksquare 4.8$

15. $0.7 \blacksquare 1.9$

16. $\dfrac{1}{4} \blacksquare \dfrac{1}{3}$

17. $-0.75 \blacksquare -0.36$

18. $-\dfrac{1}{5} \blacksquare -\dfrac{1}{8}$

19. $2.5 \blacksquare 2.50$

20. $-\dfrac{5}{6} \blacksquare 0$

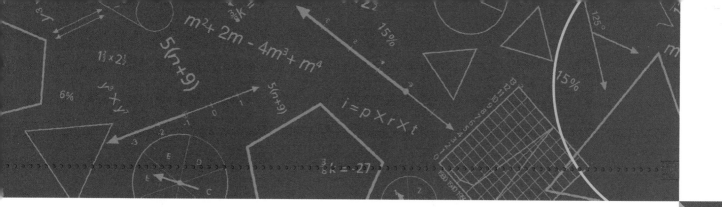

RATIONAL NUMBER OPERATIONS

Whole numbers, integers, fractions, and decimals are rational numbers. Performing basic operations with rational numbers is essential to solving many types of problems.

Examples

Addition	$4 + 7 = 11$	$\dfrac{1}{4} + \dfrac{1}{3} = \dfrac{3}{12} + \dfrac{4}{12} = \dfrac{7}{12}$
	$-6 + (-5) = -11$	$4.5 + (-1.7) = 2.8$
Subtraction	$17 - 8 = 9$	$-\dfrac{2}{3} - \dfrac{1}{8} = -\dfrac{16}{24} - \dfrac{3}{24} = -\dfrac{19}{24}$
	$14 - (-7) = 21$	$4.1 - 6.7 = -2.6$
Multiplication	$4 \cdot 11 = 44$	$\dfrac{1}{4} \cdot \left(-\dfrac{1}{2}\right) = -\dfrac{1}{8}$
	$-3 \cdot (-5) = 15$	$1.5 \cdot 4.3 = 6.45$
Division	$48 \div 4 = 12$	$\dfrac{3}{8} \div \left(-\dfrac{2}{5}\right) = \dfrac{3}{8} \cdot \left(-\dfrac{5}{2}\right) = -\dfrac{15}{16}$
	$-16 \div 8 = -2$	$-4.8 \div (-0.2) = 24$

Simplify.

21. $125 + 117$ **22.** $236 - 119$ **23.** $311 \cdot 5$ **24.** $116 \div 4$

25. $-110 + 17$ **26.** $-128 - 145$ **27.** $-19 \cdot 8$ **28.** $-225 \div (-5)$

29. $\dfrac{2}{5} + \dfrac{1}{3}$ **30.** $\dfrac{5}{8} - \left(-\dfrac{1}{2}\right)$ **31.** $\dfrac{7}{8} \cdot \dfrac{8}{9}$ **32.** $2\dfrac{1}{6} \div \left(-\dfrac{1}{3}\right)$

33. $4.18 + (-3.75)$ **34.** $7.9 - 10.67$ **35.** $2.5 \cdot (-3.15)$ **36.** $1.44 \div 8$

ORDER OF OPERATIONS

When an expression involves more than one operation, you must follow the order of operations.

Example   You can use the order of operations to simplify.

Step 1	Perform operations within parentheses.	$4^2 + (6 - 12) - 4 \cdot 5$
Step 2	Simplify exponents.	$4^2 + \quad (-6) \quad - 4 \cdot 5$
Step 3	Multiply and divide from left to right.	$16 + \quad (-6) \quad - 4 \cdot 5$
Step 4	Add and subtract from left to right.	$16 + \quad (-6) \quad - 20$
		$10 \qquad\qquad -20$
		-10

Simplify each expression using the order of operations.

37. $17 + 7(24 - 16) + 5^3$ **38.** $28 \div (9 + (-2)) \cdot 12 - 4$ **39.** $4 - (-16) \div -4^2 + 11$

40. $26 - 3 + 27 \div (-3)$ **41.** $50 \div 5^2 + 6(14 + (-10))$ **42.** $6 - 2^6 \div (8 \cdot 4) + (-12)$

Introduction to Equations

Goals ■ Understand equations and find their solutions.

Applications Recycling, Finance, Travel

Work in groups of two or three students.

Tug-of-war teams are chosen so that the total weight of each side is equal. Students and their weights are shown. Pat is unsure of his exact weight, but he knows his weight's range.

Student	Weight (pounds)	Student	Weight (pounds)
Mario	135	Mike	140
Julie	120	Ed	115
Gwen	125	Jill	125
Pao	145	Deshawn	140
Heather	105	Pat	120-135

1. How would you make up the teams so that each side has the same total weight?

2. What weight did your group assign to Pat?

3. Check with other groups. Did they make the same teams?

◤ BUILD UNDERSTANDING

An **equation** is a statement that two numbers or expressions are equal. Refer to the table above of students and their weights.

If a team of Mario and Julie compete against a team of Deshawn and Ed, the weight on each side is equal.

$$135 + 120 = 140 + 115$$

$$255 = 255 \quad \text{This equation is true.}$$

If a team of Mario and Gwen compete against a team of Pao and Julie, the weight on each side is not equal.

$$135 + 125 = 145 + 120$$

$$260 \neq 265 \quad \text{This equation is false.}$$

A team of Mike and Heather may or may not want to compete against a team of Pat and Jill. It depends upon Pat's weight.

$$140 + 105 = x + 125 \quad \text{This is an open sentence.}$$

An **open sentence** is a sentence that contains one or more variables. It can be true or false, depending on the values substituted for the variables. A value of the variable that makes an equation true is called a **solution of the equation**.

Open sentence: $140 + 105 = x + 125$
Solution: $x = 120$

Math: Who, Where, When

Around A.D. 830, the Arab mathematician and teacher Al-Khowarizmi wrote a book describing methods of solving equations. The word *algebra* is derived from the word *al-jabr*, which appeared in the title of his book.

Example 1

Tell whether the equation is *true, false,* or an *open sentence.*

a. $3(8 - 2) = 20$ **b.** $2y - 6 = 10$ **c.** $-9 + 5 = 2(7 - 9)$

Solution

a. $3(8 - 2) = 20$

$3(6) = 20$

$18 \neq 20$

The equation is *false.*

b. $2y - 6 = 10$

The equation contains a variable, so it is an *open sentence.*

c. $-9 + 5 = 2(7 - 9)$

$-4 = 2(-2)$

$-4 = -4$

The equation is *true.*

Example 2

Which value, 1 or 2, is a solution of the equation $2x + 5 = 9$?

Solution

$2x + 5 = 9$

$2(1) + 5 \overset{?}{=} 9$ Substitute 1 for x.

$2 + 5 \overset{?}{=} 9$ Use order of operations.

$7 \neq 9$

So 1 is not a solution.

$2x + 5 = 9$

$2(2) + 5 \overset{?}{=} 9$ Substitute 2 for x.

$4 + 5 \overset{?}{=} 9$

$9 = 9$

So 2 is a solution.

Example 3

SPORTS Together, Lisa and Maria scored 48 points in a basketball game. Maria scored 27 points. Lisa scored p points. Find the number of points Lisa scored. Use the equation $p + 27 = 48$, and try the values 19 and 21 for p.

Solution

Let $p = 19$.

$p + 27 = 48$

$19 + 27 \overset{?}{=} 48$

$46 \neq 48$

Let $p = 21$.

$p + 27 = 48$

$21 + 27 \overset{?}{=} 48$

$48 = 48$

Lisa scored 21 points in the basketball game.

> **Check Understanding**
>
> Write a true equation.
>
> Write a false equation.
>
> Write an open sentence.

Tell whether each equation is *true*, *false*, or an *open sentence*.

1. $-3(7-4) = 6$

2. $n - 4 = 4$

3. $-6 + 3 = -2(3) + 3$

4. $\frac{1}{2}(2-4) = 3$

5. $\frac{x}{3}(3) = \frac{9}{4}$

6. $12 - (4 \div 2) = 10$

Which of the given values is a solution of the given equation?

7. $3 - x = -1$; 4, -4

8. $2m - 1 = 9$; 3, 5

9. $\frac{x}{4} = 1$; 4, -4

10. $-4m + 9 = 1$; 2, -1

11. $3\left(\frac{y}{15}\right) = 1$; 5, -5

12. $x(12 \div 4) = -9$; 2, -3

13. A midsize car can be driven 34 mi on 1 gal of gas. This is 10 mi more than a luxury car. Use the equation $x + 10 = 34$ to find how many miles a luxury car can be driven on 1 gal of gas.

14. **TRAVEL** Toya drives 75 mi, which is half the distance from Madison, WI to Chicago, IL. Use the equation $\frac{x}{2} = 75$ to find if the total trip is 160 mi or 150 mi.

15. **WRITING MATH** Use an example to explain what it means to solve an equation.

Tell whether each equation is *true*, *false*, or an *open sentence*.

16. $8 = 9 - y$

17. $-11 = -15 - 4$

18. $24 \div (-6) = -4$

19. $4(2 + 5) = 30 - (4 - 6)$

20. $\frac{-3 + 11}{2} = 5 - (4 - 3)$

21. $8 + (7 - 5) = 3 - x + 2$

Which of the given values is a solution of the given equation?

22. $b - 5 = -1$; 4, 6

23. $9 + c = 9$; 0, 1

24. $d + 7 = -1$; -6, -8

25. $1 - e = 4$; -3, -5

26. $-6 = 6 + h$; 0, -12

27. $8 = f - 4$; 4, 12

28. $2k = 2$; 1, -1

29. $8m - 2 = -10$; -8, -1

30. $17 = 2n$; $8\frac{1}{2}$, $9\frac{1}{2}$

31. $19 = 10 - 2p$; 4.5, -4.5

32. $\frac{p}{4} = 12$; 8, 48

33. $7 + k = 9$; 0, 2

34. $2x - 2 = 0$; 1, -1

35. $3c + 3 = 2c + 7$; 2, 4

36. $\frac{d+1}{8} = \frac{1}{2}$; 1, 3

37. $\frac{5-x}{3} = 2$; 0, -1

38. $\frac{y-8}{5} = -2$; 2, -2

39. $\frac{9+y}{2y} = 2$; 2, 3

Use mental math to solve each equation.

40. $x + 5 = 7$

41. $n - 4 = 3$

42. $2 = 10 - k$

43. $3 + h = 0$

44. $-4m = 20$

45. $-3 = 18 \div m$

46. $-3y = -33$

47. $n - 12 = 12$

48. $-21 \div t = -3$

49. **FINANCE** Pete's earnings equal the sum of Doug and Sam's earnings. Pete earns \$82 and Sam earns \$39. Use the equation $39 + x = 82$, where x represents the amount Doug earns, and try these values for x: 37, 43, 45. Determine the amount Doug earns.

50. A stack of nickels is worth \$1.35. Use the equation $0.05n = 1.35$ and these values for n: 26, 27, 28. Find n, the number of nickels in the stack.

51. YOU MAKE THE CALL Tia uses mental math to solve the equation $\frac{x}{3} = -12$. She said $x = 36$. Is her answer correct? Explain.

52. A radio is on sale for $112, which is $\frac{2}{3}$ of its regular price. Use the equation $\frac{2}{3}r = 112$ and these values for r: 130, 165, 168. Find r, the regular price of the radio.

RECYCLING In 1973, Nathaniel Wyeth patented the recycled plastic, PET (polyethylene terephthatlate). Seventy-eight years earlier New York City started the first residential waste program in the United States.

53. Let y represent the year New York City started its residential waste program. Write an equation to represent this situation.

54. Use these values for y: 1890, 1895, 1900 to determine the year New York City started its residential waste program.

▰ EXTENDED PRACTICE EXERCISES

55. A certain rectangle has a length of 12 cm and a perimeter of 36 cm. Use the formula $P = 2l + 2w$, where l represents the length and w represents the width. Use these values for w: 5, 6, 7. Determine the width of the rectangle.

WRITING MATH Write an equation that has each of the following solutions.

56. solution: 4   **57.** solution: -1 **58.** solution: 0

Use mental math to find two solutions of the equation.

59. $x^2 = 25$ **60.** $y^2 = 49$ **61.** $h^2 - 4 = 5$

62. CRITICAL THINKING Verify that 4 is a solution of the equation $-2 + k^2 = 14$. Can you think of any other values of k that are solutions? Explain.

▰ MIXED REVIEW EXERCISES

Use the frequency table shown. (Lesson 1-5)

63. How many more people chose sales manager than artist as a career choice?

64. As many people chose newspaper columnist as chose which two possible pairs of career choices together?

65. How many fewer people chose engineer than accountant as a career choice?

Career choices	Tally	Frequency
Sales manager	ⵏⵏ ⵏⵏ |||	13
Politician	|||	3
Doctor	||	2
Engineer	|||	3
Accountant	||||	4
Newspaper columnist	ⵏⵏ |||	8
Artist	ⵏⵏ	5

66. What are the outliers for this data set? (Lesson 1-3)

67. Where does the data cluster? (Lesson 1-3)

68. Add: $1.2 + \frac{1}{2} + \frac{2}{5} - \frac{3}{10}$. (Basic Skills)

5-2 Add or Subtract to Solve Equations

Goals ■ Use addition and subtraction to solve equations.

Applications Retail, Sports, Hobbies

Work in a small group using Algeblocks.

To solve an equation with Algeblocks, use a Sentence Mat. A Sentence Mat is two Basic Mats with an equal symbol between them. Follow the steps to solve the following equation.

$$x - 2 = 5$$

1. Sketch or show the equation on the Sentence Mat.

2. To isolate the x-block, create zero pairs by adding the opposite of -2, or 2, to each side.

3. Simplify each side of the equation to find the solution.

4. Read the solution, $x = 7$, from the mat.

5. How would you solve the equation $x + 2 = 5$ using Algeblocks? Explain.

Sentence Mat

$x - 2 \quad = \quad 5$

$x - 2 + 2 \quad = \quad 5 + 2$

◣ BUILD UNDERSTANDING

When you **solve an equation**, you find all the values of the variable that make the equation true.

Use the inverse operation to undo each operation performed on the variable. The reason for performing the inverse operation is to isolate the variable on one side. In that way, what is unknown (the variable) can be expressed by numbers that are known. Recall that the inverse operation of addition is subtraction. The inverse operation of subtraction is addition.

Keep the equation in balance by performing the same operations on both sides. It helps to think of the equation as a balance scale. Any term added or subtracted from one side must be added or subtracted to the other. Continue until the variable is alone on one side of the equation.

Check your solution by substituting the value for the variable in the original equation.

Example 1

Solve each equation. Check the solution.

a. $x + 7 = \dfrac{11}{2}$

b. $-3 = n - 5$

Solution

a.

$x + 7 = \dfrac{11}{2}$	Undo addition with subtraction.	
$x + 7 - 7 = \dfrac{11}{2} - 7$	Keep the equation in balance.	
$x + 0 = \dfrac{11}{2} - \dfrac{14}{2}$	Write with common denominator.	
$x = \dfrac{-3}{2} = -1\dfrac{1}{2}$	The solution is correct.	

Check

$$x + 7 = \dfrac{11}{2}$$
$$-1\dfrac{1}{2} + 7 \stackrel{?}{=} \dfrac{11}{2}$$
$$5\dfrac{1}{2} = \dfrac{11}{2} \quad \checkmark$$

b.

$-3 = n - 5$	Undo addition with subtraction.	
$-3 + 5 = n - 5 + 5$	Keep the equation in balance.	
$2 = n + 0$		
$2 = n$	The solution is correct.	

Check

$$-3 = n - 5$$
$$-3 \stackrel{?}{=} 2 - 5$$
$$-3 = -3 \quad \checkmark$$

Example 2

a. What equation is modeled on the Sentence Mat?

b. Describe how to solve this equation using Algeblocks.

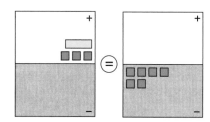

Solution

a. $x + 3 = -6$

b. Create a zero pair to get the x-block by itself. Do this by adding the opposite of 3, or -3, to both sides of the Sentence Mat. Simplify each side. Read the solution from the Sentence Mat.
$x = -9$

Check Understanding

Zero pairs are opposites whose sum is zero. Which of the following pairs of numbers are zero pairs?

4 and 4

-4 and 4

Example 3

RETAIL After Eriq spent $145.29 to buy in-line skates, he had $23.21 left. How much money did he have before he bought the skates?

Solution

Let x represent the amount that Eriq had originally. Subtract $145.29 from x. This equals $23.21, the amount remaining.

$$x - 145.29 = 23.21$$
$$x - 145.29 + 145.29 = 23.21 + 145.29$$
$$x = 168.50$$

Eriq had $168.50 before he bought the in-line skates.

Solve each equation. Check the solution.

1. $y + 8 = -3$
2. $h - 6 = -1$
3. $x - 7 = 2$
4. $4 + t = -5$
5. $2 + y = -8$
6. $k - 5 = 2$
7. $4 = h - 6$
8. $5 - z = -3$
9. $-15 + a = 19$

10. **WRITING MATH** Write an equation involving addition that has 4 as its solution. Explain how to solve your equation.

11. The sum of the measures of two angles is 180°. One of the angles measures 64°. What is the measure of the other angle?

Solve each problem by writing an equation.

12. **RECYCLING** Serena and Asha collect extra recycle bins so that they can put one in each classroom at school. They collect 14 bins. There are 48 classrooms. How many more recycle bins do they need?

13. **HOBBIES** Sandra gives her brother 16 baseball cards, leaving her with 42 cards. How many cards did she have originally?

14. **MODELING** What equation is modeled on the mat?

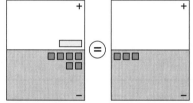

Solve each equation. Check the solution.

15. $c - 6 = -3$
16. $d + 9 = -2$
17. $4 = a - 4$
18. $3 + f = -5$
19. $g - 1.2 = 2.8$
20. $n - 5 = 6$
21. $y + 9 = -3$
22. $2.7 = m - 0.4$
23. $0 = k + 7$
24. $\frac{3}{4} + m = -1$
25. $-4.5 = n - 2.7$
26. $-6 = y - 6$
27. $-8 + k = 3$
28. $h - 1 = -1$
29. $-7 = d + 0$
30. $x + 25 = -50$
31. $-15 + r = 15$
32. $-1.3 + a = 4.8$
33. $\frac{1}{2} + y = -\frac{3}{4}$
34. $x - 0.9 = 10.8$
35. $t + \frac{3}{8} = 2\frac{3}{4}$
36. $p - \frac{3}{4} = \frac{3}{8}$
37. $-4.6 = 35 + c$
38. $\frac{4}{5} + b = \frac{7}{10}$

Solve each problem by writing an equation.

39. Eugene rides his bike for $3\frac{1}{2}$ h on Saturday. If he rides $1\frac{1}{4}$ h in the morning, how long did he ride the rest of the day?

40. Janine reads 230 pages of a novel. She read 39 more pages than Harrison has read. How many pages has Harrison read?

41. **SPORTS** This year Jay's best height in the high jump was 6′ 3″. This jump is 6″ more than his best jump last year. What was the height of Jay's best jump last year?

42. **RETAIL** A sweater is on sale for $47.10. The sale price is $16 lower than the regular price. Find the regular price.

43. **CALCULATOR** Describe how you might use a calculator to solve Exercise 13.

44. Supplementary angles are two angles whose sum is 180°. The figure represents supplementary angles. Write and solve an equation to determine the measure of the missing angle.

117° / x

DATA FILE For Exercises 45 and 46, use the U.S. roller coaster data on page 521. A roller coaster designer wants to design a roller coaster that is similar to the Millenium Force. However, this new roller coaster will be 250 ft taller and 10 mi/h faster.

45. Write and solve an equation for the height of the new roller coaster.

46. Write and solve an equation for the speed of the new roller coaster.

47. WRITING MATH Write an equation involving subtraction that has −6 as its solution. Explain how to solve your equation.

Write an equation for each phrase.

48. y added to three equals eleven.

49. n increased by five is twelve.

50. Fifteen decreased by x is eight.

51. Six subtracted from s is one.

52. Negative three added to h is six.

53. k less than two equals three.

EXTENDED PRACTICE EXERCISES

Solve each equation. Check the solution.

54. $x - 8 = -2(4 - 5)$

55. $2 + 3(4) = k + 5$

56. $\frac{8(9)}{3(2)} = n + 17$

57. $6 + p + 4 = -2 + 8$

58. $\frac{1}{2}(12 - 8) = x + 5$

59. $x + 2.5 + 4.7 = 8.1 - 5.3$

60. CRITICAL THINKING During an 11-h period, Evelina spent $1\frac{3}{4}$ h at track practice, $\frac{1}{2}$ h at lunch, $2\frac{1}{2}$ h doing homework, 1 h on the phone, and the remaining time in class. Write and solve an equation to find the amount of time she spent in class.

61. MODELING Use Algeblocks to model and solve the equation $x - 4 = -2$.

62. STATISTICS Recall from Chapter 1 that the range of a set of numbers is the difference between the greatest number and the smallest number. The range of a certain set of numbers is 64. The smallest number of the set is 23. Write and solve an equation to determine the highest number of the set.

MIXED REVIEW EXERCISES

Simplify. Round to the nearest hundredth when necessary. (Lesson 3-9)

63. $\sqrt{114}$ **64.** $\sqrt{136}$ **65.** $\sqrt{125}$ **66.** $\sqrt{200}$ **67.** $\sqrt{576}$

68. Cameron hiked a mountain path. He started at an elevation of 1575 ft. In the first hour, his elevation increased by 50 ft. The next hour he hiked down 15 ft and back up 28 ft. After breaking for lunch, he hiked for another 2 hours which increased his elevation another 87 ft. At what elevation was Cameron at the end of his hike? (Lesson 3-1)

Review and Practice Your Skills

Tell whether each equation is *true*, *false*, or an *open sentence*.

1. $-14 = -4 - (-10)$ **2.** $-8 = 2 + m$ **3.** $4 \cdot 8 = 2^3$

4. $2 - c + 1 = -7$ **5.** $24 \div (8 - 5) = 6 - 3$ **6.** $8 - 13 = -1 \cdot 5$

7. $-5 + 5 \cdot 3 = 8 - (-2)$ **8.** $6 - (-2w) = 3 + w$ **9.** $18 - 3 \cdot 4 = (9 - 7)^3$

Which of the given values is a solution of the given equation?

10. $y + 3 = -1; -4, -2, 2$ **11.** $-5b = 5; -25, -1, 1$ **12.** $2 = \dfrac{x}{8}; 2, 4, 16$

13. $\dfrac{d}{3} = -6; -18, -3, -2$ **14.** $-1 = t - 9; -10, -8, 8$ **15.** $-18 = 3w; -54, -6, 6$

16. $5 - j = 10; -15, -10, -5$ **17.** $4 = s + 7; -11, -3, 11$ **18.** $q - 3 = -6; -9, -6, -3$

19. $-2z = -12; -6, 6, 24$ **20.** $\dfrac{n}{-4} = 2; -8, -2, 8$ **21.** $-2 + a = 7; -9, 5, 9$

22. $3d + 2 = -4; -2, 1, 2$ **23.** $11 = 3 - 4p; -4, -2, -1$ **24.** $2 + 3n = -10; -4, -3, -2$

25. $\dfrac{4 - m}{-6} = 0; -4, 0, 4$ **26.** $\dfrac{e + 2}{3} = -3; -11, -9, 7$ **27.** $\dfrac{2 + h}{5} = -1; -7, 3, 7$

Use mental math to solve each equation.

28. $d - 2 = 12$ **29.** $-8 = 4y$ **30.** $-1 = q + 3$

31. $-5w = 20$ **32.** $4 = a - 7$ **33.** $n - 6 = -1$

34. $-3 + g = -10$ **35.** $k \div 3 = -5$ **36.** $-2f = 12$

Solve each equation. Check the solution.

37. $d + 6 = -2$ **38.** $4 - k = 8$ **39.** $12 = -3 - g$

40. $10 = n - (-3)$ **41.** $-5 + x = -7$ **42.** $a + 6 = 2$

43. $-2 = 12 + y$ **44.** $-4 = 2 - w$ **45.** $3 = -1 + h$

46. $-1 - z = 5$ **47.** $m + 3 = -1$ **48.** $-2 - b = -9$

49. $7 + d = -3$ **50.** $f - 10 = 3$ **51.** $-8 = p - 8$

52. $-3.6 + c = 2.1$ **53.** $0.8 = 2 - s$ **54.** $1.9 - d = -4.5$

55. $6.5 = -3.3 + w$ **56.** $a + (-1.9) = -0.2$ **57.** $r - 0.4 = -1.7$

58. $\dfrac{1}{3} - m = -\dfrac{4}{3}$ **59.** $y + \dfrac{1}{2} = \dfrac{7}{8}$ **60.** $-\dfrac{1}{4} = \dfrac{1}{8} - k$

61. $g + \dfrac{2}{5} = -\dfrac{7}{10}$ **62.** $\dfrac{1}{10} = -q + \dfrac{1}{2}$ **63.** $b - 1\dfrac{2}{3} = \dfrac{5}{6}$

Tell whether each equation is *true, false*, or an *open sentence*. (Lesson 5-1)

64. $-11 = 9 - 20$
65. $-7 \cdot (-10) = -70$
66. $5 - (-17) = -3 \cdot (-4)$

67. $14 - p = 2 \cdot (-6)$
68. $4^2 - 1 = -5 \cdot (-3)$
69. $5 \cdot (6 - 10) = 3 \cdot 6 + 2$

Which of the given values is a solution of the given equation? (Lesson 5-1)

70. $-3q = -9$; $-3, 3, 27$
71. $-4 = 2 - n$; $-6, -2, 6$
72. $-3 + y = -10$; $-13, -7, 7$

73. $w + 6 = -1$; $-7, -5, 5$
74. $\dfrac{h}{-3} = -12$; $-36, 4, 36$
75. $4 = -12g$; $-3, -\dfrac{1}{3}, \dfrac{1}{3}$

76. $5 - 2x = 3$; $-3, -1, 1$
77. $\dfrac{v + 4}{2} = -3$; $-10, -1, 2$
78. $5 = \dfrac{6 - d}{3}$; $-21, -9, 21$

Solve each equation. Check the solution. (Lesson 5-2)

79. $-4 + m = -11$
80. $3 - g = -6$
81. $5 = b - (-1)$

82. $5 = d + (-2)$
83. $x + 2 = -8$
84. $-1 = -5 + y$

85. $w - 8.1 = 2.2$
86. $6.5 = -3.2 - m$
87. $-4.8 + x = -2.3$

88. $\dfrac{1}{4} + r = \dfrac{4}{5}$
89. $\dfrac{1}{10} - z = \dfrac{3}{5}$
90. $-\dfrac{1}{6} = a + \dfrac{2}{3}$

Math*Works* Career – Solid Waste Disposal Staff
Workplace Knowhow

S olid waste disposal staff collects recyclable materials at site facilities and curbside. They make sure that customers properly dispose of hazardous household waste (HHW). Typical HHW's include antifreeze, automotive batteries, insecticide, and paint thinner. HHW's should not be placed in regular trash collection. Adding the costs of HHW programs to regular garbage collection is a common way to fund an HHW program. One way to calculate the additional charge is to use the formula $m = p \div h \div 12$, where m is the monthly charge, p is the program cost and h is the number of households being served.

1. One program serves 25,000 households and costs $60,000. What is the monthly charge to one customer?

2. Another program serves 100,000 households and has an annual cost of $75,000. What is the yearly cost to one consumer?

3. Suppose a new program will cost consumers an additional $.05/mo. The new program will serve 35,000 households. What is the total cost of the new program?

4. One program charges customers $45/yr and there are 132,000 people being served. What is the program cost?

5-3 Multiply or Divide to Solve Equations

Goals ■ Solve equations using multiplication and division.

Applications Fitness, Travel

In gym class, 12 students divide into equal teams. How many equal teams could be formed and what would be the number of people on each team?

One possibility is modeled by the Sentence Mat shown.

1. What expression is shown on each side of the mat?

2. You can solve the problem by dividing the blocks on the right side into 4 equal groups. Sketch how you arrange the blocks to solve the problem.

3. What equation represents this problem?

4. Repeat this process with another possible number of teams.

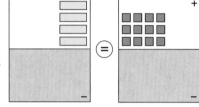

▼ BUILD UNDERSTANDING

Like addition and subtraction, multiplication and division are inverse operations. You can solve equations involving multiplication or division by using the inverse operation.

> **Think Back**
>
> Recall that the inverse operation of multiplication is division.
>
> The inverse operation of division is multiplication.

Example 1

Solve each equation. Check the solution.

a. $4m = -52$

b. $\dfrac{k}{8} = -2$

Solution

a. $4m = -52$

$\dfrac{4m}{4} = \dfrac{-52}{4}$ Undo multiplication with division.

$1m = -13$

$m = -13$

Check

$4m = -52$

$4(-13) \overset{?}{=} -52$

$-52 = -52$ ✔

b. $\dfrac{k}{8} = -2$

$8\left(\dfrac{k}{8}\right) = 8(-2)$ Undo division with multiplication.

$1k = -16$

$k = -16$

Check

$\dfrac{k}{8} = -2$

$\dfrac{-16}{8} \overset{?}{=} -2$

$-2 = -2$ ✔

Another way to solve an equation involving multiplication or division is to use reciprocals. **Reciprocals** are two numbers whose product is 1. For example, 4 and $\frac{1}{4}$ are reciprocals because $4 \cdot \frac{1}{4} = 1$. Also, $\frac{2}{3}$ and $\frac{3}{2}$ are reciprocals because $\frac{2}{3} \cdot \frac{3}{2} = 1$.

Example 2

Solve each equation using reciprocals. Check the solution.

a. $12p = 60$

b. $\frac{2}{5}y = 6$

Problem Solving Tip

Always check the solution when solving equations.

Solution

a. $12p = 60$

$\left(\frac{1}{12}\right)12p = \left(\frac{1}{12}\right)60$ The reciprocal of 12 is $\frac{1}{12}$.

$1p = \frac{60}{12}$

$p = 5$

Check

$12p = 60$

$12 \cdot 5 \stackrel{?}{=} 60$

$60 = 60$ ✔

b. $\frac{2}{5}y = 6$

$\left(\frac{5}{2}\right)\frac{2}{5}y = \left(\frac{5}{2}\right)6$ The reciprocal of $\frac{2}{5}$ is $\frac{5}{2}$.

$1y = \frac{30}{2}$

$y = 15$

Check

$\frac{2}{5}y = 6$

$\frac{2}{5} \cdot 15 \stackrel{?}{=} 6$

$6 = 6$ ✔

Example 3

Write an equation for each sentence.

a. The product of p and three is nine.

b. The quotient of k and negative two is twelve.

Solution

a. $3p = 9$

b. $\frac{k}{-2} = 12$

Example 4

PART-TIME JOB Kirk earned $83.75 working for an hourly wage of $6.70 at the movie theater. Write and solve an equation to determine how many hours he worked.

Solution

Let h equal the number of hours that Kirk worked.

$6.70h = 83.75$

$\frac{6.70h}{6.70} = \frac{83.75}{6.70}$

$1h = 12.5$

$h = 12.5$

Kirk worked $12\frac{1}{2}$ h. Be sure to check solution.

Solve each equation. Check the solution.

1. $3t = 9$

2. $\dfrac{p}{-3} = 9$

3. $-4x = 16$

4. $\dfrac{x}{4} = -32$

5. $2c = -12$

6. $-5h = -15$

7. $\dfrac{2}{3}a = -10$

8. $42 = 21z$

9. $11 = \dfrac{1}{4}d$

Write an equation for each sentence.

10. The quotient of q and three is seven.

11. The product of six and g is four.

12. MODELING What equation does the model show?

13. April bought 5 pens for \$2.75. What did each pen cost?

14. A rectangular garden has an area of 105 ft². The length of the garden is 12 ft. Write and solve an equation to determine the width of the garden.

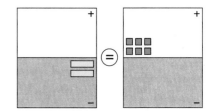

15. WRITING MATH How would you solve the equation $6x = 30$? How would you solve the equation $6 + x = 30$? How are the methods alike? How are they different?

Solve each equation. Check the solution.

16. $228 = 4x$

17. $-10 = \dfrac{t}{2}$

18. $-12p = 3$

19. $-8w = 10$

20. $5.2h = -15.6$

21. $\dfrac{4}{3}x = 6$

22. $6k = 6$

23. $-2 = \dfrac{c}{2}$

24. $12r = -18$

25. $\dfrac{3}{4}y = \dfrac{-2}{5}$

26. $-x = 7$

27. $\dfrac{n}{4} = -8$

28. $14a = 35$

29. $\dfrac{g}{-3.2} = -1.3$

30. $\dfrac{2}{7}y = \dfrac{-5}{14}$

31. $\dfrac{y}{-6} = -2$

32. $\dfrac{n}{-9} = -8$

33. $14 = -3.5m$

34. $1 = \dfrac{h}{5}$

35. $\dfrac{m}{-3} = \dfrac{4}{9}$

36. $30r = 18$

CALCULATOR What keystrokes will you enter in your calculator to solve each equation?

37. $-12x = -42$

38. $\dfrac{d}{17.8} = 160.2$

39. $0.0165y = 0.495$

Write an equation for each sentence.

40. The product of five and n is negative ten.

41. The quotient of negative six and y is six.

42. Negative seven times t equals fourteen.

43. The quotient of eight and r is three.

44. Tanya receives 26 paychecks a year and each check is for \$827.43. Write and solve an equation to find Tanya's yearly net pay. (*Net pay* is the amount of the check. It is the amount that a person is paid after all taxes and deductions.)

45. TRAVEL Landrea traveled 1045 mi during 16 h. What was her average speed? Write and solve an equation for the situation.

DATA FILE For Exercises 46–48, refer to the data on the maximum speeds of animals on page 518. Find the amount of time (in seconds) that it would take for each of the following animals to travel 2 mi if they can run at their maximum speed for that distance. Use the formula $d = rt$.

46. lion
47. elephant
48. chicken

FITNESS Use the data in the table. Let m represent the number of minutes a person does a given activity.

49. Rebecca burned 750 calories playing basketball. Write and solve an equation to determine the number of minutes she spent playing basketball.

50. Thomas burned 604 calories playing tennis. Write and solve an equation to determine the number of minutes he spent playing tennis.

51. Would Juan burn more calories by walking for 35 min or dancing for 30 min?

Calories Burned in Activities

Activity	Calories/min
Basketball	12.5
Cross-country skiing	11.7
Dancing	10.5
Walking (5 mi/h)	9.3
Tennis (singles)	7.1

52. TALK ABOUT IT In solving the equation $-3x = 27$, your classmate says the solution is 9. How would you show your classmate the error and how to correct it?

Solve each equation. Check the solution.

53. $-4(-5) = 8n$

54. $\dfrac{k}{3} = 2(6 - 8 + 12)$

55. $\dfrac{64.8}{3} = -2.4x$

EXTENDED PRACTICE EXERCISES

56. INDUSTRY Each year 320,000,000 metric tons of waste are generated in the U.S. by various industries. This is 8% of the total metric tons of waste produced in the U.S. each year. Write and solve an equation to determine the total number of metric tons generated each year.

57. MODELING Use Algeblocks to model and solve the equation $-2x = 8$.

Describe all solutions of each equation.

58. $2n = 0$

59. $0n = 2$

60. CRITICAL THINKING What is the result if you add 0 to each side of an equation? What is the result if you multiply each side of an equation by 0?

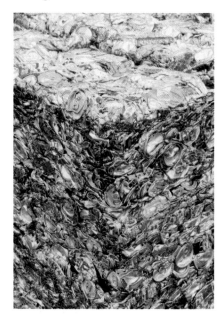

MIXED REVIEW EXERCISES

Write the name of each term described. (Lessons 4-1 and 4-2)

61. an object with one vertex

62. a polygon with opposite congruent sides parallel and no right angles

63. an object that can be named by two endpoints

64. a quadrilateral with only one pair of parallel sides

5-4 Solve Two-Step Equations

Goals ■ Solve two-step equations.

Applications Photography, Sports, Recycling

On the map shown, each line is equal to $\frac{1}{2}$ mi and north is to the top.

1. If you tell your friends how to go from Memorial Union to Kay Gymnasium, then from Kay Gymnasium to Wilson Museum, what directions (N-S-E-W) and distance do you tell them?

2. If they backtrack, what directions and distance do they travel?

3. Describe the relationship between your answers in Questions 1 and 2.

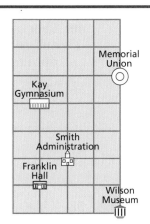

◢ BUILD UNDERSTANDING

You have solved *one-step equations* by applying a single inverse operation. To solve *two-step equations*, apply more than one inverse operation.

The order of operations for $2x + 5 = 7$ is to start with x, multiply by 2 and add 5. The result is 7.

Order of Operations: | x | multiply by 2 | add 5 | 7 |

To solve the equation, undo the steps. Start with 7, subtract 5, then divide by 2 to find x.

Solve the equation: | 7 | subtract 5 | divide by 2 | x |

Example 1

Solve each equation. Check the solution.

a. $4x + 7 = 23$

b. $7(y + 4) = 21$

Solution

a. $4x + 7 = 23$

$4x + 7 - 7 = 23 - 7$ Undo addition.

$4x = 16$

$\dfrac{4x}{4} = \dfrac{16}{4}$ Undo multiplication.

$x = 4$

Check $4x + 7 = 23$

$4(4) + 7 \overset{?}{=} 23$

$16 + 7 \overset{?}{=} 23$

$23 = 23$ ✔

b. $7(y + 4) = 21$

$7y + 28 = 21$ distributive property

$7y + 28 - 28 = 21 - 28$ Undo addition.

$7y = -7$

$\dfrac{7y}{7} = \dfrac{-7}{7}$ Undo multiplication.

$y = -1$

Check $7(y + 4) = 21$

$7(-1 + 4) \overset{?}{=} 21$

$7(3) \overset{?}{=} 21$

$21 = 21$ ✔

Example 2

Use Algeblocks to solve $4x - 1 = 7$.

Solution

Model the equation.

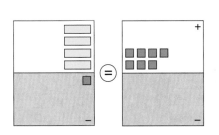

$$4x - 1 = 7$$

Add 1 to both sides. Remove zero pairs.

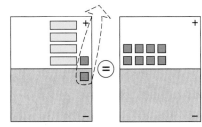

$$4x - 1 + 1 = 7 + 1$$
$$4x + 0 = 8$$

Divide both sides by 4.

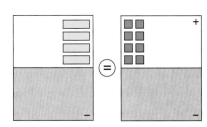

$$\frac{4x}{4} = \frac{8}{4}$$

Read the solution, $x = 2$.

To solve a proportion that has an unknown term, use cross-products.

Example 3

Solve the proportion. $\dfrac{n}{16} = \dfrac{15}{24}$

Solution

$$n \cdot 24 = 16 \cdot 15 \qquad \text{Write the cross-products.}$$

$$24n = 240$$

$$\frac{24n}{24} = \frac{240}{24}$$

$$n = 10 \qquad \text{Be sure to check solution.}$$

> ### Think Back
>
> If a statement is a proportion, the *cross-products* of the terms are equal.
>
> If $\frac{a}{b} = \frac{c}{d}$, then $ad = bc$.

Example 4

RETAIL Rudy bought a new water bottle and two rolls of handlebar tape for his bike. The water bottle cost $2.99. The total cost was $15.97. What was the cost of a roll of tape?

Solution

Write and solve an equation that represents the situation.

Let r = the cost of one roll of tape.

$$2r + 2.99 = 15.97$$

$$2r + 2.99 - 2.99 = 15.97 - 2.99$$

$$2r = 12.98$$

$$r = 6.49$$

The cost of a roll of tape was $6.49.

Solve each equation. Check the solution.

1. $-3d + 2 = -10$

2. $2k - 4 = -2$

3. $4p - 8 = -4$

4. $4k - 2 = -18$

5. $2(y - 2) = 6$

6. $-5(x + 3) = 15$

Solve each proportion. Check the solution.

7. $\frac{4}{3} = \frac{a}{45}$

8. $\frac{x}{8} = \frac{21}{56}$

9. $\frac{9}{m} = \frac{15}{10}$

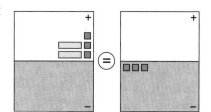

10. **MODELING** What equation is modeled on the Sentence Mat shown?

11. A bike rents for $4.50/h plus a $3 fee. Pedro pays $25.50 to rent a bike. For how many hours did he rent the bike?

12. Sarah buys a case of motor oil and a quart of transmission fluid for a total of $18.37. The transmission fluid sells for $1.17 per quart. There are 12 quarts of motor oil in a case. How much does one quart of oil cost?

13. **WRITING MATH** How do you use the order of operations when solving a two-step equation? Explain.

Solve each equation. Check the solution.

14. $\frac{y}{-7} - 3 = 4$

15. $2 = 8x + 5$

16. $\frac{13}{10} = \frac{52}{x}$

17. $8 + 3n = -1$

18. $\frac{z}{39} = \frac{4}{12}$

19. $9 = -6 + \frac{k}{2}$

20. $6p - 7 = 5$

21. $11 = -2x + 5$

22. $\frac{k}{-2} + 4 = -8$

23. $\frac{4.5}{6} = \frac{a}{20}$

24. $3 = 3h + 6$

25. $\frac{78}{m} = \frac{18}{12}$

26. $0 = 7 + \frac{r}{6}$

27. $9n - 1.5 = 10.2$

28. $\frac{0.9}{3.6} = \frac{1.2}{b}$

29. $34 = \frac{f}{-2} + 47$

30. $\frac{x}{6} = \frac{77}{42}$

31. $5(k + 3) = 30$

32. $\frac{2.4}{32} = \frac{y}{16}$

33. $-7(p - 9) = 14$

34. $\frac{3}{5} + 3c = 3$

35. $26 = 8(m + 7)$

36. $\frac{0.004}{0.12} = \frac{x}{1.8}$

37. $\frac{x}{7} - 15 = -4$

38. The distance around a rectangle is 48 cm. The length of the rectangle is 12 cm. Write and solve an equation to determine the width of the rectangle.

39. **PHOTOGRAPHY** Mark uses color film for 48 out of every 72 photographs. If he takes 156 photographs, how many will be on color film?

40. **RECYCLING** In the U.S., 68% of aluminum cans are recycled. The percent of glass containers recycled in the U.S. is one more than half the percent of recycled aluminum cans. Write and solve an equation to determine the percent of recycled glass containers in the U.S.

Use the figure for Exercises 41–43.

41. The length of this rectangle is 3 times the width. Write an equation to model this situation.

42. Solve the equation to determine the value of x.

43. Use the value of x to find the width, perimeter and area of the rectangle.

44. **TALK ABOUT IT** Evelin says to solve the equation $9x - 6 = 12$, you should first divide both sides of the equation by 9. Parvis says that you should add 6 to both sides of the equation first. Which method would you use? Explain.

45. Jessica has $1.05 in change to use a pay phone. The call costs 35¢ for the first minute and 20¢ for each additional minute. Write and solve an equation to determine the length of call she can make.

46. **WRITING MATH** Write a two-step equation that has a solution of -3. Explain how to solve your equation.

47. **SPORTS** The length of a football field is 200 ft longer than the width. The perimeter of the field is 1040 ft. Find the length and width of a football field.

DATA FILE For Exercises 48–51, refer to the data on the climate of U.S. cities on page 535. Convert the average temperatures to Celsius by using the following formula, where F represents Fahrenheit temperature and C represents Celsius temperature. $F = 1.8C + 32$

48. April average for Honolulu

49. October average for New York

50. January average for Fairbanks

51. July average for St. Louis

▧ EXTENDED PRACTICE EXERCISES

Solve each equation. Check the solution.

52. $\dfrac{-45}{-9} + (5p - 3p) = 4 + 7 \cdot 3$

53. $9\left(x - \dfrac{1}{3}\right) = -21$

54. **MODELING** Use Algeblocks to solve $3x + 4 = -5$. Make sketches to record each step.

Write and solve an equation for each sentence.

55. Seven times a number minus three is 39.

56. A number divided by negative six added to two is negative four.

57. **CHAPTER INVESTIGATION** Use the MSW equation on page 205 to find the total MSW recycled in each of the following years: 1970, 1980, 1990, and 2001.

▧ MIXED REVIEW EXERCISES

Find the surface area and volume of each figure. Round to the nearest hundredth.
(Lessons 4-6 and 4-7)

58.
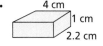
4 cm
1 cm
2.2 cm

59.
6.5 cm
3.7 cm

60.
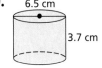
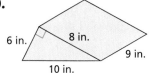
6 in. 8 in.
9 in.
10 in.

Review and Practice Your Skills

PRACTICE ◣ LESSON 5-3

Solve each equation. Check the solution.

1. $-96 = 6p$

2. $\dfrac{5}{4}a = 15$

3. $8y = -2$

4. $\dfrac{m}{-3} = -6$

5. $\dfrac{w}{3} = \dfrac{1}{-12}$

6. $1 = \dfrac{k}{9}$

7. $-9 = \dfrac{3}{5}w$

8. $9y = -18$

9. $0.3n = 8.1$

10. $3 = -g$

11. $-4p = -36$

12. $\dfrac{2}{9} = \dfrac{-t}{3}$

13. $\dfrac{2}{1.5} = \dfrac{c}{0.3}$

14. $6 = 0.2d$

15. $10 = \dfrac{d}{-2}$

16. $\dfrac{m}{-6} = \dfrac{-3}{2}$

17. $4 = \dfrac{-h}{20}$

18. $5a = 80$

19. $-3p = 15$

20. $\dfrac{-5}{8} = \dfrac{e}{4}$

21. $-4.8 = \dfrac{v}{4}$

22. $\dfrac{b}{3} = -5$

23. $2 = -6d$

24. $-\dfrac{2}{7}z = 10$

25. $-0.5 = -7.5f$

26. $\dfrac{n}{6} = \dfrac{10}{3}$

27. $\dfrac{1}{-2} = \dfrac{z}{12}$

Write an equation for each statement.

28. The quotient of twelve and y is three.

29. Negative three times w equals one.

30. The product of two and m is negative five.

31. The quotient of one-third and g equals two.

32. Four times h equals negative one-half.

33. The quotient of negative two and d is eight.

PRACTICE ◣ LESSON 5-4

Solve each equation. Check the solution.

34. $4 = -3a + 2$

35. $-1 = 2p + 3$

36. $3j - 1 = -5$

37. $1 - 5m = -3$

38. $-2f + 1 = -3$

39. $4 = 2t + 7$

40. $5 = \dfrac{h}{-2} + 3$

41. $-2 + \dfrac{d}{3} = -1$

42. $5d - 1 = -3$

43. $\dfrac{3}{-8} = \dfrac{z}{5}$

44. $\dfrac{2}{9} = \dfrac{-r}{5}$

45. $-2y + 3 = -8$

46. $3s - 1 = -8$

47. $-3 = 2v - 5$

48. $\dfrac{k}{6} = \dfrac{-1}{5}$

49. $4 = 3g - 8$

50. $5 - \dfrac{b}{3} = -6$

51. $\dfrac{w}{-2} + 7 = 1$

52. $-5q + 1 = 4$

53. $8 = -n + 2$

54. $-6m + 3 = -1$

55. $6 + 3r = -1$

56. $\dfrac{5}{2} = \dfrac{f}{-3}$

57. $6 = 2g + 4$

58. $\dfrac{c}{-4} = \dfrac{-2}{9}$

59. $4 = 3h - 2$

60. $5 + \dfrac{c}{-3} = 2$

61. $5w - 3 = 2$

62. $-3 = 2 + \dfrac{x}{-4}$

63. $1 - 2m = -6$

64. $7 + 2b = 3$

65. $-1 - 3t = 4$

66. $7y + 2 = -4$

Which of the given values is a solution of the given equation? (Lesson 5-1)

67. $s + 8 = -3; -11, -5, 5$

68. $\dfrac{r}{-2} = -10; \dfrac{1}{5}, 5, 20$

69. $7 - c = -6; -13, 1, 13$

70. $-4 = -3b; \dfrac{3}{4}, \dfrac{4}{3}, 12$

71. $10 - 3g = -2; -4, 2, 4$

72. $\dfrac{m-2}{-1} = 5; -5, -3, 3$

Solve each equation. Check the solution. (Lesson 5-2)

73. $-3 + y = -1$

74. $5 = -2 + g$

75. $6 = x - (-1)$

76. $1.2 - m = 0.3$

77. $-1.4 = n + 3.6$

78. $d + \dfrac{3}{8} = -\dfrac{3}{4}$

Solve each equation. Check the solution. (Lesson 5-3)

79. $-45 = 3g$

80. $\dfrac{w}{-3} = -1$

81. $\dfrac{1}{5}d = -4$

82. $9 = \dfrac{c}{3}$

83. $6n = 2.5$

84. $\dfrac{3}{8} = \dfrac{t}{-16}$

Solve each equation. Check the solution. (Lesson 5-4)

85. $5 = 3p + 2$

86. $4e - 1 = -8$

87. $7 - \dfrac{a}{2} = 3$

88. $-1 = 4 + \dfrac{r}{-3}$

89. $\dfrac{3}{-7} = \dfrac{k}{2}$

90. $\dfrac{c}{3} = \dfrac{-1}{8}$

Mid-Chapter Quiz

Tell whether each equation is true, false, or open. (Lesson 5-1)

1. $-8 = 3 - 11$

2. $2 = 7 + x$

3. $5(2 + 7) = 20 - (6 - 3)$

4. $-\dfrac{12}{6} = 2$

Write the inverse of each step. (Lessons 5-2 and 5-3)

5. Add -4.

6. Subtract 1.2.

7. Divide by -4.

8. Multiply by 0.4.

Solve and check. (Lessons 5-1, 5-3, and 5-4)

9. $-7 + m = -2$

10. $2.1 = t - 3.4$

11. $5x = 35$

12. $\dfrac{n}{-2} = 4.3$

13. $45n = -90$

14. $2.4 = \dfrac{j}{3.2}$

15. $\dfrac{8}{2.4}x = -100$

16. $2(y - 3) = -20$

17. $\dfrac{v}{6} - 2 = 7$

18. $4 = 3x + 16$

19. $4a - 3 = 12$

20. $\dfrac{c}{-5} + 3 = 7$

21. Robine is 3 yr older than 4 times Karim's age. If Robine is 19, how old is Karim? (Lesson 5-4)

22. Solve the equation. Check the solution. (Lesson 5-4)

$$-28 = -4(2y - 3)$$

5-5 Combine Like Terms

Goals ◼ Solve an equation by combining like terms.

Applications Safety, Market research, Food service

Study the Sentence Mat shown.

1. What equation is modeled by the Algeblocks?

2. Determine a simpler way to write the left side of the mat.

3. Determine a simpler way to write the right side of the mat.

4. Redraw the simplified equation on a Sentence Mat. Use the Algeblocks to solve the simplified equation.

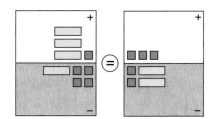

◣ BUILD UNDERSTANDING

The parts of a variable expression that are separated by addition or subtraction signs are called **terms**. The variable expression $x + 3y + 2x - 4y^2$ contains four terms: x, $3y$, $2x$ and $-4y^2$. The terms x and $2x$ are **like terms** because they have the same variable raised to the same power. The terms x, $3y$ and $4y^2$ are **unlike terms** because they have different variable parts.

To simplify a variable expression, use the properties from Chapter 3 to combine like terms.

Check Understanding

Identify the following terms as *like* or *unlike*.

a. 3x, 3y   b. 7k, −3k   c. 2p, 2pt

Example 1

Simplify.

a. $3m + 5m$

b. $8x - 3x + 2y + 4$

c. $4k - 2h + 3k$

Solution

a. $3m + 5m = (3 + 5)m$ Use the distributive property.
$\qquad\qquad = 8m$

b. $8x - 3x + 2y + 4 = (8 - 3)x + 2y + 4$ Use the distributive property.
$\qquad\qquad\qquad = 5x + 2y + 4$

c. $4k - 2h + 3k = 4k + 3k - 2h$ Rewrite the expression using the commutative property.
$\qquad\qquad = (4 + 3)k - 2h$ Use the distributive property.
$\qquad\qquad = 7k - 2h$

In some equations, it is necessary to first simplify both sides of the equation by combining like terms. Then solve for the variable.

Example 2

Solve $5x - 7 = 3x + 13$. Check the solution.

Solution

$$5x - 7 = 3x + 13$$

$5x - 3x - 7 = 3x - 3x + 13$ Subtract $3x$ from both sides.

$2x - 7 = 13$ Simplify.

$2x - 7 + 7 = 13 + 7$ Add 7 to both sides.

$$2x = 20$$

$$\frac{2x}{2} = \frac{20}{2}$$ Divide both sides by 2.

$$x = 10$$

Check

$$5x - 7 = 3x + 13$$

$$5(10) - 7 \stackrel{?}{=} 3(10) + 13$$

$$50 - 7 \stackrel{?}{=} 30 + 13$$

$$43 = 43 \quad ✔$$

Example 3

Write and solve an equation for the following sentence.

Three more than a number is four less than twice the number.

Solution

Let n represent the number.

$$n + 3 = 2n - 4$$

$$n + 3 - n = 2n - n - 4$$

$$3 = n - 4$$

$$3 + 4 = n - 4 + 4$$

$$7 = n$$

> **Check Understanding**
>
> How would you use Algeblocks to represent and solve the equation in Example 3?

TRY THESE EXERCISES

Simplify.

1. $2y + 13y$

2. $5z + 4 - 3z$

3. $7n + 3m - m + 2n$

4. $-2x + 4 + 4x$

5. $-5f - 2f + 3 + 3f$

6. $\frac{-1}{7} + \frac{2}{3}y - \frac{1}{2} - \frac{1}{2}y$

7. Jesse worked $2p$ hours in the morning, $5r$ hours in the afternoon, and $4p$ hours in the evening. Write and simplify an expression for the total number of hours he worked.

Solve each equation. Check the solution.

8. $4n - n + 7 = 2$

9. $9x - 6x = -15$

10. $7c - 3c - 3 = 5$

11. $8m - 2m = 18$

12. $-6x + 20 = 180 + 2x$

13. $1.5y - 0.7 = 0.4y + 1.5$

14. MODELING What equation is modeled on the Sentence Mat shown?

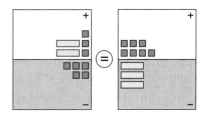

Write and solve an equation for the following sentence.

15. Twice a number added to 45 is 180.

PRACTICE EXERCISES • For Extra Practice, see page 563.

Simplify.

16. $4m + 8m$

17. $8s + 5t - 3s$

18. $x + 11x$

19. $3x + 5 - 9 - 5x$

20. $n + 5n + n^2 - 3n$

21. $a + 2a + 3a$

22. $7y + 3y - 4$

23. $14n - 6x + (-5n) + 8x$

24. $-12r - 21s + r - 15r$

25. $3(x - 5) + 2(x + 2)$

26. $-3x - 3x + 3xy$

27. $-(p + 3) + p^2 + 4p$

Solve each equation. Check the solution.

28. $5x - 2x - 7 = 2$

29. $3n - 2n - 4 = 3$

30. $4n + 9 - n = -3$

31. $8r - 11 - 6r = -6$

32. $3n - 10 + 2n = 5$

33. $n - 5 - 5n = 1$

34. $-3c - 4 = -5c + 6$

35. $5x - 21 = -2x + 28$

36. $y - 3 = -y + 5$

37. $9b + 3 = 10b - 5$

38. $-8 + 3h = 2h - 10$

39. $6f + 20 = -2f - 4$

40. $24 - 9t = -13t + 8$

41. $7x - 8 = 10x + 1$

42. $\frac{1}{4}x = 2x + 17.5$

43. $7x = 10(x - 1.5)$

44. $2y + y - 10 = -2y + 15$

45. $3(5.5 - m) = 8m$

46. $3(b - 2) - b = 5b + 12$

47. $3n + 5(n - 2) = 3(n - 5)$

48. $\frac{3}{4}(x + 12) = \frac{5}{8}(x + 8)$

Write and solve an equation for each sentence.

49. Seven is a number divided by the sum of negative three and five.

50. Five times a number added to six is twice that number less 15.

51. Three times a number less six is eight more than negative four times the number.

SAFETY A traffic court judge fines speeding offenders $75 plus $2 for every mile per hour they exceed the speed limit. Use s to represent the speed a person travels. Write and solve an equation to determine the person's speed based on their fine.

52. $95 fine; Speed limit: 45 mi/h

53. $129 fine; Speed limit: 65 mi/h

PART-TIME JOB Kesha babysits after school Monday through Friday. The table shows the number of hours that she babysits each day. She earns $78/wk.

Day	Mon.	Tues.	Wed.	Thurs.	Fri.
Hours	2	3	3	3	1

54. Write and solve an equation to determine how much money Kesha earns in 1 h.

55. FOOD SERVICE Carlos works as a restaurant server. He worked 4 h on Friday and 3 h on Saturday. Over the 2 days he earned $35 in tips. The total amount of money he earned working Friday and Saturday was $81.20. Write and solve an equation to determine Carlos' hourly wage.

Use the figure shown.

56. Write an equation to model the sum of the angles in the triangle. (Hint: The sum of the measures of the angles in a triangle is 180°.)

57. Solve the equation. Check the solution.

58. Use the value for x to determine the measure of each angle.

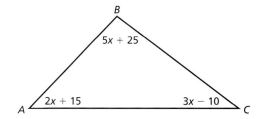

59. WRITING MATH Write a word problem leading to an equation with variables on both sides of the equal sign. Be sure your problem has a solution that makes sense.

■ EXTENDED PRACTICE EXERCISES

MARKET RESEARCH Bell Cellular offers a payment plan with a monthly fee of $40 and $0.30/min. TALK Cellular offers a payment plan with a monthly fee of $25 and $0.50/min.

60. Write an expression for the cost of a cellular phone service from Bell Cellular if you speak m min per month.

61. Write an expression for the cost of a cellular phone service from TALK Cellular if you speak m min per month.

62. For what number of minutes are the costs equal?

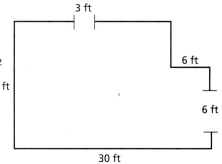

Copy the table at the right. Choose 3 numbers and write in spaces provided.

63. Complete steps 1 through 5 for all four columns.

64. Simplify the expression in step 5 of the fourth number column. What is the result?

65. CRITICAL THINKING Explain why you obtained the unusual results in the table.

	First number	Second number	Third number	Fourth number
1. Add 3				
2. Multiply by 4				
3. Subtract 6				
4. Divide by 2				
5. Subtract twice the original number				

■ MIXED REVIEW EXERCISES

Evaluate each expression. Let $g = 7$ and $h = -1$. (Lesson 3-5)

66. $g - h$ **67.** $h + (-g)$ **68.** $3h - g$ **69.** $\dfrac{g}{h}$

70. $4h + (-2g)$ **71.** $4gh + g - 5h$ **72.** $g - 4h + 17$ **73.** $h^2 - g^2$

74. Michael needs to purchase quarter-round molding which he will place at the base of every wall in this room. Molding cost $2.25 for each 6-ft length. How much will it cost Michael to purchase the molding he needs? How much will be left over? (Lesson 2-3)

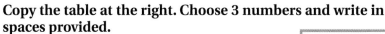

5-6 Use Formulas to Solve Problems

Goals ■ Use formulas to solve problems.

Applications Weather, Retail, Biology, Fitness

The five figures form a pattern.

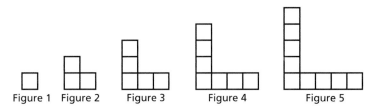

Figure 1 Figure 2 Figure 3 Figure 4 Figure 5

1. Copy and complete the table to find the perimeter of each figure. Each side of each individual square is 1 unit.

Figure number	1	2	3	4	5
Perimeter					

2. Without drawing a picture, describe what the sixth figure will look like and predict its perimeter.

3. If you continued this pattern, what would be the perimeter of the 35th figure?

4. Explain how the perimeter of each figure is related to its figure number.

5. Using the variables n for the figure number and P for the perimeter, write an equation for the relationship in Question 4.

BUILD UNDERSTANDING

A **formula** is an equation stating a relationship between two or more variables. For example, the number of square units in the area (A) of a rectangle is equal to the number of units of length (l) multiplied by the number of units of width (w). Therefore, the formula for the area of a rectangle is $A = lw$.

Sometimes you can evaluate a variable in a formula by using the given information.

In the figure shown, the length is 9 units and the width is 5 units.

$A = lw$

$A = 9 \cdot 5$

$A = 45$

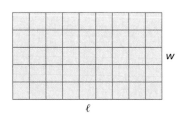

The area is 45 square units, or 45 units^2.

At other times, you must use your knowledge of equations to solve for a variable in a formula.

Example 1

A car traveled 120 mi in $2\frac{1}{2}$ h. Find its average rate.

Problem Solving Tip

When solving equations using multiplication or division, convert mixed numbers to improper fractions.

Solution

Use the formula $d = rt$, where d = distance, r = rate and t = time.

$$d = r \cdot t$$
$$120 = r\left(2\frac{1}{2}\right) \qquad \text{Substitute the values of the variables into the formula.}$$
$$120 = r\left(\frac{5}{2}\right) \qquad \text{Change } 2\frac{1}{2} \text{ to improper fraction.}$$
$$120\left(\frac{2}{5}\right) = r\left(\frac{5}{2}\right)\left(\frac{2}{5}\right) \qquad \text{Multiply both sides by } \frac{2}{5}.$$
$$48 = r$$

The car's average rate was 48 mi/h.

Example 2

WEATHER The formula $F = \frac{9}{5}C + 32$ relates Fahrenheit and Celsius temperatures, where F = Fahrenheit temperature and C = Celsius temperature. If the outside temperature is 86°F, find the temperature in degrees Celsius.

Solution

$$F = \frac{9}{5}C + 32$$
$$86 = \frac{9}{5}C + 32 \qquad \text{Substitute.}$$
$$86 - 32 = \frac{9}{5}C + 32 - 32 \qquad \text{Subtract to get } C \text{ term alone.}$$
$$54 = \frac{9}{5}C$$
$$\left(\frac{5}{9}\right)54 = \left(\frac{5}{9}\right)\left(\frac{9}{5}\right)C \qquad \text{Multiply to get } 1C.$$
$$30 = C$$

The temperature is 30°C.

Sometimes you must solve a formula for a variable that is not alone on one side of the equal symbol.

Example 3

In the formula $F = ma$, where F = force, m = mass and a = acceleration, solve for m.

Solution

$$F = ma$$
$$\frac{F}{a} = \frac{ma}{a} \qquad \text{Divide both sides by } a \text{ to get } m \text{ alone on one side.}$$
$$\frac{F}{a} = m$$

The new formula indicates that mass is found by dividing force by acceleration.

1. **FITNESS** Jolene jogs 8 mi at an average rate of 6 mi/h. Use the formula $d = rt$ to determine how long she jogs.

2. A formula for the perimeter of a rectangle is $P = 2l + 2w$, where P = perimeter, l = length and w = width. A rectangle has a width of 7 cm and a perimeter of 46 cm. Find the length of the rectangle.

Solve each formula for the indicated variable.

3. $A = lw$, for w 4. $3w - m = r$, for w 5. $H = \dfrac{15N}{60}$, for N

6. $ab + 7 = c$, for b 7. $9a = C$, for a 8. $3x + 7 = y$, for x

9. **WRITING MATH** Describe the steps you used to solve for x in Exercise 8.

10. The formula for the area of a triangle is $A = \frac{1}{2}bh$, where b = length of the base and h = height. A triangle has a 13 cm base and an area of 91 cm². Find its height.

11. The area of a trapezoid is $A = \frac{1}{2}(b_1 + b_2)h$, where b_1 and b_2 are the bases and h is the height. A trapezoid has a base that measures 3 in., another base that measures 5 in. and an area of 62 in.² Find its height.

BUSINESS A car salesperson earns a base salary of $300 a week plus $\frac{1}{2}$% of the purchase price as commission for each car she sells.

12. Write a formula for the salesperson's weekly earnings (E) when she sells cars for d dollars.

13. One week she sells vehicles valued at a total of $125,000. What are her earnings for that week?

14. What is the total value of vehicles she must sell for her weekly earnings to be $825?

RETAIL Jones' Department Store is having a 25% off sale.

15. Write a formula for calculating sale price, represented by s, based on the original price, represented by p.

16. What is the sale price of an item whose original price is $39.99?

17. What is the original price of an item whose sale price is $25.50?

BIOLOGY Crickets increase their rate of chirping as the temperature increases. One formula for estimating the temperature from crickets' chirps is $F = \frac{n}{4} + 32$, where F = Fahrenheit temperature and n = number of chirps per minute.

18. Find the temperature when a cricket chirps 188 chirps/min.

19. Find the chirping frequency at 70°F.

Solve each formula for the indicated variable.

20. $I = prt$, for t

21. $E = mc^2$, for m

22. $A = p + i$, for i

23. $P = 3s$, for s

24. $P = 2l + 2w$, for w

25. $P = 2l + 2w$, for l

26. $A = bh$, for h

27. $F = \dfrac{9}{5}C + 32$, for C

28. $y = mx + b$, for x

29. $O = \dfrac{R + M}{2}$, for M

30. $T = p + prt$, for r

31. $v = r + at$, for a

32. $ex - 2y = 3z$, for x

33. $ex - 2y = 3z$, for y

DATA FILE For Exercises 34–36, refer to the data on champion trees on page 529. Use the formula for the circumference of a circle to find the approximate radius of the following trees at 4.5 ft. Use $\pi = 3.14$.

34. sitka spruce

35. coast redwood

36. common bald cypress

37. TRAVEL The formula for finding gas mileage is $m = \dfrac{d}{g}$, where m is miles per gallon, d is the distance traveled, and g is the number of gallons of gas used. If your car averages 31 mi/gal and the gas tank contains 8 gal of gas, can you drive 275 mi on one tank of gas? Explain.

◤ EXTENDED PRACTICE EXERCISES

Determine if the variable underlined was isolated correctly. If not, what is the correct rewritten formula?

38. $r = p(220 - \underline{a})$ $a = \dfrac{r - 220}{p}$

39. $y = \underline{m}x + b$ $m = \dfrac{y - b}{x}$

40. RECYCLING Americans throw away enough office and writing paper each year to build a wall 12 ft high stretching from Los Angeles to New York City. This distance is 2824 mi. Find the approximate area of one face of this wall. (*Hint:* 1 mi = 5280 ft)

41. CRITICAL THINKING The formula for the surface area of a cylinder is $SA = 2\pi rh + 2\pi r^2$ where r is the radius and h is the height of the cylinder. A 12-oz soda can has a radius of 1.3 in. and an approximate surface area of 51 in.^{2} What is the approximate height of a 12-oz can of soda?

CHAPTER INVESTIGATION Use the formula on page 205 to find the percent of increase of the total MSW recycled for the following time periods.

42. 1980 to 1990

43. 1990 to 1995

44. 1995 to 2001

◤ MIXED REVIEW EXERCISES

Use the drawing to answer the questions.
(Lessons 2-3 and 2-9)

45. Find the perimeter and the area, to the nearest tenth. Use 3.14 for π.

46. Carpet is sold by the square yard. How many square yards are needed to cover this floor? (Lesson 2-2)

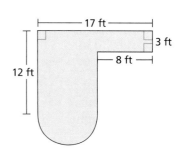

Review and Practice Your Skills

Simplify.

1. $2d - 3d$

2. $6 + a - 3a$

3. $5n + 2m - 3n$

4. $4 - 3m + m$

5. $-p + 3q - 4p + q$

6. $-4w - w + 3w$

7. $4(b + 2) - 3b$

8. $6x - 2y + x - z$

9. $7k + 3(1 - k)$

Solve each equation. Check the solution.

10. $3w + w - 3 = -1$

11. $-a + 3 - 2a = 6$

12. $5 - z + 3z = 2$

13. $-4d + 3 + 2d = 4$

14. $7n + 2n - 1 = -3$

15. $c - 3c + 6c = 4$

16. $2 - h = 3h + 4$

17. $3y + 4 = y - 1$

18. $-4 + 3k = -2k - 3$

19. $5 + 2m = -m - 1$

20. $-g - 3 = 2 + 3g$

21. $5 - 4b = 3 + b$

22. $-2t + 3 = 11 - 4t$

23. $6 + 3q = 1 - 2q$

24. $1 - e = 4e + 3$

25. $8 - 3b = 2(3 - 2b)$

26. $k + 3(1 + 2k) = 3k$

27. $2(p + 1) = 3p - 6$

28. $4s - 1 = 3 + 4(1 - s)$

29. $5 - 2d = -(3 - d) + 1$

30. $5w - 3 = -2(1 - 2w)$

Solve each formula for the indicated variable.

31. $A = \frac{1}{2}bh$, for b

32. $F = ma$, for a

33. $V = lwh$, for w

34. $s = \frac{d}{t}$, for t

35. $A = lw$, for w

36. $V + F - E = 2$, for E

Find the area or the missing length of each rectangle.

37. area = 16 in.^{2},
length = 4 in.
Find the width.

38. width = 2.5 m,
length = 0.5 m
Find the area.

39. area = 8 yd^2,
width = 3 yd
Find the length.

40. width = 150 m,
area = 30 m^2
Find the length.

41. area = 20 in.^{2},
length = 8 in.
Find the width.

42. length = 4.5 ft,
width = 3 ft
Find the area.

Find the area or the missing length of each rectangle.

43. area = 3.2 mm^2,
base = 4 mm
Find the height.

44. height = 40 in.,
area = 10 in.^{2}
Find the base.

45. area = 72 ft^2
base = 4 ft
Find the height.

46. area = 15 in.^{2},
height = 30 in.
Find the base.

47. base = 10 yd,
height = 3 yd
Find the area.

48. area = 51 ft^2
height = 6 ft
Find the base.

49. base = 3.2 km,
area = 90 km^2
Find the height.

50. area = 15 in.^{2},
base = 4 in.
Find the height.

51. height = 3 m
base = 4.5 m
Find the area.

Which of the given values is a solution of the given equation? (Lesson 5-1)

52. $d - 3 = -6$; $-9, -6, -3$ **53.** $4 = 2m + 8$; $-4, -2, 2$ **54.** $\dfrac{3 + a}{-5} = 1$; $-8, -2, 2$

Solve each equation. Check the solution. (Lessons 5-2 and 5-3)

55. $-1 + c = 8$ **56.** $-3 = f - 6$ **57.** $m + 0.3 = -4.7$

58. $-2.4 = 0.8k$ **59.** $\dfrac{d}{-2} = -7$ **60.** $\dfrac{n}{3} = \dfrac{-4}{9}$

Solve each equation. Check the solution. (Lessons 5-4 and 5-5)

61. $-3 + 2g = -7$ **62.** $2 = \dfrac{-m}{3} + 1$ **63.** $\dfrac{w}{-3} = \dfrac{11}{4}$

64. $3 - 5k + 2k = -6$ **65.** $7 - 3y = -y + 4$ **66.** $q + 2 = -3(2q + 1)$

Solve each formula for the indicated variable. (Lesson 5-6)

67. $C = \dfrac{5}{9}(F - 32)$, for F **68.** $A = \dfrac{1}{2}bh$, for h

MathWorks
Workplace Knowhow
Career – Financial Analyst

Financial analysts for a collection facility manage and calculate the budget for reusable materials. Collection facilities sell recycled plastics and other materials. One such plastic, PET is shredded and used as filler in sleeping bags and parkas. Currently 50% of plastics are recyclable using existing technology. The table shows the percent of solid waste and selling price of recyclable items.

Component	Solid waste (percent)	Selling price (cents/pound)
Glass bottles	8%	2.0
Newspaper	6%	1.0
Plastic bottles	2%	6.0
Steel cans	2%	0.5
Aluminum cans	1%	40.0

A formula to calculate the amount of money a facility earns on the resale of recycled items is $m = pt\left(\dfrac{s}{100}\right)$, where p is the percent of solid waste, t is the number of pounds of solid waste collected and s is the selling price.

Calculate the amount of money earned from each component when a facility collects 100 T of solid waste.

1. Glass bottles   2. Newspaper   3. Plastic bottles
4. Steel cans   5. Aluminum cans
6. What is the total amount of money earned?

Problem Solving Skills:
Work Backwards

Some problems tell you the outcome of a series of steps and ask you to find where the steps began. To solve such a problem, start at the end and **work backwards** to the beginning. For example, when solving a mystery, a detective might work backwards to discover all the details. A mechanic might work backwards to find the cause of a certain problem on an automobile.

Problem Solving Strategies

Guess and check

Look for a pattern

Solve a simpler problem

Make a table, chart or list

Use a picture, diagram or model

Act it out

✔ Work backwards

Eliminate possibilities

Use an equation or formula

Problem

Jarret purchased a $925 clarinet using his $6.20/h earnings as a cashier. While he was saving, his uncle donated $150 toward his goal. How many hours did Jarret work to earn enough money to buy the clarinet?

Solve the Problem

■   · $6.20   + $150   = $925

number of hours worked	hourly wages	uncle's contribution	total

The diagram shows the steps and operations leading to the $925 total. To solve, work backwards from $925.

Start with the total.	925
Subtract the contribution.	$\begin{array}{r} -150 \\ \hline 775 \end{array}$
Divide by the hourly wage.	$775 \div 6.20 = 125$

Jarret worked 125 h.

Check: $(125 \cdot 6.20) + 150 = 925$

◤ TRY THESE EXERCISES

Write a variable expression for each situation.

1. In April the price of milk increased by 15%.

2. Phillipe's dad is three times as old as Phillipe.

3. The sale price of the shoes is $\frac{1}{4}$ the original price.

4. Jonica spends $\frac{1}{2}$ her free time reading and $\frac{1}{4}$ her free time shopping.

5. All clearance items at a department store are 30% off the original price. The store has a sale where all clearance items are an additional 50% off.

Five-step Plan

1 Read
2 Plan
3 Solve
4 Answer
5 Check

6. A new car has a sticker price of $14,586. This price includes the base price plus $1360 in options, $180 dealer preparation fee, and $66 license fee. Find the base price.

7. During April the price of an apple tripled. In May the price dropped $0.18. In June it halved. In July the price rose $0.06 to $0.30/apple. Find the price at the beginning of April.

8. Find the value of w in the diagram. The total area is 40 cm².

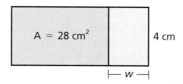

$A = 28$ cm²   4 cm

$\vdash w \dashv$

9. **RECYCLING** In 2002, there were 1767 landfills in operation. This is 1324 fewer than there were in 1998. In 1998, there were 3357 more landfills than in 1988. Find how many landfills were in operation in 1988.

10. The formula for the volume of a rectangular prism is $V = lwh$, where l is the length, w is the width and h is the height. Find the length of a rectangular prism with width 4 cm, height 13 cm and volume 416 cm³.

11. The formula for the volume of a cylinder is $V = \pi r^2 h$, where r is the radius of the base and h is the height. Find the height of a cylinder whose volume is 129.8 in.³ and radius is 5 in. Round to the nearest hundredth.

12. Lee wants to save $900 for his vacation. When he has saved four times as much as he already has saved, he'll need $44 more to reach his goal. How much has he saved?

13. Julie multiplied her age by 2, subtracted 5, divided by 3 and added 9. The result was 20. How old is she?

14. **WRITING MATH** Write a problem that can be solved by working backwards.

15. The formula for the area of a trapezoid is $A = \frac{1}{2}(b_1 + b_2)h$, where b_1 and b_2 are the lengths of the bases and h is the height. Lee wishes to buy the trapezoidal plot of land shown. The length of the land bordering the river is twice as long as the side bordering the street. The total area of the land is 14,040 ft². How long is the border along each side?

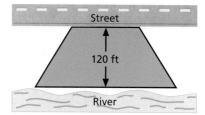

Street

120 ft

River

16. **CHAPTER INVESTIGATION** Contact a local recycling center to find the total MSW recycled and the MSW generated in your area. Calculate the MSW recycling rate.

MIXED REVIEW EXERCISES

Simplify. (Lessons 3-3, 3-7 and 3-1)

17. $4 - 5^2 + 2 \cdot 6 \div 3$

18. $28(14 + (-29)) \div (3 \cdot (-5))$

19. $112 + (-144) \div 2^2 - 8 \cdot (-4)$

20. 4.907×10^8

21. 1.16×10^{-5}

22. 6×10^{-4}

23. What is the sign on the product of an even number of negative numbers?

24. What is the sign on the product of an odd number of negative numbers?

5-8 Graph Open Sentences

Goals ■ Graph open sentences on a number line.

Applications Retail, Sports, Environment

Work in groups of two or three students.

1. List the numbers in order from least to greatest.

$$-1\frac{1}{2} \quad 2^2 \quad \pi \quad \sqrt{3} \quad -\frac{2}{9} \quad 4\frac{1}{4} \quad -3$$

2. Draw this number line. Place a solid dot to represent each number. Label each dot with its number.

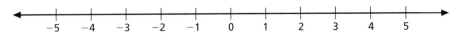

3. Using one of the symbols in the table, write four true statements using different pairs of numbers above.

4. Compare your statements with your classmates. Are they the same? Explain.

Symbol	Meaning
>	is greater than
<	is less than
≥	is greater than or equal to
≤	is less than or equal to

▼ BUILD UNDERSTANDING

A mathematical sentence that contains one of the symbols $<, >, \le$ or $\ge$ is called an **inequality**.

Inequality: $5 < 7$ $7 \ge y - 2$

In words: 5 is less than 7   7 is greater than or equal to $y - 2$

In Lesson 5-1, you learned that an open sentence is a sentence that contains one or more variables. An open sentence may be either an equation or an inequality. A **solution of an open sentence** is a value of the variable that makes the equation or inequality true.

Equation: $x + 7 = 11$ Inequality: $k > 3$

Solution: 4   Sample solutions: 4, 3.5, 7, π

The inequality $k > 3$ has an infinite number of solutions, since any real number greater than 3 is a solution.

You can graph the solutions to an equation or an inequality on a number line. A number line consists of the set of real numbers. You graph points by using a *solid dot*. To show that a point is not included you use an *open dot*. The table at the right may help you remember when to use an open dot and when to use a solid dot.

> **Think Back**
>
> *Real numbers* are the sets of rational and irrational numbers.
>
> *Rational numbers* can be expressed in the form $\frac{a}{b}$, where a is any integer and b is any integer except 0.
>
> *Irrational numbers* are non-terminating, non-repeating decimals.

Symbol	Meaning	Graph
>	is greater than	o
<	is less than	o
≥	is greater than or equal to	●
≤	is less than or equal to	●

Example 1

Solve the equation $m + 9 = 14$ and graph the solution.

Solution

$$m + 9 = 14$$
$$m + 9 - 9 = 14 - 9$$
$$m = 5$$

The equation has one solution. Graph the solution on a number line by drawing a solid dot at the point 5.

To check a solution of an inequality, substitute it for the variable and check that the resulting sentence is true.

Example 2

Graph each inequality.

a. $p > -2$ **b.** $k \leq 4$

Solution

a. The solutions are all real numbers greater than -2. Draw a solid arrow beginning at -2, pointing right. Since -2 is not part of the solution, draw an open dot at -2.

Check. Let $p = 0$.
$$p > -2$$
$$0 > -2$$

b. The solutions are all real numbers less than or equal to 4. Draw a solid arrow beginning at 4, pointing left. Since 4 is part of the solution, draw a solid dot at 4.

Check. Let $k = 1$.
$$k \leq 4$$
$$1 \leq 4$$

Example 3

RETAIL Every frame in the store is priced under \$35. Write an inequality to describe the situation. Name three possible frame prices.

Solution

Let c represent the cost of a frame. $c < 35$

There are variously priced frames. Three sample prices are \$34.99, \$30.00 and \$12.50.

Graph each open sentence on a number line.

1. $x - 7 = 3$

2. $h \geq -5$

3. $a < 1$

4. $p > -9$

5. $m + 2 = -3$

6. $n \leq 3$

Write an inequality to describe each situation.

7. There were more than 300 students in the gym.

8. On Tuesday, the price of unleaded gas was less than or equal to $0.94/gal.

9. **WRITING MATH** Explain the difference between $x > 7$ and $x \geq 7$.

10. **SPORTS** Gregor makes at least 8 foul shots in every basketball game. Write and graph an inequality to describe the situation.

▼ **PRACTICE EXERCISES** • For Extra Practice, see page 564.

Graph each open sentence on a number line.

11. $x > 4$

12. $m < -1$

13. $h + 5 = 2$

14. $y \leq 0$

15. $p \geq \dfrac{1}{2}$

16. $-5 = a - 6$

17. $n > -8$

18. $y \geq 3.5$

19. $m + 10 = 5$

20. $b \geq 12$

21. $x - 6 = 3$

22. $s < 5\dfrac{1}{2}$

23. $8s + 9 = -23$

24. $9h + 6 = -66$

25. $d \geq -10$

Write three solutions to the following inequalities.

26. $x \leq -2$

27. $p > 18$

28. $f \geq -6$

29. $m > -12$

30. $w \leq -2.4$

31. $k < -\dfrac{1}{5}$

Write an inequality to describe each situation.

32. The budget must not exceed $4500.

33. The stadium can hold up to 52,000 people.

34. Throughout the city, there were at least 5 in. of snowfall.

35. Joaquin's height is greater than or equal to 2 m.

36. Americans throw away more than 242 million automotive tires each year.

Write an open sentence for each graph.

37.

38.

39.

40.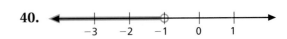

SAFETY Write an inequality to represent the allowable driving speed *s* on each sign.

41.
```
SCHOOL
SPEED
LIMIT
20
```

42.
```
SPEED
LIMIT
45
```

ENVIRONMENT In this decade, Americans will throw away the following items. Write an inequality to describe each situation.

43. over 1 million tons of aluminum cans and foil

44. more than 11 million tons of glass bottles and jars

45. over $4\frac{1}{2}$ million tons of office paper

46. at most 10 million tons of newspaper

47. WRITING MATH Write a situation to describe the graph shown.

EXTENDED PRACTICE EXERCISES

DATA FILE For Exercises 48 and 49, refer to the data on Broadway Runs on page 522.

48. How many more performances could *Rent* have and still have fewer performances than *Cats*?

49. How many more performances could *The Lion King* have and still have fewer performances than *Grease*?

Write an inequality to describe each situation.

50. The cost of renting a car is $75 per day plus $0.30/ mi. The total cost for one day cannot exceed $100. Let *m* represent the number of miles.

51. Double the distance is at least 500 miles. Let *d* represent the distance.

52. One-third of the area is no more than 24 ft^2. Let *a* represent this area.

MIXED REVIEW EXERCISES

Find the value of each variable. (Lesson 3-8)

53. $4^8 \cdot 4^t = 4^{16}$ **54.** $9^g \div 9^{12} = 9^{-3}$ **55.** $(15^3)^r = 15^{15}$

Decide whether each statement is true or false. (Lesson 3-4)

56. The associative property can be used when subtracting.

57. The distributive property can be used when subtracting.

Review and Practice Your Skills

PRACTICE ■ LESSON 5-7

Write an expression to represent each situation.

1. Art spends 40% of his homework time on math.

2. The cost of our electricity went up 5% last month.

3. Jose is one-half as old as his father.

4. I paid for $\frac{1}{3}$ of the stereo.

5. The price of the computer was 10% less than last year.

6. The sweater was $\frac{3}{4}$ of the original price.

7. Kelley plans to save twice as much money this month as last month.

8. Gail saved 20% when she bought the coat on sale.

9. My average speed was three times as great on the second half of my trip.

PRACTICE ■ LESSON 5-8

Graph each open sentence on a number line.

10. $w + 2 = -3$ 11. $y > -2\frac{1}{2}$ 12. $r \leq 5$

13. $a < -6$ 14. $1 = 5 + b$ 15. $w > -1$

16. $2d - 3 = -1$ 17. $n \leq 3$ 18. $c \geq -4.5$

19. $t \geq 3.5$ 20. $p < -1.5$ 21. $q \leq \frac{4}{3}$

22. $10 + 3e = 1$ 23. $5 - m = 8$ 24. $k > -7$

25. $6 + p = 2$ 26. $g \geq -4$ 27. $v < 0.25$

Write three different solutions for each inequality.

28. $p > -6$ 29. $m \leq 4$ 30. $b < 3$

31. $c \leq 5$ 32. $x < -1$ 33. $p \geq 0$

34. $a \geq 3.5$ 35. $y < 6.2$ 36. $w \leq -0.5$

37. $h > \frac{1}{3}$ 38. $q \geq -\frac{7}{8}$ 39. $f < -3\frac{1}{10}$

Write an inequality to describe each situation. Let *n* represent "a number."

40. A number can be any value up to 25.

41. A number is at least 10.

42. A number cannot be less than 50.

43. A number must be less than -100.

44. A number must not exceed 450.

45. A number must be more than 6.

Which of the given values is a solution of the given equation? (Lesson 5-1)

46. $-2 = e - 9$; $-11, 7, 11$

47. $3 = \dfrac{w - 1}{4}$; $-13, 11, 13$

48. $-1 = 5r + 9$; $-2, \dfrac{1}{5}, 2$

Solve each equation. Check the solution. (Lessons 5-2)

49. $a + (-4) = 1$

50. $8 = d - (-4)$

51. $10 - s = -2$

52. $3.2 = b - 6.1$

53. $4.5 + h = -0.8$

54. $x + \dfrac{1}{3} = \dfrac{1}{5}$

Solve each equation. Check the solution. (Lesson 5-3)

55. $\dfrac{r}{4} = -1$

56. $0.4c = -3.6$

57. $\dfrac{x}{5} = \dfrac{-3}{10}$

58. $2 = -8p$

59. $-3 = \dfrac{m}{-9}$

60. $\dfrac{3}{2} = \dfrac{-w}{8}$

Solve each equation. Check the solution. (Lesson 5-4)

61. $-1 = 2m + 3$

62. $3x - 4 = -7$

63. $4 + \dfrac{h}{2} = -3$

64. $2 = \dfrac{a}{3} - 5$

65. $4 - 3p = -6$

66. $\dfrac{6}{-5} = \dfrac{d}{2}$

Solve each equation. Check the solution. (Lesson 5-5)

67. $d + 4d - 3 = -8$

68. $4 = 2m + 1 - m$

69. $3p + 2 = 7 - p$

70. $-1 + 3y = 2y + 6$

71. $2(e - 5) = -3e$

72. $5h + 2 = -3(1 - 2h)$

Solve each formula for the indicated variable. (Lesson 5-6)

73. $y = mx + b$, for m

74. $d = st$, for t

75. $I = prt$, for r

The formula for the area of a rectangle is $A = lw$, where l is the length and w is the width. Find the area or the missing length. (Lesson 5-6)

76. area = 4 ft²,
length = 8 ft
Find the width.

77. width = 15 yd,
length = 3 yd
Find the area.

78. area = 8 m²,
width = 0.2 m
Find the length.

Write an expression to represent each situation. (Lesson 5-7)

79. The population decreased 35% since 1990.

80. The speed of the winning car was 15% greater than last year.

81. The class raised $\dfrac{1}{8}$ of the money needed for the new school parking lot.

Graph each open sentence on a number line. (Lesson 5-8)

82. $p \geq -5$

83. $a \leq \dfrac{3}{2}$

84. $e > -5$

85. $2 + n = -8$

86. $g \geq -1$

87. $4 = -3 - b$

88. $z < 1.5$

89. $3 = 2g + 9$

90. $h < -4\dfrac{1}{2}$

5-9 Solve Inequalities

Goals ■ Solve inequalities with one variable.

Applications Finance, Transportation, Hobbies

Work with a partner. The integers shown are ordered from least to greatest. The following eight steps involve a single operation.

| −8 | −4 | 0 | 4 | 12 |

Step 1 add 6

Step 2 add −4

Step 3 subtract 8

Step 4 subtract −3

Step 5 multiply by 5

Step 6 multiply by −1

Step 7 divide by 2

Step 8 divide by −4

1. Perform Step 1 on the five integers. List your results.

2. Repeat Question 1 for Steps 2–8.

3. List the Steps in which the operations you performed resulted in a set of integers ordered from least to greatest.

4. List the Steps in which the operations you performed resulted in a set of integers ordered from greatest to least.

◤ BUILD UNDERSTANDING

Many equations have only one solution. As you discovered in Lesson 5-8, an inequality may have an infinite number of solutions. To find the solutions of an inequality containing several operations, follow the rules for solving equations.

Example 1

Solve $k + 3 \geq -4$ and graph the solution.

Solution

$$k + 3 \geq -4$$

$$k + 3 - 3 \geq -4 - 3 \quad \text{Undo the addition.}$$

$$k \geq -7 \quad \text{Simplify.}$$

Draw the graph.

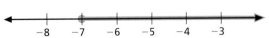

The solution is all real numbers greater than or equal to −7.

The following rule is the only difference between solving an equation and solving an inequality.

Inequality Rule	When multiplying or dividing both sides of an inequality by a negative number, reverse the direction of the inequality symbol.

Example 2

Solve $-3n - 5 < 1$ and graph the solution.

Solution

$$-3n - 5 < 1$$

$$-3n - 5 + 5 < 1 + 5 \qquad \text{Undo the subtraction.}$$

$$-3n < 6$$

$$\frac{-3n}{-3} > \frac{6}{-3} \qquad \text{Reverse the direction of the inequality since you are dividing by } -3.$$

$$n > -2$$

Draw the graph.

The open circle shows that -2 is not a solution.

Choose a point to check your solution.

The solution is all real numbers greater than -2.

Check Understanding

Is -3 a solution of each inequality?

1. $x + 1 \leq 3$ 2. $\frac{x}{6} > 0$

3. $2x + 12 > -3$ 4. $-\frac{9}{x} \geq 4$

Example 3

TRANSPORTATION The pilot of a small plane wants to stay at least 700 ft beneath the storm clouds, which are at 5500 ft. The plane is ascending at a rate of 640 ft/min. How long after taking off can the pilot ascend before reaching the clouds?

Solution

Write and solve an inequality that represents the situation.

Let m = number of minutes ascending at 640 ft/min.

$$640m + 700 \leq 5500$$

$$640m + 700 - 700 \leq 5500 - 700$$

$$\frac{640m}{640} \leq \frac{4800}{640}$$

$$m \leq 7.5$$

The pilot can climb for up to 7.5 min.

Solve and graph each inequality.

1. $x - 9 \geq -5$

2. $5y < -45$

3. $-2p \leq 10$

4. $p + 9 < 6$

5. $-7 < 2k + 3$

6. $5n - 1 \geq 14$

7. RETAIL Kaitlyn receives $75 for her birthday. She wants to buy a sweater that costs $41 and two CDs. Use the inequality $2c + 41 \leq 75$ to find how much she can spend on each CD.

8. Members of the orchestra hope to raise at least $1925 from the sale of tickets to the spring concert. They estimate that 350 people will attend the concert. Let c represent the amount they will charge per ticket. Write and solve an inequality to determine the cost of each ticket.

9. WRITING MATH Explain the similarities and differences between the solution of $x + 6 = 3$ and $x + 6 < 3$.

PRACTICE EXERCISES • For Extra Practice, see page 565.

Solve and graph each inequality.

10. $w - 6 < -2$

11. $3p \geq -12$

12. $x + 9 \leq 3$

13. $5n > 0$

14. $-\dfrac{y}{3} \geq 1$

15. $4 < \dfrac{2}{3}h$

16. $15 + t > 17$

17. $12 \leq -4a$

18. $\dfrac{1}{7}m > -1$

19. $13 \leq c + 8$

20. $-x < 2$

21. $k - (-3) > -2$

22. $6w + 2 > -4$

23. $4 + 5y < 29$

24. $7n + 4 \geq -10$

25. $4\left(x - \dfrac{1}{2}\right) < 6$

26. $2y - 5y \geq 12$

27. $-2 + 5(2) > -2(x - 1)$

28. $\dfrac{2}{3}x - \dfrac{3}{4}x \leq 1$

29. $-1 < 4n - 5n$

30. $15(a + 4) \leq 5(12 - 5a)$

Write and solve an inequality.

31. RETAIL Michelle is saving money to buy a new leather jacket. The jacket she wants is $349. She has saved $116. Determine the least amount that Michelle must still save to be able to buy the jacket.

32. FINANCE Miyoshi borrows $1500 for a new car stereo system. She will pay $50 a month until the loan is repaid. Determine the number of months it will take before her debt will be less than $200.

33. The current math department budget will allow no more than $6000 to be spent on technology. The math department decides to buy one computer that will cost $2400. The rest will be spent on graphing calculators, which cost $80 each. Determine the number of calculators that the math department can purchase.

34. Kathy earns $7/h and double that for any time over 40 h. This week she wishes to earn at least $350. Determine the number of hours that she should work.

35. Ashanti scores an 85, 96, and 87 on her first three math tests. Determine the score she must get on her fourth math test to have an average of at least 90 for all the tests.

36. HOBBIES On an ocean dive, Tracy wants to stay at least 7 m above her deepest previous depth, 40 m. How long can she descend at a rate of 2.2 m/min?

37. YOU MAKE THE CALL Olisa translated the sentence "*Marta spent no more than $24*" as $m > 24$. Ryan translated it as $m \leq 24$. Who is correct? Explain.

38. Write a problem that could be solved using the inequality $t - 5 > 9$.

39. Write a problem that could be solved using the inequality $\frac{x}{2} < 32$.

40. WRITING MATH Explain why $x > 11$ and $11 < x$ are equivalent statements.

EXTENDED PRACTICE EXERCISES

Find all integers that are solutions of both inequalities.

41. $x < -4$ and $x > -8$ **42.** $x \geq 3$ and $x < 9$ **43.** $x < -2$ and $x > -5$

Write and solve an inequality for each statement.

44. Four times a number is greater than or equal to -32.

45. The sum of four and a number is less than seven.

46. Seven times a number, added to four, is greater than negative ten.

47. Ted and Tom worked more than 14 h painting a garage. Ted worked 2 h more than Tom. Find the least whole number of hours each of them might have worked.

48. CRITICAL THINKING The product of an integer and 8 is less than 43. Find the greatest integer that meets this condition.

MIXED REVIEW EXERCISES

49. RECYCLING A company charges $20.67 per month to remove waste and recyclable items from your home. They also charge a $15 one-time fee for a recycling bin. The first bill was $97.68. Write and solve an equation to determine the number of months the bill included.

Find the value of the variable in each equation.

50. $17 + s = -25$ **51.** $\frac{3}{4}d = -21$ **52.** $28 - k = 32$

53. $10x - 15 = 25$ **54.** $\frac{(y - 14)}{-2} = 21$ **55.** $\frac{3}{8}t + 6\frac{1}{4} = -5\frac{5}{6}$

56. TECHNOLOGY Use a graphing utility to graph $n > 3$ and $3 > n$. How do the graphs differ?

Chapter 5 Review

VOCABULARY ◢

Choose the word from the list that best completes each statement.

1. When you __?__, you find all values of the variable that makes the equation true.

2. A(n) __?__ is the value of the variable that makes the equation or inequality true.

3. The parts of a variable expression that are separated by addition or subtraction signs are called __?__.

4. A mathematical sentence that contains one of the symbol's $<$, $>$, $\leq$, or $\geq$ is called a(n) __?__.

5. A(n) __?__ is a statement that two numbers or expressions are equal.

6. The fractions $\frac{1}{8}$ and $\frac{8}{1}$ are called __?__.

7. The terms $3x^2$ and $2x$ are __?__.

8. In the equation $3y + 2 = 8$, 2 is the __?__.

9. A math sentence that relates two or more variables is called a(n) __?__.

10. The terms $2m$ and $3m$ are __?__.

a.	equation
b.	formula
c.	inequality
d.	like terms
e.	reciprocals
f.	solution of an open sentence
g.	solution of the equation
h.	solve an equation
i.	terms
j.	unlike terms
k.	variables
l.	work backwards

LESSON 5-1 ◢ Introduction to Equations, p. 208

▶ An **equation** is a statement that two numbers or expressions are equal.

▶ An **open sentence** is a sentence that contains one or more variables.

▶ A value that makes an equation true is called a **solution of the equation.**

Tell whether each equation is true, false, or an open sentence.

11. $k - 7 = -1$ 12. $-4 + 6 = -6 + 4$ 13. $-2(3 + 5) = -16$

14. $3(2 + 6) = 6 + 6$ 15. $2x + 5 = 15$ 16. $8 - 12 = 10 - 6$

17. $3m - 5m = 16$ 18. $-5 + 7 = 4 - 2$ 19. $18(3 - 5) = 4 - 40$

Which of the given values is a solution of the given equation?

20. $36 = -18k$; 2, -2, 18   21. $c + 1\frac{1}{2} = 2$; $\frac{1}{2}$, $3\frac{1}{2}$ 22. $3.42 = 2p$; 3.4, 6.84, 1.71

23. $d + 5 = 12$; 7, -7 24. $6.4g = 16$; 3.1, 2.5   25. $54 = -6j$; 9, -9, 12

26. $88 = -4r$; -22, 22   27. $t - 12 = 32$; 20, 44   28. $8.5 = -2k$; -4.25, 4.25

LESSON 5-2 ◢ Add or Subtract to Solve Equations, p. 212

▶ To **solve an equation,** undo an operation by using the inverse operation. Addition and subtraction are inverse operations.

Solve each equation. Check the solution.

29. $x - 5 = -2$

30. $y + 8 = -3$

31. $8 - r = 10$

32. $m + 6 = 17$

33. $12 - b = -15$

34. $p - 5.6 = 12.7$

35. $3.5 - r = 6$

36. $w + 23 = 56$

37. $z + 6 = -19$

38. Ramon bought a CD for $13, leaving him with $12. How much money did he have with him before he bought the CD? Write an equation and solve.

39. Paul gave his sister 18 cards, leaving him 29 cards. How many cards did he have to begin with? Write an equation and solve.

LESSON 5-3 ◣ Multiply or Divide to Solve Equations, p. 218

▶ Solve equations involving multiplication and division using the inverse operation.

▶ Reciprocals are two numbers whose product is 1.

Solve each problem. Check the solution.

40. $330 = 33m$

41. $\dfrac{x}{-2} = 21$

42. $12 = \dfrac{1}{4}d$

43. $\dfrac{y}{3} = 4$

44. $-x = 15$

45. $2t = -34$

46. $3d = -15$

47. $-\dfrac{1}{3}t = 24$

48. $220 = 11x$

Write an equation for each sentence.

49. The product of six and h is negative twenty-four.

50. The quotient of four and z is two.

LESSON 5-4 ◣ Solve Two-Step Equations, p. 222

▶ To solve *two-step equations,* undo addition and subtraction first. Then undo multiplication and division.

▶ To solve a proportion that has an unknown term, use cross products.

Solve each problem. Check the solution.

51. $7r + 12 = 5$

52. $6(s - 3) = 12$

53. $3.2x - 2.1 = 4.3$

54. $\dfrac{y}{6} + 7 = 1$

55. $3x + 3 = 15$

56. $2y - 6 = -12$

57. $5t - 7 = 18$

58. $6(2 - r) = 6$

59. $4 - \dfrac{w}{3} = 2$

60. $5(c + 3) = 20$

61. $4.4v + 3 = 11.8$

62. $3.6y - 8 = 6.4$

LESSON 5-5 ◣ Combine Like Terms, p. 228

▶ In some equations it is necessary to first simplify both sides of the equation by combining **like terms.**

Simplify.

63. $3m + 4m$

64. $8x - 2x + 3d$

65. $12(d - 2) + 3(2 - d)$

66. $4t - 6t$

67. $5(3z - 2z)$

68. $3(12 - n) - 4(n - 5)$

69. $3y - 8y + 8x$

70. $7(j - k)$

71. $12(4 + w) - 3(w + 2r)$

LESSON 5-6 ◼ Use Formulas to Solve Problems, p. 232

▶ A **formula** is an equation stating a relationship between two or more variables.

Solve each formula for the indicated variable.

72. $I = prt$, for p **73.** $y = mx + b$, for m **74.** $P = 2l + 2w$, for w

75. $d = rt$, for t **76.** $A = \frac{1}{2}bh$, for h **77.** $V = \frac{1}{3}\pi r^2 h$, for h

78. Use the formula $d = r \cdot t$. A car traveled 252 miles in 4.5 h. Find its average rate.

LESSON 5-7 ◼ Problem Solving Skills: Work Backwards, p. 238

▶ To solve some problems start at the end and work backwards to the beginning.

79. During March the price of snow shovels doubled. In June the price dropped $10.00. In August it halved. In September the price rose $3.00 to $16/shovel. Find the price at the beginning of March.

80. A certain bacteria doubles its population every 12 hours. After 3 full days, there are 1600 bacteria in a culture. How many bacteria were there at the beginning of the first day?

LESSON 5-8 ◼ Graph Open Sentences, p. 240

▶ **Inequalities** can be graphed on a number line. A solid dot indicates the number is part of the solution. An open circle indicates that the number is not part of the solution.

Graph each open sentence on a number line.

81. $m < -1$ **82.** $-7 = p - 2$ **83.** $m \geq 2$

Write an open sentence for each graph.

84.
```
  ←——+——+——+——+——●——+——→
     -1  0  1  2  3  4
```

85.
```
  ←——+——+——○——+——+——+——→
        -3 -2 -1  0  1  2
```

86.
```
  ←——+——+——+——+——+——+——●——+——+——+——+——+——→
    -5 -4 -3 -2 -1  0  1  2  3  4  5  6
```

87.
```
  ←——+——○——+——+——+——+——+——+——+——+——+——→
    -5 -4 -3 -2 -1  0  1  2  3  4  5  6
```

LESSON 5-9 ◼ Solve Inequalities, p. 246

▶ Follow all steps for solving equations except for the following situation: When multiplying or dividing both sides of an inequality by a negative number, reverse the direction of the inequality.

Solve and graph each inequality.

88. $x - 4 < -3$ **89.** $-3n \geq 12$ **90.** $\dfrac{h}{-6} + 5 > -2$

CHAPTER INVESTIGATION

EXTENSION The EPA has a goal for each area of the U.S. to have an MSW recycling rate of 25%. How close is your local area to this goal? Use a bar graph to display the MSW recycling rate for each time period and for your area.

Chapter 5 Assessment

Tell whether the equation is *true*, *false* or an *open sentence*.

1. $v + 8 = -6$

2. $3(4 - 5) = 3$

3. $-8 + 6 = 2(4 - 3)$

Solve each equation. Check the solution.

4. $e - 5 = 2$

5. $-6d = 420$

6. $m + 4.4 = 12.2$

7. $\dfrac{x}{3} = -5$

8. $5m - 3 = 17$

9. $19 = 2c + 3$

10. $\dfrac{-5}{6}y = \dfrac{1}{5}$

11. $\dfrac{3}{2}x + 4 = 2$

12. $2(y - 3) = 14$

13. $5d + 5 = 2d - 4$

14. $28 = \dfrac{x}{-2} + 34$

15. $-3(m - 1) = 2(8 + m) - 4$

Solve.

16. Convert a temperature of 167°F to Celsius scale. Use the formula $F = \dfrac{9}{5}C + 32$ (F = Fahrenheit, C = Celsius).

17. Solve the formula $P = 24 + d$ for d.

Simplify.

18. $x + 2x + 3x^2 + 5x$

19. $7(a + b - c) - 3(a + 2b - 4c)$

Graph each open sentence on a number line.

20. $f \leq -2$

21. $2x - 5 = -7$

22. $2 < 0.5x + 1.5$

23. $-4p + 9 \leq -3$

Write an open sentence for each graph.

24.

25.

Write and solve an equation for the situation.

26. For a bike-a-thon, Wendy had pledges that totaled $34.35 for each kilometer she rode. She collected $64.55 in other donations. If she collected $751.55 in all, how far did she ride?

27. Marcellian earns $6.90 per hour. His weekly salary before taxes is $241.50. How many hours does he work per week?

Write and solve an inequality.

28. Mr. Cintron needs at least $325 as a down payment for a couch he wants to purchase. He has $123 saved so far. Determine the least amount of money he needs to save to purchase the couch.

29. Gretchen earns $6.90/h and double time for any time over 40 h. She works a whole number of hours each week. This week she wishes to earn at least $500. Determine the number of hours she should work.

Standardized Test Practice

Part 1 | Multiple Choice

Record your answers on the answer sheet provided by your teacher or on a sheet of paper.

1. The scores on Tammy's first four math tests are 76, 88, 96, and 90. What score must she earn on the fifth test to have a test average of 90? (Lesson 1-2)
   - Ⓐ 95
   - Ⓑ 96
   - Ⓒ 98
   - Ⓓ 100

2. Which unit is most appropriate to find the amount of water to fill a bathtub? (Lesson 2-1)
   - Ⓐ cup
   - Ⓑ gallon
   - Ⓒ pint
   - Ⓓ quart

3. Marco is making cookies for his classmates. The recipe makes 36 cookies and calls for $1\frac{1}{3}$ c of flour. How much flour is needed to make 48 cookies? (Lesson 2-6)
   - Ⓐ $2\frac{3}{4}$ c
   - Ⓑ $2\frac{1}{6}$ c
   - Ⓒ $1\frac{7}{9}$ c
   - Ⓓ $1\frac{1}{2}$ c

4. At 7:00 A.M., the temperature was 15°F. By 3:00 P.M., the temperature was −10°F. By how many degrees did the temperature drop? (Lesson 3-1)
   - Ⓐ 5°F
   - Ⓑ 10°F
   - Ⓒ 15°F
   - Ⓓ 25°F

5. A video store has previously viewed DVDs on sale at 3 for $20, with a limit of 3 at the sale price. Additional DVDs are available at the regular price of $9.95 each. Which expression could be used to find the cost of 6 DVDs? (Lesson 3-3)
   - Ⓐ 6 · 9.95
   - Ⓑ 20 + 3 · 9.95
   - Ⓒ 3 · 20 + 3 · 9.95
   - Ⓓ 20 + 6 · 9.95

6. Which net could be folded into a cube? (Lesson 4-3)
   - Ⓐ
   - Ⓑ
   - Ⓒ
   - Ⓓ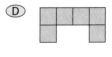

7. What is the volume of the cylinder? Use 3.14 for π. (Lesson 4-7)

   3 cm
   5 cm
   - Ⓐ 47.1 cm³
   - Ⓑ 94.2 cm³
   - Ⓒ 135.7 cm³
   - Ⓓ 141.3 cm³

8. Tanesha is covering a small box with fabric. What is the surface area of the box? (Lesson 4-9)

   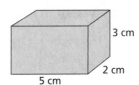

   3 cm
   2 cm
   5 cm
   - Ⓐ 30 cm²
   - Ⓑ 42 cm²
   - Ⓒ 50 cm²
   - Ⓓ 62 cm²

9. Tamika's family is returning from vacation. If the distance is 378 mi, how long will it take to drive home traveling at 54 mi/h? (Lesson 5-6)
   - Ⓐ 5.6 h
   - Ⓑ 7 h
   - Ⓒ 8.6 h
   - Ⓓ 9 h

Test-Taking Tip

Question 6
Use estimation to help eliminate answer choices that are unreasonably large or small.

Part 2 Short Response/Grid In

Record your answers on the answer sheet provided by your teacher or on a sheet of paper.

10. The length of a board is 12 ft 8 in. The board is to be cut into 4 pieces of equal length. How long is each piece? (Lesson 2-2)

11. Use the law of exponents to simplify $(6^3)^2$. (Lesson 3-8)

12. Classify the triangle. (Lesson 4-2)

13. Name a polyhedron with only one base. (Lesson 4-3)

14. How many cubes are there in the isometric drawing? (Lesson 4-5)

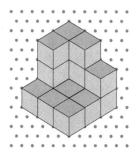

15. Eight less than ten times a number is 82. What is the number? (Lesson 5-5)

16. Matt volunteered to stuff 1000 envelopes for a nonprofit organization. If he can stuff 12 envelopes in one minute and 112 envelopes are already finished, how many minutes will it take to finish the task? (Lesson 5-4)

17. A car's fuel economy E is given by the formula $E = \frac{m}{g}$, where m is the number of miles driven and g is the number of gallons of fuel used. If a car has an average fuel consumption of 30 mi/gal while using 9.5 gal, how far can the car go? (Lesson 5-6)

18. Katie used half of her allowance to buy a ticket to the class play. Then she spent $1.75 for an ice cream cone. Now she has $2.25 left. How much is her allowance in dollars? (Lesson 5-7)

Part 3 Extended Response

Record your answers on a sheet of paper. Show your work.

19. Toby is constructing a model of a rocket. He used a cylinder for the base and a cone for the top. (Lessons 4-7 and 4-8)

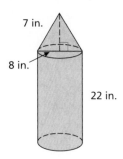

7 in.

8 in.

22 in.

a. What is the volume of the cone to the nearest cubic inch?

b. What is the volume of the cylinder to the nearest cubic inch?

c. What is the total volume of the rocket once it is assembled? Round to the nearest cubic inch.

20. A popcorn company is deciding between two types of packaging for its product. The first is a rectangular prism, and the second is a cylinder. (Lessons 4-7 and 4-9)

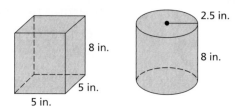

8 in.

5 in.

5 in.

2.5 in.

8 in.

a. Find the surface area of each container.

b. Find the volume of each container.

c. Which container uses less packaging? Which container holds more product?

Math Online mathmatters1.com/standardized_test

Equations
and Percents

THEME: Living on Your Own

In this chapter, you will solve problems that involve percents, discounts, taxes, commission, and simple interest. These terms are used in many aspects of daily life such as retail, employment and finances.

Living on your own involves managing money and making daily decisions. Taxes, fees, and interest charges can help you decide where to live, how to save money and which items to purchase. If you understand equations and percents, you are better prepared to live on your own.

- **Financial counselors** (page 269) advise customers on how to manage and invest their money and assets. They help their clients establish budgets, invest their money, and manage spending.

- **Political scientists** (page 289) study government and how the general population interacts and participates in it. They survey people about their opinions, research legal issues, and study election results.

Math
Online

mathmatters1.com/chapter_theme

Average Apartment Rent and Square Footage

Type	Monthly rent (dollars)	Square footage
Studio	472	475
One bedroom	624	733
Two bedroom	739	1066
Three bedroom	951	1240

Where Americans Live

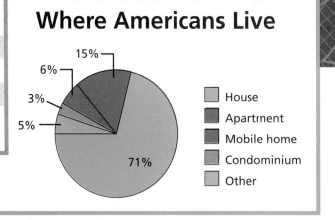

- House
- Apartment
- Mobile home
- Condominium
- Other

15%
6%
3%
5%
71%

Data Activity: Where Americans Live

Use the table and circle graph for Questions 1–5.

1. If 1,500,000 people were polled for the data in the circle graph, how many of them live in a house?

2. How many of them live in an apartment?

3. How many of them live in a mobile home?

4. If you and a roommate rent a two-bedroom apartment and split the rent equally, how much do you pay in monthly rent?

5. If you and 2 roommates rent a two-bedroom apartment and split the rent equally, how much do you pay in monthly rent?

CHAPTER INVESTIGATION

Credit cards often charge miscellaneous fees and finance charges. Before you begin using credit cards, you need to understand terms like grace period, annual percentage rate (APR) and annual fees. Once you know the language of credit cards, you can save money and manage debt.

Working Together

Your first purchase on a credit card is a new sofa for $875.00. Your credit card has no annual fee, and an annual interest rate of 18.7%. Use the Chapter Investigation icons throughout the chapter to help you find the most economical way to pay off the credit card bill.

6 Are You Ready?

Refresh Your Math Skills for Chapter 6

The skills on these two pages are ones you have already learned. Use the examples to refresh your memory and complete the Exercises. For additional practice on these and more prerequisite skills, see pages 536–544.

PERCENT

You may remember that "percent" means "per 100." The percent form of a decimal number has been multiplied by 100. Converting from percent to decimal and decimal to percent involves multiplying and dividing by 100.

Examples

$57.6\% = 57.6 \div 100 = 0.576$ $7.1 = 7.1 \times 100 = 710\%$

$0.03\% = 0.03 \div 100 = 0.0003$ $0.08 = 0.08 \times 100 = 8\%$

Change each percent to a decimal and each decimal to a percent.

1. 38%
2. 5.7%
3. 0.62%
4. 126%
5. 1.6%
6. 0.09%
7. $\frac{1}{2}\%$
8. $\frac{2}{5}\%$
9. 4.9
10. 0.6
11. 0.483
12. 0.035
13. 1.001
14. 0.0081
15. $\frac{1}{2}$
16. $\frac{2}{5}$

WRITING EQUATIONS

English terms can be "translated" to math terms to write an equation.

Examples   What is the sum of 47 and 8?   The product of a number and 7 is $3\frac{1}{8}$

$x = 47 + 8$ $7x = 3\frac{1}{8}$

Write an equation for each statement.

17. The number x is the difference of 27.9 and $\frac{2}{6}$.
18. The number x is the quotient of $4\frac{1}{5}$ and $\frac{2}{3}$.
19. The sum of 39 and x is 12.
20. Twice the difference of a number and 15 is 32.
21. The number x is the product of 7.1 and 3.2.
22. Half the quotient of 42 and a number x is $3\frac{1}{2}$.
23. The product of x and $2x$ is 2.
24. The sum of the length and the width is 12.
25. The length is two times the width.
26. A triangle has a base that is three times its height.

258   Chapter 6 **Equations and Percents**

FORMULAS

You have used formulas to find perimeter, area, and volume. Formulas can help you solve problems involving percents, too.

Example   Find the area of a triangle with a base of 7.1 in. and a height of 6.2 in.

$$A = \frac{1}{2}b \cdot h$$

$$A = \frac{1}{2}(7.1)(6.2)$$

$$A = 22.01 \text{ in.}^2$$

Choose the appropriate formula and solve each problem.

27. Find the perimeter of a square of width 1.7 cm.

28. Find the area of a rectangle that measures 1.6 in. by 2.75 in.

29. Find the volume of a 5-ft cube.

30. Find the volume of a cone with radius 1 m and height 1 m.

31. Find the area of a triangle with a base of 1.7 yd and a height of 3 yd.

32. Find the area of a trapezoid with bases that measure 18 cm and 25 cm and has a height of 10 cm.

33. Find the volume of a rectangular prism with a length of 15 cm, width of 7.6 cm, and height of 14.7 cm.

34. Find the volume of a square pyramid that measures 5 ft tall and whose base measures 2 ft on each side.

$$P = 2l + 2w$$

$$P = 4s$$

$$A = l \cdot w$$

$$A = \frac{1}{2}b \cdot h$$

$$A = \frac{1}{2}(b_1 + b_2)h$$

$$V = l \cdot w \cdot h$$

$$V = B \cdot h$$

$$V = \frac{1}{3}B \cdot h$$

$$V = s^3$$

SOLVING PROPORTIONS

You can solve proportions by finding the cross-products.

Examples

$$\frac{5}{7} = \frac{x}{14}$$

$$5 \cdot 14 = 7x$$

$$\frac{70}{7} = \frac{7x}{7}$$

$$10 = x$$

$$\frac{14}{23} = \frac{27}{x}$$

$$14x = 23 \cdot 27$$

$$\frac{14x}{14} = \frac{621}{14}$$

$$x = 44\frac{5}{14} \text{ or about } 44.4$$

Find the value of x in each proportion. Round to the nearest tenth.

35. $\dfrac{1}{7} = \dfrac{x}{9}$

36. $\dfrac{2}{17} = \dfrac{5}{x}$

37. $\dfrac{42}{51} = \dfrac{x}{100}$

38. $\dfrac{x}{11} = \dfrac{12}{50}$

39. $\dfrac{14}{x} = \dfrac{5}{7}$

40. $\dfrac{x}{21} = \dfrac{17}{100}$

41. $\dfrac{4}{x} = \dfrac{2.5}{7}$

42. $\dfrac{x}{16} = \dfrac{7}{12}$

43. $\dfrac{14}{27} = \dfrac{x}{11}$

44. $\dfrac{28}{4.9} = \dfrac{41}{x}$

45. $\dfrac{2.1}{0.8} = \dfrac{6}{x}$

46. $\dfrac{23}{14.5} = \dfrac{x}{17}$

6-1 Percents and Proportions

Goals ■ Use a proportion to solve problems involving percents.

Applications Retail, Weather, Real estate, Travel, Living on your own

Work with a partner.

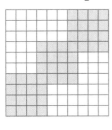

Grid A   Grid B   Grid C

1. Estimate the ratio of the number of shaded parts to the total number of squares for each grid shown.

2. Count the grid squares to find the actual ratio for each grid.

3. How do your estimates compare with the actual ratios?

4. Draw a design on a 10 by 10 grid. Estimate the ratio of the shaded squares to the total number of squares. Then count to find the actual ratio.

◤ BUILD UNDERSTANDING

A **percent** is a special ratio that compares a number to 100. Percent means "per one hundred." To find the percent of a number, you can write a proportion using the following relationship.

$$\frac{\text{part}}{\text{whole}} = \frac{\text{part}}{\text{whole}}$$

In one of the ratios, the percent is *always* the part and 100 is *always* the whole.

Example 1

What number is 60% of 45?

Solution

Let x = the part. The whole is 45. Write and solve a proportion.

$$\begin{array}{c} \text{part} \rightarrow \\ \text{whole} \rightarrow \end{array} \frac{60}{100} = \frac{x}{45} \begin{array}{c} \leftarrow \text{part} \\ \leftarrow \text{whole} \end{array}$$

$$60 \cdot 45 = 100 \cdot x \quad \text{Use the cross-products.}$$

$$2700 = 100x$$

$$\frac{2700}{100} = \frac{100x}{100}$$

$$27 = x$$

■ 27 is 60% of 45.

> ### Problem Solving Tip
>
> Calculators may require different keystrokes to find the percent of a number.
>
> To find 40% of 360, one calculator may require this key sequence:
>
> 360 ☒ 40 %
>
> Another calculator may not have a % key and will require this key sequence:
>
> 360 ☒ 0.40 ENTER

Example 2

What percent of 65 is 13?

Solution

Let y = the percent. Since you are to find the percent, y is over 100.

$$\frac{y}{100} = \frac{13}{65} \quad \leftarrow \text{part} \atop \leftarrow \text{whole}$$

$$y \cdot 65 = 100 \cdot 13 \quad \text{Use the cross-products.}$$

$$65y = 1300$$

$$\frac{65y}{65} = \frac{1300}{65}$$

$$y = 20$$

■ 13 is 20% of 65.

Example 3

20 is $25\frac{1}{4}$% of what number?

Solution

Let n = the number (whole value).

$$\frac{25.25}{100} = \frac{20}{n} \qquad 25\frac{1}{4}\% = 25.25\%$$

$$25.25 \cdot n = 100 \cdot 20$$

$$25.25n = 2000$$

$$\frac{25.25n}{25.25} = \frac{2000}{25.25}$$

$$n \approx 79.2$$

■ 20 is approximately $25\frac{1}{4}$% of 79.2.

> ### Estimation Tip
>
> One way to estimate a percent of a number is to round to simpler numbers and use mental math. For example, when finding 34% of 1175, use 33%, or $\frac{1}{3}$, and round 1175 to 1200.
>
> $\frac{1}{3}$ of 1200 is 400.
>
> So, 34% of 1175 is about 400.

Example 4

RETAIL Keiko buys a camera for 20% off the original price. The original price of the camera is $79. How much money does she save?

Solution

Let p = the amount Keiko saves. Write and solve a proportion.

$$\frac{20}{100} = \frac{p}{79}$$

$$20 \cdot 79 = 100 \cdot p$$

$$1580 = 100p$$

$$\frac{1580}{100} = \frac{100p}{100}$$

$$15.80 = p$$

■ Keiko saves $15.80 on the camera.

Write and solve a proportion.

1. What number is 87% of 80?

2. What number is $33\frac{1}{3}$% of 189?

3. What percent of 200 is 68?

4. What percent of 120 is 18?

5. 25% of what number is 6?

6. 42 is 60% of what number?

7. What number is 90% of 30?

8. What percent of 64 is 32?

9. Seven percent of Paul's weekly salary is deducted for taxes. If his salary is $315 per week, how much is deducted?

10. Regina purchases a new VCR for $215. The sales tax is $11.83. What is the sales tax rate?

11. **CALCULATOR** List the keystrokes used to solve Exercise 10 on your calculator.

12. In Ms. Garrett's English class, 80% of the students were present last Thursday. She had 20 students in class on Thursday. What is the total number of students enrolled in Ms. Garrett's English class?

13. **WRITING MATH** Explain how you can use mental math to solve Exercise 10.

◤ **PRACTICE EXERCISES** • **For Extra Practice, see page 565.**

Use mental math to estimate each percent.

14. 42% of 495

15. 70% of 189

16. 21% of 204

17. 32% of 63

18. 11% of 375

19. 49% of 93

Write and solve a proportion.

20. 63% of 215 is what number?

21. 12 is what percent of 15?

22. What percent of 56 is 8.4?

23. 64% of 88 is what number?

24. 12 is 5% of what number?

25. What percent of 20 is 7?

26. 60 is 45% of what number?

27. 9 is what percent of 27?

28. 2% of what number is 3?

29. 84 is what percent of 350?

30. 34 is what percent of 80?

31. What percent of 125 is 87.5?

32. 4.68 is 18% of what number?

33. What number is $7\frac{1}{2}$% of 60?

34. $\frac{1}{2}$% of what number is 3?

35. 250 is what percent of 625?

36. 11.5 is $12\frac{1}{2}$% of what number?

37. What number is 72% of 90?

38. 2.4 is what percent of 200?

39. $33\frac{1}{3}$% of what number is $29\frac{1}{3}$?

40. **WEATHER** Last April it rained 9 out of 30 days. What percent of days did it not rain last April?

41. **REAL ESTATE** A real estate agent receives a $9900 commission, which is 6% of the selling price. At what price does the agent sell the house?

TRAVEL During one weekend at a mountain resort, 81 out of 125 guests ski on Saturday and 80% of the 105 visitors ski on Sunday.

42. Which day had the greatest percentage of skiers?

43. Which day had the greatest number of skiers?

DATA FILE For Exercises 44 and 45, refer to the data on top-rated TV shows on page 523.

44. What method could you use to compare the number of households watching any two given TV shows? Is it possible for one TV show to have a higher rating than another, yet have fewer people watching? Explain.

45. Use the method you developed in Exercise 44 to compare the number of households watching *The Cosby Show* in 1986–1987 to *Friends* in 2001–2002.

Use the table for Exercise 46.

46. There are 9500 students in a freshman class. According to the data, how many of these students will study each field?

Intended Field of Study for College Freshman

Field of study	Percent	Field of study	Percent
Business studies	18	Social service	8
Professional studies	18	Biological studies	4
Engineering	10	Technical	4
Education	9	Physical service	2
Arts and humanities	8	Other	19

◣ EXTENDED PRACTICE EXERCISES

47. CRITICAL THINKING If 15% of 50 is 7.5, would 30% of 50 be 15?

LIVING ON YOUR OWN To be approved for a Standard Conventional Mortgage to buy a house, you can spend a certain amount of your gross monthly income on your mortgage. If you do not have any debt, such as school loans or a car payment, you can spend 28% or less of your monthly income on your mortgage.

48. Together, Mr. and Mrs. Quan earn $65,000. Without any debt, how much can they spend each month on their mortgage?

49. Juanita earns $12.75/hr and works 40 hr/wk. How much can she spend in a year on her mortgage if she has no debt?

50. Together, Mr. and Mrs. Benkert earn $48,500 and do not have any debt. They want to buy a house that will have a mortgage of about $1300/mo. With a Standard Conventional Mortgage, will they be able to afford a monthly payment of this amount? Explain.

◣ MIXED REVIEW EXERCISES

Graph the solution for each equation or inequality on a number line. (Lesson 5-8)

51. $x + 17 = -4$

52. $4y + 6 = 8$

53. $5m - 6 \geq 9$

54. $38 - w < 10$

Write an inequality and solve. (Lesson 5-9)

55. Each student donated $5 towards a class trip. The school gave the class an additional $100. How many students made a donation if the total collection was at least $350?

6-2 Write Equations for Percents

Goals ■ Write equations to solve problems involving percents.

Applications Entertainment, Retail, Sports, Finance, Living on your own

Copy the table shown.

1. Determine the total number of minutes you are at school.

2. Complete the table by writing the number of minutes in a typical school day that you spend doing each activity. Make sure the total in the number of minutes column equals your answer to Question 1.

3. Find the percent of your school day that is spent on each activity.

4. Compare your percentages with a classmate. Are your percentages similar?

Activity	Number of minutes
Doing homework	
Eating	
Mathematics class	
Reading	
Socializing with friends	
Talking with teachers	
Walking between classes	
Other activities	

◤ BUILD UNDERSTANDING

You can use an equation to find the percent of a number, find what percent one number is of another, and find a number when a percent of it is known. Translate the phrase into an equation. Then solve for the unknown. "Is" means *equal.* "Of" means *multiply.*

Example 1

What number is 40% of 75?

Solution

Find the percent of a number by writing an equation using a decimal or fraction.

What number is 40% of 75?

$$x = 0.40 \cdot 75$$
$$x = 30$$

What number is 40% of 75?

$$x = \frac{2}{5} \cdot 75$$
$$x = 30$$

30 is 40% of 75.

Mental Math

You may find it helpful to memorize these equivalent percents, decimals and fractions.

$$12\tfrac{1}{2}\% = 0.125 = \tfrac{1}{8}$$

$$20\% = 0.2 \ = \tfrac{1}{5}$$

$$25\% = 0.25 = \tfrac{1}{4}$$

$$33\tfrac{1}{3}\% \approx 0.33 \approx \tfrac{1}{3}$$

$$40\% = 0.4 \ = \tfrac{2}{5}$$

$$50\% = 0.5 \ = \tfrac{1}{2}$$

$$60\% = 0.6 \ = \tfrac{3}{5}$$

$$66\tfrac{2}{3}\% \approx 0.67 \approx \tfrac{2}{3}$$

$$75\% = 0.75 = \tfrac{3}{4}$$

$$80\% = 0.8 \ = \tfrac{4}{5}$$

Example 2

What percent of 56 is 14?

Solution

Write an equation. Let b = the percent.

What percent of 56 is 14?

$$b \cdot 56 = 14$$

$$56b = 14$$

$$\frac{56b}{56} = \frac{14}{56}$$

$$b = 0.25 \qquad 0.25 = 25\% \text{ Convert to a percent.}$$

25% of 56 is 14.

Check Understanding

50% of a number is half of that number.

100% of a number is that number.

200% of a number is twice that number.

What is 200% of 5?

What is 300% of 5?

Example 3

68% of what number is 17?

Solution

68% of what number is 17?

$$0.68 \cdot x = 17$$

$$0.68x = 17$$

$$\frac{0.68x}{0.68} = \frac{17}{0.68}$$

$$x = 25$$

68% of 25 is 17.

Example 4

ENTERTAINMENT Tamike bought 4 concert tickets for a total of $132. She paid $9.90 in sales tax. What is the sales tax rate?

Solution

Let the sales tax rate be s. Write and solve an equation.

What percent of $132 is $9.90?

$$s \cdot 132.00 = 9.90$$

$$132.00s = 9.90$$

$$\frac{132.00s}{132.00} = \frac{9.90}{132.00}$$

$$s = 0.075 \qquad 0.075 = 7.5\%$$

The sales tax rate is 7.5%.

Write and solve an equation. If necessary, round to the nearest tenth.

1. 36 is what percent of 80?

2. 60 is 40% of what number?

3. 70% of what number is 147?

4. What number is 64.5% of 200?

5. 75% of 64 is what number?

6. What percent of 80 is 12?

7. What number is $\frac{3}{4}$% of 80?

8. What percent of 96 is 7.68?

9. **CALCULATOR** Explain how to find 54.3% of 300 using your calculator.

10. The Alvarez family has a total annual income of $62,000. They save 10% of their annual income. How much money does the family save in a year?

11. **RETAIL** A 6-volume CD anthology was marked down from $120 to $96. What percent of the original number was the sale price?

▼ **PRACTICE EXERCISES** • For Extra Practice, see page 566.

Write and solve an equation.

12. 20% of what number is 21?

13. 6 is what percent of 42?

14. 90% of 45 is what number?

15. 35% of what number is 50?

16. 5 is what percent of 75?

17. 75% of 24 is what number?

18. 42 is what percent of 70?

19. 30 is what percent of 45?

20. 3% of 72 is what number?

21. 6% of what number is 16,450?

22. What percent of 224 is 28?

23. 9 is what percent of 27?

24. 120% of what number is 36?

25. 24 is $2\frac{1}{2}$% of what number?

26. 12% of what number is 32?

27. 35 is what percent of 7?

28. 20 is what percent of 15?

29. 24 is 7.5% of what number?

30. What number is $12\frac{1}{2}$% of 92?

31. 3 is what percent of $4\frac{1}{2}$?

32. 18 is $2\frac{1}{2}$% of what number?

33. What number is $66\frac{2}{3}$% of 60?

34. What number is 0.1% of 150?

35. 0.144 is 0.4% of what number?

SPORTS Write and solve an equation to find the percent of field goals made out of field goals attempted for each player in the table. Round your answer to the nearest tenth of a percent.

36. Gilson

37. Juang

38. McNance

39. Parker

40. Pitell

41. Rodriguez

Player	Field goals attempted	Field goals made
Gilson	993	560
Juang	1307	790
McNance	1062	585
Parker	937	557
Pitell	1057	603
Rodriguez	894	529

42. List the players in order from greatest to least field goal percentage.

43. Thomas correctly answered 49 out of 55 questions on a math test. What percent of the questions did Thomas answer correctly?

44. WRITING MATH Write an equation that you can use to find your percentage grade on a test with 30 questions.

45. FINANCE Ming has $452 in her savings account. Last month she added $32 to the account. The amount she added is what percent of her total savings?

46. Last month 685 books and videos were checked out from the library. The books account for 80% of the items checked out. How many books have been checked out? How many videos have been checked out?

47. If you photocopy an image at 90%, are you reducing or enlarging the image?

LIVING ON YOUR OWN The table shows Grant's monthly expenses. He earns $1525/mo. Round to the nearest tenth.

Rent	$425
Car payment	$177
Insurance	$85
Phone	$37
Electric	$30
Heat	$60
Food	$100

48. What percent of Grant's monthly income does he spend on rent?

49. What percent of Grant's monthly income does he spend on phone, electric, and heat?

50. Grant spends $125/mo on entertainment. After paying all his bills, what is the maximum amount of money Grant can put in savings each month?

51. What other expenses might Grant encounter throughout the year? What percent of the savings do you think Grant will spend on these other expenses?

◣ EXTENDED PRACTICE EXERCISES

DATA FILE For Exercises 52 and 53, refer to the contents of a garbage can data on page 529.

52. One week Queisha's family threw away 150 lb of garbage. According to the circle graph, how many pounds of each type of waste was in their garbage?

53. If Queisha's family threw away 14 lb of food, how many pounds did their total garbage weigh?

54. CRITICAL THINKING Is it sometimes easier to use an equation rather than a proportion to solve problems involving percents? Explain.

55. CHAPTER INVESTIGATION Find the monthly interest rate, rounded to the nearest hundredth. Multiply this times the credit card balance to find the interest charge. Add this to the balance to find the total amount owed. How much do you pay the first month if you owe 2% of the bill?

◣ MIXED REVIEW EXERCISES

56. A map is made to the scale of 1 in. = 5 mi. What is the actual distance if the distance between two cities on the map is $2\frac{1}{4}$ in.? (Lesson 2-8)

57. An architect's drawing is made to the scale of 1 cm = 10 m. How long will the front of a 58 m-building be in the drawing? (Lesson 2-8)

58. Find the perimeter of a square courtyard that measures $16\frac{1}{3}$ yd on each side. (Lesson 2-3)

Review and Practice Your Skills

Write and solve a proportion.

1. 35% of 40 is what number?

2. 1 is 20% of what number?

3. What percent of 72 is 6?

4. 65% of what number is 247?

5. What number is 4% of 30?

6. What percent of 500 is 5?

7. 16 is $\frac{1}{4}$% of what number?

8. 22% of 70 is what number?

9. 15% of 150 is what number?

10. 14 is what percent of 50?

11. What number is 73% of 320?

12. What number is $66\frac{2}{3}$% of 12?

13. 10 is what percent of 15?

14. 60 is 5% of what number?

15. 2% of what number is $10\frac{1}{2}$?

16. 35 is what percent of 140?

17. What percent of 92 is 58.5?

18. 75% of what number is 22.5?

19. 80% of what number is 4?

20. What number is 25% of 6?

21. 40% of 240 is what number?

22. 42 is what percent of 230?

23. What percent of 80 is 5?

24. 55 is 40% of what number?

Write and solve an equation.

25. 31.5 is what percent of 90?

26. What number is 14% of 60?

27. What percent of 60 is 1.8?

28. 112 is 5% of what number?

29. 18 is 16% of what number?

30. 1% of 38 is what number?

31. 3% of 130 is what number?

32. $33\frac{1}{3}$% of what number is 41?

33. 35% of what number is 70?

34. 6 is what percent of 50?

35. What number is 15% of 54?

36. What number is 80% of 250?

37. What percent of 75 is 15?

38. What percent of 48 is 32?

39. $87\frac{1}{2}$% of 480 is what number?

40. 70 is what percent of 200?

41. 5 is what percent of $2\frac{1}{2}$?

42. What number is 0.3% of 45?

43. 120 is 25% of what number?

44. 75% of 360 is what number?

45. What percent of 25 is 10?

46. 12% of what number is 60?

47. 70% of what number is 35?

48. 12 is 3% of what number?

49. 16% of what number is 2?

50. $2\frac{1}{2}$ is what percent of 15?

51. 48 is 60% of what number?

52. What percent of 350 is 140?

53. What number is 8% of 230?

54. What number is 34% of 60?

Write and solve a proportion. (Lesson 6-1)

55. 42% of 90 is what number?

56. 14 is 70% of what number?

57. What percent of 30 is 6?

58. $33\frac{1}{3}$% of what number is 55?

59. 15 is 60% of what number?

60. What number is 2% of 190?

61. $5\frac{1}{2}$% of what number is 26?

62. 88% of 450 is what number?

63. 120 is what percent of 150?

64. What percent of 75 is 30?

65. What number is 18% of 60?

66. 40 is what percent of 60?

Write and solve an equation. (Lesson 6-2)

67. $\frac{1}{4}$% of 600 is what number?

68. 21 is 70% of what number?

69. What percent of 90 is 1.8?

70. 6% of what number is 15?

71. 18 is what percent of 72?

72. 28% of 320 is what number?

MathWorks
Workplace Knowhow

Career – Financial Counselor

Financial counselors help people decide how to use their money wisely. They teach their clients how to spend more carefully and how to invest for future profits. Financial counselors help their customers establish *budgets*, which are plans to meet regular expenses, pay off debt and invest in savings.

Answer Questions 1–8 using the circle graph.

A financial counselor gave a client this circle graph describing how to budget his income. If the client brings home $1200.00 each month, how many dollars should he spend in each category?

1. housing
2. food
3. car loan and maintenance
4. utilities
5. phone
6. clothing
7. entertainment
8. The client spends $450 on housing, $200 on food, $350 on a car, $100 on utilities, $125 on phone, $50 on clothing and $100 on entertainment. For which categories did he stay within the budget? For which categories did he exceed the budget?

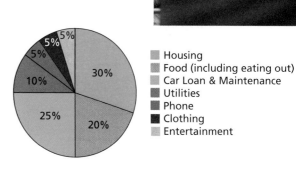

Housing
Food (including eating out)
Car Loan & Maintenance
Utilities
Phone
Clothing
Entertainment

Discount and Sale Price

Goals
- Find the discount and sale price of items.
- Solve problems involving discount and sale price.

Applications Retail, Living on your own, Travel

Use the advertisements shown for Questions 1–3.

Betsy is going to buy a new pair of gym shoes. She finds two different advertisements for the Excelsior Walking Shoe.

1. Find the price of the Excelsior Walking Shoe at The Happy Foot.

2. Which store is offering a better sale?

3. Find the difference in the shoe price at The Shoe Company and The Happy Foot.

The Shoe Company
featuring
The Excelsior Walking Shoe   Regular Price $79   Sale Price $59.99

The Happy Foot
Best Prices Around
The Excelsior Walking Shoe
15% OFF Regular Price
Regular Price $79

◢ BUILD UNDERSTANDING

When an item is on sale, you can save money buying it at less than regular price. The **discount** is the amount that the regular price is reduced. The **sale price** is the regular price minus the discount.

You can use a proportion or an equation to solve problems involving discount and sale price.

Example 1

Find the discount and the sale price of a mountain bike with a regular price of $298.95 and a 20% discount rate.

Solution

Use an equation.

discount = percent of discount · regular price

$d = 20\% \cdot \$298.95$

$d = 0.20 \cdot 298.95$ Convert percent to decimal.

$d = 59.79$

The discount is $59.79.

Use a proportion.

$$\frac{d}{298.95} = \frac{20}{100}$$

$d \cdot 100 = 298.95 \cdot 20$ Use the cross-products.

$100d = 5979$

$$\frac{100d}{100} = \frac{5979}{100}$$

$d = 59.79$

Find the sale price.

sale price = regular price − discount

$s = 298.95 - 59.79$

$s = 239.16$

The sale price is $239.16.

Problem Solving Tip

Another method to find the sale price of an item is to subtract the percent of discount from 100. Then multiply this number times the regular price.

How could you use this method for Example 1?

To find the percent of discount, first find the amount of the discount. Then divide by the regular price.

Example 2

A CD player is on sale for $174.30. The regular price of the CD player is $249.00. Find the percent of discount.

Solution

Find the discount.

discount = regular price − sale price

$$d = 249.00 - 174.30$$

$$d = 74.70$$

The discount is $74.70.

Find the percent of discount.

$$\text{percent of discount} = \frac{\text{discount}}{\text{regular price}}$$

$$p = \frac{74.70}{249.00}$$

$$p = 0.3 \quad \textit{Convert decimal to a percent.}$$

The percent of discount is 30%.

Example 3

Jonte went to a 25%-off sale at a fruit and vegetable market. The regular price of a pepper is $0.48. Find the sale price.

Solution

Use mental math.

$$25\% = \frac{1}{4}$$

$\frac{1}{4}$ of 0.48 is 0.12.

$$0.48 - 0.12 = 0.36$$

The sale price is $0.36.

◤ TRY THESE EXERCISES

Find the discount and the sale price. Round your answers to the nearest cent.

1. Regular price: $415.38
Percent of discount: 15%

2. Regular price: $63.45
Percent of discount: 25%

3. Regular price: $89.95
Percent of discount: 50%

4. Regular price: $369.99
Percent of discount: 20%

5. Regular price: $115.18
Percent of discount: 3%

6. Regular price: $3880.00
Percent of discount: 12.5%

Find the percent of discount.

7. Regular price: $149.78
   Sale price: $104.85

8. Regular price: $2450.00
   Sale price: $1470.00

9. Regular price: $19.75
   Sale price: $16.79

10. Regular price: $347.80
    Sale price: $313.02

11. Francie bought a sweater on sale for 20% off the regular price of $49.95. What is the sale price?

12. Chago went to an end-of-the-season ski sale. Every item was 40% off. How much did he pay for $1580 worth of regular-priced merchandise?

13. Rosario bought a new outfit on sale for $73.36. The original price was $109.50. What was the percent of discount on the outfit?

14. **WRITING MATH** Explain the difference between *regular price*, *discount*, and *sale price*.

▼ PRACTICE EXERCISES • For Extra Practice, see page 566.

Find the discount and the sale price. Round your answers to the nearest cent.

15. Regular price: $150.50
    Percent of discount: 10%

16. Regular price: $79.50
    Percent of discount: 20%

17. Regular price: $689.99
    Percent of discount: 18.5%

18. Regular price: $1450.39
    Percent of discount: 25%

19. Regular price: $24.00
    Percent of discount: 25%

20. Regular price: $495.55
    Percent of discount: 11%

21. Regular price: $511.09
    Percent of discount: 13%

22. Regular price: $48.88
    Percent of discount: 3%

23. Regular price: $1385.00
    Percent of discount: 5%

24. Regular price: $8175.00
    Percent of discount: 16%

Find the percent of discount.

25. Regular price: $56.99
    Sale price: $42.74

26. Regular price: $2480.00
    Sale price: $1984.00

27. Regular price: $785.00
    Sale price: $392.50

28. Regular price: $349.25
    Sale price: $232.84

29. Regular price: $92.00
    Sale price: $59.80

30. Regular price: $4258.95
    Sale price: $1703.58

31. **TRAVEL** The regular price of a Caribbean cruise is $995. For a limited time, the price is reduced by 28%. How much will you save with the reduced price?

32. **WRITING MATH** Explain the difference between 25% *of* $90 and 25% *off* $90.

33. Jarad bought a mountain bike on sale for $355. This sale price is $50 off the original price. What is the percent of discount on the bike?

34. Which is cheaper, a $29 wallet that is 15% off or a $40 wallet that is 35% off?

35. LIVING ON YOUR OWN Rashaun is looking for a one-bedroom apartment. At Blue Sky Apartments, rent for the first two months is 20% off. The one-bedroom rate at Blue Sky is $525/mo. At City View Apartments, the first month is 50% off. The one-bedroom rate at City View is $550/mo. Which apartment will be cheaper for the first two months? Explain.

RETAIL The Nelsons purchased a new television that was on sale for 15% off the original price. The original price is $459.99.

36. How much did the Nelsons save?   **37.** How much did they pay?

The Smart Dresser discounts all merchandise that has been on the racks for six weeks. Any merchandise remaining another four weeks has an additional discount. Complete the table to show the final sale price for Exercises 38–41.

38. $89.00

39. $384.50

40. $25.75

41. $2853.27

Regular price	First discount	Second discount	Final sale price
$89.00	10%	15%	■
$384.50	5%	20%	■
$25.75	25%	10%	■
$2853.27	30%	5%	■

■ EXTENDED PRACTICE EXERCISES

42. CRITICAL THINKING A pair of shoes is reduced by 15%. Two weeks later, the same pair of shoes is reduced by an additional 20%. After the second markdown, will the shoes cost the same if they had been reduced 35% at the start of the sale? Use an example to explain your answer.

YOU MAKE THE CALL Vernon bought a VCR at 15% off the original price of $199.99. Benjamin bought the same VCR from a different store at $30 off the original price of $189.99.

43. How much did Vernon pay for his VCR?

44. How much did Benjamin pay for his VCR?

45. Who bought the VCR at a better discount rate?

46. Who bought the VCR with a greater amount of discount?

■ MIXED REVIEW EXERCISES

Find the number of faces, edges, and vertices for each solid. (Lesson 4-2)

47. 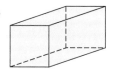   **48.**   **49.**

50. Find the surface area of a cylinder that has a height of 5.5 ft and a diameter of 1 ft. Use 3.14 for π. (Lesson 4-7)

6-4 Solve Problems Using Tax Rates

Goals ■ Use proportions to solve tax problems.
■ Use equations to solve tax problems.

Applications Retail, Buying a house, Living on your own, Part-time job

Work with a partner.

Tasha goes to a bike store. She wants to spend less than $300.00 for a new bike and helmet. The bike she wants to buy costs $219.99. The helmet she prefers costs $59.99. The sales tax is 6.5%.

1. If Tasha buys the bike and helmet, does she stay within the budgeted amount?

2. How much over or under budget is she?

▼ BUILD UNDERSTANDING

There are various taxes that people must pay. Some examples are sales tax, property tax and income tax. A **tax** is a charge, usually a percentage, that is imposed by an authority, generally local, state or federal government.

You can set up a proportion or equation to solve problems involving taxes.

Example 1

The price of a pair of shoes is $35.98. The sales tax is 6%. Find the amount of the sales tax and total cost of the shoes.

Solution

Find the sales tax.

sales tax = percent of sales tax · regular price

sales tax = 0.06 · 35.98

sales tax = 2.1588   Round 2.1588 to $2.16.

Find the total cost.

total cost = regular price + sales tax

total cost = 35.98 + 2.16

total cost = $38.14

The amount of sales tax on the shoes is $2.16 and the total cost is $38.14.

Problem Solving Tip

Another way to find the total cost of an item is to add the sales tax percent to 100%. Then multiply this new percent by the regular price.

For example, if the regular price is $35.98 and the sales tax is 6%, add 6% to 100%. Then find 106% of $35.98.

1.06 · 35.98 = 38.14

The total cost is $38.14.

People who are employed in the U.S. must pay taxes based on their income. These taxes are called **income taxes**. The **net pay**, or take-home pay, is the amount of money that a person is paid after taxes are subtracted.

Example 2

Wanda earns $950 every two weeks. She pays 25% in income taxes. Calculate the amount of income tax she pays and her net pay.

Solution

Find the income tax amount using a proportion or an equation.

$$\frac{x}{950} = \frac{25}{100}$$ Use a proportion. $x = 950 \cdot 0.25$ Use an equation.

$x \cdot 100 = 950 \cdot 25$ Use the cross-products. $x = 237.5$

$$\frac{100x}{100} = \frac{23{,}750}{100}$$

$x = 237.50$

Find Wanda's net pay. $950.00 - \$237.50 = \$712.50$

Wanda pays $237.50 for income tax, and her net pay is $712.50.

Homeowners pay **property taxes** based on the value of their house and property. These taxes help pay for services such as schools, libraries, and a police force.

Example 3

The Cole's home is valued at $93,000. The property tax rate for their home is about 3%. Estimate how much the Coles will pay in property taxes.

Solution

$$\frac{x}{93{,}000} = \frac{3}{100}$$ Use a proportion. $x = 93{,}000 \cdot 0.03$ Use an equation.

$x \cdot 100 = 93{,}000 \cdot 3$ Use the cross-products. $x = 2790$

$$\frac{100x}{100} = \frac{279{,}000}{100}$$

$x = 2790$

The Coles will pay about $2790 in property tax.

Example 4

Tyrell purchases a pair of jeans for $29.99. The total amount of the purchase is $31.79. Find the sales tax rate.

Solution

Find the amount of tax. $31.79 - \$29.99 = \$1.80$

Find the tax rate. $$\frac{x}{100} = \frac{1.80}{29.99}$$ Use a proportion. $1.80 = 29.99 \cdot x$ Use an equation.

$x \cdot 29.99 = 100 \cdot 1.80$ $$\frac{1.80}{29.99} = \frac{29.99x}{29.99}$$

$$\frac{29.99x}{29.99} = \frac{180}{29.99}$$ $0.06 \approx x$

$x \approx 6$

The sales tax rate is 6%.

Find the amount of the sales tax and total cost of each item.

1. Shoes: $45
Sales tax rate: 5.5%

2. Calculator: $25.99
Sales tax rate: 7%

3. Computer disks: $10.99
Sales tax rate: 6.5%

Find the income tax and net pay for each.

4. Income: $800/wk
Income tax rate: 25%

5. Income: $26,500/yr
Income tax rate: 20%

6. Income: $45,000/yr
Income tax rate: 22%

Find the property tax paid by each homeowner.

7. Home value: $72,500
Property tax rate: 2.5%

8. Home value: $205,000
Property tax rate: 3%

9. Home value: $127,500
Property tax rate: 4%

Find the sales tax rate.

10. Price: $109.99
Total cost: $116.04

11. Price: $37.50
Total cost: $39.75

12. Price: $875.00
Total cost: $936.25

13. PART-TIME JOB Jolene is paid $125/wk working at a fast-food restaurant.
The income tax rate is 11%. What is Jolene's net pay?

Find the amount of the sales tax and total cost of each item.

14. Ski jacket: $125
Sales tax rate: 7%

15. Speakers: $135
Sales tax rate: 6.5%

16. Suit: $155
Sales tax rate: 6.5%

17. Computer software: $49.99
Sales tax rate: 6%

18. Radio: $138
Sales tax rate: 5%

19. T-shirt: $12.99
Sales tax rate: 7.5%

Find the income tax and net pay for each.

20. Income: $545/wk
Income tax rate: 29%

21. Income: $68,000/yr
Income tax rate: 22%

22. Income: $395/wk
Income tax rate: $21\frac{1}{2}$%

23. Income: $22,000/yr
Income tax rate: 19%

24. Income: $31,400/yr
Income tax rate: 23%

25. Income: $750/wk
Income tax rate: 17.5%

Find the property tax paid by each homeowner.

26. Home value: $102,000
Property tax rate: $2\frac{1}{2}$%

27. Home value: $69,000
Property tax rate: 3%

28. Home value: $175,000
Property tax rate: 2%

29. Home value: $95,000
Property tax rate: 3%

30. Home value: $117,500
Property tax rate: $2\frac{1}{3}$%

31. Home value: $136,000
Property tax rate: $3\frac{1}{4}$%

Find the sales tax rate.

32. Price: $32.45
Total cost: $34.88

33. Price: $1195.00
Total cost: $1266.70

34. Price: $562.99
Total cost: $599.58

35. Price: $117.99
Total cost: $123.89

36. Price: $78.50
Total cost: $84.78

37. Price: $210.20
Total cost: $225.44

LIVING ON YOUR OWN This year Javiera bought a house. The value of the house is $98,500. The property tax rate is $3\frac{1}{2}\%$. Javiera has an annual income of $35,900. Her income tax rate is 23%.

38. How much does Javiera pay in property taxes each year?

39. How much does Javiera pay in income taxes each year?

40. Javiera pays $725/mo on her mortgage. What percent of her monthly net income goes towards her mortgage payment?

SPREADSHEETS This spreadsheet represents a sale at a furniture store. Find the values in the empty cells.

41. cell C2

42. cell B3

43. cell E4

44. cell D5

	A	B	C	D	E
1	Item	Regular price	Discount	Discount price	Percent of discount
2	Queen bed	$899.99	■	$810.00	10
3	Sleeper sofa	■	$250.00	$1000.00	20
4	Chair with ottoman	$679.89	$101.98	$577.91	■
5	Lamp	$109.00	$27.25	■	25

■ EXTENDED PRACTICE EXERCISES

45. **RETAIL** Silvia is purchasing a bracelet at 15% off the original price of $129.99. The sales tax is $5\frac{1}{2}\%$. Including sales tax, how much does the bracelet cost?

46. **WRITING MATH** Write an equation to find the sales tax rate on an item priced at $54.75 that sells for $58.31 with tax. Explain each step.

47. **CRITICAL THINKING** When calculating a percent, why does a proportion give the correct answer but in an equation you must move the decimal?

48. **RETAIL** Malcolm purchased a $29.99 book at 20% off the original price. He is a member of the book-of-the-month club in which he receives an additional $5 off any purchase. The sales tax is 5.5%. How much does Malcolm owe the bookstore?

■ MIXED REVIEW EXERCISES

Find the surface area and volume of each solid. Use 3.14 for π. Round to the nearest hundredth. (Lessons 4-7 and 4-9)

49.
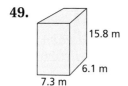
15.8 m
6.1 m
7.3 m

50.
1 in.
5 in.

51.
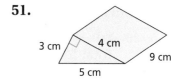
3 cm   4 cm   9 cm
5 cm

Divide. (Lesson 3-6)

52. $-56 \div 7$

53. $64 \div (-8)$

54. $-48 \div (-6)$

55. $49 \div (-7)$

Review and Practice Your Skills

Find the discount and the sale price. Round your answers to the nearest cent.

1. Regular price: $89.90
   Percent of discount: 20%

2. Regular price: $120
   Percent of discount: 35%

3. Regular price: $45.50
   Percent of discount: 5%

4. Regular price: $76.89
   Percent of discount: 18%

5. Regular price: $124.99
   Percent of discount: 60%

6. Regular price: $2459
   Percent of discount: 45%

Find the percent of discount. Round to the nearest tenth of a percent.

7. Regular price: $78.90
   Sale price: $69.95

8. Regular price: $320
   Sale price: $299

9. Regular price: $114
   Sale price: $98

10. Regular price: $49
    Sale price: $45

11. Regular price: $24.50
    Sale price: $17.99

12. Regular price: $82.50
    Sale price: $75.59

PRACTICE ■ LESSON 6-4

Find the amount of the sales tax and total cost of each item.

13. Price: $32.80
    Sales tax rate: 8.5%

14. Price: $380.50
    Sales tax rate: 4.5%

15. Price: $68.99
    Sales tax rate: 7.5%

16. Price: $125.49
    Sales tax rate: 6%

17. Price: $29.95
    Sales tax rate: 7%

18. Price: $723
    Sales tax rate: 5%

Find the income tax and net pay for each.

19. Income: $875/wk
    Income tax rate: 24%

20. Income: $485/wk
    Income tax rate: 16%

21. Income: $730/wk
    Income tax rate: 31%

22. Income: $32,000/yr
    Income tax rate: $23\frac{1}{2}$%

23. Income: $46,500/yr
    Income tax rate: 28%

24. Income: $26,700/yr
    Income tax rate: 19%

Find the property tax paid by each homeowner.

25. Home value: $178,000
    Property tax rate: 3%

26. Home value: $89,500
    Property tax rate: $2\frac{3}{4}$%

27. Home value: $240,000
    Property tax rate: 3%

Find the sales tax rate. Round to the nearest tenth of a percent.

28. Price: $118.89
    Total cost: $124.83

29. Price: $28.50
    Total cost: $30.92

30. Price: $249.99
    Total cost: $264.99

31. Price: $58.49
    Total cost: $62.58

32. Price: $73.40
    Total cost: $78.17

33. Price: $37.65
    Total cost: $40.66

Solve each problem. (Lessons 6-1 and 6-2)

34. 20% of 340 is what number?

35. What number is 40% of 116?

36. What percent of 60 is 5?

37. 13 is what percent of 65?

38. $1\frac{1}{2}$% of what number is 30?

39. 7 is 80% of what number?

Find the discount and the sale price. (Lesson 6-3)

40. Regular price: $158.90
Percent of discount: 12%

41. Regular price: $49.95
Percent of discount: 30%

Find the percent of discount. (Lesson 6-3)

42. Regular price: $245
Sale price: $229

43. Regular price: $54.99
Sale price: $49.95

Find the amount of the sales tax and total cost of each item. (Lesson 6-4)

44. Price: $64.50
Sales tax rate: 6.5%

45. Price: $1475
Sales tax rate: 8%

46. Price: $329.95
Sales tax rate: 7%

Find the income tax and net pay for each. (Lesson 6-4)

47. Income: $678/wk
Income tax rate: 14%

48. Income: $36,500/yr
Income tax rate: 23%

49. Income: $1250/wk
Income tax rate: 17%

Mid-Chapter Quiz

Write as a percent. (Are You Ready?, Lessons 6-1 and 6-2)

1. 0.85

2. 0.9

3. 0.002

4. $\frac{3}{5}$

5. What percent of 280 is 70?

6. What percent of 650 is 260?

7. 55 is what percent of 275?

8. 60 is what percent of 200?

Calculate. Round answers to the nearest tenth. (Lessons 6-1 and 6-2)

9. 25% of 40

10. 94% of 625

11. 8.4% of 230

12. $15\frac{1}{4}$% of 96

13. 48% of what number is 192?

14. 276 is 35% of what number?

15. $33\frac{1}{3}$% of what number is 43?

16. 0.144 is 0.4% of what number?

Find the discount and the sale price. Round answers to the nearest cent. (Lesson 6-3)

17. Regular price: $224.75 Discount: 15%

18. Regular price: $4837.50 Discount: 16.5%

19. Bill bought a $57 pair of shoes for 20% off. What was the sale price?

20. The regular price of a $89.98 sweat suit is reduced by 35%. How much would you save if you bought the sweat suit at the reduced price? (Lesson 6-3)

6-5 Simple Interest

Goals ■ Calculate simple interest, interest rate and amount due.

Applications Finance, Business, Travel, Living on your own

Copy and complete each chart.

Nakia and her brother Jerome both save money. Jerome collects $0.50/wk in a jar. Nakia deposits $200 in a bank account. In the bank, Nakia's money increases 3.5% every year.

1. Calculate Jerome's savings by multiplying the number of years, 52 wk, and $0.50.

Number of years	Jerome's savings
1	1 · 52 · 0.50 = $26
2	2 · 52 · 0.50 = $52
3	
4	
5	
10	
15	
20	

2. Calculate Nakia's savings by multiplying the number of years by 3.5% of $200. Add the result to 200.

Number of years	Nakia's savings
1	(1 · 0.035 · 200) + 200 = $207
2	(2 · 0.035 · 200) + 200 = $214
3	
4	
5	
10	
15	
20	

3. After 5 years, who has the better plan? Explain.

4. After 15 years, who has the better plan? Explain.

5. After 20 years, who has the better plan? Explain.

◤ BUILD UNDERSTANDING

Interest is money paid to an individual or institution for the privilege of using the money. For example, if you put money in a savings account, a bank will pay you interest for the use of that money over a given period of time. If you borrow money from a bank, the bank will charge you interest for the use of its money for a given period of time.

The amount of money that is earning interest or that you are borrowing is called the **principal**. **Simple interest** is paid or received only on the principal.

The **rate** is the percent charged per year for the use of money over a given period of time. The **time** is how long the principal remains in the bank. In the case of borrowing, the time is how long the principal remains unpaid.

The interest (I) earned or paid on a given principal (p) at a given rate (r) over a period of time (t) is given by a formula.

Simple Interest	$I = prt$ where I is the interest earned or paid, p is the principal, r is the rate and t is the period of time.

Example 1

Find the interest on $3000 borrowed for 2 yr at a rate of 8%/yr.

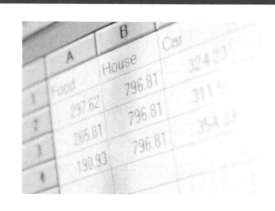

Solution

Use the formula $I = prt$.

$I = prt$

$I = \$3000 \cdot 8\% \cdot 2$ $p = \$3000, r = 8\%$ and $t = 2$

$I = 3000 \cdot 0.08 \cdot 2$ Convert percent to decimal.

$I = 480$

The interest is $480.

Example 2

Savings of $5000 are invested for 4 yr. This investment returns $1000 in interest. Find the rate of interest on the money invested.

Solution

$I = prt$ Use the formula $I = prt$ to solve.

$1000 = 5000 \cdot r \cdot 4$ $I = \$1000, p = \$5000$ and $t = 4$

$1000 = 20,000r$

$\dfrac{1000}{20,000} = \dfrac{20,000r}{20,000}$

$0.05 = r$

The money was invested at a rate of 5%.

When you repay borrowed money, the **amount due** is equal to the principal plus the interest accumulated over the time.

Check Understanding

Often the amount of time is not given in years. Write each amount of time in terms of years.

3 mo = ___?___ yr

4 mo = ___?___ yr

6 mo = ___?___ yr

9 mo = ___?___ yr

18 mo = ___?___ yr

28 mo = ___?___ yr

36 mo = ___?___ yr

42 mo = ___?___ yr

Example 3

Kendra borrows $4200 at an annual interest rate of 11.5%. The term of the loan is 36 mo. Find the simple interest and total amount due on the loan.

Solution

Find the simple interest.

$I = prt$

$I = 4200 \cdot 0.115 \cdot 3$ Change 36 mo to 3 yr.

$I = 1449$

Find the amount due.

$\$4200 + \$1449 = \$5649$

Kendra paid $1449 in simple interest. The total amount due is $5649.

Find the interest.

1. Principal: $1000
Rate: 8%/yr
Time: 2 yr

2. Principal: $895
Rate: 17.5%/yr
Time: 3 yr

3. Principal: $1300
Rate: 5.4%/yr
Time: 48 mo

Find the rate of interest.

4. Principal: $475
Interest: $14.25
Time: 6 mo

5. Principal: $900
Interest: $135
Time: 2 yr

6. Principal: $3450
Interest: $448.50
Time: 1 yr

Find the interest and amount due.

7. Principal: $650
Rate: 4%/yr
Time: 5 yr

8. Principal: $980
Rate: 7.5%/yr
Time: 3 yr

9. Principal: $1500
Rate: 15%/yr
Time: 36 mo

10. Steadman borrows $6400 from a bank at an interest rate of 14%/yr. How much interest does he owe after 4 yr?

11. WRITING MATH Explain the difference between the principal of a loan and the principal of a savings account.

PRACTICE EXERCISES • For Extra Practice, see page 568.

Find the interest and the amount due.

12. $410.00

13. $35.00

14. $1,575.00

15. $4,853.00

16. $3,000.00

17. $8,957.00

18. $505.50

19. $2,200.75

20. $775.00

21. $8,000.00

22. $10,100.00

23. $69.75

Principal	Annual rate	Time	Interest (dollars)	Amount due
$410.00	5%	6 mo	■	■
$35.00	7%	2 yr	■	■
$1,575.00	12.5%	3 yr	■	■
$4,853.00	8%	48 mo	■	■
$3,000.00	5.4%	5 yr	■	■
$8,957.00	4%	12 mo	■	■
$505.50	10.5%	1 yr	■	■
$2,200.75	11%	8 yr	■	■
$775.00	15%	36 mo	■	■
$8,000.00	16.5%	5 yr	■	■
$10,100.00	6.5%	24 mo	■	■
$69.75	8.5%	4 yr	■	■

Find the rate of interest.

24. Principal: $3500
Interest: $577.50
Time: 18 mo

25. Principal: $1300
Interest: $624
Time: 4 yr

26. Principal: $14,500
Interest: $3262.50
Time: 54 mo

LIVING ON YOUR OWN Use the table shown for Exercises 27–29.

Average Yearly College Expenses

Expense	Private	In-state public	Out-of-state public
Tuition and fees	$14,508	$3,243	$8,417
Books and supplies	$667	$662	$662
Housing and food	$5,765	$4,530	$4,530
Transportation	$547	$612	$612

27. Marcus borrows money from a bank to pay for his education at a private college. He borrows enough money for 1 yr of tuition and books. The interest rate on his loan is 7.5%. Find the interest and the total amount due on the loan.

28. Kim takes out a loan of $26,000 to go to an in-state public university. How many years of tuition, books, housing and food will this loan pay for?

29. Kim's loan has an interest rate of $6\frac{1}{4}$%/yr. Find the amount of interest that Kim pays if she takes 10 yr to pay off the loan.

30. **FINANCE** Gina opens a savings account with a deposit of $1100. What is the total amount of money in her account at the end of 2 yr if her money earned 6%/yr and she made no other deposits or withdrawals?

31. **TRAVEL** Roland took out a loan of $3000 to pay for a trip to Europe. He paid $3585 over 3 yr to pay off the loan. What was the interest rate on the loan?

Big Ben, London

EXTENDED PRACTICE EXERCISES

32. Suka deposits $150 in a savings account. The value of the savings account doubles in 25 yr. What is the interest rate?

33. **BUSINESS** Chuck borrows $18,000 to expand his courier business. He takes out two different loans. One loan is for $10,000 at 12%/yr. The other loan is for $8000 at $9\frac{1}{4}$%/yr. Both loans are for 5 yr. How much money must he repay at the end of 5 yr?

34. **CRITICAL THINKING** Suppose you earn an annual salary of $32,000. You are offered a one-time raise of 10.5% for a 4-yr period, or you can have a 3% raise each year for 4 yr. Which is the better offer?

35. **CHAPTER INVESTIGATION** Suppose you paid $100 the first month. Find the next month's new balance including the interest charge. If you pay $100/mo until the bill is paid off, how many months will you have to pay? How much will you pay in interest?

MIXED REVIEW EXERCISES

Write each number in scientific notation. (Lesson 3-7)

36. 10,400

37. 4,500,000

38. 100,000,000,000

39. 607,040,000

40. 0.015

41. 0.0006

42. 0.00000105

43. 0.000090

Write an equation and solve. (Lesson 5-5)

44. Susan is thinking of a number that when multiplied by 6.5 and divided by 13 is 2. What number is she thinking of?

Sales Commission

Goals ■ Calculate commission, commission rate and total income.

Applications Real estate, Retail, Living on your own

Work with a partner.

Seneca was hired by Megahertz Computer Store. She has to decide how to be paid. There are two plans from which to choose. Plan 1 pays a base salary of $200/wk plus $25 for every piece of hardware or software she sells. Plan 2 pays a base of $150/wk plus 7% of her total sales. To make a decision, Seneca comes up with a sample week of sales.

Hardware/software item	Price
17" color monitor	$399.00
FAX machine	$97.99
Color scanner	$145.99
Color printer	$279.95
Laptop computer	$1799.00
Internet software	$89.99
Digital camera	$799.95

1. Based on this week, how much will she earn using Plan 1?

2. Based on this week, how much will she earn using Plan 2?

3. Which plan would you advise Seneca to choose?

4. Develop a sample week of your own in which Plan 1 is the better choice.

◢ BUILD UNDERSTANDING

Many salespeople work on commission. This means they earn an amount of money that is a percent of their total sales. The percent is the **commission rate**. The amount of money they receive is the **commission**. Often, their income is a combination of salary (base pay) plus commission.

You can use either a proportion or an equation to solve problems involving commission.

Example 1

Find the commission on a sale of $6500 if the commission rate is 4%.

Solution

$$\frac{x}{6500} = \frac{4}{100}$$ Use a proportion. $x = \$6500 \cdot 4\%$ Use an equation.

$$x \cdot 100 = 6500 \cdot 4$$ Use the cross-products. $x = 6500 \cdot 0.04$ $4\% = 0.04$

$$\frac{100x}{100} = \frac{26{,}000}{100}$$ $x = 260$

$$x = 260$$

The commission is $260.

Example 2

A salesperson receives a base salary of $325/wk and a commission rate of 3%. Find her weekly income if her sales are $1250.

Solution

Find the commission.

$$\frac{x}{1250} = \frac{3}{100} \quad \text{Use a proportion.}$$

$$x \cdot 100 = 1250 \cdot 3$$

$$\frac{100x}{100} = \frac{3750}{100}$$

$$x = 37.5$$

The commission is $37.50.

Find the weekly income.

$$\$37.50 + \$325.00 = \$362.50$$

The weekly income is $362.50.

$$x = \$1250 \cdot 3\% \quad \text{Use an equation.}$$

$$x = 1250 \cdot 0.03$$

$$x = 37.5$$

Example 3

A car salesperson receives a commission of $540 for selling a car at $12,000. What is the salesperson's commission rate?

Solution

Let x = the commission rate.

$$\frac{x}{100} = \frac{540}{12,000} \quad \text{Use a proportion.}$$

$$x \cdot 12,000 = 100 \cdot 540$$

$$\frac{12,000x}{12,000} = \frac{54,000}{12,000}$$

$$x = 4.5$$

The commission rate is 4.5%.

Manufacturer's Car Lot

$$\$540 = \$12,000 \cdot x \quad \text{Use an equation.}$$

$$\frac{540}{12,000} = \frac{12,000x}{12,000}$$

$$0.045 = x$$

Check Understanding

In the solution of Example 3, explain why $x = 4.5$ when you use a proportion and $x = 0.045$ when you use an equation.

◥ TRY THESE EXERCISES

Find the commission.

1. Total sale: $450
Commission rate: 7%

2. Total sale: $1345
Commission rate: 3.5%

Find the total income.

3. Base salary: $375/wk
Total sales: $2850/wk
Commission rate: 6%

4. Base salary: $880/mo
Total sales: $13,500/mo
Commission rate: 6.2%

Find the commission rate.

5. Total sale: $345.00
   Commission: $31.05

6. Total sale: $7125
   Commission: $855

7. One week Alberto sells $14,000 worth of computers. His commission is $910. Find his commission rate.

8. Ming earns a 7.5% commission rate on sales from her used car lot. Find her commission if she sells a car for $7255.

9. **WRITING MATH** What are the benefits of working on commission? What are the drawbacks of working on commission?

▼ **PRACTICE EXERCISES** • For Extra Practice, see page 568.

SPREADSHEETS Find the values in the empty cells.

10. cell C2
11. cell A3
12. cell B4
13. cell C5
14. cell B6
15. cell A7
16. cell C8
17. cell A9
18. cell C10
19. cell B11
20. cell A12

	A	B	C
1	Sale	Commission rate	Commission
2	$1300.00	2%	■
3	■	4%	$2135.00
4	$650.00	■	$22.75
5	$320.00	4.5%	■
6	$750.00	■	$11.25
7	■	5%	$32.00
8	$1255.00	7%	■
9	■	$2\frac{1}{2}$%	$11.75
10	$3575.00	3%	■
11	$398.00	■	$35.82
12	■	$5\frac{1}{4}$%	$393.75

Find the total income.

21. Base salary: $400/wk
    Total sales: $3250/wk
    Commission rate: 6%

22. Base salary: $950/mo
    Total sales: $14,000/mo
    Commission rate: 6.2%

23. Base salary: $400/wk
    Total sales: $1450/wk
    Commission rate: 8%

24. Base salary: $1000/mo
    Total sales: $8000/mo
    Commission rate: 3%

25. Base salary: $13,500/yr
    Total sales: $175,000/yr
    Commission rate: $5\frac{1}{2}$%

26. Base salary: $190/wk
    Total sales: $3795/wk
    Commission rate: 4.5%

27. Shirley's commission rate for selling cosmetics is 15%. Her sales total $225 one week. How much commission does she receive?

28. **RETAIL** One week, Alberto sells $42,000 worth of computers. His commission is $1260. Find his commission rate.

29. Edmund receives a base salary of $265/wk and a commission rate of $3\frac{1}{2}$% of sales. How much does he earn in a week when his sales are $4859?

30. Refer to Exercise 29. Estimate Edmund's annual earnings assuming that his sales continue at the same rate.

31. LIVING ON YOUR OWN Robin sells her house for $113,500. Of the selling price, she pays 3% to the real estate agent that listed her house, and 3% to the real estate agent that sells her house. How much money does Robin pay to each real estate agent?

32. Lemuel sells cars on commission. He receives 5% for new cars and 4% for used cars. One day he sells 2 used cars for a total of $9120 and 1 new car for $12,156. How much commission does he earn?

33. Anthony receives a commission of 25% on sales of merchandise in excess of $2000. One week his sales are $7500. What is the amount of his commission?

 34. WRITING MATH Write a problem in which you have to find the amount of a sale, given the commission rate and the commission. Include the solution to the problem you write.

EXTENDED PRACTICE EXERCISES

REAL ESTATE When a real estate agent sells a house, the agent receives a commission of 3% of the cost of the house. Of this 3%, the agent keeps 60% and 40% goes to the agent's employer.

35. Dale sells a house for $127,500. How much of the commission does Dale receive?

36. In Exercise 35, how much money does the company earn on the house that Dale sells?

Some salespersons are paid a **graduated commission**. This means their rate of commission increases as their sales increase. For example, the rate may be 3% for the first $10,000 of sales, 4% on the next $6000 and 5% on sales over $16,000.

37. Markita is paid 5% commission on the first $8000 of sales and 9% on sales over $8000. Last month her sales were $26,550. What was her commission?

38. Jomei is paid $2\frac{1}{2}$% commission on the first $5000 of sales, 5% on the next $8000, and $7\frac{1}{2}$% on sales over $13,000. In June, his sales totaled $32,000. What was his commission for the month?

MIXED REVIEW EXERCISES

Solve. Round to the nearest hundredth. (Lesson 6-1)

39. What percent of 16 is 32?

40. Find 74.5% of 250.

41. What percent of 56 is 8?

42. Find 250% of 72.

43. What percent of 1 is 0.25?

44. 32% of a number is 16. What's the number?

45. What is the ratio of even exercise numbers to odd exercise numbers on this page? (Lesson 2-6)

Review and Practice Your Skills

PRACTICE ■ LESSON 6-5

Find the interest.

1. Principal: $2400
   Rate: 6%/yr
   Time: 3 yr

2. Principal: $675
   Rate: 5.5%/yr
   Time: 5 yr

3. Principal: $14,000
   Rate: 7%/yr
   Time: 18 mo

4. Principal: $4700
   Rate: 3.8%/yr
   Time: 4 yr

5. Principal: $950
   Rate: 11%/yr
   Time: 30 mo

6. Principal: $6000
   Rate: 7.5%/yr
   Time: 10 yr

Find the rate of interest.

7. Principal: $7500
   Interest: $900
   Time: 2 yr

8. Principal: $640
   Interest: $36
   Time: 9 mo

9. Principal: $1450
   Interest: $464
   Time: 4 yr

Find the interest and amount due.

10. Principal: $1800
    Rate: 4.5%/yr
    Time: 3 yr

11. Principal: $460
    Rate: 16%/yr
    Time: 24 mo

12. Principal: $3500
    Rate: 8%/yr
    Time: 1 yr

13. Principal: $670
    Rate: 15%/yr
    Time: 2 yr

14. Principal: $30,000
    Rate: 7.2%/yr
    Time: 6 yr

15. Principal: $1590
    Rate: 6%/yr
    Time: 6 mo

PRACTICE ■ LESSON 6-6

Find the commission.

16. Total sale: $758
    Commission Rate: 4.2%

17. Total sale: $45
    Commission Rate: 6%

18. Total sale: $128
    Commission Rate: 3.5%

19. Total sale: $275
    Commission Rate: 8%

20. Total sale: $1460
    Commission Rate: 1.4%

21. Total sale: $640
    Commission Rate: 5%

Find the total income.

22. Base salary: $450/wk
    Total sales: $3800/wk
    Commission Rate: 4%

23. Base salary: $1300/mo
    Total sales: $15,800/mo
    Commission Rate: 3%

24. Base salary: $950/wk
    Total sales: $1300/wk
    Commission Rate: 2.5%

25. Base salary: $2800/mo
    Total sales: $15,800/mo
    Commission Rate: 8.2%

26. Base salary: $465/wk
    Total sales: $3200/wk
    Commission Rate: 1.5%

27. Base salary: $1790/mo
    Total sales: $48,800/mo
    Commission Rate: 6%

Find the commission rate.

28. Total sale: $18,500
    Commission: $592

29. Total sale: $178
    Commission: $7.12

30. Total sale: $4300
    Commission: $279.50

31. Total sale: $4590
    Commission: $321.30

32. Total sale: $1300
    Commission: $52

33. Total sale: $490
    Commission: $58.80

Solve each problem. (Lessons 6-1 and 6-2)

34. What percent of 200 is 3?

35. What number is 3% of 250?

36. 14% of what number is $10\frac{1}{2}$?

37. 16 is 40% of what number?

Find the missing quantities. (Lessons 6-3 and 6-4)

38. Regular price: $85.90
Percent of discount: 25%
Amount of discount: ?
Sale price: ?

39. Price: $235.59
Sales tax rate: 7.5%
Sales tax: ?
Total cost: ?

40. Price: $142.50
Sales tax rate: ?%
Sales tax: ?
Total cost: $151.05

Find the total income. (Lesson 6-6)

41. Base salary: $3200/mo
Total sales: $73,500/mo
Commission Rate: 2.5%

42. Base salary: $650/wk
Total sales: $1380/wk
Commission Rate: 6%

MathWorks Career – Political Scientist
Workplace Knowhow

People living on their own are often more aware of political issues and policies. Local, state and federal policies impact taxes, legal rights and insurance. Political scientists study how governments develop, operate, and interact. Political scientists examine laws, public opinion polls, election results and special interest groups.

A certain government representative is interested in public opinion in her district. The results of the survey are in the table.

Results from District 1	
Support raising the driving age	841
Support lowering the driving age	532
Support keeping the current driving age	1256

1. Find the percentage of people surveyed who support raising the driving age.

2. Find the percentage of people surveyed who support lowering the driving age.

3. Find the percentage of people surveyed who support keeping the current driving age.

4. If there are about 175,000 people in the representative's district, how many people can you expect to support raising the driving age?

5. How many can you expect to support lowering the driving age?

6. How many can you expect to support keeping the current driving age?

7. Based on the results of the survey, should the representative support lowering the driving age?

6-7 Percent of Increase and Decrease

Goals
- Calculate percent of increase and decrease.
- Solve problems involving percent of increase and decrease.

Applications Fitness, Retail, Real estate, Living on your own

The figure shown is a geoboard with a rectangle outlined using a rubberband.

1. What is the area of the rectangle?

2. Draw a similar figure whose area is 50% larger than this figure.

3. Draw a similar figure whose area is 25% larger than this figure.

4. Suppose that the figure shown is 75% of another figure. What would the other figure look like?

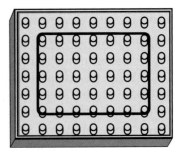

◥ BUILD UNDERSTANDING

The **percent of increase** tells what percent the amount of increase is of the original number.

To find the percent of increase, express a ratio of the amount of increase to the original number as a percent.

Percent of Increase	$\dfrac{\text{amount of increase}}{\text{original number}} \cdot 100$

Example 1

Find the percent of increase.

Original amount: 70
New amount: 84

Solution

Use the percent of increase formula.

$$r = \frac{\text{amount of increase}}{\text{original number}} \cdot 100$$

$$r = \frac{84 - 70}{70} \cdot 100$$

$$r = \frac{14}{70} \cdot 100$$

$$r = 0.2 \cdot 100$$

$$r = 20$$

The percent of increase from 70 to 84 is 20%.

The **percent of decrease** tells what percent the amount of decrease is of the original number.

To find the percent of decrease express a ratio of the amount of decrease to the original number as a percent.

Percent of Decrease	$\dfrac{\text{amount of decrease}}{\text{original number}} \cdot 100$

Example 2

Find the percent of decrease.

Original amount: $200
New amount: $175

Solution

Use the percent of decrease formula.

$$r = \frac{\text{amount of decrease}}{\text{original number}} \cdot 100$$

$$r = \frac{200 - 175}{200} \cdot 100$$

$$r = \frac{25}{200} \cdot 100$$
$$r = 0.125 \cdot 100$$

$$r = 12.5$$

The percent of decrease from $200 to $175 is 12.5%.

Example 3

FITNESS At the beginning of Barbara's exercise program, she walks 40 min/day. After six weeks she walks 55 min/day. Find the percent of increase in the amount of time she walks per day.

Solution

Use the percent of increase formula.

$$r = \frac{\text{amount of increase}}{\text{original number}} \cdot 100$$

$$r = \frac{55 - 40}{40} \cdot 100$$

$$r = \frac{15}{40} \cdot 100$$

$$r = 0.375 \cdot 100$$

$$r = 37.5$$

The percent of increase from 40 to 55 is 37.5%.

Example 4

TRAVEL In May, the round-trip airfare from Denver to Newark was $520. In November, the airlines reduced the fare to $442 round trip. Find the percent of decrease.

Solution

$$r = \frac{520 - 442}{520} \cdot 100 \qquad r = \frac{\text{amount of decrease}}{\text{original number}} \cdot 100$$

$$r = \frac{78}{520} \cdot 100$$

$$r = 0.15 \cdot 100 = 15$$

The percent of decrease in the airfare price is 15%.

◥ TRY THESE EXERCISES

Find the percent of increase.

1. Original price of refrigerator: $600
   New price: $630

2. Original number of vacation days: 14
   New number of vacation days: 21

3. Original weight: 104 lb
   New weight: 195 lb

4. Original salary: $450
   New salary: $513

Find the percent of decrease.

5. Original population: 840
   New population: 504

6. Original price: $250
   New price: $175

7. Original beats per minute: 80
   New beats per minute: 72

8. Original airfare: $680
   New airfare: $646

9. **REAL ESTATE** The Jacksons bought a house for $145,000 four years ago. They sell the house this year for $174,000. What is the percent of increase?

10. **DATA FILE** Refer to the gasoline prices data on page 528. What is the percent of increase in prices from 1996 to 2002? Round to the nearest tenth.

◥ PRACTICE EXERCISES • For Extra Practice, see page 569.

Find the percent of increase or decrease.

11. Original rent: $500
    New rent: $550

12. Original score: 185
    New score: 148

13. Original price: $0.25
    New price: $1.50

14. Original weight: 200 lb
    New weight: 144 lb

15. Original price: $120
    New price: $132

16. Original weight: 125 lb
    New weight: 135 lb

17. Original number of teams: 80
    New number of teams: 60

18. Original fare: $75
    New fare: $99

19. Original price: $500
    New price: $410

20. Original airfare: $800
    New airfare: $750

21. Original number: 472
    New number: 590

22. Original population: 320
    New population: 208

23. LIVING ON YOUR OWN Last year Monique's rent was $460/mo. This year her rent increased to $506. What is the percent of increase in her rent?

24. Lance bought a used car for $5500. He sells it two years later for $2475. What is the percent of decrease?

 25. WRITING MATH An owner of a music store pays the manufacturer for items that he sells in his store. What must he take into consideration when deciding the percent of markup for the items he sells?

DATA FILE For Exercises 26–34, refer to the life expectancy data on page 531. Copy and complete the table below to determine the percent of increase in life expectancy for males and females every ten years from 1930 to 2000. Round your answers to the nearest tenth.

26. 1930–1940

27. 1940–1950

28. 1950–1960

29. 1960–1970

30. 1970–1980

31. 1980–1990

32. 1990–2000

Time period	Percent of increase (male)	Percent of increase (female)
1930-1940	■	■
1940-1950	■	■
1950-1960	■	■
1960-1970	■	■
1970-1980	■	■
1980-1990	■	■
1990-2000	■	■

33. Which decade saw the greatest increase in life expectancy for females?

34. Which decade saw the smallest increase in life expectancy for males?

EXTENDED PRACTICE EXERCISES

 35. GEOBOARDS The enclosed area represents 75% of another area on the geoboard. Use a geoboard or draw a diagram of a geoboard to represent 100% of the area.

36. CRITICAL THINKING Does the percent of increase from 50 to 65 equal the percent of decrease from 65 to 50? Explain.

 37. WRITING MATH If an original amount is positive, explain why a 15% increase followed by a 15% decrease is less than the original amount.

 38. CHAPTER INVESTIGATION Suppose you pay $185/mo. How many months will you have to pay? How much will you pay in interest?

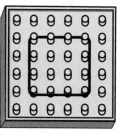

MIXED REVIEW EXERCISES

For which equation(s) does $x = 1.7$ make a true statement? (Lesson 3-5)

39. $3x - 11.7 = 5$

40. $x + (-6) = 4.3$

41. $5x + 3 = 11.5$

42. $2x - 2.4 = 1$

Find the volume of each cylinder. Use 3.14 for π. (Lesson 4-7)

43. $r = 4$ in., $h = 19.7$ in.

44. $h = 7$ cm, $d = 3.5$ cm

Problem Solving Skills: Sales and Expenses

An effective problem solving strategy for many students is to **make a table, chart or list**. Organize and classify the given data to lead you to a solution.

Problem Solving Strategies

Guess and check

Look for a pattern

Solve a simpler problem

✔ Make a table, chart or list

Use a picture, diagram or model

Act it out

Work Backwards

Eliminate possibilities

Use an equation or formula

Problem

RETAIL Shaun, Whitney, Kioshi and Tina are all salespersons at a computer store. Each sale is recorded with the salesperson's first initial, whether computer hardware (H) or software (S) is sold and the method of payment: cash ($), check (✔) or charge (C). The record of sales at the end of one weekend is shown.

S-H-C	K-S-$	K-H-✔	T-S-$	W-S-$
S-H-$	T-H-C	T-H-✔	W-H-✔	T-S-$
K-S-C	S-S-✔	T-H-✔	K-H-C	K-S-$
T-H-C	W-H-✔	K-H-C	T-H-✔	S-H-✔
W-S-$	S-H-C	S-S-✔	W-H-$	T-H-C

a. Make a table to represent the sales at the end of one week.

b. Which employee had the greatest percent of cash customers?

c. What percent of all the sales were hardware items?

Solve the Problem

a. In the table, label the columns Employee, Hardware item, Software item, Cash, Check and Charge. Then use the data from above to complete the table.

Employee	Hardware item	Software item	Cash	Check	Charge
Shaun	4	2	1	3	2
Whitney	3	2	3	2	0
Kioshi	3	3	2	1	3
Tina	6	2	2	3	3

b. Calculate the percent of cash customers for each employee.

Shaun had 6 customers and 1 paid with cash. His percent of cash customers is $\frac{1}{6}$ or 16.67%.

Whitney had 5 customers and 3 paid with cash. Her percent of cash customers is $\frac{3}{5}$ or 60%.

Kioshi had 6 customers and 2 paid with cash. His percent of cash customers is $\frac{2}{6}$ or $33\frac{1}{3}$%.

Tina had 8 customers and 2 paid with cash. Her percent of cash customers is $\frac{2}{8}$ or 25%.

Whitney had the greatest percent of cash customers.

c. There was a total of 25 sales for the month and 16 of the sales were hardware items. So, the percent of hardware sales is $\frac{16}{25}$ or 64%.

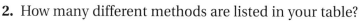

TRY THESE EXERCISES

Five-step Plan
1 Read
2 Plan
3 Solve
4 Answer
5 Check

1. Make a table to represent the different ways you can make change for a 50-cent piece using quarters, dimes and nickels. Label the columns Method, Quarter, Dime and Nickel. Label the rows 1, 2, 3 and so on.

2. How many different methods are listed in your table?

3. **CHECK YOUR WORK** Compare your table to a classmate's table. Did you leave out any possibilities? Explain.

4. In the table, what percent of the methods require 2 dimes?

PRACTICE EXERCISES

FOOD SERVICE Last Saturday, The Pizza Palace sold the following pizzas.

Small—3 cheese; 5 cheese and sausage; 8 cheese and pepperoni; 3 cheese, sausage, pepperoni, and mushrooms; 1 the works

Medium—4 cheese; 7 cheese and sausage; 2 cheese and pepperoni; 8 cheese, sausage, pepperoni, and mushrooms; 10 the works

Large—4 cheese; 5 cheese and sausage; 8 cheese and pepperoni; 12 cheese, sausage, pepperoni, and mushrooms; 8 the works

5. Organize the data from The Pizza Palace into a table. For the column headings, use the types of pizzas. For the row headings, use the different sizes available.

6. Which type of pizza is most popular?

7. Which size pizza sold the most?

8. What percent of pizzas sold are large?

9. What percent of medium pizzas are cheese?

10. **WRITING MATH** If you are manager of The Pizza Palace, how would you use this information to order supplies and ingredients?

LIVING ON YOUR OWN Every month, Jentine uses her checkbook to create a table of monthly expenses.

11. Use the checkbook shown to create a table of her monthly expenses.

12. What percent of her paycheck deposits does she spend on rent?

13. Did she spend a greater percent on groceries or other bills?

RECORD ALL CHARGES OR CREDITS THAT AFFECT YOUR ACCOUNT

NUMBER	DATE	DESCRIPTION OF TRANSACTION	PAYMENT DEBT (−)		✔ T	FEE (IF ANY) (−)	PAYMENT CREDIT (+)		BALANCE	
									$134	27
	5/1	paycheck deposit					+650	00	784	27
213	5/1	rent	-425	00					359	27
214	5/6	car payment	-167	00					192	27
215	5/11	car insurance	-72	00					120	27
216	5/12	groceries	-37	41					82	86
217	5/14	gas and electric	-34	50					48	36
	5/16	paycheck deposit					+650	00	698	36
ATM	5/17	spending money	-100	00					598	36
218	5/19	water	-11	72					586	64
219	5/19	phone	-39	43					547	21
220	5/21	credit card payment	-90	00					457	21
221	5/26	groceries	-43	66					413	55

MIXED REVIEW EXERCISES

Find the area of each shape to the nearest hundredth. (Lesson 2-9)

14.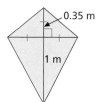
0.35 m
1 m

15.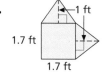
1 ft
1.7 ft
1.7 ft

16.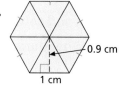
0.9 cm
1 cm

Chapter 6 Review

VOCABULARY

Choose the word from the list that best completes each statement.

1. The percent of sales a salesperson receives as income is __?__ .

2. The percent of the regular price you save when you buy an item on sale is the __?__ .

3. A(n) __?__ is a ratio that compares a number to 100.

4. A(n) __?__ is a percentage charge imposed by an authority.

5. The __?__ is the money paid to an individual or institution for the privilege of using their money.

6. The equation $I = prt$ is the formula for __?__ .

7. The amount of money that a person takes home from their job after the taxes are paid is called their __?__ .

8. Homeowners pay __?__ taxes based on the value of their house and land.

9. The amount of money you borrow from a financial institution for a loan is called the __?__ .

10. Amounts deducted from your paycheck by federal, state, and local governments based on your income are __?__ taxes.

a. bargain

b. commission rate

c. discount

d. income

e. interest

f. loan

g. net pay

h. percent

i. principal

j. property

k. simple interest

l. tax

LESSON 6-1 ■ Percents and Proportions, p. 260

▶ **Percent** means *per one hundred.*

$$\frac{part}{whole} = \frac{part}{whole}$$

Write and solve a proportion.

11. 75% of 64 is what number?   12. What percent of 80 is 12?   13. 5 is what percent of 30?

14. Last August there were 12 out of 31 days when the temperature was above 85°F. About what percent of days was the temperature 85°F or less last August?

15. A financial group charges 20% for an advance on your income tax refund. If your refund is $250, how much do they receive?

16. Sam saved 30% on a television. If Sam saved $128, what was the original price of the television to the nearest dollar?

LESSON 6-2 ■ Write Equations for Percents, p. 264

▶ To solve a problem involving percent, translate the phrase into an equation, then solve for the unknown. "Is" means *equal.* "Of" means *multiply.*

Write and solve an equation.

17. What number is 7% of 80?

18. What percent of 86 is 215?

19. 1 is what percent of 40?

20. 60 is 40% of what number?

21. 70% of what number is 147?

22. 80% of 16 is what number?

LESSON 6-3 ◣ Discount and Sale Price, p. 270

▶ The **discount** is the amount that the regular price is reduced.

▶ The **sale price** is the regular price minus the discount.

Find the discount and the sale price. Round your answers to the nearest cent.

23. Regular price: $899.98
    Discount: 25%

24. Regular price: $510.20
    Discount: 10%

Find the percent of discount.

25. Regular price: $24.50
    Sale price: $17.15

26. Regular price: $3457.00
    Sale price: $2592.75

Find the regular price.

27. Sale price: $750
    Discount rate: 25%

28. Sale price: $6.30
    Discount rate: 10%

LESSON 6-4 ◣ Solve Problems Using Tax Rates, p. 274

▶ People employed in the U.S. must pay an **income tax** based on their income. The **net pay,** or take-home pay, is the amount of money that a person is paid after taxes are subtracted.

▶ Homeowners pay **property taxes** based on the value of their house and property.

Find the income tax and net pay for each.

29. Income: $35,850/yr
    Income tax rate: 12%

30. Income: $830/wk
    Income tax rate: 23%

31. Find the amount of sales tax and the total cost for a jacket priced at $45. The sales tax rate is 6%.

32. Find the sales tax rate on paint priced at $65 that sells for $68.58 with tax.

33. Find the property tax on a home valued at $216,000 with a tax rate of 3.1%.

34. What is the tax rate on a $135,000 home if the property tax is $3700?

35. What is the value of a home if its tax rate is 4% and the property tax is $6240?

LESSON 6-5 ◣ Simple Interest, p. 280

▶ The amount of money that is earning interest or that you are borrowing is called the **principal**.

| Simple interest | $I = prt$ |

▶ The **amount due** is equal to the principal plus the accrued interest.

36. Find the interest and the amount due when the principal is $600, the rate is 7%/year, and the time is 2 years.

37. Find the rate of interest when the principal is $9000, the interest is $5040, and the time is 4 years.

You are buying a car and need to finance $15,000 of the cost. The dealership offers you two different loan options: (1) 5-year loan at 3% and (2) 3-year loan at 5%.

38. Find the total cost of each loan plan for financing $15,000.

39. Which plan would you choose if you wanted the lowest monthly car payment? Explain your choice.

LESSON 6-6 ◣ Sales Commission, p. 284

▶ You can use either a proportion or an equation to solve problems involving **commission**.

40. A real estate company sells four houses in one month for a total of $654,320. How much commission does the company earn if the commission rate is 6%? $

41. What rate of commission does a salesperson earn if a total sales of $536 results in a $13.40 commission?

42. Find the total weekly income of a salesperson who earns a 5% commission, whose base salary is $400/wk, and whose total sales in one week is $2500.

LESSON 6-7 ◣ Percent of Increase and Decrease, p. 290

▶ The **percent of increase** tells what percent the amount of increase is of the original amount.

▶ The **percent of decrease** tells what percent the amount of decrease is of the original amount.

Percent of increase	$\dfrac{\text{amount of increase}}{\text{original number}} \cdot 100$
Percent of decrease	$\dfrac{\text{amount of decrease}}{\text{original number}} \cdot 100$

43. Find the percent of increase if the original price was $96 and the new price is $108.

44. Find the percent of decrease if the original population was 800 and the new population is 716.

45. A new car listed at $26,450 went on sale for $23,805. What was the percent of decrease in the price of the car?

46. A newspaper increased its subscription rate by $25 per year. If the original rate was $168, what was the percent of increase in the subscription rate?

LESSON 6-8 ◣ Problem Solving Skills: Sales and Expenses, p. 294

47. DATA FILE Use the data on the ten most popular U.S. roller coasters on page 521. If you wanted to plan a vacation in order to ride the most popular roller coasters, which state would you visit?

CHAPTER INVESTIGATION

EXTENSION You have been pre-approved for two credit cards. You want to accept one card to purchase a new microwave for $109.00 and a television for $328.00. One credit card has no annual fee. Its interest rate is 19.2%. The other card has an annual fee of $69.95 and an interest rate of 17.5%. Suppose you will pay $75 each month. Which credit card should you choose? Why? Make a presentation of your data to support your choice.

Chapter 6 Assessment

Write and solve a proportion.

1. 17% of 50 is what number?

2. What percent of 800 is 280?

3. 28 is what percent of 140?

4. Write 13 out of 20 as a percent.

Write and solve an equation.

5. What percent of 65 is 29.9?

6. What number is 52% of 95?

7. 84 is what percent of 350?

8. 300% of what number is 27?

9. What percent of 50 is 10?

10. What number is 87% of 3?

11. What number is 5.8% of 70?

12. 62% of what number is 254.2?

Find the discount and the sale price. Round your answers to the nearest cent.

13. Regular price: $419
    Discount: 12%

14. Regular price: $53.69
    Discount: 25%

Find the percent of discount.

15. Regular price: $18.90
    Sale price: $16.63

16. Regular price: $329.00
    Sale price: $230.30

Find the income tax and net pay for each.

17. Income: $250/wk
    Income tax rate: 8%

18. Income: $50,000/yr
    Income tax rate: 18%

Solve.

19. Find the amount of sales tax and the total cost of a stereo that is priced at $200. The sales tax rate is 7%.

20. Find the property tax paid by the owner of a home valued at $120,000. The property tax rate is 3%.

21. Find the sales tax rate on a painting that is priced at $130 and sells for $136.50.

22. What rate of commission does a salesperson earn if a total sales of $6000 results in a $240 commission?

23. Find the total weekly income of a salesperson who earns a 6% commission, whose base pay is $500/wk, and whose total sales in one week is $800.

Find the missing values.

24. Principal: $170
    Rate: 6%/yr
    Time: 2 years
    Interest: ?
    Amount due: ?

25. Principal: $8100
    Rate: ?
    Time: 36 months
    Interest: $1822.50
    Amount due: ?

26. Principal: $517
    Rate: 12%/yr
    Time: ?
    Interest: $31.02
    Amount due: ?

Find the percent of increase or decrease.

27. Original amount: $90
    New amount: $144

28. Original amount: $500
    New Amount: $320

Standardized Test Practice

Part 1 Multiple Choice

Record your answers on the answer sheet provided by your teacher or on a sheet of paper.

1. Which measure of central tendency best represents the most requested music videos? (Lesson 1-2)

- (A) mean
- (B) median
- (C) mode
- (D) quartile

2. What is the area of a square whose side measures $2\frac{1}{2}$ ft? (Lesson 2-4)

- (A) 6.25 ft
- (B) 6.25 ft^2
- (C) 10 ft
- (D) 10 ft^2

3. Find the circumference to the nearest tenth of a centimeter. Use 3.14 for π. (Lesson 2-7)

5.5 cm

- (A) 379.94 cm
- (B) 94.985 cm
- (C) 34.54 cm
- (D) 17.27 cm

4. A community garden is 10.5 yd wide and 12 yd long. The extension office recommends that 1 lb of compost be added for every 10 yd^2 of garden. About how many pounds of compost are needed for the community garden? (Lesson 2-5)

- (A) 4.52
- (B) 12.6
- (C) 126
- (D) 1260

5. $-7 + (-15) = ?$ (Lesson 3-1)

- (A) -22
- (B) -8
- (C) 8
- (D) 22

6. Which number should be substituted for the ■ in the following expression? (Lesson 3-4)

$$(7 \times ■) \times 3 = 7 \times (5 \times 3)$$

- (A) 3
- (B) 5
- (C) 15
- (D) 35

7. A solid figure has two circular bases that are congruent and parallel. Name the figure. (Lesson 4-3)

- (A) cylinder
- (B) prism
- (C) circular pyramid
- (D) sphere

8. What is the volume of the prism? (Lesson 4-7)

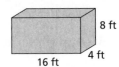

8 ft

16 ft   4 ft

- (A) 112 ft^3
- (B) 192 ft^3
- (C) 256 ft^3
- (D) 512 ft^3

9. What is the solution of $1\frac{3}{4}n = 17\frac{1}{2}$? (Lesson 5-3)

- (A) $n = 10$
- (B) $n = 15\frac{3}{4}$
- (C) $n = 19\frac{1}{4}$
- (D) $n = 30\frac{5}{8}$

10. Which inequality is represented by the graph? (Lesson 5-8)

- (A) $x > -1$
- (B) $x \geq -1$
- (C) $x > -2$
- (D) $x \geq 3$

$-4\ -3\ -2\ -1\ \ 0\ \ 1\ \ 2\ \ 3$

11. How is $\frac{2}{5}$ expressed as a percent? (Lesson 6-1)

- (A) 2.5%
- (B) 0.40%
- (C) 25%
- (D) 40%

Test-Taking Tip

(A) (B) (C) (D)

Exercise 6

Check your solution by replacing the shaded box with your answer. If this results in a true statement, your solution is correct.

Part 2 Short Response/Grid In

Record your answers on the answer sheet provided by your teacher or on a sheet of paper.

12. If the mean score for a class on a math test is 80, what would a student who was absent have to score on a makeup test for the class mean to remain the same? (Lesson 1-2)

13. Complete. Write the answer in simplest form. (Lesson 2-2)
$$1.9 \text{ km} - 1.25 \text{ km} = \underline{\ ?\ } \text{ m}$$

14. Solve $3 : 4 = b : 16$ for b. (Lesson 2-8)

15. What property of real numbers is shown by this equation? (Lesson 3-4)
$$(9 + 3) + 5 = 9 + (3 + 5)$$

16. When simplified, the exponent of $x^8 \cdot x^2$ would be what number? (Lesson 3-7)

17. Draw a net for a cube. (Lesson 4-4)

18. How many cubes are in the isometric drawing? (Lesson 4-5)

19. What is the volume of the prism to the nearest tenth? (Lesson 4-7)

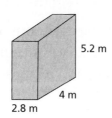

5.2 m

4 m

2.8 m

20. Solve $5x \times 11 > -4$. (Lesson 5-9)

21. Find the property tax paid by the owner of a home valued at $75,000 if the property tax rate is 2.5%. (Lesson 6-4)

22. What is the amount of interest earned in one year on $900 if the rate of interest is 9.5% per year? (Lesson 6-5)

23. Charmaine receives a 4% commission on her furniture sales. Last week, she sold $2800 worth of furniture. What was the amount of her commission? (Lesson 6-6)

24. The price per share of an Internet-related stock decreased from $90 per share to $36 per share early in 2001. What was the percent of decrease? (Lesson 6-7)

Part 3 Extended Response

Record your answers on a sheet of paper. Show your work.

25. Consider the following methods of gathering data. (Lesson 1-1)

- personal interviews
- records of performance
- questionnaire
- telephone interviews

a. Which method should Trendy Tires use to gather data about the quality of their Consumer Service Department?

b. Which method would be the best way to determine the baseball player with the best batting average?

c. Which method should a campaign office use to find out whether a government official will be relected to office in the next election?

26. A car salesperson states that a new SUV model in the showroom can go 232 mi on a tank of gas. The formula for gas mileage is $m = \frac{d}{g}$, where m is miles per gallon, d is the distance driven, and g is number of gallons used. (Lesson 5-6)

a. If the tank holds 16 gal, how many miles per gallon will the SUV get?

b. How many tanks of gas would it take on a trip that was 448 mi?

c. How far can the SUV go on 6 gal of gas?

Functions and Graphs

THEME: Velocity

When two sets of data are related, you can use them to describe relationships, determine trends, and make predictions. One set of data depends on the other set. For example, the cost of mailing a letter depends on its weight, the color of light depends on its wavelength, and a baseball's velocity depends on the amount of force applied to it.

Velocity is the rate that an object or particle changes position with respect to time. Velocity can describe the speed of cars, trains, planes, light, sound, animals, communications, and manufacturing. The mathematics used to calculate and measure velocity uses functions and graphs.

- **Truck drivers** (page 313) transport goods from one point to another. They must keep accurate records of their mileage, speed, and expenses for each load they haul.

- **Ship captains** (page 333) plan the route and speed of the ship, and keep records of the ship's travels and cargo.

Math Online

mathmatters1.com/chapter_theme

Speed of Sound in Various Media

Substance	Temp (° C)	Speed (m/sec)
Gases		
Carbon Dioxide	0	259
Air	0	331
Air	21	343
Liquids		
Ethanol	20	1162
Mercury	20	1450
Water	20	1482
Solids		
Lead	–	1960
Copper	–	3810
Glass	–	5640

Data Activity: Speed of Sound

Sound travels at different speeds depending on what it is traveling through. Although air is the most common medium, sound can travel through any type of matter. Sound travels slowest through gases, faster through liquids and fastest through solids. As the temperature of the medium rises, so does the speed of sound.

Use the table for Questions 1–5.

1. Through which substances does sound travel between 1000 m/sec and 2000 m/sec?

2. How fast does sound travel through water?

3. About how long would it take sound to be transmitted through 10 km of copper wire?

4. Many animals in the ocean rely on sound waves to communicate over long distances. Why do you think this is so?

5. Why do you think tin can telephones work?

CHAPTER INVESTIGATION

Velocity is a rate of change influenced by various factors. For instance, ship captains must account for wind and its effects. Engineers can improve the quality of sound by adjusting the physical surroundings. An animal's speed is affected by muscular structure, health and habitat. A car coasting down a ramp is influenced by the slope and length of the ramp.

Working Together

Use functions and graphs to determine which is more important to the speed of a car coasting down a ramp, the slope of the ramp or the length of the ramp. The Chapter Investigation icons will help you find the answer.

7 Are You Ready?

Refresh Your Math Skills for Chapter 7

The skills on these two pages are ones you have already learned. Use the examples to refresh your memory and complete the exercises. For additional practice on these and more prerequisite skills, see pages 536–544.

COORDINATE GRAPHS

You have plotted points on a coordinate plane. Sometimes analyzing data on a graph can help you make conclusions or set a rule.

Examples The ordered pair $(3, -4)$ can be shown on the coordinate plane. Start at zero $(0, 0)$. Move 3 to the right and down 4. Point C lies at $(3, -4)$.

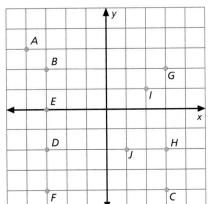

Name the point that corresponds to each point.

1. A 2. H

3. J 4. G

5. F 6. E

7. I 8. B

Plot each ordered pair on a coordinate plane.

9. $S\,(0, -1)$ 10. $T\,(-3, -4)$ 11. $U\,(4, 2)$ 12. $V\,(4, -1)$

13. $W\,(-2, 0)$ 14. $X\,(3, 3)$ 15. $Y\,(-1, -2)$ 16. $Z\,(0, 2)$

EVALUATE EXPRESSIONS

You can find the value of an expression by substituting values for the variables.

Evaluate each expression. Use $x = -4$, $y = 1$, and $z = -\dfrac{1}{2}$.

17. $x + 7$ 18. $y - 11$ 19. $4x$ 20. $14z$

21. $x + y$ 22. $3x - 7$ 23. $27 - xy$ 24. $y - \dfrac{3}{4}x$

25. $x + y - z$ 26. $xz + y$ 27. $\dfrac{1}{2}x + \dfrac{1}{2}z$ 28. $\dfrac{2}{3}y - 5z$

SQUARES AND SQUARE ROOTS

While there are many applications for squares and square roots, in this chapter you will use them to solve equations involving right triangles.

Simplify.

29. 16^2 30. 14^2 31. 21^2 32. 18^2

33. $\sqrt{225}$ 34. $\sqrt{784}$ 35. $\sqrt{961}$ 36. $\sqrt{289}$

PYTHAGOREAN THEOREM

You can use the formula $a^2 + b^2 = c^2$ to find the length of the sides of a right triangle.

Example Find the measure of the unlabeled side. Variables a and b represent the measures of the two legs, and c represents the hypotenuse. Replace the variables in the formula with the known values. Then simplify.

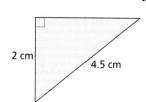

$$a^2 + b^2 = c^2$$
$$2^2 + b^2 = (4.5)^2$$
$$4 + b^2 = 20.25$$
$$4 - 4 + b^2 = 20.25 - 4$$
$$b^2 = 16.25$$
$$b = \sqrt{16.25} \approx 4.03$$

The remaining side measures about 4.03 cm.

Use the Pythagorean Theorem to find the measure of each unlabeled side. Round to the nearest hundredth.

37.

6 in.

6 in.

38. 4 cm

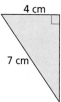

7 cm

39. 4.2 ft

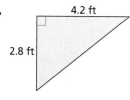

2.8 ft

EQUATIONS

You have found the solutions for equations. Sometimes it is necessary to solve an equation for a variable in terms of another variable.

Example Solve $x + 3y = -6$ for y.

$$x - x + 3y = -x - 6$$
$$\frac{3y}{3} = \frac{-x - 6}{3}$$
$$y = -\frac{1}{3}x - 2$$

Solve each equation for y.

40. $4x + y = 19$ **41.** $9y = x + 3$ **42.** $3y - 4x = 12$ **43.** $18 - 2y = -6x$

7-1 Problem Solving Skills: Qualitative Graphing

A graph can be used as a form of communication. For example, a graph can describe the speed of an object, the growth of a child, or even a ride on a rollercoaster. For such situations, instead of creating an exact picture, you can use a **qualitative graph**.

When you must interpret a qualitative graph to solve a problem, a good strategy for understanding is to **act it out**. Acting out the problem or possible solutions can mean that you physically go through the motions described or that you use objects to represent the elements of the problem.

Problem Solving Strategies

Guess and check

Look for a pattern

Solve a simpler problem

Use a picture, diagram or model

Make a table, chart or list

✔ Act it out

Work backwards

Eliminate possibilities

Use an equation or formula

Problem

The graph shown provides information about water in a bathtub.

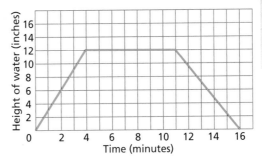

a. What is represented by the horizontal axis? the vertical axis?

b. What does the line moving upwards represent?

c. What does the horizontal segment of the graph represent?

d. What does the line moving downwards represent?

e. What is the total time spent from turning on the water until the bathtub is empty?

f. What could be represented by the graph shown at the right?

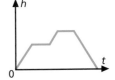

Solve the Problem

a. The horizontal axis represents time in minutes. The vertical axis represents the water height in inches.

b. The line moving upwards represents filling the bathtub with water.

c. The horizontal segment represents the time between filling and draining the tub with nothing else added to it.

d. The line moving downwards represents draining the water from the bathtub.

e. The total time is 16 min.

f. The new graph could represent filling the bathtub with water, then a person bathing, and finally draining the water from the tub.

TRY THESE EXERCISES

Match each situation with one of the graphs shown. A graph can be used more than once.

1. The speed of a rollercoaster as it goes steadily up, then faster down a hill.

2. The speed of a subway train as it pulls into a station, stops, then pulls out.

3. Your speed on a bike as you ride up a hill, rest, then ride down the other side.

4. Your distance from the ground as you swing on a swing.

5. The amount of popcorn in a bowl during a movie.

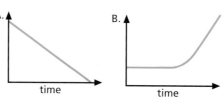

Five-step Plan

1 Read
2 Plan
3 Solve
4 Answer
5 Check

PRACTICE EXERCISES

Carla and Seishi walk to school together. One day Seishi forgot her lunch so she walked back home to get it. Her mom then drove her to meet up with Carla.

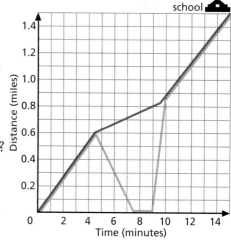

6. What is represented by the horizontal and vertical axes?

7. Who is represented by the red line? by the blue line?

8. What is represented by the upward lines on the graph?

9. What is represented by the downward line on the graph?

10. What is represented by the horizontal blue line?

11. How long did it take them to get to school?

12. How many minutes did Seishi walk?

LITERATURE Write a short story that can be represented by each graph.

13.

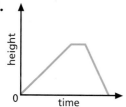

14.

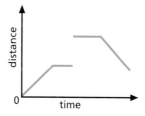

15. **WRITING MATH** Suppose you read a news article that included a qualitative graph. Make a list of at least six possible topics that the news article could be about. Name the labels on the axes for each topic.

MIXED REVIEW EXERCISES

Calculate the interest. (Lesson 6-4)

16. $p = \$5000$, $r = 3.5\%$
    $t = 3$ years

17. $p = \$2500$, $r = 3.0\%$
    $t = 5$ years

18. $p = \$10,000$, $r = 2.75\%$
    $t = 9$ months

7-2 The Coordinate Plane

Goals
- Identify points on the coordinate plane.
- Graph ordered pairs on the coordinate plane.

Applications Travel, Navigation, Velocity

NAVIGATION A location on a map is specified with a letter followed by a number. On the map shown, Topeka, Kansas, is located in column C and row 2. This region is written as C2.

1. Which cities are located in region C4?

2. In which region is Omaha, Nebraska located?

3. Through which three regions might you travel if you go from Fargo, North Dakota, to Sioux Falls, South Dakota?

4. If you were to travel from St. Louis, Missouri, to Omaha, Nebraska, in which region would your trip begin? In which region would it end?

▼ BUILD UNDERSTANDING

The system used to identify locations on a map is similar to the mathematical system called a **coordinate plane**.

On a coordinate plane, two number lines are drawn perpendicular to each other. The horizontal number line is called the **x-axis**. The vertical number line is called the **y-axis**. The axes separate the plane into four regions, called **quadrants**. The axes intersect at their zero points, called the **origin**.

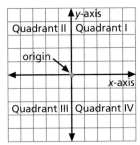

In the graph at the right, point *S* is located 3 units to the right of the origin and 4 units up from the origin. This point is identified as *S*(3, 4). The *x*-coordinate is 3. The *y*-coordinate is 4. The location of point *S*(3, 4) is different than the location of point *T*(4, 3).

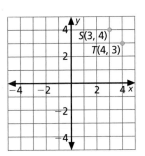

Each point on the coordinate plane can be identified by an ordered pair. An **ordered pair** is a set of two numbers (*x*, *y*) where *x* is the *x*-coordinate and *y* is the *y*-coordinate of the point. Every point on the coordinate plane is associated with a unique ordered pair.

When you read coordinates and graph points in the plane, you need to be aware of directions. The table shows you which direction to move to locate a point based on the sign of each coordinate in the ordered pair.

	x-coordinate	*y*-coordinate
Positive	move right →	move up ↑
Negative	move left ←	move down ↓

Example 1

Find the coordinates.

a. point D

b. two points in Quadrant III

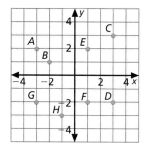

Solution

a. Point D is 3 units to the right of the origin and 2 units down from the origin. The coordinates of D are $(3, -2)$.

b. Points G and H are in Quadrant III. Their coordinates are $G(-3, -2)$ and $H(-1, -3)$.

Example 2

Graph points $X(0, 3)$, $Y(2, 0)$ and $Z(-2, -2)$ on a coordinate plane.

Solution

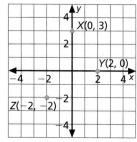

Point X is 0 units to the right or left of the origin and 3 units up from the origin.

Point Y is 2 units to the right of the origin and 0 units up or down.

Point Z is 2 units to the left of the origin, then 2 units down.

Check Understanding

If the x-coordinate of a point is 0, what can you tell about the location of the point?

If the y-coordinate of a point is 0, what can you tell about the location of the point?

Example 3

On a coordinate plane, sketch square $JKLM$ with a diagonal having endpoints at $J(3, 3)$ and $L(-3, -3)$.

Solution

Begin by graphing $J(3, 3)$ and $L(-3, -3)$ on a coordinate plane.

Graph point M directly to the left of J and above L. The point $M(-3, 3)$ is a vertex of the square.

Graph point K directly below J and to the right of L. The point $K(3, -3)$ is another vertex of the square.

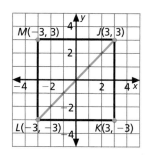

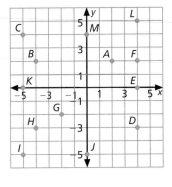

Use the coordinate plane shown. Give the coordinates of each.

1. point *A*   2. point *D*   3. point *M*

4. point *H*   5. origin   6. a point on the *x*-axis

7. a point in Quadrant IV

8. two points with the same *x*-coordinate

Graph each point on a coordinate plane.

9. *A*(4, 0)   10. *B*(−3, 4)   11. *C*(3, −2)

12. **ERROR ALERT** Perico graphed *A*(−2, 3) in Quadrant III. What mistake do you think he made?

Sketch each figure on a coordinate plane.

13. triangle *XYZ* with vertices *X*(3, 2), *Y*(5, 6) and *Z*(2, −4)

14. square *ABCD* with a diagonal having endpoints at *B*(5, −5) and *D*(−5, 5)

15. **WRITING MATH** Given point (2, −3), explain how its coordinates tell its location.

▼ **PRACTICE EXERCISES** • **For Extra Practice, see page 569.**

Use the coordinate plane shown. Give the coordinates of each of the following.

16. point *A*   17. point *B*

18. point *E*   19. point *N*

20. point *I*   21. point *R*

22. two points on the *x*-axis

23. two points with the same *y*-coordinate

24. a point in Quadrant I

25. a point in Quadrant III

26. a point whose *x*- and *y*-coordinates are the same number

27. a point whose coordinates are opposites

Graph each point on a coordinate plane.

28. *T*(3, 2)   29. *V*(4, −1)   30. *W*(0, 4)   31. *X*(−7, 8)

32. *Y*(−4, 0)   33. *Z*(3, −5)   34. *R*(−8, −4)   35. *O*(0, 0)

Sketch each figure on a coordinate plane.

36. rectangle *QRST* with a diagonal with endpoints at *R*(3, 2) and *T*(−3, −2)

37. triangle *ABC* with vertices *A*(−2, 2) *B*(4, −3) and *C*(−3, −1)

38. Which pair of points is closer together, (−2, 7) and (5, 7) or (−1, 3) and (−1, −1)? Explain your answer.

39. Graph $L(1, 2)$. Move 3 units to the right and 2 units up to locate point M. What are the coordinates of M?

All but one of these points suggest a pattern: $A(-7, -4)$, $B(-3, -2)$, $C(3, 1)$, $D(-5, -3)$, $E(1, 0)$ and $F(7, 4)$.

40. Graph these points.

41. Write the coordinates of the one point that does not fit the pattern.

42. Explain why it does not fit the pattern.

Graph each set of points on a coordinate plane. Then join the points in order. Identify the geometric figure and find its area.

43. $A(0,0)$, $B(0, 5)$, $C(4, 5)$ and $D(4, 0)$

44. $E(-7, 1)$, $F(-3, 1)$, $G(-3, -3)$ and $H(-7, -3)$

EXTENDED PRACTICE EXERCISES

TRAVEL To get home from work, Kara drives 6 mi west, 3 mi south, 2 mi west, 4 mi south and 3 mi east. Her husband Phil works at an office 4 mi east of Kara's office. To get home from work, he drives 5 mi south, 5 mi west, 2 mi south and 2 mi west.

45. On a coordinate plane, use two different colored pens to draw the route that both Kara and Phil take home from work.

46. How many miles does Kara drive home from work?

47. How many miles does Phil drive home from work?

48. According to your graph, whose workplace is a closer distance to home?

49. VELOCITY Kara drives at an average speed of 42 mi/h, and Phil drives at an average speed of 37 mi/h. If they both leave work at the same time, who generally makes it home quicker?

50. CHAPTER INVESTIGATION Use a marble to model the car. A ruler with a groove down the center will be the ramp. Secure the ruler to a stack of three books so that the slant of the ruler will not change. Roll the marble down the ruler from the top. Measure the different distances traveled using a meter stick or tape measure. Record three rolls of the marble.

MIXED REVIEW EXERCISES

Calculate the discount and sale price for each item. (Lesson 6-2)

51. Original price: $55, Discount rate: 15%

52. Original price: $125, Discount rate: 20%

53. Original price: $215, Discount rate: 25%

54. Find the commission on a sale of $13,425 if the commission rate is 3%. (Lesson 6-6)

Review and Practice Your Skills

PRACTICE ◣ LESSON 7-1

Both graphs show bicycle trips. Use the graphs to answer the questions.

1. On both graphs, what is shown by the horizontal axis?

2. On both graphs, what is shown by the vertical axis?

3. On Graph 1, how far did the rider go in the first hour?

4. On Graph 1, how many miles had been covered when the rider stopped for lunch?

5. On Graph 1, did the rider go faster before or after lunch?

6. On Graph 2, what does the downward part of the graph represent?

7. On Graph 2, how far did the rider go in the first 45 minutes?

8. The rider for Graph 2 stopped for lunch after one and one-half hours. How long did she stop for lunch?

9. How many miles away from home was the rider for Graph 2 when she stopped for lunch?

10. During which time period did the rider for Graph 2 go the fastest?

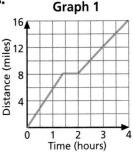

Graph 1

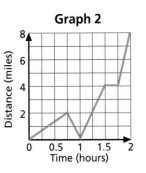

Graph 2

PRACTICE ◣ LESSON 7-2

Use the coordinate plane shown. Give the coordinates of each point.

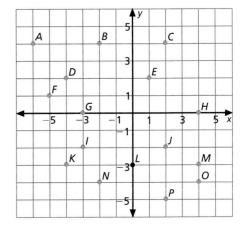

11. point A

12. point J

13. point C

14. point L

15. point F

16. point N

17. point P

18. point B

19. point K

20. point D

21. point M

22. point H

23. a point with the same x-coordinate as point B

24. a point with the same y-coordinate as point N

In which quadrant is each of these points?

25. point K

26. point E

27. point M

28. point F

Graph each point on a coordinate plane.

29. $D(-1, 3)$

30. $M(3, -1)$

31. $K(-2, -2)$

32. $J(-2, 0)$

33. $X(4, -2)$

34. $A(4, 4)$

35. $R(0, 3)$

36. $W(-1, -3)$

37. $L(3, 0)$

38. $H(-1, 1)$

39. $C(0, -1)$

40. $P(2, -3)$

The graph shows a bicycle trip. (Lesson 7-1)

41. How far did the rider go in the first two hours?

42. After two hours, the rider stopped for lunch. How long did he stop for lunch?

43. How long did it take the rider to get home after lunch?

44. Did the rider go faster before or after lunch?

Use the coordinate plane shown. Give the coordinates of each point. (Lesson 7-2)

45. point F **46.** point K **47.** point B

48. point E **49.** point A **50.** point J

51. a point with the same y-coordinate as point D

Graph each point on a coordinate plane. (Lesson 7-2)

52. $H(-2, -2)$ **53.** $M(0, 3)$ **54.** $D(2, -4)$ **55.** $P(-4, 2)$

56. $Y(1, 4)$ **57.** $B(-2, 3)$ **58.** $L(-3, 0)$ **59.** $F(3, -1)$

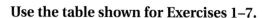

MathWorks Career – Truck Driver
Workplace Knowhow

Truck drivers transport goods from one point to another. They work for themselves or as an employee of a shipping or trucking company. Some drivers own their own trucks, while others rent or lease trucks. They must keep accurate records of their mileage and expenses for each load they ship.

Use the table shown for Exercises 1–7.

Hours	Miles
1	45
2	103
3	164
4	167
5	221
6	279

1. Plot and connect the data on a coordinate grid.

2. What is represented by the horizontal axis?

3. What is represented by the vertical axis?

4. What intervals did you choose for each axis?

5. What do you think the almost horizontal line from hours $3-4$ represents?

6. What is the truck driver's average speed for hours $1-3$?

7. What is the truck driver's average speed for the entire trip?

7-3 Relations and Functions

Goals ■ State the domain, range and whether a relation is a function.
■ Evaluate a function by using a function rule.

Applications Part-time job, Velocity, Manufacturing, Number sense

TRAVEL While driving on vacation, the Carmonas kept track of the total distance they traveled after each hour of the trip. This data is presented in the table.

1. Use a full sheet of graph paper to plot each set of data as a point on the coordinate plane. For example, the first point you would plot is (1, 33), which represents 33 mi after 1 h.

2. What does the graph look like?

3. For the sixth hour, the ordered pair is (6, 274). Explain possible reasons why the distance is unchanged from the fifth hour.

Number of hours	Miles driven
1	33
2	95
3	155
4	219
5	274

◥ BUILD UNDERSTANDING

A **relation** is a set of paired data, or ordered pairs. The **domain** consists of all first values (*x*-coordinates) of the ordered pairs in a relation. All of the second values (*y*-coordinates) of the ordered pairs in a relation are called the **range**. If each value of the domain is paired with one and only one value of the range, then the relation is a **function**. The above table is a function.

A *mapping* is a visual way to show the pairing of the domain and range. The domain and range can be numbers, names, geometric figures or items.

> **Problem Solving Tip**
>
> If a mapping shows a domain value paired with more than one value in the range, then the relation is not a function.

Example 1

Does the mapping show that the relation is a function?

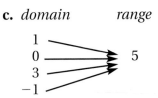

a. *domain*   *range*
Hamburger $4.00
Pizza $2.75
Chicken $3.50
Salad $2.50

b. *domain*   *range*
California → San Francisco
              Houston
Texas   → Dallas
              Los Angeles

c. *domain*   *range*
1
0   → 5
3
−1

Solution

a. Function. Each item (domain) relates to only one price (range).

b. Not a function. Each state relates to more than one city (range).

c. Function. Each domain value relates to the same range value, 5.

A relation is graphed on the coordinate plane as sets of ordered pairs. In the ordered pair (*x*, *y*), *x* is from the domain and *y* is from the range.

You can tell if a graphed relation is a function by using the **vertical line test**. Visualize a vertical line that moves left to right through the coordinate plane. If the vertical line ever goes through the plotted graph more than once, then the relation is *not* a function.

Example 2

Use the vertical line test to determine if each relation is a function.

a. 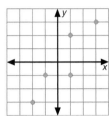   **b.** 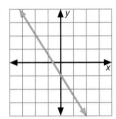   **c.**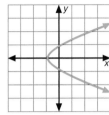

Solution

a.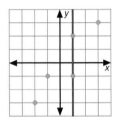

This is not a function. The vertical line intersects the relation at two points at the same time.

b.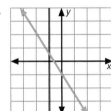

This is a function. The vertical line intersects the relation at only one point at a time.

c.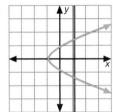

This is not a function. The vertical line intersects the relation at two points at the same time.

Example 3

Use the relation to determine the following.

$\{(3, 5), (4, 8), (5, 10), (6, 8), (7, 9)\}$

a. domain   **b.** range   **c.** Is the relation a function?

Solution

a. domain: 3, 4, 5, 6, 7   **b.** range: 5, 8, 9, 10

c. Yes, this relation is a function since each different value in the domain has only one value in the range.

The description of a function is a **function rule**. For example, if x is the domain, then the *add 5 rule* becomes $x + 5$. A symbol like $f(x)$ or $g(x)$ denotes the range assigned to x. The function is written $f(x) = x + 5$, and the ordered pair is $(x, f(x))$.

The function $f(x) = x + 5$ is read "*f* of *x* equals *x* plus 5." To evaluate a function at a given value of the domain, substitute the value into the function.

If $x = 1$, then $f(1) = 1 + 5$, or 6.   If $x = 4$, then $f(4) = 4 + 5$, or 9.

Example 4

PART-TIME JOB Martina charges a base fee of $3/h when she babysits and an additional $1/h for each child. This is represented by the function $f(x) = 3 + x$, where x is the number of children she is babysitting and $f(x)$ is the amount she earns per hour.

a. Evaluate $f(x)$ to find Martina's hourly rate when she babysits 1, 2, 3 and 4 children.

b. Martina babysits 2 children for 5 h. How much does she earn?

Solution

a. Set up a table to represent the function.

b. For 2 children, Martina earns $5/h. So, for 5 h, she will earn $5 \cdot 5$, or $25.

x	f(x)
1	3 + 1 = 4
2	3 + 2 = 5
3	3 + 3 = 6
4	3 + 4 = 7

◥ TRY THESE EXERCISES

Does the mapping show that the relation is a function?

1. Ed ⟶ 135
Sue ⟶ 110
Ana ⟶ 123
Ali ⟶ 118

2. 1987 ⟶ New York Times
1988 ⟶ Philadelphia Inquirer
1989 ⟶ Seattle Times
1990

3. 6 ⟶
4 ⟶ 0
12 ⟶ 1
20 ⟶

Use the vertical line test to determine if each relation is a function.

4.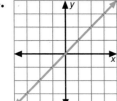

5.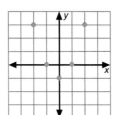

6.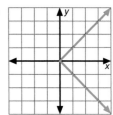

Use the relation $\{(-1, 3), (0, 6), (4, -1), (7, 2)\}$ to determine the following.

7. the domain

8. the range

9. Is the relation a function?

10. WRITING MATH Write "f of x equals x minus four" in function form.

11. VELOCITY For a car traveling at a constant rate of 45 mi/h, the distance is a function of time: $d(t) = 45t$. Find $d(3)$. Explain the meaning of your answer.

◥ PRACTICE EXERCISES • For Extra Practice, see page 570.

Does the mapping show that the relation is a function?

12. Cincinnati ⟶ Reds
Braves
Atlanta Bengals
Falcons

13. −10 ⟶ −1
15 ⟶ 2
5 ⟶ −3
−25 ⟶ 5

14. shoes
accessories ⟶ 25%
juniors ⟶ 20%
dresses

15. YOU MAKE THE CALL Hasina says that the relation $\{(2, 4), (0, 2), (-4, 2)\}$ is a function. Eugene disagrees. Is Hasina or Eugene correct? Explain why.

Use the vertical line test to determine if each relation is a function.

16.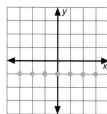
17.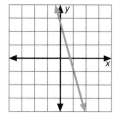
18.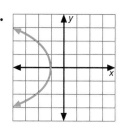
19.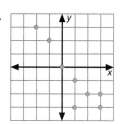

State the domain and range of each relation.

20. {(6, 3), (5, 4), (4, 5), (3, 6)}

21. {(0.1, 3.6), (1.1, 3.4), (2.1, 3.2), (1.2, 0.5)}

22. $\left\{\left(1, \frac{1}{2}\right), \left(-2, \frac{2}{3}\right), \left(4, \frac{1}{2}\right)\right\}$

23. {(2, −4), (3, −2), (2, 0), (3, 2), (2, 4)}

24–27. State whether each relation in Exercises 20–23 is a function.

28. **DATA FILE** Refer to the data on champion trees on page 529. Let the height represent the domain and let the crown spread represent the range. Is this relation a function? Explain.

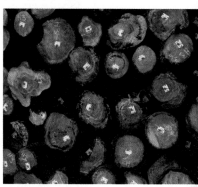

MANUFACTURING The total cost *C* of producing gizmos depends on the number of gizmos and is given by the function $C(n) = 1.85n + 400$. Find the cost of producing the following number of gizmos.

29. 1 gizmo
30. 50 gizmos
31. 450 gizmos
32. 1000 gizmos

EXTENDED PRACTICE EXERCISES

FINANCE Deontae buys a CD player for $96. He pays in installments, making a down payment of $15, then payments of $18/mo.

33. Write a function for the cost *C*(*m*), where *m* is the number of months.

34. About how many months will it take Deontae to pay for the CD player?

NUMBER SENSE Consider a new function called the * function. The relationship of the * function for some numbers are given below.

*(5) = 17   *(9) = 29   *(12) = 38   *(20) = 62

35. What is the * function rule?

36. Find *(7).
37. Find *(19).
38. Find *(40).

MIXED REVIEW EXERCISES

Write the formula needed to find the area of each shape. Then find the area.
(Lessons 2-4, 2-7, 5-6)

39.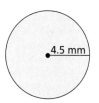
4.5 mm

40.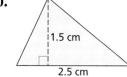
1.5 cm
2.5 cm

41.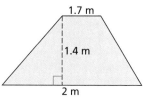
1.7 m
1.4 m
2 m

42. If the area of a circle is 4.58 m², what is the measure of its diameter, to the nearest hundredth of a meter? (Lessons 2-7, 5-7)

Linear Graphs

Goals ■ Find solutions and intercepts of equations.
         ■ Graph functions.

Applications Fitness, Food service, Velocity, Business, Health

---

Work with a partner.

1. Copy and complete the table.

2. Plot each point in your table on a coordinate plane.

3. What can you tell about the points?

4. Use the vertical line test to determine if this is a function.

$f(x) = 3x + 2$		
x	**3x + 2**	**(x, y)**
−3	$3(-3) + 2 = -9 + 2 = -7$	(−3, −7)
−1		
0		
2		
4		

◤ BUILD UNDERSTANDING

A **linear function** is a function that can be represented on a graph by a straight line. It can also be represented by a **linear equation in two variables**. The linear equation can be written in the form $y = mx + b$, where x and y are the variables and m and b are the constants.

The solutions to a linear equation are written as ordered pairs. To determine solutions of an equation with two variables, first choose any value for the first variable, x.

Then substitute that value into the equation for x and solve to find the corresponding value of y. Do this for at least three different values of x. Make a table to organize the ordered pairs that are solutions of the equation.

Example 1

Find three solutions of the equation $y = x - 3$.

Solution

To solve $y = x - 3$, first choose three values for x. You can choose a negative number, zero and a positive number for these values.

$y = x - 3$ Let $x = -2$.	$y = x - 3$ Let $x = 0$.	$y = x - 3$ Let $x = 3$.
$y = (-2) - 3$	$y = 0 - 3$	$y = 3 - 3$
$y = -5$	$y = -3$	$y = 0$

Write the solutions in a table.

x	y	(x, y)
−2	−5	(−2, −5)
0	−3	(0, −3)
3	0	(3, 0)

When x is −2, y is −5.
When x is 0, y is −3.
When x is 3, y is 0.

A linear equation has an infinite number of solutions. Graph at least three ordered pairs on a coordinate plane to find a line that represents the solutions.

Although you can find the line by graphing only two ordered pairs, graphing a third point verifies your solution. All points on the line are solutions to the linear equation.

Think Back

Infinite means "goes on forever." You are not able to list all ordered pairs that are solutions to a linear equation.

Example 2

Graph the equation $y = -2x + 4$.

Solution

Make a table of at least three solutions of the equation. Graph the ordered pairs. Draw a line through the points.

All the points on the line are solutions of the equation $y = -2x + 4$.

x	y	(x, y)
−1	6	(−1, 6)
0	4	(0, 4)
1	2	(1, 2)

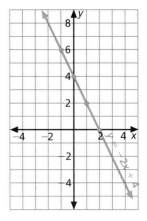

The points where a line crosses an axis is called the **intercept**.

The **x-intercept** is the x-coordinate of the point $(x, 0)$ where the graph crosses the x-axis. In Example 2, the x-intercept is at $(2, 0)$.

The **y-intercept** is the y-coordinate of the point $(0, y)$ where the graph crosses the y-axis. In Example 2, the y-intercept is at $(0, 4)$.

Technology Tip

You can use a graphing utility to find the x- and y-intercepts.

After graphing the equation, use the zoom feature to find where the line crosses an axis. Then use the trace feature to identify the coordinates.

Example 3

Find the x-intercept and y-intercept of $y = \frac{2}{3}x + 4$.

Solution

To find the x-intercept, let $y = 0$.

$$0 = \frac{2}{3}x + 4$$

$$-4 = \frac{2}{3}x \qquad \text{Subtract 4 from both sides.}$$

$$\left(\frac{3}{2}\right)(-4) = \left(\frac{2}{3}x\right)\left(\frac{3}{2}\right) \qquad \text{The reciprocal of } \frac{2}{3} \text{ is } \frac{3}{2}.$$

$$-6 = x$$

The x-intercept is -6.

To find the y-intercept, let $x = 0$.

$$y = \frac{2}{3} \cdot 0 + 4$$

$$y = 0 + 4$$

$$y = 4$$

The y-intercept is 4.

Example 4

FITNESS Jahari walks on a treadmill at an average rate of 4 mi/h. The distance he walks is a function of the amount of time he walks. Write an equation for this function. Graph the function.

Solution

Let x = the time and y = the distance. So $y = 4x$ because distance = rate · time ($d = r \cdot t$).

Make a table of solutions. Use the ordered pairs to graph the function.

x	y	(x, y)
1	4	(1, 4)
2	8	(2, 8)
3	12	(3, 12)

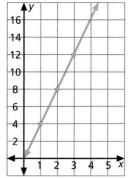

TRY THESE EXERCISES

Find three solutions of each equation. Write the solutions in a table.

1. $y = 4x$

2. $y = 5x - 1$

3. $y = 2x + 2$

Make a table of three solutions for each equation. Then graph the equation.

4. $y = 2x - 1$

5. $y = -3x$

6. $y = x + 4$

Find the x-intercept and y-intercept of each equation.

7. $y = 3x - 4$

8. $y = 4x + 8$

9. $y = \frac{3}{4}x - 2$

FOOD SERVICE Chicken cutlets cost \$3/lb. This is expressed by the equation $y = 3x$ where y is the cost and x is the number of pounds.

10. Complete the table of solutions for the function.

11. Graph the function.

12. WRITING MATH Explain why it is a good idea to use at least three different values of x when you graph a linear equation.

x	y	(x, y)
0	■	(0, ■)
1	■	(1, ■)
2	■	(2, ■)
3	■	(3, ■)

PRACTICE EXERCISES • For Extra Practice, see page 570.

Make a table of three solutions for each equation. Then graph the equation.

13. $y = x + 2$

14. $y = 3x - 2$

15. $y = x - 1$

16. $y = -2x - 4$

17. $y = 2x + 4$

18. $y = -2x$

19. $y = 4x$

20. $y = -2x + 1$

21. $y = \frac{1}{2}x + 5$

22. $y = -3x + 4$

23. $y = -x - 1$

24. $y = -\frac{1}{3}x + 1$

Find the x-intercept and y-intercept of each equation.

25. $y = 5x - 3$

26. $y = \frac{2}{3}x - 2$

27. $y = -3x + 2$

28. $y = -\frac{1}{2}x - 2$

29. $y = 3x - 1$

30. $y = \frac{2}{3}x - 8$

VELOCITY The speed of a certain ball thrown into the air is found by the formula $V = 20 - 9.8T$ where V is the velocity in m/sec and T is the time in seconds. When V is positive, the ball is going up. When V is negative, the ball is falling.

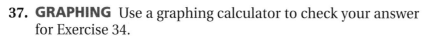

31. Evaluate the function for times of 0 to 6 sec.

32. How fast is a ball traveling in 1.5 sec?

BUSINESS A cell phone carrier charges $1.10 per long-distance call plus $0.25 for each minute.

33. Write a function to represent the cost of a telephone call lasting x minutes.

34. Graph the function.

35. What does the y-intercept represent for this function?

36. How much will a 5-min telephone call cost?

37. GRAPHING Use a graphing calculator to check your answer for Exercise 34.

38. On a coordinate plane, graph the ordered pairs $(-2, -1)$ and $\left(1, \frac{1}{2}\right)$. Draw a line through the points. Make a table of these ordered pairs and three other points on the line. Write an equation for the line.

39. A person's age 5 yr from now (y) is a function of the person's present age (x). Write an equation that represents this function. Then graph the function.

40. YOU MAKE THE CALL Fernando thinks the ordered pair $(-1, 3)$ is a solution of $y = 2x - 5$. Is he correct? Explain.

■ EXTENDED PRACTICE EXERCISES

HEALTH The formula $h = 17 - \frac{a}{2}$ where h is the hours of sleep and a is the age, is used to find the amount of sleep needed by people under the age of 18.

41. Graph the function to calculate the hours of sleep needed for anyone under age 18.

42. According to your graph, at what age would someone not need any sleep?

43. How much sleep does someone your age need?

CRITICAL THINKING Solve for y in terms of x. Then graph the equations.

44. $y - 3x + 1 = 0$ **45.** $4x + 2y = 8$

■ MIXED REVIEW EXERCISES

Simplify. (Lessons 3-1, 3-2, 3-3, 3-5)

46. $3 + -7 \div (-2)$ **47.** $|-5| - 6 \cdot (-2)$ **48.** $4(7-10) \div (-3)$ **49.** $-5^2 - |6| + 30$

50. The distance to a comet is 4.15×10^{20} kilometers from Earth. Write the distance in standard form. (Lesson 3-7)

Review and Practice Your Skills

PRACTICE ◣ LESSON 7-3

Does the mapping show that the relation is a function?

1.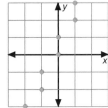

2. mammals 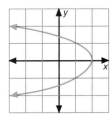 turtle
   dog
   reptiles horse
   snake

3.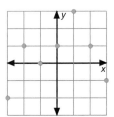

Use the vertical line test to determine if each relation is a function.

4.
5.
6.
7.

State the domain and range of each relation.

8. $\{(-3, 11), (2, -4), (-1, 5), (0, 2)\}$

9. $\{(-4, 0), (-2, 0), (2, 1), (4, 1), (6, 1)\}$

10. $\{(-3, -9), (-2, -6), (-1, -3), (0, 0), (1, 3)\}$

11. $\{(5, -2), (-2, -1), (-3, 0), (-2, 1), (5, 2)\}$

12. $\{(4, -2), (4, -1), (4, 0), (4, 1), (4, 2)\}$

13. $\{(-10, 5), (-6, 3), (-2, 1), (4, -2), (8, -4)\}$

14. $\{(0.1, 5), (0.2, 10), (0.1, 10), (0.3, 5)\}$

15. $\{(1, 1), (1, 2), (2, 1), (2, 2), (3, 1), (3, 2)\}$

16. $\{(-1.5, 7.25), (-1, 6), (0, 5), (1.5, 7.25)\}$

17. $\{(-3, 2), (2, 5), (-1, 0), (-3, -1), (-2, 3)\}$

18–27. State whether each relation in Exercises 8–17 is a function.

PRACTICE ◣ LESSON 7-4

Make a table of three solutions for each equation. Then graph the equation.

28. $y = -x - 3$

29. $y = -3x + 3$

30. $y = 2x - 3$

31. $y = -\frac{1}{4}x - 1$

32. $y = \frac{1}{2}x + 1$

33. $y = -x + 2$

34. $y = -2x + 2$

35. $y = x - 2$

36. $y = -\frac{1}{2}x - 1$

37. $y = -4x + 1$

38. $y = \frac{1}{3}x + 1$

39. $y = 2x + 2$

Find the x-intercept and y-intercept of each equation.

40. $y = -2x - 1$

41. $y = x + 3$

42. $y = -3x - 12$

43. $y = -x - 5$

44. $y = -\frac{4}{3}x + 4$

45. $y = 2x + 7$

46. $y = \frac{1}{5}x + 2$

47. $y = -5x + 15$

48. $y = x - 9$

49. $y = 4x - 1$

50. $y = 3x - 6$

51. $y = -2x + 3$

The graph shows a long walk. (Lesson 7-1)

52. How long did it take the person to walk the first 3 miles?

53. How long did the person stop for lunch?

54. How far from home was the person at the end of the walk?

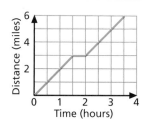

Graph each point on a coordinate plane. (Lesson 7-2)

55. $K(0, -3)$ **56.** $A(-1, 3)$ **57.** $X(3, -1)$ **58.** $E(-2, -3)$

State the domain and range of each relation. (Lesson 7-3)

59. $\{(0.2, -5), (0.1, -3), (0.2, 0), (0.1, 3)\}$ **60.** $\{(6, -2), (5, 0), (4, 2), (3, 4), (2, 6)\}$

61–62. State whether each relation in Exercises 59–60 is a function.

Make a table of three solutions. Then graph the equation. (Lesson 7-4)

63. $y = -x + 3$ **64.** $y = \dfrac{3}{2}x + 1$ **65.** $y = 2x - 1$

Find the *x*-intercept and *y*-intercept of each equation. (Lesson 7-4)

66. $y = -3x + 2$ **67.** $y = -\dfrac{1}{5}x - 1$ **68.** $y = x + 4$

Mid-Chapter Quiz

Plot each ordered pair on the same coordinate plane. (Lesson 7-2)

1. $A(-3, -1)$ **2.** $B(4, 0)$ **3.** $C(-5, -2)$

4. $D(-3, 3)$ **5.** $E(5, 5)$ **6.** $F(0, -1)$

Refer to the coordinate plane used for Exercises 1–6. (Lesson 7-2)

7. Which point has equal coordinates?

8. Which points have the same *x*-coordinate?

9. Which point lies on the *y*-axis?

10. Which point lies in Quadrant II?

Use the relation $\{(1, 3), (-1, 6), (1, 9), (-1, 12), (1, 15)\}$**.** (Lesson 7-3)

11. State the domain.   **12.** State the range.   **13.** Is the relation a function?

Find the *x*-intercept and *y*-intercept of each equation. (Lesson 7-4)

14. $y = -4x + 1$ **15.** $y = 2x - 5$ **16.** $y = 8x + 1$

7-5 Slope of a Line

Goals
- Find the slopes of lines.
- Identify slopes as positive, negative, zero or undefined.

Applications Travel, Velocity, Carpentry

Use grid paper to complete Questions 1–7.

1. Plot the point $A(1, 2)$ on a coordinate plane.

2. From point A, move up two units and then to the right three units. Label this point B.

3. Repeat Question 2, but start at point B. Label the new point C.

4. Repeat Question 2 again. This time start at point C. Label the new point D.

5. What do you notice about these four points?

6. From point D, move up three units and then to the right two units. Label this point E.

7. What do you notice about point E in relation to the other four points?

▼ BUILD UNDERSTANDING

As you move from one point to another along a straight line, the change in position has two components. There is a vertical change in position, called the **rise**, and a horizontal change in position, called the **run**. The ratio of the rise to the run is called the **slope** of the line.

You can find the slope of a line graphed on a coordinate plane by counting the units of change between the coordinates of two points on the line.

Slope of a Line	$\dfrac{\text{rise}}{\text{run}} = \dfrac{\text{difference of } y\text{-coordinates}}{\text{difference of } x\text{-coordinates}}$

Example 1

Find the slope of the line shown.

Solution

Choose two points on the line, such as $A(1, 4)$ and $B(2, 1)$. Find the number of units of change from point A to point B.

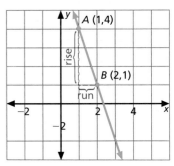

$$\text{slope} = \frac{\text{rise}}{\text{run}} = \frac{3 \text{ units down}}{1 \text{ unit right}} = \frac{-3}{1} = -3 \qquad \text{A move down or to the left is represented by a negative number.}$$

To find the slope of a line, use the coordinates of any two points on the line. The rise is found by subtracting the *y*-coordinate of the first point from the *y*-coordinate of the second point. The run is found by subtracting the *x*-coordinate of the first point from the *x*-coordinate of the second point.

Example 2

Find the slope of the line that passes through each pair of points.

a. $M(0, 2)$ and $N(-2, -1)$

b. $S(2, 4)$ and $T(5, -1)$

c. $A(1, 3)$ and $B(-4, 3)$

d. $X(6, -3)$ and $Y(6, 2)$

Problem Solving Tip

Before you find the slope of two points, label one ordered pair as (x_1, y_1) and the other as (x_2, y_2).

Then use the formula

$\dfrac{y_1 - y_2}{x_1 - x_2}$ or $\dfrac{y_2 - y_1}{x_2 - x_1}$.

Solution

Be sure to subtract the *y*- and *x*-coordinates in the same order.

a. slope $= \dfrac{2 - (-1)}{0 - (-2)} = \dfrac{3}{2}$ The slope is $\dfrac{3}{2}$.

b. slope $= \dfrac{4 - (-1)}{2 - 5} = \dfrac{5}{-3} = -\dfrac{5}{3}$ The slope is $-\dfrac{5}{3}$.

c. slope $= \dfrac{3 - 3}{1 - (-4)} = \dfrac{0}{5} = 0$ There is no change in the rise. The slope is zero.

d. slope $= \dfrac{-3 - 2}{6 - 6} = -\dfrac{5}{0}$ There is no change in the run. The slope is undefined because division by zero is undefined.

If a line slants upward from left to right, it has a **positive slope**. If a line slants downward from left to right, it has a **negative slope**. Any horizontal line has a **zero slope**. The slope of any vertical line is **undefined**.

Example 3

Identify the slope of each line as positive, negative, zero or undefined.

a.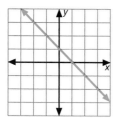

b.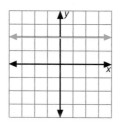

c.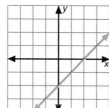

d.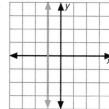

Solution

a. The slope of this line is negative since it slants downward from left to right.

b. The slope of this line is zero since it is a horizontal line.

c. The slope of this line is positive since it slants upward from left to right.

d. The slope of this line is undefined since it is a vertical line.

Check Understanding

For each graph in Example 3, choose two points on the line. Use these points to find the slope.

Verify that the value of the slope corresponds to the description of the slope.

Example 4

TRAVEL Signs along parts of the highway often indicate the slope of upcoming hills so drivers can adjust their speed. Find the slope of a hill that increases 792 ft in 3 mi.

Solution

First convert 3 mi to feet. $3 \text{ mi} \cdot \dfrac{5280 \text{ ft}}{1 \text{ mi}} = 15{,}840 \text{ ft}$

Then find the slope. $\text{slope} = \dfrac{\text{rise}}{\text{run}} = \dfrac{792 \text{ ft}}{15{,}840 \text{ ft}} = \dfrac{1}{20}$

The road rises 1 ft vertically for every 20 ft horizontally.

◥ TRY THESE EXERCISES

Find the slope of each line on the coordinate plane shown.

1. line AB

2. line EF

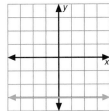

Find the slope of the line that passes through each pair of points.

3. $A(4, 5)$ and $B(3, 7)$

4. $J(1, 6)$ and $K(3, 9)$

5. $P(9, -1)$ and $Q(0, 4)$

Identify the slope of each line as positive, negative, zero, or undefined.

6.

7.

8.

9.

10. WRITING MATH List examples of when someone might need to know the slope of something. Why is it helpful to use numbers in these instances?

11. TRAVEL Find the slope of a hill whose height increases 704 ft in 2 mi.

◥ PRACTICE EXERCISES • For Extra Practice, see page 571.

Identify the slope of each line as positive, negative, zero, or undefined.

12. a

13. b

14. c

15. d

16. e

17. f

Find the slope of each line shown.

18. a

19. b

20. c

21. d

22. e

23. f

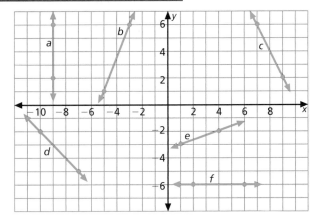

Find the slope of the line that passes through each pair of points.

24. $X(-4, 2)$ and $Z(3, -2)$ **25.** $M(3, 2)$ and $N(1, 2)$ **26.** $R(-5, -7)$ and $S(8, -2)$

27. $M(-3, 4)$ and $N(-8, 1)$ **28.** $X(6, -5)$ and $Y(6, -3)$ **29.** $R(1, 1)$ and $S(-1, -2)$

30. $A(2, 5)$ and $B(5, 5)$ **31.** $I(-1, -2)$ and $J(5, -6)$ **32.** $G(3, -3)$ and $H(3, 2)$

On a coordinate plane, graph the point (1, 1). Then graph lines through (1, 1) with each of the following slopes.

33. $\frac{2}{3}$ **34.** $\frac{3}{2}$ **35.** $-\frac{2}{3}$ **36.** $-\frac{3}{2}$

DATA FILE Refer to the data on maximum heart rate on page 531.

37. Graph the set of data.   **38.** What is the slope of the data?

Find the slope of the following.

39. A hill that decreases 990 ft in 1.5 mi.   **40.** A hill that increases 1144 ft in 3.25 mi.

41. VELOCITY Which runner wins the race in the graph shown? Explain.

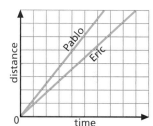

■ EXTENDED PRACTICE EXERCISES

CRITICAL THINKING Find the slope of the sides of a quadrilateral with the given vertices. Determine if the quadrilateral is a parallelogram or a trapezoid.

42. $A(-2, 6)$, $B(4, 6)$, $C(1, 2)$, $D(-5, 2)$

43. $E(-5, -3)$, $F(1, 1)$, $G(4, 6)$, $H(-8, -2)$

44. CARPENTRY The slope of the handrail on a staircase is the ratio of the riser to the tread. If the riser is 26 cm and the tread is 34 cm, find the slope of the handrail.

45. CHAPTER INVESTIGATION Secure the ruler to five books, and roll the marble three times from the top. Compare your results to the earlier results of rolling the marble.

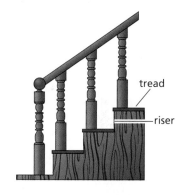

■ MIXED REVIEW EXERCISES

Evaluate for $x = -5$. (Lesson 3-5)

46. $3x + 2$ **47.** $x - 7$

48. $-4x + -1$ **49.** $|-x| + 20$

50. $\frac{2}{5}x - 3$ **51.** $-\frac{4}{5}x + (-12)$

52. $-|2x + 1|$ **53.** $3x^2 + (-10)$

54. DATA FILE Refer to the data on total and partial eclipses of the moon on page 524. What percent of the time was the moon in a total eclipse during the two events in 2004? Round to the nearest tenth of a percent.

7-6 Slope-Intercept Form of a Line

Goals
- Find the slope and *y*-intercept of a line.
- Write an equation and graph a line using slope-intercept form.

Applications Fitness, Sports, Shipping

Use grid paper to complete Questions 1–4.

1. When given a slope of $-\frac{2}{3}$, describe two different movements of rise and run that you can use to plot points on the line.

2. Plot the point $(0, 3)$. Given a slope of $-\frac{1}{3}$, plot two more points on the line, one to the right and one to the left of $(0, 3)$.

3. When given a slope of $\frac{3}{2}$, describe two different movements of rise and run that you can use to plot points on the line.

4. Plot the point $(0, -2)$. Given a slope of $\frac{1}{2}$, plot two more points on the line, one to the right and one to the left of $(0, -2)$.

◤ BUILD UNDERSTANDING

When a linear equation is written with *y* alone on one side of the equal symbol, it is in the form $y = mx + b$. This is called the **slope-intercept form** of an equation. The *m* (the coefficient of *x*) is the slope of the line. The *b* is the *y*-intercept.

Example 1

Find the slope and *y*-intercept of $y = -\frac{1}{2}x + 4$. Then graph the line.

Solution

$y = mx + b$ slope-intercept form

$y = -\frac{1}{2}x + 4$ The slope is *m*, or $-\frac{1}{2}$. The *y*-intercept is *b*, or 4.

Plot the *y*-intercept at $(0, 4)$.

Use the slope to count rise and run units. Locate two more points. Draw a line through the points.

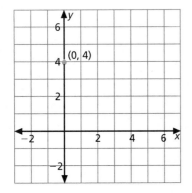

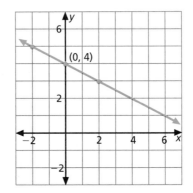

You can use a graphing calculator to graph a line or to check your work. For Example 1, you can check the graph. Enter the equation $y = -\frac{1}{2}x + 4$ on the equation entry screen. Use the graph feature to verify that this graph matches the solution for Example 1.

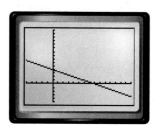

Example 2

Write an equation of the line shown.

Solution

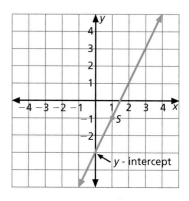

The line intersects the y-axis at $(0, -3)$, so the y-intercept is -3. Therefore $b = -3$.

Find the slope by counting rise and run units from the y-intercept to point S.

slope $= \dfrac{\text{rise}}{\text{run}} = \dfrac{2}{1} = 2$ The slope is 2. Therefore $m = 2$.

The equation of the line is $y = 2x - 3$.   Slope-intercept form: $y = mx + b$.

Example 3

Name the slope and y-intercept of each line. Write an equation of the line in slope-intercept form.

a.
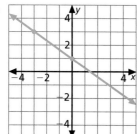

b.
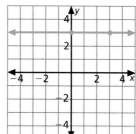

c.
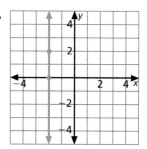

Solution

a. The line intersects the y-axis at $(0, 1)$, so the y-intercept is 1. Select points $(0, 1)$ and $(-3, 3)$ to find the slope.   slope $= \dfrac{1 - 3}{0 - (-3)} = -\dfrac{2}{3}$

   The equation of the line is $y = -\dfrac{2}{3}x + 1$.

b. The line intersects the y-axis at $(0, 3)$, so the y-intercept is 3. Select points $(0, 3)$ and $(3, 3)$ to find the slope.   slope $= \dfrac{3 - 3}{0 - 3} = \dfrac{0}{-3} = 0$

   The equation of the line is $y = 0 \cdot x + 3$, or $y = 3$. Notice that the value 3 is the y-coordinate of all points on the line.

c. The line does not intersect the y-axis, so it does not have a y-intercept. Select points $(-2, 0)$ and $(-2, 2)$ to find the slope.   slope $= \dfrac{0 - 2}{-2 - (-2)} = -\dfrac{2}{0}$

   The slope is undefined. The equation of any vertical line can be written by using the x-coordinate of any point on the line. The value -2 is the x-coordinate of all points on the line. So the equation of the line is $x = -2$.

Find the slope and *y*-intercept of each line. Then graph the line.

1. $y = \frac{1}{3}x - 2$

2. $y = 2x + 6$

3. $y = \frac{3}{5}x - 3$

Name the slope and *y*-intercept of each line. Write an equation of the line.

4.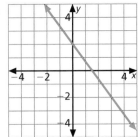

5.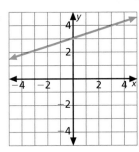

6.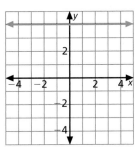

FITNESS A health club charges a fee of $50 when a member first joins. Thereafter, a member is charged $135/yr. This is represented by the equation $y = 135x + 50$ where *x* is the number of years as a member and *y* is the total amount paid.

7. Name the slope of the equation.

8. Name the *y*-intercept of the equation.

9. Graph the equation.

10. How much will a 6-yr member pay in total?

11. How many years has a person been a member if they have paid $455 in total?

 12. WRITING MATH Do all lines with the same *y*-intercept have the same slope? Do all lines with the same slope have the same *y*-intercept? Provide examples.

Find the slope and *y*-intercept of each line. Then graph the line.

13. $y = 4x + 1$

14. $y = -\frac{1}{2}x - 2$

15. $y = -2x - 1$

16. $y = -x + 4$

17. $y = \frac{3}{2}x - 5$

18. $y = -\frac{5}{4}x + 3$

19. $y = \frac{1}{4}x - 4$

20. $y = -3x$

21. $y = 2$

Name the slope and *y*-intercept of each line. Write an equation of the line.

22.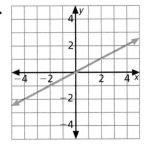

23.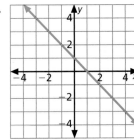

24.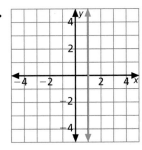

25. YOU MAKE THE CALL Erica graphed the equation $y = -\frac{2}{3}x + 3$ on the grid shown. Is her graph correct? If not, explain her error.

SHIPPING A store charges a fixed amount of $5.00 to ship each package and an additional charge of $2.20/lb. This is represented by the equation $y = 2.20x + 5.00$ where x is the number of pounds and y is the cost of shipping the package.

26. Name the slope of the equation.

27. Name the y-intercept of the equation.

28. Graph the equation.

29. How much will a 3-lb package cost?

30. How many pounds are in a package that costs $14.20 to ship?

SPORTS Admission to a high school soccer game is charged on a per-vehicle basis. For each vehicle, there is a fixed parking fee of $3 and an additional $2 for each person in the car.

31. Write an equation for this situation.

32. Graph the equation.

■ EXTENDED PRACTICE EXERCISES

Graph each equation on the same coordinate plane.

33. $y = \frac{1}{2}x + 3$ **34.** $y = \frac{1}{2}x - 1$

35. What do you notice about these two lines?

36. CRITICAL THINKING Write a generalization about the slopes of parallel lines.

37. Write an equation of the line that has a y-intercept of 2 and is parallel to the line whose equation is $y = \frac{1}{2}x + 1$.

38. Write an equation of a line that has a slope of 0.

■ MIXED REVIEW EXERCISES

Use $a^2 + b^2 = c^2$ to find the value of the missing variable. Round to the nearest hundredth. (Lesson 5-6)

39. $a = 3; b = 4$ **40.** $b = 17; c = 19$ **41.** $a = 21; c = 25$

42. MANUFACTURING The circle shown is an enlarged drawing of a part used in manufacturing. It is drawn to the scale of 1 cm = 0.5 mm. The diameter of the circle in the drawing is 3.2 cm. Find the circumference and area of the drawing and the circumference and area of the actual part to the nearest ten thousandth. Use 3.14 for π. (Lessons 2-7, 2-8, 5-7)

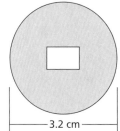

3.2 cm

Review and Practice Your Skills

PRACTICE ◤ LESSON 7-5

Find the slope of each line.

1.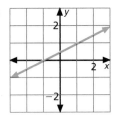

2.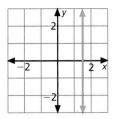

3.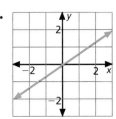

4.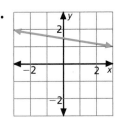

5.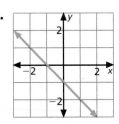

6.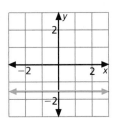

7.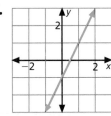

8.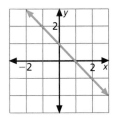

Find the slope of the line that passes through each pair of points.

9. $X(-1, 0)$ and $Y(3, 2)$

10. $S(5, 1)$ and $T(-1, -3)$

11. $K(3, -2)$ and $L(-2, 3)$

12. $P(2, -2)$ and $Q(-5, 1)$

13. $L(-1, 4)$ and $M(2, 4)$

14. $R(-4, -1)$ and $S(-2, -3)$

15. $C(6, 4)$ and $D(0, -3)$

16. $J(0, 3)$ and $K(-1, -1)$

17. $F(2, -3)$ and $G(2, 5)$

PRACTICE ◤ LESSON 7-6

Find the slope and y-intercept of each line. Then graph the line.

18. $y = -3x + 2$

19. $y = x + 3$

20. $y = -x - 2$

21. $y = -x + 2$

22. $y = 3x - 1$

23. $y = \frac{2}{3}x + 1$

24. $y = 2x - 2$

25. $y = -\frac{3}{2}x - 2$

26. $y = -2x + 3$

27. $y = -\frac{1}{2}x - 1$

28. $y = 2x + 1$

29. $y = -\frac{1}{3}x - 1$

Name the slope and y-intercept of each line. Write an equation for the line.

30.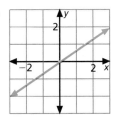

31.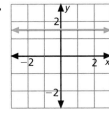

32.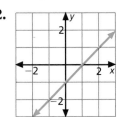

33.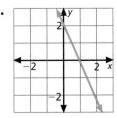

The graph shows a long walk. (Lesson 7-1)

34. How far did the person walk in the first hour?

35. What does the downward part of the graph represent?

Graph each point on a coordinate plane. (Lesson 7-2)

36. $P(-3, 1)$ **37.** $D(-1, 0)$ **38.** $N(2, -3)$

State the domain and range of each relation. (Lesson 7-3)

39. $\{(3, -5), (4, -2), (5, -1), (6, -2), (7, -5)\}$ **40.** $\{(-1, 2), (3, 0), (-2, 2), (-4, 2), (1, 0)\}$

41–42. State whether each relation in Exercises 43–44 is a function.

Find the slope of the line that passes through each pair of points. (Lesson 7-5)

43. $X(-2, -1)$ and $Y(3, -1)$ **44.** $S(2, 4)$ and $T(-1, -1)$ **45.** $K(3, -2)$ and $L(-6, 0)$

Find the slope and y-intercept of each line. Then graph the line. (Lesson 7-6)

46. $y = -\dfrac{1}{3}x + 1$ **47.** $y = x - 4$ **48.** $y = -3x + 2$

Math*Works* **Career – Ship Captain**
Workplace Knowhow

A ship captain determines the course and speed, monitors the vessel's position, directs the crew and maintains logs of the ship's movements and cargo. An *anemometer* is important to ship captains. It is an instrument that measures wind speed. An anemometer has three or four cups attached to short rods. The rods are connected to a vertical shaft. As the wind blows, the cups rotate around the center shaft. The number of turns per minute determines the wind speed.

Use the table for Exercises 1–7.

1. Use a coordinate plane to plot the data in the table. Let the x-axis be the number of revolutions per minute. Label the x-axis in intervals of 100. Let the y-axis be the wind velocity. Label the y-axis in intervals of 5.

2. Is this relation linear? If so, draw the line connecting the points.

3. Find the slope of the line.

4. Find the y-intercept of the line.

5. Write an equation of the line.

6. What is the wind velocity when the anemometer spins at a rate of 1200 revolutions per minute?

Number of revolutions per minute	Wind velocity (mi/hr)
100	3.6
500	18.0
1000	36.0
1500	54.0
2000	72.0
2500	90.0
3000	108.0
3500	126.0

Distance and the Pythagorean Theorem

Goals
- Use distance formula to find distance between two points.
- Use the Pythagorean theorem to find distance.

Applications Hobbies, Architecture, Travel, Safety

Use the ski slope shown and a coordinate plane.

ski slope

300 m

625 m

1. Label the *x*-axis and *y*-axis with intervals of 25 m. Then label (0, 0) point *A*.

2. Plot point *B*(625, 0) to represent the base of the ski slope. Then plot point *C*(0, 300) to represent the height of the ski slope.

3. Draw $\overline{AB}$, $\overline{AC}$ and $\overline{BC}$ so that you have a right triangle on your grid.

4. Label the lengths of $\overline{AB}$ and $\overline{AC}$.

5. Find the approximate length of $\overline{BC}$ by using the edge of another piece of grid paper with the same size squares. Recall that the length of each square on your grid represents 25 m.

◢ BUILD UNDERSTANDING

To find the distance between two points on a coordinate plane, you can draw a right triangle with the points and then apply a formula.

To construct a right triangle, first draw a segment connecting the given points. Then draw a vertical line through one point and a horizontal line through the other point. The intersection of these two lines forms the right angle of the right triangle.

In a right triangle, the side opposite the right angle is the **hypotenuse**. The other sides are the **legs**. There is a special relationship between the legs of a right triangle and the hypotenuse.

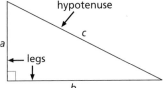

hypotenuse

c

a

← legs

b

The sum of the squares of the lengths of the legs is equal to the square of the length of the hypotenuse. This relationship is called the **Pythagorean Theorem**.

Pythagorean Theorem	If a right triangle has legs of length *a* and *b* and a hypotenuse of length *c*, then $a^2 + b^2 = c^2$.

The Pythagorean Theorem is named after the Greek mathematician Pythagoras. He discovered the theorem at least 3000 years ago.

Think Back

Refer to the lesson introduction. Use the Pythagorean Theorem to verify the length of the ski slope.

Example 1

Use the Pythagorean Theorem to find the distance from point $D(-1, 2)$ to point $E(2, 6)$.

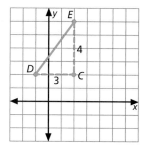

Solution

When no numbers are shown on the x- and y-axis, you can assume that each grid is 1 unit long on each side. First draw right triangle CDE with $\overline{DE}$ as the hypotenuse. Draw a horizontal line through point D and a vertical line through point E. The lines intersect at point $C(2, 2)$.

Since $\triangle CDE$ is a right triangle, use the Pythagorean Theorem to find the length of $\overline{DE}$.

$DE^2 = CD^2 + CE^2$ $c^2 = a^2 + b^2$

$DE^2 = 3^2 + 4^2$ The length of $\overline{CD}$ is 3, and the length of $\overline{CE}$ is 4.

$DE^2 = 25$

$DE = \sqrt{25} = 5$ Use only the positive solution since distance cannot be a negative value.

The distance from D to E is 5 units.

You can also use the distance formula to find the distance between two points on a coordinate plane. This formula can be calculated using the Pythagorean Theorem and any two points. Use points $P(x_1, y_1)$ and $Q(x_2, y_2)$ to represent any two ordered pairs.

$c^2 = a^2 + b^2$

$PQ^2 = (x_2 - x_1)^2 + (y_2 - y_1)^2$ The difference represents the length of each leg of the triangle.

$PQ = \sqrt{(x_2 - x_1)^2 + (y_2 - y_1)^2}$

> ### Check Understanding
>
> In Example 1, could you have drawn a horizontal line through E and a vertical line through D? Find the distance from $F(1, 1)$ to $G(3, 4)$.

> **Distance Formula** $PQ = \sqrt{(x_2 - x_1)^2 + (y_2 - y_1)^2}$ for points $P(x_1, y_1)$ and $Q(x_2, y_2)$.

Example 2

Given points $M(-3, 5)$ and $N(5, -4)$, use the distance formula to find the length of $\overline{MN}$.

Solution

Let $M(-3, 5)$ be the ordered pair (x_1, y_1). Let $N(5, -4)$ be the ordered pair (x_2, y_2).

$MN = \sqrt{(x_2 - x_1)^2 + (y_2 - y_1)^2}$

$MN = \sqrt{[5 - (-3)]^2 + (-4 - 5)^2}$

$MN = \sqrt{8^2 + (-9)^2}$

$MN = \sqrt{64 + 81}$

$MN = \sqrt{145} \approx 12$ The length of $\overline{MN}$ is about 12 units.

Calculate the length of each segment.

1.   **2.**   **3.** 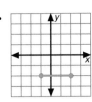 **4.**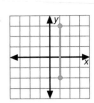

Graph each pair of points on a coordinate plane. Then use the Pythagorean Theorem to find the distance between each pair of points.

5. $A(1, 2)$ and $B(4, 6)$ **6.** $C(1, 6)$ and $D(6, -6)$ **7.** $E(1, -2)$ and $F(7, 6)$

8. $R(-4, -3)$ and $S(2, 5)$ **9.** $T(-2, 4)$ and $U(10, -1)$ **10.** $V(0, 2)$ and $W(2, 0)$

Use the distance formula to find the distance between each pair of points.

11. $J(1, 4)$ and $K(2, 1)$ **12.** $L(1, -2)$ and $M(3, 5)$ **13.** $P(-1, -2)$ and $Q(-3, -7)$

14. SPORTS A rope is attached from the top of a sailboat mast to a point 7 ft from the base of the mast. If the mast is 16 ft high, how long is the rope to the nearest tenth?

Graph each set of points on a coordinate plane. Then use the Pythagorean Theorem to find the distance between each pair of points. Round distances to the nearest tenth.

15. $G(1, 2)$ and $H(3, 5)$ **16.** $A(-1, 2)$ and $B(3, -1)$

17. $C(3, 2)$ and $D(-3, -6)$ **18.** $E(1, -3)$ and $F(2, 4)$

19. $G(1, -2)$ and $H(3, -3)$ **20.** $J(2, 5)$ and $K(-3, -7)$

21. $L(-3, 5)$ and $M(1, 1)$ **22.** $X(0, 0)$ and $Y(-1, 4)$

23. $P(-3, -4)$ and $Q(2, 5)$ **24.** $P(-8, 1)$ and $M(-5, 0)$

Use the distance formula to find the distance between each pair of points. Round distances to the nearest tenth.

25. $P(2, 5)$ and $Q(3, 7)$ **26.** $R(0, 0)$ and $S(-3, -1)$

27. $R(4, -2)$ and $S(-3, 1)$ **28.** $T(-1, -2)$ and $U(-3, 3)$

29. $V(1, 2)$ and $W(2, 5)$ **30.** $X(1, 2)$ and $Y(4, 8)$

31. $Z(-4, -5)$ and $N(4, 5)$ **32.** $C(-6, 0)$ and $D(0, -4)$

33. $G(3, -2)$ and $H(-4, 4)$ **34.** $Q(5, 2)$ and $R(-3, 2)$

35. ARCHITECTURE A cable at the top of a bridge support is attached to a hook in the ground 12 ft from the base of the support. If the support is 16 ft tall, how long is the cable?

36. TRAVEL Mandi leaves her home and travels 2 mi west, 3 mi south, 4 mi east and 5 mi north to the orthodontist. What is the shortest distance between the orthodontist and Mandi's home?

37. WRITING MATH Explain how you can find the length of a leg of a right triangle if you know the length of one leg and the hypotenuse.

Use the Pythagorean Theorem to find each unknown measurement.

38. length of the ladder

39. height of the kite above the ground

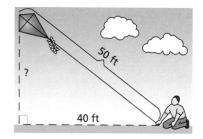

40. SAFETY Safety experts recommend that the bottom of a 12-ft ladder should be at least 4 ft from the base of the wall. How high can the ladder safely reach?

■ EXTENDED PRACTICE EXERCISES

Find the perimeter of each triangle formed by the points given. Is the triangle isosceles or equilateral?

41. $A(1, 1)$, $B(5, 0)$, $C(9, 1)$

42. $D(0, 1)$, $E(-4, 4)$, $F(-8, 1)$

CRITICAL THINKING Find the lengths of the sides of the triangle formed by the points given. Use the Pythagorean Theorem to find the side lengths and determine if it is a right triangle.

43. $G(1, 1)$, $H(-2, -2)$, $K(4, -2)$

44. $P(-1, 1)$, $Q(5, 0)$, $R(2, -1)$

45. CHAPTER INVESTIGATION Tape two rulers together end to end. Secure the long ruler to three books. Roll the marble three times. Then secure the long ruler to five books. Record three more rolls of the marble. Compare your results to the earlier results with the single ruler.

■ MIXED REVIEW EXERCISES

Graph each solution on a number line. (Lessons 5-8 and 5-9)

46. $4x = 19$

47. $16 \geq x + 12$

48. $3x - 4 < 5$

Solve. (Lesson 6-3)

49. Find the sales tax of 5.2% on a purchase of $1959. Then find the total cost.

50. Find the property tax of 2.1% on a home worth $85,000.

Solutions of Linear and Nonlinear Equations

Goals
- Determine if an ordered pair is included in the solution.
- Use the line of best fit to interpret a set of data.

Applications   Retail, Part-time job, Sports, Velocity

Use a graphing utility or grid paper to answer Questions 1-7.

1. Graph the equation $y = x$. Does this line represent a function?

2. What do you know about the points $(3, 3)$ and $(-2, -2)$ in relation to the equation graphed in Question 1?

3. What do you know about the point $(-2, 5)$ in relation to the equation graphed in Question 1?

4. Use a chart like the one shown to graph the equation $y = |x|$.

5. What do you know about the points $(3, 3)$ and $(-2, -2)$ in relation to the equation graphed in Question 4?

6. What do you know about the point $(-2, 5)$ in relation to the equation graphed in Question 4?

7. What are the similarities and differences in the graphs for Questions 1 and 4?

x	$y = \|x\|$	(x, y)
-4	$y = \|-4\| = 4$	$(-4, 4)$
-2		
0		
2		
4		

◥ BUILD UNDERSTANDING

Not all equations with two variables represent linear functions. A **nonlinear function** is represented by an equation whose graph is not a straight line.

The **solutions of an equation** are values that make the equation true. When the equation has two variables, there are usually an infinite number of solutions. These solutions are written as ordered pairs.

One way to determine the solutions of a nonlinear equation is to substitute an ordered pair into the equation and check whether it makes the equation true. Another method to check a solution is to determine if the ordered pair is on the graph of the function.

Check Understanding

Some examples of nonlinear functions are
$y = |x|$ $y = x^2$
$y = x^3$ $y = \dfrac{1}{x}$

Use a graphing calculator to graph these functions.

Example 1

Determine if the ordered pair is a solution.

a. $(-4, -1)$

$y = \dfrac{1}{2}x - 3$

b. $(3, 7)$

$y = |x| + 4$

c. $(2, 0)$

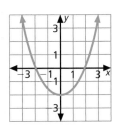

Solution

a. Test the point $(-4, -1)$ in the equation.

$$y = \frac{1}{2}x - 3 \quad \rightarrow \quad -1 \overset{?}{=} \frac{1}{2}(-4) - 3$$

$$-1 \overset{?}{=} -2 - 3$$

$$-1 \neq -5 \qquad \text{False}$$

Therefore, the ordered pair $(-4, -1)$ is not a solution of the equation.

b. Test the point $(3, 7)$ in the equation.

$$y = |x| + 4 \quad \rightarrow \quad 7 \overset{?}{=} |3| + 4$$

$$7 = 7 \qquad \text{True}$$

Therefore, the ordered pair $(3, 7)$ is a solution of the equation.

c. Locate the point $(2, 0)$ on the coordinate plane. The graph of the function goes through this point, so $(2, 0)$ is a solution of the equation for the graph even though we were not told what the equation was.

Some linear and nonlinear graphs have regions that are shaded above or below, to the right or to the left, or inside or outside of the line or curve. *Inequalities* in two variables produce these types of graphs. Ordered pairs on the line or curve as well as in the shaded regions are solutions of the inequality that produced the graph.

Example 2

State whether the ordered pair is a solution of the inequality that produced the graph.

a. $(1, -2)$

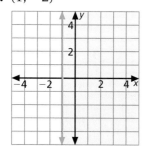

b. $(-2, 4)$

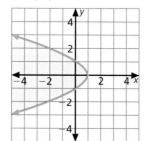

c. $(1, -3)$

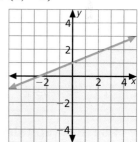

Solution

a. The ordered pair $(1, -2)$ is a solution since the point is in the shaded region.

b. The ordered pair $(-2, 4)$ is not a solution since the point is in the non-shaded region.

c. The ordered pair $(1, -3)$ is a solution since the point is in the shaded region.

A function whose graph you can draw without lifting your pencil is called a **continuous function**. So, the graph of a continuous function has no holes or gaps and its domain is the set of real numbers. The linear and nonlinear graphs that you have been studying in this chapter are continuous.

A scatter plot is a function that is not continuous, but is defined for only specific domain values. On the other hand, a line of best fit is a continuous function that best approximates a trend for the data in the scatter plot and thus, is used to summarize the data. Use the position of a point in relation to the line of best fit to interpret that data.

Example 3

PART-TIME JOB A survey is taken of students who work part-time jobs. The data graphed represent hours worked and money earned per week for ten students. Each point is labeled with the students' initials. The line of best fit is drawn.

a. Who earns the most money per hour?

b. Who earns the least money per hour?

c. Nina works 14 hr/wk and earns $75/wk. Would her data be above, below, or on the line?

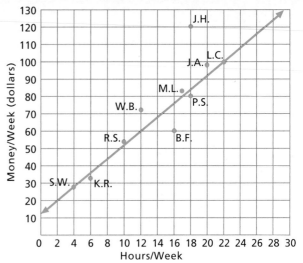

Solution

a. J.H. earns the most per hour. This point is farthest above the line of best fit.

b. B.F. earns the least per hour. This point is farthest below the line of best fit.

c. The point representing Nina would be just above the line of best fit.

◤ TRY THESE EXERCISES

Determine if the ordered pair is a solution of the equation.

1. $(-5, 15)$
$$y = -\frac{3}{5}x + 12$$

2. $(-3, -11)$
$$y = -8 + |x|$$

3. $(-2, 5)$
$$y = 0.5x + 6$$

4. $(-3.5, 0)$
$$y = -2x - 7$$

Determine if the ordered pair is a solution of the equation or inequality that produced the graph.

5. $(2, 3)$

6. $(-1, 4)$

7. $(2, -3)$

8. $(0, 4)$

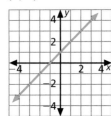
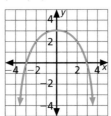
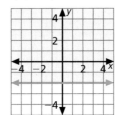
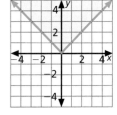

Use the graph shown for Exercises 9–11.

9. Give the coordinates of two points above the line.

10. Give the coordinates of two points below the line.

11. Would the point $(5, 3)$ appear above, below, or on the line?

12. VELOCITY If the velocity of a car between traffic lights is graphed on a coordinate plane, would this data be linear or nonlinear?

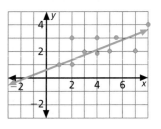

◤ PRACTICE EXERCISES • For Extra Practice, see page 573.

Determine if the ordered pair is a solution of the equation.

13. $\left(\frac{2}{3}, \frac{1}{4}\right)$
$$y = -\frac{3}{4}x + \frac{1}{2}$$

14. $(0, 5)$
$$y = 5 + |x|$$

15. $(-0.8, 5.3)$
$$y = 4.5x - 1.7$$

16. $(3, 9)$
$$y = x^2$$

Determine if the ordered pair is a solution of the equation or inequality that produced the graph.

17. (2, 2)

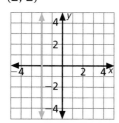

18. (2, −3)

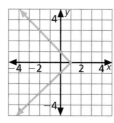

19. (2, −2)

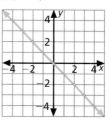

20. (1, −5)

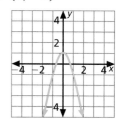

21. (0, 3)

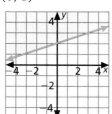

22. (1, −4)

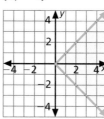

23. (−3, 0)

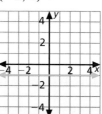

24. (−2, 4)

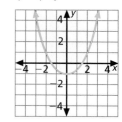

RETAIL The manager at an ice cream parlor keeps track of sales according to the temperature on a given day. The graph shown represents the sales over two weeks according to the daily temperature. The line of best fit is drawn.

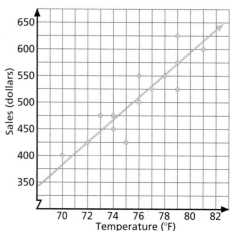

25. What was the temperature on the best day of sales?

26. What were the sales on the day with lowest temperature?

27. On a particular day the temperature was 76° F and the sales were $525. Would this data be plotted above, below, or on the line?

28. WRITING MATH What relationship do you notice between sales and temperature?

EXTENDED PRACTICE EXERCISES

DATA FILE For Exercises 29–31, refer to the data on the Summer Olympic Games on page 533.

29. Use a full sheet of grid paper to make a scatter plot displaying the number of countries and the number of competitors.

30. Draw a line of best fit on the scatter plot.

31. CRITICAL THINKING What relationship do you notice between the number of countries and the number of competitors?

MIXED REVIEW EXERCISES

For Exercises 32–34, use the relation {(1, 4), (2, −9), (−3, −2), (0, 0), (4, −3)}. (Lesson 7-3)

32. Find the domain.   **33.** Find the range.   **34.** Is the relation a function?

35. Richard earned $450 commission one month. He was responsible for $15,000 in sales. What was his rate of commission? (Lesson 6-5)

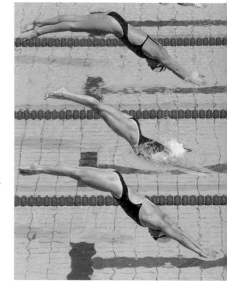

Chapter 7 Review

VOCABULARY ◢

1. When each value of a domain is paired with one and only one value of the range, then the relation is a __?__.

2. The points where a line crosses the axes are called the __?__.

3. The __?__ is the point where the *x*- and *y*-axes intersect.

4. A(n) __?__ is a function that is represented by a straight line.

5. The __?__ of a line is the ratio of the rise to the run.

6. In a relation, all of the *y*-coordinates of the ordered pairs are called __?__.

7. The slope of any vertical line is __?__.

8. The __?__ is the side opposite the right angle in a right triangle.

9. A(n) __?__ can be used to tell if a graphed relation is a function.

10. The axes separate the plane into four regions called __?__.

a.	coordinate plane
b.	domain
c.	function
d.	hypotenuse
e.	intercepts
f.	linear function
g.	origin
h.	quadrants
i.	range
j.	slope
k.	undefined
l.	vertical line test

LESSON 7-1 ◢ Problem Solving Skills: Qualitative Graphing, p. 306

▶ A **qualitative graph** can be used to describe a situation such as a rollercoaster ride.

Match each situation with one of the graphs shown.

11. Your distance from the ground as you climb down a ladder.

12. The level of the water in a glass as you add ice cubes.

For Exercises 13-15, use Graph C.

13. Suppose Graph C shows Hector's travel plans last Sunday. At which times was Hector traveling?

14. How far did Hector travel?

15. For how many minutes did Hector travel?

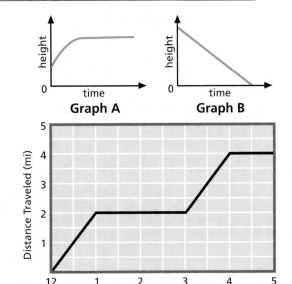

Graph A

Graph B

Graph C

LESSON 7-2 ◢ The Coordinate Plane, p. 308

▶ The **coordinate plane** is separated into 4 regions called **quadrants**. Every location on the coordinate plane is associated with a unique **ordered pair**.

16. Graph *A*(3, 2), *B*(−2, 5), *C*(−1, −4), *D*(0, 2), *E*(−2, 0), and *F*(0, 0).

Use the coordinate plane to find the coordinates of each point.

17. *A*

18. *B*

19. *C*

20. *D*

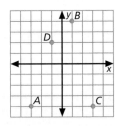

LESSON 7-3 ◣ Relations and Functions, p. 314

▶ A **relation** is a set of ordered pairs. The **vertical line test** can determine if a relation is a **function**. The description of a function is a **function rule**.

▶ In a relation, the **domain** consists of all the *x*-coordinates of the ordered pairs and the **range** consists of all the *y*-coordinates of the ordered pairs.

21. Does the mapping below show that the relation is a function?

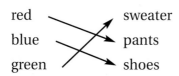

22. Use the vertical line test to determine if the relation below is a function.

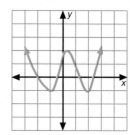

23. Find the domain and range of the relation $\{(-1, 4), (0, 8), (3, -2), (8, 0)\}$. Then tell if the relation is a function.

LESSON 7-4 ◣ Linear Graphs, p. 318

▶ A **linear equation** in two variables can be written in the form $y = mx + b$, where *x* and *y* are the variables and *m* and *b* are the constants.

▶ The **x-intercept** is the *x*-coordinate of the point $(x, 0)$ where the graph crosses the *x*-axis. The **y-intercept** is the *y*-coordinate of the point $(y, 0)$ where the graph crosses the *y*-axis.

24. Graph the equation $y = 2x + 5$.

25. Find the *x*- and *y*-intercepts of $y = -2x + 7$.

Ms. Lerman rented a car for $40 per day plus a one-time insurance cost of $20.

26. Write a function to represent the cost of renting the car for *x* days.

27. What does the *y*-intercept represent in this function?

28. How much will it cost her if she rents the car for 4 days?

LESSON 7-5 ◣ Slope of a Line, p. 324

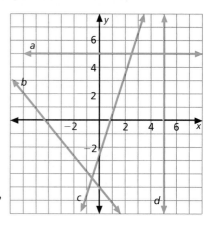

Slope of a line	$\dfrac{\text{rise}}{\text{run}} = \dfrac{\text{difference of } y\text{-coordinates}}{\text{difference of } x\text{-coordinates}}$

▶ A line slanting upward from left to right has a **positive slope**. If a line slants downward from left to right it has a **negative slope**. A horizontal line has **zero slope**. A vertical line has undefined slope.

29. Identify the slope of each line shown at the right as positive, negative, zero, or undefined.

30. Find the slope of the line that passes through $P(8, -3)$, $Q(-5, 6)$.

LESSON 7-6 ◣ Slope Intercept Form of a Line, p. 328

▶ The **slope intercept form** of an equation is $y = mx + b$, where m is the slope of the line and b is the y-intercept.

31. Name the slope and y-intercept and then write an equation of the line shown at the right.

32. Find the slope and y-intercept of $y = \frac{3}{2}x - 3$. Then graph the line.

33. Write the slope-intercept form of a line that has a y-intercept of 0 and slope of $-\frac{5}{3}$.

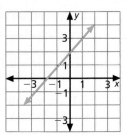

LESSON 7-7 ◣ Distance and the Pythagorean Theorem, p. 334

Pythagorean Theorem	$a^2 + b^2 = c^2$

Distance formula	$PQ = \sqrt{(x_2 - x_1)^2 + (y_2 - y_1)^2}$

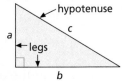

34. Graph the points $A(-3, 6)$ and $B(2, -1)$. Then use the Pythagorean Theorem to find the distance between A and B.

35. Use the distance formula to find the distance between $R(0, 6)$ and $S(-3, 5)$.

36. The size of a television screen is measured by the length of the screen's diagonal. If a television screen measures 24 in. high and 18 in. wide, what size of television is it?

LESSON 7-8 ◣ Solutions of Linear and Nonlinear Equations, p. 338

▶ A **nonlinear function** is represented by an equation whose graph is not a straight line.

▶ The **solutions of an equation** are values that make an equation true.

Determine if the ordered pair is a solution of the equation.

37. $\left(\frac{1}{2}, -\frac{1}{6}\right)$
$y = -x + \frac{1}{3}$

38. $(0, 5)$
$y = 5x + 3$

39. $(2, -3)$
$y = -\frac{1}{3}x + -\frac{1}{3}$

Determine if the ordered pair is a solution of the equation that produced the graph.

40. $(3, 8)$

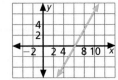

41. $(21, 21)$

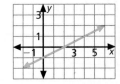

42. $(0, 21)$

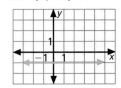

CHAPTER INVESTIGATION

EXTENSION Make a table of your results. Which situation allows the marble to roll the farthest? If you wanted a car to travel the greatest distance off a given ramp, how would you adjust the slope and length of the ramp? Explain why this information is valuable to people who design and build roads.

Chapter 7 Assessment

Write a story that can be represented by each graph.

1.
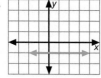

2.
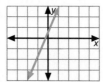

Graph each point on the same coordinate plane.

3. $Y(5, 1)$

4. $P(22, 23)$

5. $D(0, 3)$

6. a point in Quadrant IV

7. a point whose coordinates are opposites

8. Does the mapping at the right show that the relation is a function?

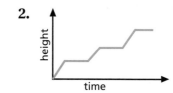

9. Use the vertical line test to determine if the relation is a function.

10. State the domain and range of $\{(2, -30), (3, -40), (2, 10), (3, 40), (2, 30)\}$. Then state whether the relation is a function.

11. Write "f of x is equal to 2 times x minus three" in function form.

12. Find three solutions of the equation $y = x + 8$.

13. Graph the equation $y = 2x - 5$.

14. Find the x- and y-intercepts of $y = \frac{3}{4}x - 2$.

15. Find the slope of the line graphed at the right.

16. Find the slope of the line that passes through $A(0, 22)$ and $B(8, 23)$.

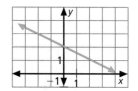

Identify the slope of each line as positive, negative, zero, or undefined.

17.

18.

19.

20.

21. Find the slope and y-intercept of $y = 2x - 1$. Then graph the line.z

22. Find the equation of the line graphed in Exercise 15.

23. Name the slope and y-intercept of the line in Exercise 18. Write an equation of the line in slope-intercept form.

24. Graph $N(6, 1)$ and $M(3, -2)$ on a coordinate plane. Then use the Pythagorean Theorem to find the distance between N and M.

25. Use the distance formula to find the distance between $A(5, 3)$ and $S(3, 5)$.

26. Kim leaves the pool and travels 3 mi west, 4 mi south, and 6 mi north to the gym. What is the shortest distance between the pool and the gym?

27. Determine if $\left(\frac{3}{4}, 5\right)$ is a solution of $y = 4x + 2$.

28. Determine if $(5, -1)$ is a solution of $y = -1$.

Standardized Test Practice

Part 1 Multiple Choice

**Record your answers on the answer sheet
provided by your teacher or on a sheet of paper.**

1. Which pair of ratios is equivalent?
(Lesson 2-6)

(A) $\dfrac{3}{5}, \dfrac{15}{25}$ (B) $\dfrac{18}{12}, \dfrac{12}{18}$

(C) $\dfrac{9}{3}, \dfrac{36}{15}$ (D) $\dfrac{4}{9}, \dfrac{16}{27}$

2. Find the surface area of the cylinder. Use
3.14 for π. Round to the nearest tenth.
(Lesson 4-9)

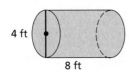

4 ft

8 ft

(A) 108.5 ft^2 (B) 125.6 ft^2

(C) 150.7 ft^2 (D) 301.4 ft^2

3. Li has a total of 92 coins in his coin collection.
This is 8 more than three times the number of
quarters in the collection. How many quarters
does Li have in her collection? (Lesson 5-4)

(A) 11   (B) 28

(C) 30   (D) 33

4. The Lions won 18 of their last 24 games. What
percent of the games did the team win?
(Lesson 6-1)

(A) 25%   (B) 43%

(C) 75%   (D) 90%

5. A $1400 refrigerator is on sale at 15% off
the regular price. What is the discount?
(Lesson 6-3)

(A) $210   (B) $390

(C) $420   (D) $1190

6. A loan of $1500 was taken out at a simple
interest rate of 6.65%. What is the total amount
due after 3 yr? (Lesson 6-5)

(A) $299.25   (B) $1200.75

(C) $1799.25   (D) $2999.25

7. As a lifeguard, you drew the diagram below to
show the positions of different attractions at
the water park. Which coordinates represent
Lion Falls? (Lesson 7-2)

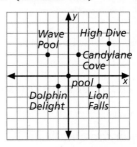

(A) $(-1, -1)$ (B) $(1, 2)$

(C) $(-2, 2)$ (D) $(3, -1)$

8. Determine which relation is a function.
(Lesson 7-3)

(A)
x	f(x)
−3	0
2	−2
−3	−5

(B)
x	f(x)
−1	9
1	8
2	6

(C)
x	f(x)
2	7
4	9
5	5

(D)
x	f(x)
−2	9
1	4
3	0

9. Find the slope of a line that passes through
$(1, -2)$ and $(1, 3)$. (Lesson 7-5)

(A) $-\dfrac{2}{5}$ (B) 0

(C) $\dfrac{2}{5}$ (D) undefined

10. Jewel has written 10 pages of a novel and
plans to write 15 additional pages per month
until she is finished. Which equation shows
the total number of pages P after any number
of months m? (Lesson 7-6)

(A) $P = 15(1 + 10m)$ (B) $P = 10 + 15m$

(C) $P = 15 + 10m$ (D) $P = 10 + (15 + m)$

11. Find the perimeter of $\triangle GHI$ with vertices
$G(-2, 3)$, $H(2, 2)$, and $I(0, -3)$. (Lesson 7-7)

(A) 11.6 units   (B) 14.3 units

(C) 15.8 units   (D) 17.0 units

Part 2 Short Response/Grid In

Record your answers on the answer sheet provided by your teacher or on a sheet of paper.

12. Identify the three-dimensional shape formed by the net. (Lesson 4-4)

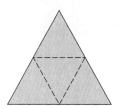

13. Find an equation for the sentence: Thirteen more than a number m is 12. (Lesson 5-2)

14. A survey asked a 324-student freshman class their preferences for pizza. The circle graph shows the results. How many students chose vegetable as their favorite? (Lesson 6-1)

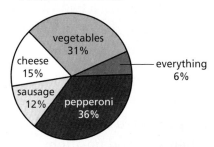

Pizza Preferences

vegetables 31%
cheese 15%
everything 6%
sausage 12%
pepperoni 36%

15. On Thursday, a baker sold 235 cookies. On Friday, she sold 198 cookies. What is the percent of decrease from Thursday to Friday? Round to the nearest tenth. (Lesson 6-7)

16. Find the x-intercept and the y-intercept of $y = 5 - 2x$. (Lesson 7-4)

17. Find the slope of the line. (Lesson 7-5)

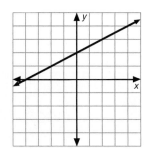

18. A taxicab company charges $2.25 plus $0.28 per mi. If this situation is represented as a line on a graph, what is the y-intercept? (Lesson 7-6)

19. Find the distance between (6, 2) and (4, 0.5). (Lesson 7-7)

Part 3 Extended Response

Record your answers on a sheet of paper. Show your work.

20. Dina and Jasmine each go for walks around a lake a few times per week. Last week Dina walked 7 mi more than Jasmine. (Lesson 5-6)

  a. If j represents the number of miles Jasmine walked, write an equation that represents the total number of miles T the two girls walked.

  b. If Jasmine walked 9 mi during the week, how many miles did Dina walk?

  c. If Jasmine walked 11 mi during the week, how many miles did the two girls walk together?

21. As a thunderstorm approaches, you see lightning as it occurs, but you hear the accompanying thunder a short time later. The distance d in miles that sound travels in t seconds is given by the equation $d = 0.21t$. (Lesson 7-4)

  a. Find three solutions of the equation. Write the solutions in a table.

  b. Graph the equation.

  c. Estimate how long it will take to hear the thunder from a storm 3 mi away.

Test-Taking Tip

Questions 20 and 21
If a problem seems difficult, reread the question slowly and carefully. Always ask yourself, "What have I been asked to find?" and, "What information will help me work out the answer?"

Relationships in Geometry

THEME: Animation

Each time you enter a room, you see geometric relationships in the lines and planes that form the walls, ceiling, and floor. Each time you watch a film, you are viewing three-dimensional objects on a two-dimensional screen. You don't have to be able to name the geometric shapes and their relationships to enjoy a film, but the creators of the film must.

One specific type of film is animation. Animation is a series of still images that when viewed in rapid succession give the illusion of motion. Animation is based on hand-drawn or computer-generated shapes and spatial relationships of the shapes, such as their size and position.

- **Animators** (page 361) draw the images for animated films, features, and commercials. They use traditional drawing tools, such as ink, paint, and rulers, and modern tools like computers to create animation.

- **Video editors** (page 379) carefully watch animation and other videos for mistakes, such as images that are jumpy or inconsistent. They may also coordinate the writing and creation of a video feature.

Math Online

mathmatters1.com/chapter_theme

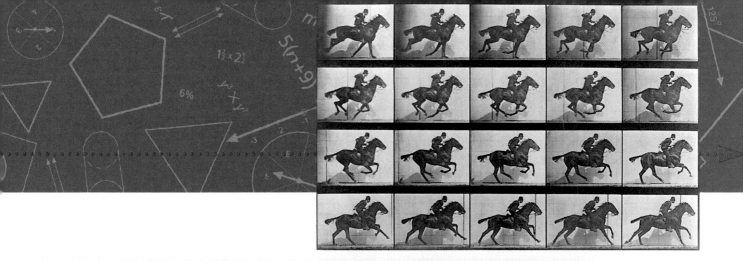

Data Activity: Motion Photography

Eadweard Muybridge studied motion, especially how motion is portrayed by photography. In 1887 he used 24 cameras to photograph this horse galloping. Each camera was triggered by a trip-wire on the course.

Use the photographs to answer Questions 1–6.

1. Muybridge proved that a horse's feet are held up under the body, and don't always touch the ground during a gallop. This was controversial, and contradicted what painters had depicted for centuries. How many photographs prove that the horse is totally in the air?

2. Could any of the photographs be taken out without losing the flow of the images?

3. Describe or sketch an image that could appear before the first photograph.

4. Describe or sketch an image that could appear after the last photograph.

5. If the photographs shown create 0.5 sec of animation, how many photographs will create 5 sec of animation?

6. How many photographs will create 5 min of animation?

CHAPTER INVESTIGATION

A fundamental tool used by animators to simulate movement and visualize how a scene or character will come to life is a flipbook. Flipbooks are series of pages containing progressively changing drawings. Each page of the book represents a *frame* in the animation. When the pages are flipped through quickly, the scene unfolds before you.

Working Together

Design a flipbook to illustrate the animation of an object or a scene. Use the Chapter Investigation icons in the chapter to help you create your flipbook.

Are You Ready?

Refresh Your Math Skills for Chapter 8

The skills on these two pages are ones you have already learned. Use the examples to refresh your memory and complete the exercises. For additional practice on these and more prerequisite skills, see pages 536–544.

GEOMETRIC TERMS

You often do not use geometric terms in other branches of math, so it is easy to forget their meanings. This matching exercise will help you remember key geometric terms you will use in this chapter.

Match each definition with the term it describes.

1. two lines that meet at 90° angles

2. a closed shape with six sides

3. a portion of a line having two endpoints

4. two lines in the same plane that never meet or intersect

5. a four-sided figure with all its sides congruent

6. a closed shape with any number of sides that are all the same length

7. two rays that share an endpoint

8. a pair of lines that are not in the same plane and do not intersect

9. a polygon having eight sides

10. a portion of a line having one endpoint and continuing forever in one direction

a.	point
b.	line
c.	line segment
d.	ray
e.	angle
f.	plane
g.	perpendicular
h.	skew
i.	parallel
j.	regular polygon
k.	irregular polygon
l.	hexagon
m.	octagon
n.	rhombus
o.	quadrilateral
p.	equilateral
q.	obtuse

GEOMETRIC SHAPES

Being able to identify geometric shapes will be helpful throughout this chapter. Refer to Chapter 4 if you need to review.

Identify each shape. Describe with as many terms as possible.

11.

12.

13.

14.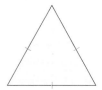

MEASURING ANGLES

Careful use of a protractor will allow you to check constructions you make in this chapter. To make an angle easier to measure, copy angles to another sheet of paper. Extend the rays until they are long enough to cross the arc of the protractor.

Use a protractor to find the measure of each angle.

15.

16.

17.

18.

19.

20.

COORDINATE PLANE

In this chapter you will work with figures on a coordinate plane. For the new skill you will learn, it is necessary that you know how to plot points and identify coordinates.

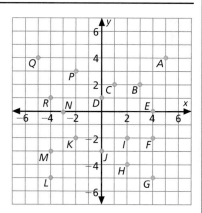

Example Remember that the *x*-coordinate tells how far left or right to move from zero. The *y*-coordinate tells how far up or down to move from zero.

Find $Q(-5, 4)$.
x-coordinate = -5
y-coordinate = 4
Start at the origin. Since the *x*-coordinate is negative, move to the left 5 units. The *y*-coordinate tells you to move up 4 units.

Name the point represented by each ordered pair.

21. $(4, 0)$ **22.** $(2, -2)$ **23.** $(3, 2)$ **24.** $(-2, -2)$

25. $(-2, 3)$ **26.** $(1, 2)$ **27.** $(5, 4)$ **28.** $(4, -2)$

Name the coordinates for each point.

29. D **30.** M **31.** J **32.** G

33. L **34.** N **35.** R **36.** H

Angles and Transversals

Goals
- Measure and classify angles.
- Explore the relationship between transversals and angles.

Applications Civil engineering, Construction, Animation

Use two straight objects such as pencils, spaghetti or straws to model lines.

1. Imagine that your desktop is a plane. Position your "lines" on the plane in as many ways as you can.

2. Position your two lines so that they are parallel to each other.

3. Position your two lines perpendicular to each other.

4. How else can the two lines be positioned?

> **Think Back**
>
> Two lines are perpendicular to each other if they intersect to form right angles.
>
> Two lines are parallel if they have no points in common, that is, they do not intersect.

◣ BUILD UNDERSTANDING

Certain pairs of angles have special relationships. Two angles are called **complementary angles** if the sum of their measures is 90°. In the figure at the right, ∠*ABC* is the *complement* of ∠*CBD*.

Two angles are called **supplementary angles** if the sum of their measures is 180°. In the figure below, ∠*LOM* is the *supplement* of ∠*MON*.

Adjacent angles lie in the same plane have a common vertex and a common side, but have no interior points in common. In the figures shown, ∠*ABC* is *adjacent* to ∠*CBD*, and ∠*LOM* is *adjacent* to ∠*MON*.

Complementary and supplementary angles are not necessarily adjacent.

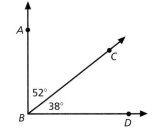

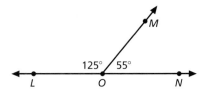

Example 1

Use the figure shown to name the following.

a. complementary angles

b. supplementary angles

c. adjacent angles

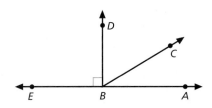

Solution

a. complementary angles: ∠*ABC* and ∠*CBD*

b. supplementary angles: ∠*ABC* and ∠*CBE*, ∠*ABD* and ∠*DBE*

c. adjacent angles: ∠*ABC* and ∠*CBD*, ∠*ABC* and ∠*CBE*, ∠*ABD* and ∠*DBE*, ∠*CBD* and ∠*DBE*

Congruent angles have the same measure. In the figure, ∠1 and ∠3 are congruent. This is written as ∠1 ≅ ∠3.

Intersecting lines are lines that have exactly one point in common. When two lines intersect in a plane, the angles that are not adjacent to each other are called **vertical angles**. In the figure shown, ∠1 and ∠3 form a pair of vertical angles, and ∠2 and ∠4 form a pair of vertical angles.

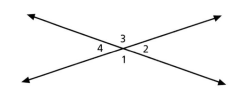

Vertical angles have the same measure. In this figure, $m\angle 1 = m\angle 3$ and $m\angle 2 = m\angle 4$.

Example 2

In the figure, $m\angle 1 = 145°$. Find the angle measurements.

a. $m\angle 3$ **b.** $m\angle 2$

Solution

a. ∠1 and ∠3 are vertical angles, so they have the same measure.
$m\angle 1 = m\angle 3$, or $m\angle 3 = 145°$.

b. ∠2 and ∠3 are supplementary angles, so the sum of their measures is 180°.

$$m\angle 2 + 145° = 180°$$
$$m\angle 2 = 180° - 145°$$
$$m\angle 2 = 35°$$

A **transversal** is a line that intersects two or more lines in a plane at different points. In the figure shown, $\overleftrightarrow{AB}$ is a transversal intersecting $\overleftrightarrow{CD}$ and $\overleftrightarrow{EF}$. A transversal such as $\overleftrightarrow{AB}$ forms several pairs of angles with the lines it intersects.

Corresponding angles are angles that are in the same position relative to the transversal and the lines being intersected. In the figure, there are four pairs of corresponding angles.

∠1 and ∠5, ∠2 and ∠6, ∠3 and ∠7, ∠4 and ∠8

When two parallel lines are cut by a transversal, corresponding angles are congruent.

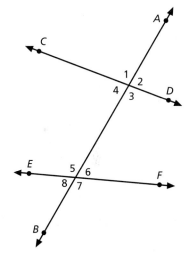

Example 3

In the figure, $\overleftrightarrow{AB} \parallel \overleftrightarrow{CD}$ and $m\angle 1 = 118°$. Find each measure.

a. $m\angle 5$ **b.** $m\angle 8$ **c.** $m\angle 7$

Solution

a. ∠5 and ∠1 are corresponding angles, so $m\angle 5 = m\angle 1$. So $m\angle 5 = 118°$.

b. ∠8 and ∠5 are vertical angles, and $m\angle 8 = m\angle 5$. So $m\angle 8 = 118°$.

c. ∠7 and ∠8 are supplementary angles, so $m\angle 7 + m\angle 8 = 180°$. So $m\angle 7 = 180° - 118° = 62°$.

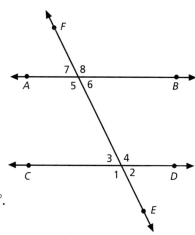

In the figure, $\overleftrightarrow{AB} \parallel \overleftrightarrow{CD}$.

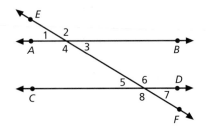

1. Which angles have the same measure as ∠5?

2. Which angles have the same measure as ∠6?

3. Identify a pair of vertical angles.

4. Identify two pairs of corresponding angles.

5. **ANIMATION** A cartoonist is drawing a pair of scissors cutting a piece of paper. How should the angle of the handles relate to the angle of the blades?

In the figure, $\overleftrightarrow{RS} \perp \overleftrightarrow{TU}$. Find each measure.

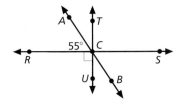

6. $m\angle ACT$ 7. $m\angle UCB$

8. $m\angle SCB$ 9. $m\angle RCB$

Find each measure.

10. $m\angle 2$ 11. $m\angle 3$ 12. $m\angle 4$

13. **MODELING** Fold a piece of paper in half lengthwise. Open it and fold it in half in the opposite direction. What is the relationship between the creases? The folds create four angles. How are these angles related?

In the figure, $\overleftrightarrow{AB} \parallel \overleftrightarrow{CD}$. Find each measure.

14. $m\angle 1$ 15. $m\angle 2$ 16. $m\angle 3$ 17. $m\angle 4$

18. $m\angle 5$ 19. $m\angle 6$ 20. $m\angle 7$ 21. $m\angle 8$

22. **YOU MAKE THE CALL** Devon claims that in the figure, if $m\angle BCA = m\angle DCE$, then $\overleftrightarrow{AB}$ and $\overleftrightarrow{CD}$ must be parallel because $\angle BCA$ and $\angle DCE$ are vertical angles. Is he correct? Explain.

23. **WRITING MATH** In the definition of a transversal, why is it necessary to include the phrase *at different points*?

24. **ENGINEERING** Sofie is a civil engineer designing a parking lot for a new shopping mall. She wants lines *a*, *b*, and *c* to be parallel to each other so that the parking spaces are the same size. What angle relationships must be true for lines *a*, *b*, and *c* to be parallel?

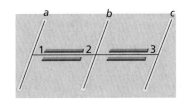

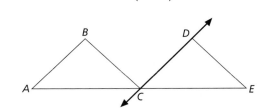

25. Find $m\angle 1$, $m\angle 2$, $m\angle 3$, and $m\angle 4$.

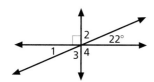

26. $\overleftrightarrow{LM} \parallel \overleftrightarrow{NO}$. Find $m\angle 5$ and $m\angle 6$.

27. CONSTRUCTION A carpenter makes two parallel cuts to create the board shown. Find the measures of $\angle 1$, $\angle 3$, and $\angle 4$.

EXTENDED PRACTICE EXERCISES

Use the figures to find the given angle measures.

28. Find $m\angle ATS$ and $m\angle STB$.

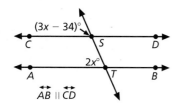

$\overleftrightarrow{AB} \parallel \overleftrightarrow{CD}$

29. Find $m\angle QRN$ and $m\angle PRN$.

30. CRITICAL THINKING Do you think that $\overleftrightarrow{AB}$ and $\overleftrightarrow{CD}$ are parallel to each other? Explain why or why not.

31. *Alternate interior angles* are pairs of angles that are interior to the lines and on alternate sides of the transversal. In Exercise 26, $\angle 3$ and $\angle 6$ are alternate interior angles. Name a pair of alternate interior angles in the figure for Exercises 14–21.

32. *Alternate exterior angles* are pairs of angles that are exterior to the lines and on alternate sides of the transversal. In Exercise 26, $\angle 1$ and $\angle 8$ are alternate exterior angles. Name a pair of alternate exterior angles in the figure for Exercises 14–21.

33. WRITING MATH Suppose two parallel lines are intersected by a transversal. How are the alternate interior angles related? Alternate exterior angles?

MIXED REVIEW EXERCISES

Draw each geometric figure. Then write a symbol for each figure. (Lesson 4-1)

34. ray AB

35. line segment RT

36. line GH

37. angle XYZ

38. plane $\mathscr{L}$

39. point V

Use the relation $\{(-4, -2), (-2, -1), (-1, 0), (0, 1), (1, 2)\}$ to determine the following. (Lesson 7-3)

40. the domain

41. the range

42. Is the relation a function?

Beginning Constructions

Goals
- Construct line segments and copies of angles.
- Construct angle bisectors and perpendicular bisectors.

Applications Machinery, Sports, Animation

Use paper and a straightedge for the activity.

1. Use the straightedge to draw a line segment on the paper. Label the endpoints A and B.

2. Fold the paper so that point A overlaps B. Open the paper and label the intersection of the crease and $\overline{AB}$ point M.

3. What do you notice about the lengths of $\overline{AM}$ and $\overline{MB}$?

◤ BUILD UNDERSTANDING

There are methods of constructing congruent figures and copies of other figures without using measuring devices such as rulers or protractors. Constructions can be made with a compass and straightedge, paper folding and geometry software.

Example 1

PAPER FOLDING Construct a copy of each figure.

a. a line segment   **b.** an angle

> ### Think Back
>
> Two figures that have the same size and shape are said to be congruent.
>
> The symbol ≅ is read as "is congruent to."

Solution

a. First, draw $\overline{AB}$. Then make a fold just below or above the segment. Trace the image of $\overline{AB}$ and label it $\overline{CD}$. $\overline{CD}$ is a copy of $\overline{AB}$.

b. To copy an angle, draw an angle on paper. Fold just below or above the original angle and trace the image. The new angle is a copy of the original.

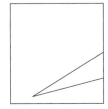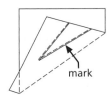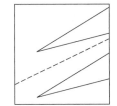

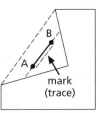

The **midpoint** of a line segment is the point that separates it into two line segments of equal length. To bisect means "to divide into two equal parts." The **perpendicular bisector** of a line segment is a line, ray or line segment that is perpendicular to another line segment at its midpoint. An **angle bisector** is a ray that separates an angle into two congruent adjacent angles.

Example 2

Use a compass and a straightedge to construct the perpendicular bisector of $\overline{JK}$.

Solution

Step 1 Place the point of the compass at *J*. Open the compass a little more than half the length of $\overline{JK}$. Draw one arc above and another below $\overline{JK}$.

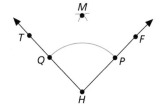

$\overleftrightarrow{LN}$ is the perpendicular bisector of $\overline{JK}$.

Step 2 Place the compass point at *K*. With the setting used in Step 1, draw arcs above and below $\overline{JK}$. Label the points of intersection *L* and *N*. Draw $\overleftrightarrow{LN}$. Label the midpoint *M*.

Example 3

SPORTS On a baseball diamond, the angle formed by first base, home plate and third base is a right angle. The pitcher's mound and second base both lie on the bisector of this angle. Draw ∠ *THF* and construct the angle bisector using a compass and a straightedge.

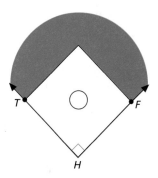

Solution

Step 1 With the compass point at *H*, draw an arc that intersects $\overrightarrow{HT}$ and $\overrightarrow{HF}$. Label the intersection points *P* and *Q*.

Step 2 With the compass point at *P*, adjust the compass so that the opening is a little greater than arc *PQ*. Draw an arc inside ∠ *THF*.

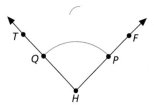

Step 3 Place the compass point at *Q*. With the setting used in Step 2, draw an arc that intersects the last arc at *M*.

Step 4 Draw $\overrightarrow{HM}$, the angle bisector.

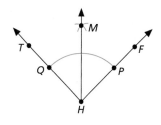

Checking with a protractor reveals that ∠ *THM* and ∠ *MHF* are both 45° angles.

Example 4

Construct a line perpendicular to $\overleftrightarrow{RS}$ through point A.

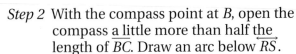

Step 1 With the compass point at A, draw an arc that intersects $\overleftrightarrow{RS}$ at B and C.

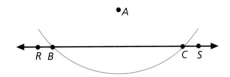

Step 2 With the compass point at B, open the compass a little more than half the length of $\overline{BC}$. Draw an arc below $\overleftrightarrow{RS}$.

Step 3 Place the compass point at C. With the setting used in Step 2, draw an arc below $\overleftrightarrow{RS}$ that intersects the first arc. Label the point of intersection D.

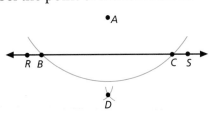

Step 4 Draw $\overleftrightarrow{AD}$.

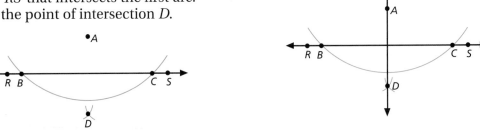

$\overleftrightarrow{AD}$ is perpendicular to $\overleftrightarrow{RS}$ through point A.

◥ TRY THESE EXERCISES

Using a protractor, draw each angle. Then construct each angle bisector.

1. $m\angle ABC = 132°$ **2.** $m\angle DEF = 48°$ **3.** $m\angle GHI = 90°$

Trace each segment. Then copy it and construct the perpendicular bisector.

4.    **5.** C————D   **6.** E————F

Trace each figure. Then construct a line perpendicular to $\overleftrightarrow{AB}$ through point P.

7.    **8.**    **9.**

10. WRITING MATH Compare the construction of an angle bisector to the construction of a segment bisector.

◥ PRACTICE EXERCISES • For Extra Practice, see page 574.

Using a protractor, draw each angle. Then construct each angle bisector.

11. $m\angle JKL = 32°$ **12.** $m\angle MNO = 85°$ **13.** $m\angle PQR = 170°$

Trace each segment. Then copy it and construct the perpendicular bisector.

14. G———H   **15.** J————K   **16.** L—————————M

Trace each figure. Then construct a line perpendicular to $\overrightarrow{AB}$ through point P.

17.

18.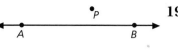

19.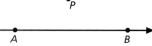

20. Two line segments can be added to construct a longer line segment. Construct two line segments on your paper. Name them $\overline{AB}$ and $\overline{CD}$. Then construct a line segment of length $AB + CD$. This is called *segment addition*.

21. Two angles can be added to construct a larger angle. Draw two angles. Then construct the angle that would represent their sum to model *angle addition*. Write a mathematical statement to express this addition.

22. **MACHINERY** A machine cuts copper pipe into 4 equal sections. Draw a line segment about 4 or 5 in. long on your paper to model a copper pipe. Then divide it into four equal sections. Check your work with a ruler.

Trace each triangle. Construct the perpendicular bisectors through each side. Label the intersection point P.

23.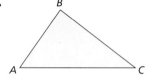

24.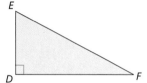

25.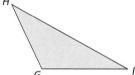

26. **CRITICAL THINKING** Measure the distances from each point P to each of the triangle vertices. What do you observe? Explain.

EXTENDED PRACTICE EXERCISES

27. **GEOMETRY SOFTWARE** Use geometry software to draw three triangles. Construct the three perpendicular bisectors of each triangle. Label the intersection point. Then construct a circle with the intersection point as the center, and one vertex as a point on the circle. What do you observe about each circle's relationship to each triangle?

28. **ANIMATION** Animators often sketch their ideas on paper. How do you think constructions are used in animation? Give an example.

29. **CHAPTER INVESTIGATION** Decide what you will animate in your flip book. You might choose a ball bouncing, a person's face changing, a setting sun or a child releasing a balloon. Be creative. Make a few sketches of your subject.

MIXED REVIEW EXERCISES

Draw a net for each three-dimensional figure. (Lesson 4-4)

30. a cylinder

31. a cube

Find the slope and y-intercept of each line. (Lesson 7-6)

32. $y = \dfrac{4}{5}x - 6$

33. $y = -3x + 7$

34. $y = -\dfrac{1}{2}x - \dfrac{2}{3}$

35. $y = 7x + -3$

Review and Practice Your Skills

PRACTICE ■ LESSON 8-1

Use Figure A. Find each measure.

1. $m\angle 1$ 2. $m\angle 2$ 3. $m\angle 3$

In Figure B, $\overleftrightarrow{XY} \parallel \overleftrightarrow{PQ}$. Find each measure.

4. $m\angle 1$ 5. $m\angle 2$ 6. $m\angle 3$

7. $m\angle 4$ 8. $m\angle 5$ 9. $m\angle 6$

10. $m\angle 7$ 11. $m\angle 8$ 12. $m\angle 9$

Use Figure C. Find each measure.

13. $m\angle 1$ 14. $m\angle 2$ 15. $m\angle 3$

In Figure D, $\overleftrightarrow{AB} \parallel \overleftrightarrow{ST}$. Find each measure.

16. $m\angle 1$ 17. $m\angle 2$ 18. $m\angle 3$

19. $m\angle 4$ 20. $m\angle 5$ 21. $m\angle 6$

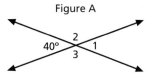

Figure A

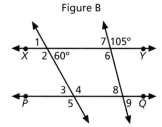

Figure B

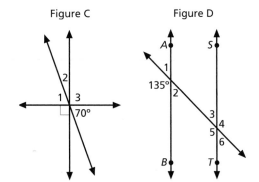

Figure C   Figure D

PRACTICE ■ LESSON 8-2

Using a protractor, construct each angle. Then bisect the angle.

22. $m\angle ABC = 58°$ 23. $m\angle PQR = 126°$ 24. $m\angle XYZ = 24°$

Trace each figure below. Copy it and construct the perpendicular bisector.

25. D ●————————————● E

26. K ●————————————————————● M

Trace each figure below. Construct a perpendicular to the line from point P.

27.

28.

PRACTICE ◣ LESSON 8-1–LESSON 8-2

In Figure A, $\overleftrightarrow{EF} \parallel \overleftrightarrow{JK}$. Find each measure. (Lesson 8-1)

29. $m\angle 1$ **30.** $m\angle 2$ **31.** $m\angle 3$

32. $m\angle 4$ **33.** $m\angle 5$ **34.** $m\angle 6$

Use Figure B. Find each measure. (Lesson 8-1)

35. $m\angle 1$ **36.** $m\angle 2$ **37.** $m\angle 3$

Using a protractor, construct each angle. Then bisect the angle. (Lesson 8-2)

38. $m\angle MNR = 174°$ **39.** $m\angle FGH = 85°$ **40.** $m\angle CDE = 110°$

41. Trace the figure below. Copy it and construct the perpendicular bisector.

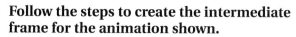

42. Trace the figure below. Construct a perpendicular to the line from point *P*.

Figure A   Figure B

MathWorks — Career – Animator
Workplace Knowhow

Animators produce images that are transferred to film or tape. These images are shown in rapid succession to create the illusion of motion.

Traditionally, animators drew or painted images that were filmed one frame at a time. Now, computer animation can create special effects that are impossible with other techniques. Computer animation can produce images from data to predict storms, reconstruct accidents, and demonstrate particle collisions.

A storyboard is a scene-by-scene illustration of the events in an animated feature. The storyboard identifies important animation frames. Computers can create the frames that fill in the action between the key scenes.

Follow the steps to create the intermediate frame for the animation shown.

1. Measure $\angle 1$.

2. Measure $\angle 2$.

3. Add $m\angle 1$ and $m\angle 2$.

4. Bisect the sum of $m\angle 1$ and $m\angle 2$. This is $m\angle 3$ in the in-between frame.

5. Draw the in-between frame, and indicate $\angle 3$.

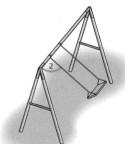

Diagonals and Angles of Polygons

Goals
■ Determine the number of diagonals in a polygon.
■ Explore the sum of the internal angles in a polygon.

Applications Safety, Number sense, Animation

A tangram is an ancient Chinese puzzle made up of seven pieces which can be combined to form different figures. Use a set of tangram pieces or trace the figure and cut them out to make your own.

1. Use any number of pieces to form each of these figures.

2. Record which pieces were used to make each figure.

3. Compare the pieces you used to make each figure with a classmate. Is there more than one way to make each figure? Explain.

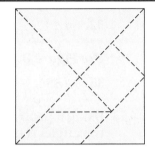

◥ BUILD UNDERSTANDING

A **diagonal** of a polygon is a line segment that joins two vertices and is not a side. There is a rule for determining the number of diagonals in a polygon that has *n* sides.

Diagonals of a Polygon	The number of diagonals in a polygon with *n* sides is given by the formula $\dfrac{n(n-3)}{2}$.

Example 1

Find the number of diagonals for the following figures.

a. a pentagon   **b.** a decagon   **c.** a triangle

Solution

a. A pentagon has five sides. So applying the formula gives
$$\frac{5(5-3)}{2} = \frac{5(2)}{2} = 5 \text{ diagonals.}$$

b. A decagon has ten sides. So applying the formula gives
$$\frac{10(10-3)}{2} = \frac{10(7)}{2} = 35 \text{ diagonals.}$$

c. A triangle has three sides. So applying the formula gives
$$\frac{3(3-3)}{2} = \frac{3(0)}{2} = 0 \text{ diagonals.}$$

Polygons can be separated into nonoverlapping triangular regions by drawing all the diagonals from one vertex. Since the sum of the angles of any triangle is 180°, the sum of the measures of a polygon's interior angles is the product of the number of triangles formed and 180°.

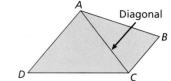

The sum of the angles of △ABC equals 180°.
The sum of the angles of △ACD equals 180°.

Example 2

SAFETY Find the sum of the angle measures of a stop sign.

Solution

A stop sign is an octagon, so it has 8 sides. When all the diagonals are drawn from one vertex, there are 6 triangular regions. Since 6 · 180° = 1080°, the sum of the angle measures of a stop sign is 1080°.

Example 3

Find $m\angle D$ in the polygon shown.

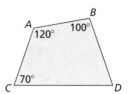

Solution

Since the polygon can be separated into two triangles, the sum of the angle measures is 2 · 180° = 360°.

$$m\angle A + m\angle B + m\angle C + m\angle D = 360°$$

$$120° + 100° + 70° + m\angle D = 360°$$

$$290° + m\angle D = 360°$$

$$m\angle D = 70°$$

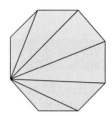

◣ TRY THESE EXERCISES

Find the total number of diagonals in each polygon. Then find the sum of the angle measures for each.

1.

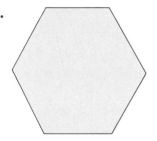

2.

3.

4. How many diagonals do each of the polygons in Exercises 1–3 have from one vertex?

5. The angles of the scalene triangle have the measures shown. What is the measure of $\angle x$?

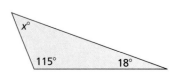

Find the unknown angle measure for each polygon.

6.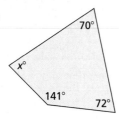

7.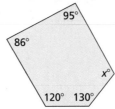

8.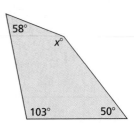

In the chart, information is complete for triangles and quadrilaterals. Complete the information for the following polygons.

9. pentagon

10. hexagon

11. heptagon

12. octagon

13. nonagon

14. decagon

Polygon	Number of sides	Number of triangles formed by drawing diagonals from one vertex	Sum of the angle measures	Total number of diagonals
Triangle	3	1	180°	0
Quadrilateral	4	2	2 • 180° = 360°	2
Pentagon	5	■	■	■
Hexagon	6	■	■	■
Heptagon	7	■	■	■
Octagon	8	■	■	■
Nonagon	9	■	■	■
Decagon	10	■	■	■

15. What is the measure of each angle of a regular hexagon? (Recall that a regular polygon is both equilateral and equiangular.)

Trace each polygon. Then draw all of the diagonals from one vertex.

16.

17.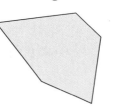

18.

 19. **WRITING MATH** Write a conclusion about the number of diagonals that a polygon with *n* sides has from one vertex. How many nonoverlapping triangles can be formed in a polygon with *n* sides?

Find the unknown angle measure in each figure.

20.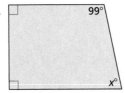

21.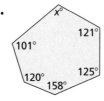

22.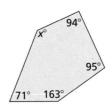

23. NUMBER SENSE How many diagonals can be drawn from one vertex in a 100-sided polygon? How many nonoverlapping triangles would these diagonals create?

24. What is the sum of the angle measures of the figure in Exercise 23?

25. MODELING Eight teenagers who live in the same neighborhood decide to install a telephone and intercom system. The system will connect each house with each of the other seven houses. How many wires will the system require? Draw a model of the system to solve the problem.

26. ANIMATION Juang is a cartoon animator who is designing a shield in the shape of a regular octagon for a character. He begins by sketching a model of the shield on a square piece of paper. At what angle must he cut along adjacent edges of the paper?

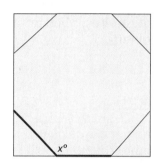

▰ Extended Practice Exercises

Find the measure of each angle.

27.

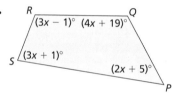

28.

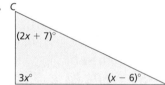

29.

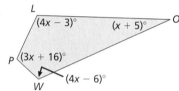

MODELING Use tangram pieces to create each figure. Then calculate the sum of each figure's angles and the total number of diagonals.

30.

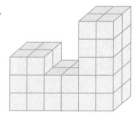

31.

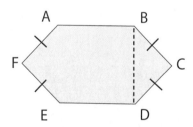

▰ Mixed Review Exercises

How many cubes are in each isometric drawing? (Lesson 4-5)

32.

33.

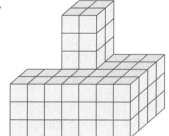

34.

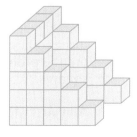

35. Make three isometric drawings of an object in the classroom. Each drawing should be made from a different angle.

Problem Solving Skills: Modeling Problems

Many problems in mathematics and life are too complex or involve too many steps to be solved mentally or on paper. A strategy for understanding such a problem and arriving at its solution is to **act it out**.

Acting out the problem means that you physically go through the motions described or use objects to represent elements of the problem.

Problem Solving Strategies

Guess and check

Look for a pattern

Solve a simpler problem

Make a table, chart or list

Use a picture, diagram or model

✔ Act it out

Work backwards

Eliminate possibilities

Use an equation or formula

Problem

If one penny is placed next to another and then is rolled halfway around the other, will Lincoln's head be right-side-up or upside down?

Solve the Problem

It is difficult to visualize mentally what the solution to this problem is, so a good way to solve it is to act it out. Take two pennies, place them as shown, and then roll the right penny halfway around the left penny. Lincoln's head is right-side-up.

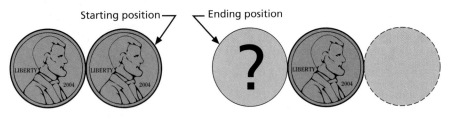

Starting position — Ending position

◤ TRY THESE EXERCISES

1. Make three squares by moving exactly three toothpicks in the pattern shown. Act out the problem and draw a picture of the three squares.

2. Starting with the original pattern in Exercise 1, make three squares by moving exactly four toothpicks. Act out the problem and draw a picture of the three squares.

3. **ANIMATION** Work with a partner. Record the time in seconds that it takes your partner to make a paper airplane. One second of animation requires 24 images. How many images are required to animate your partner making a paper airplane?

PAPER FOLDING Trace and cut out square *ABDC*. Label the corners as shown. What figure is formed with each move? Check by acting it out.

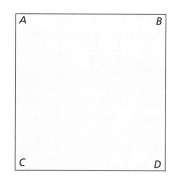

4. *A* folded onto *B*

5. *A* folded onto *D*

6. *A* folded onto *B*, then *B* folded onto *C*

7. *A* folded onto *D*, then *B* folded onto *C*

Use your cutout of square *ABDC* to solve.

8. Fold *A* onto *B* and then *B* onto *C*. The perimeter of the resulting figure is what fractional part of the perimeter of the original square?

9. Fold *A* onto the midpoint of side *AB*. If the area of the resulting figure is 12 cm², what was the area of the original square?

10. A heavy log is being moved by rolling it on cylinders. If the circumference of each cylinder is 6 ft, how far will the log move for each revolution of the cylinders? Use cardboard or paper to make a model.

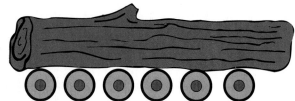

11. **MODELING** Write your name on paper so when you hold it up to a mirror you can read it in the reflection. Use a reflective device.

12. **DATA FILE** Use the Data Index on pages 516–517 to find basic data on the planets. Make a visual of the sun and the planets. Draw the Sun at the edge of your paper. Let the length of a paper clip represent 100 million km. Draw the location of Mercury, Venus, Earth and Mars in relation to the sun. What is the distance between Earth and the Sun in terms of paper clips?

13. **WRITING MATH** Write your own problem that can be solved by acting out the situation. Exchange your problem with a partner and solve.

■ **MIXED REVIEW EXERCISES**

Use the distance formula to find the distance between each pair of points. Round to the nearest hundredth. (Lesson 7-7)

14. *J*(3, −4) and *K*(−2, −1)   15. *G*(6, −10) and *H*(−3, 5)   16. *L*(0, 9) and *M*(−4, −5)

17. Find the perimeter of a triangle formed by the points *A*(1, 3), *B*(0, −5) and *C*(−2, 4). State whether the triangle is scalene, isosceles or equilateral.

Classify each of the following triangles as scalene, isosceles or equilateral and as either obtuse, right or acute. (Lesson 4-2)

18.

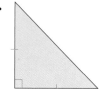

19.

20.

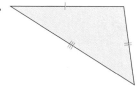

Review and Practice Your Skills

PRACTICE ▰ LESSON 8-3

Trace each polygon. Then draw all of the diagonals from one vertex and find the sum of the angle measures for each polygon.

1.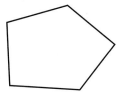

2.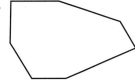

3.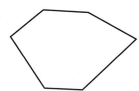

4.

5.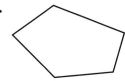

6.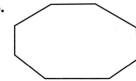

Find the unknown angle measure in each polygon.

7.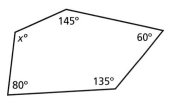

8.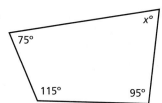

9.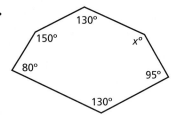

PRACTICE ▰ LESSON 8-4

Use toothpicks for Exercises 10–12.

10. Remove 4 toothpicks to make 5 congruent squares.

11. Move 4 toothpicks to make 4 congruent rhombuses.

12. Move 6 toothpicks to make a six-pointed star.

Cut out square *WXYZ* so that it is 8 in. on each side.
Fold the square in half twice to find the center point *P*.
Use your square for these problems.

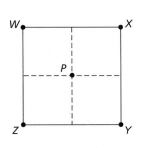

13. Fold *X* onto *P* and *Z* onto *P*. Describe the figure formed and find its area.

14. Fold *W* onto *Y*, *X* onto *P*, and *Z* onto *P*. Describe the figure formed and find its area.

368 | Chapter 8 **Relationships in Geometry**

PRACTICE ◤ LESSON 8-1–LESSON 8-4

Use Figure A in which $\overrightarrow{BY} \parallel \overrightarrow{KG}$. Find each measure. (Lesson 8-1)

15. $m\angle 1$ **16.** $m\angle 2$ **17.** $m\angle 3$ **18.** $m\angle 4$

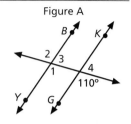

Use a compass and straightedge for these exercises. (Lesson 8-2)

19. Construct $\angle ABC$ and then construct its bisector.

20. Draw line *EF* at least two inches long. Mark point *P* above the line. Construct a perpendicular to the line from point *P*.

Figure B

Use Figure B for these exercises. (Lesson 8-3)

21. Trace the polygon. Draw all of the diagonals from one vertex.

22. Find the sum of the angle measures for this polygon.

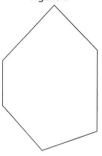

Use toothpicks to solve these problems. (Lesson 8-4)

23. Remove 6 toothpicks to make 3 squares of different sizes.

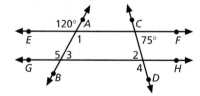

24. Move 2 toothpicks to make 6 equilateral triangles.

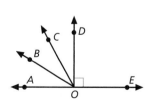

Mid-Chapter Quiz

Use the figure in which $\overleftrightarrow{EF} \parallel \overleftrightarrow{GH}$. Find each angle measure. (Lesson 8-1)

1. $\angle 1$ **2.** $\angle 2$

3. $\angle 3$ **4.** $\angle 4$

5. $\angle 5$

Use the figure to identify each angle. (Lesson 8-1)

6. adjacent complementary angles

7. adjacent supplementary angles

8. adjacent angles that are neither supplementary nor complementary

9. adjacent angles, one of which is acute and one of which is obtuse

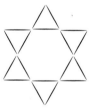

10. Find the number of triangles formed by drawing diagonals from one vertex, the sum of the angle measures, and the total number of diagonals of an 18-sided figure. (Lesson 8-3)

8-5 Translations in the Coordinate Plane

Goals ■ Identify and draw translations.

Applications Interior design, Animation, Health

Obtain grid paper, a straightedge and a small cube, such as a green Algeblock.

1. Draw a vertical line down the middle of your paper.

2. Place the block on either side of the line so that its sides line up with the lines of the grid paper. Trace it.

3. Slide the block horizontally along the grid lines until it passes over the vertical line. Keep the block lined up with its original position.

4. Trace the block's new position, and draw an arrow from the first image to the second.

5. How many units did you slide the block across the paper? How did the block change as you slid it? How did it stay the same?

◣ BUILD UNDERSTANDING

A **translation**, or slide, of a figure produces a new figure that is exactly like the original. As a figure is translated, you imagine all its points sliding along a plane the same distance and in the same direction.

The sides and angles of the new figure are equal in measure to the sides and angles of the original, and each side of the new figure is parallel to the corresponding side of the original.

A move such as a translation is called a **transformation** of the figure. The new figure is called the **image** of the original, and the original is called the **preimage** of the new figure.

Reading Math

An image is said to be graphed "under" a given translation of the preimage.

This does not mean that the image is actually located under the preimage.

Example 1

Graph the image of the point $C(-5, 3)$ under a translation 4 units right and 3 units down.

Solution

Add 4 to the x-coordinate and subtract 3 from the y-coordinate.

$$C(-5, 3) \rightarrow C'(-5 + 4, 3 - 3) \rightarrow C'(-1, 0)$$

C', which is read "C prime," is the translated image of C.

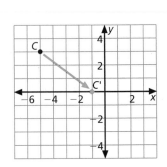

Example 2

Graph the image of △*ABC* with vertices *A* (−3, 3), *B* (−5, −1) and *C* (−2, 0) under a translation 5 units right and 4 units up.

Problem Solving Tip

The translation image can be traced directly from the preimage once the coordinates of one vertex have been found.

Solution

Add 5 to the *x*-coordinate of each vertex.
Add 4 to the *y*-coordinate of each vertex.

$$A (−3, 3) \rightarrow A' (−3 + 5, 3 + 4)$$
$$\rightarrow A' (2, 7)$$

$$B (−5, −1) \rightarrow B' (−5 + 5, −1 + 4)$$
$$\rightarrow B' (0, 3)$$

$$C (−2, 0) \rightarrow C' (−2 + 5, 0 + 4)$$
$$\rightarrow C' (3, 4)$$

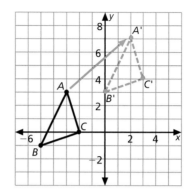

Example 3

INTERIOR DESIGN Annette, her brother Jon and their parents are preparing a nursery for a new arrival. The pattern of the wallpaper they have chosen is shown. The top row of the sheet they have just hung is shown below.

Top of sheet hung

How should the next sheet be positioned to the right of this sheet?

Solution

Align the left edge of the next sheet along the right edge of the sheet just hung. Translate the new sheet upward until the right part of the butterfly matches its left half on the sheet just hung.

◥ TRY THESE EXERCISES

1. Graph the image of the point *M* (3, −3) under a translation 5 units left and 6 units down.

2. Graph the image of △*DEF* with vertices *D* (−4, 6), *E* (1, 7) and *F* (1, 2) under a translation 0 units right or left and 3 units down.

3. Graph the image of △*XYZ* with vertices *X* (−2, 4), *Y* (5, 6) and *Z* (1, −3) under a translation 3 units right and 2 units up.

4. Graph the image of quadrilateral *MNOP* with vertices *M* (0, 5), *N* (4, 7), *O* (5, 1) and *P* (0, −2) under a translation 4 units left and 1 unit down.

Copy each preimage onto grid paper. Then graph each image under a translation 6 units left and 3 units up.

5.   6. 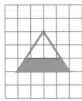 7.  8.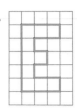

PRACTICE EXERCISES • For Extra Practice, see page 575.

For Extra Practice, see page 575.

9. Graph the image of the point $Q(0, 4)$ under a translation 4 units left and 1 unit down.

10. Graph the image of the point $P(-2, -5)$ under a translation 3 units right and 4 units up.

11. Graph the image of $\triangle ABC$ with vertices $A(-3, -4)$, $B(-1, 6)$ and $C(0, -2)$ under a translation 3 units left and 2 units down.

12. Graph the image of $\triangle XYZ$ with vertices $X(4, 2)$, $Y(2, -6)$ and $Z(0, -2)$ under a translation 5 units right and 3 units up.

13. On a coordinate plane, graph trapezoid $DEFG$ with vertices $D(-4, 3)$, $E(6, 3)$, $F(4, -3)$ and $G(-2, -3)$. Then graph its image under a translation 4 units left and 0 units up or down.

Copy each set of figures on a coordinate plane. Then graph the image of each figure under the given translation.

14. 5 units right, 3 units up

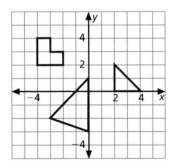

15. 1 unit left, 4 units down

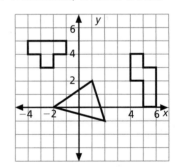

16. Which of the lettered figures are translations of the shaded figure? Describe each translation.

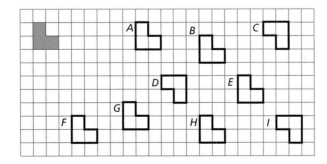

Determine the direction and number of units of the translation for each image.

17.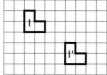

18.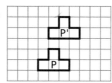

19.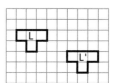

20. ANIMATION In sequencing pictures to illustrate motion, animators often use translations of images from frame to frame. Suppose an animator wishes to use a translation of 1 unit left and 1 unit up on this image of an airplane. Draw the next two frames.

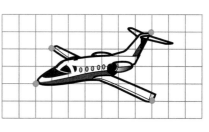

Translations were used to make each design. Describe how each was made.

21.

22.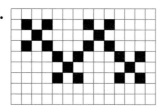

23.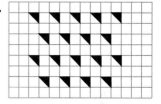

24. WRITING MATH Describe a translation by which △ABC is transformed into △A′B′C′. Can you describe the movement in more than one way? If so, give a second description.

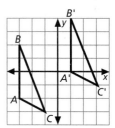

EXTENDED PRACTICE EXERCISES

25. DATA FILE Refer to the data on the predicted maximum heart rate by age on page 531. Plot the data on grid paper as a line graph with age on the *x*-axis and heart rate on the *y*-axis. Suppose experts predict that in five years the maximum heart rate will be 5 beats greater on average for all ages. Translate your line graph to reflect this trend.

26. HEALTH The diagram represents a typical heartbeat over one cycle of pumping. Copy the diagram, and then sketch the next heartbeat.

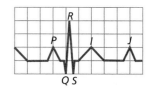

27. CHAPTER INVESTIGATION Decide how you want your animation to begin and end, so you can plan how much your subject needs to change from frame to frame. Draw your first frame. Keep it simple because you will need to make several drawings for your flip book.

MIXED REVIEW EXERCISES

Make an orthogonal drawing of each stack of cubes, showing top, front and side views.
(Lesson 4-6)

28.

29.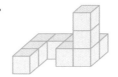

30.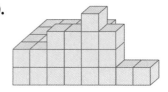

8-6 Reflections and Line Symmetry

Goals ■ Identify and draw reflections.
■ Identify and use lines of symmetry.

Applications Physics, Landscaping, Animation

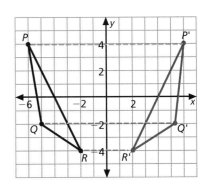

Use patty paper and a pencil.

1. Draw a simple figure on the left side of the paper.

2. Draw a line down the middle of the paper. Make sure that the line does not cross the figure you have drawn. Fold the paper along this line.

3. Trace the figure from Step 1 on the folded side of the paper. Then unfold. What do you observe?

4. Describe the two images on the paper.

◥ BUILD UNDERSTANDING

Another type of transformation is a reflection. A **reflection** is a transformation in which a figure is flipped, or reflected, over a *line of reflection.* When you think of a reflection, you probably think of a mirror.

In the diagram, point P' is the image of point P under a reflection across $\overleftrightarrow{EF}$. The two points, P and P', are the same distance from $\overleftrightarrow{EF}$. If line PP' were drawn, it would be perpendicular to $\overleftrightarrow{EF}$.

The x-axis and y-axis can be used as lines of reflection for figures drawn on a coordinate plane.

Example 1

Graph the image of $\triangle PQR$ with vertices $P(-6, 4)$, $Q(-5, -2)$ and $R(-2, -4)$ under a reflection across the y-axis.

Solution

The y-coordinates will remain the same, but the x-coordinates are all opposite. Multiply the x-coordinate of each vertex by -1.

$$P(-6, 4) \rightarrow P'(-6 \cdot -1, 4) \rightarrow P'(6, 4)$$

$$Q(-5, -2) \rightarrow Q'(-5 \cdot -1, -2) \rightarrow Q'(5, -2)$$

$$R(-2, -4) \rightarrow R'(-2 \cdot -1, -4) \rightarrow R'(2, -4)$$

The reflected image of $\triangle PQR$ across the y-axis is $\triangle P'Q'R'$.

Example 2

The vertices of △*KLM* are *K* (−4, 3), *L* (6, 5) and *M* (0, −2). Graph the image of △*KLM* under a reflection across the *x*-axis.

Solution

The *x*-coordinates will remain the same, but the *y*-coordinates are all opposite. Multiply the *y*-coordinate of each vertex by −1.

$$K(-4, 3) \rightarrow K'(-4, 3 \cdot -1) \rightarrow K'(-4, -3)$$

$$L(6, 5) \rightarrow L'(6, 5 \cdot -1) \rightarrow L'(6, -5)$$

$$M(0, -2) \rightarrow M'(0, -2 \cdot -1) \rightarrow M'(0, 2)$$

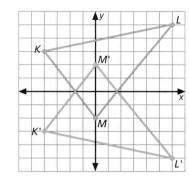

The reflected image of △*KLM* across the *x*-axis is △*K'L'M'*.

A figure has **line symmetry** if there is a line that can be drawn through it that divides the figure into two matching parts.

In the figure shown, if the isosceles triangle is folded along line *PQ*, one side fits exactly over the other. So the triangle has line symmetry, and line *PQ* is called a *line of symmetry*.

Math: Who, Where, When

Many properties of reflections are found in the Japanese art of paper folding called *origami*.

Books with diagrams and instructions for folding paper decorations were published as long ago as the early 1700s.

Example 3

Trace each figure and draw all the lines of symmetry.

a.

b.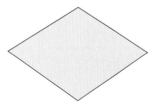

c.

Solution

a.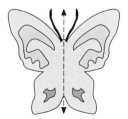

1 line of symmetry

b.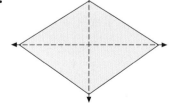

2 lines of symmetry

c.

1 line of symmetry

Give the coordinates of the image of each point under a reflection across the given axis.

1. $(4, -5)$; x-axis
2. $(-3, 8)$; y-axis
3. $(-2, 0)$; y-axis
4. $(7, 9)$; x-axis

5. Graph the image of $\triangle GHI$ with vertices $G(1, -3)$, $H(3, -2)$ and $I(5, -4)$ under a reflection across the y-axis.

6. Graph the image of $\triangle JKL$ with vertices $J(-5, 1)$, $K(-3, 6)$ and $L(-1, 2)$ under a reflection across the x-axis.

Trace each figure and draw all the lines of symmetry.

7.

8.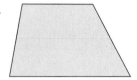

9.

Give the coordinates of the image of each point under a reflection across the given axis.

10. $(4, 1)$; x-axis
11. $(-2, 5)$; x-axis
12. $(-4, 1)$; y-axis
13. $(7, -2)$; y-axis

Graph the image of each figure under a reflection across the y-axis.

14.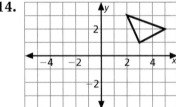

15.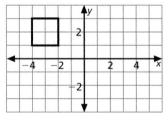

16. Graph the image of the figure in Exercise 14 under a reflection across the x-axis.

17. Graph the image of the figure in Exercise 15 under a reflection across the x-axis.

Tell whether the dashed line is a line of symmetry.

18.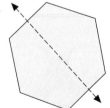

19.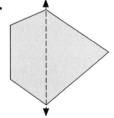

20.

 21. WRITING MATH Write a short explanation of how to draw the reflection of $\triangle ABC$ across a vertical line m. Include a diagram with your explanation.

22. PHYSICS The diagram shows the path of a soccer ball when it is kicked into the air. Copy the diagram and draw a line of symmetry.

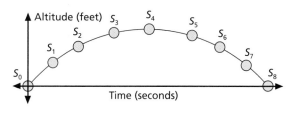

23. The capital letter H is a reflection of itself across two different lines of symmetry. What other capital letters have this property?

Draw a triangle with the given number of lines of symmetry.

24. 1 line **25.** 3 lines **26.** 0 lines

27. YOU MAKE THE CALL Erika claims that the reflection of $(-2, 0)$ across the x-axis is $(2, 0)$. Pico thinks that the reflection is actually the same point $(-2, 0)$. Who is correct and why?

28. ANIMATION Give two examples of when animators use reflections in cartoons and movies.

29. MODELING Draw a simple figure on a piece of paper. Use a reflective device to draw its reflection. Where is the line of reflection?

■ EXTENDED PRACTICE EXERCISES

Copy each diagram onto grid paper. Then sketch the image of each set of squares under a reflection across line *l*.

30.

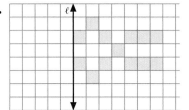

31.

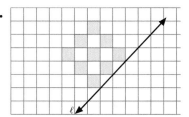

32. DATA FILE Refer to the data on the ten tallest buildings in the world on page 521. Find and copy pictures of three of the buildings, and draw any lines of symmetry that they have.

33. LANDSCAPING Mia is designing the layout of a rectangular garden and wants it to appear symmetrical. Make a sketch of a possible designs using reflection and symmetry to position the different regions. Include the types and colors of flowers and plants in each region.

■ MIXED REVIEW EXERCISES

Determine if the ordered pair is a solution. (Lesson 7-8)

34. $(6, -8)$
$y = -\dfrac{2}{3}x + 4$

35. $(-2, 8)$
$y = |-x| + 6$

36. $(5, 10)$
$y = 5x + (-15)$

37. $(5, -6.5)$
$y = -6 + 0.1x$

38. Graph the equation $y = -\dfrac{3}{4}x - 12$ on a coordinate plane. (Lessons 7-2 and 7-8)

For Exercises 39–42, use the graph in Exercise 38.

39. Find the slope. (Lesson 7-5)

40. Find the y-intercept. (Lesson 7-4)

41. Is this relation a function? (Lesson 7-3)

42. Is $\left(-6, -7\dfrac{1}{2}\right)$ on the graph? (Lesson 7-8)

Review and Practice Your Skills

PRACTICE ◣ LESSON 8-5

Use grid paper.

1. Graph the image of point $A(2, -3)$ under a translation 2 units left and 4 units down.

2. Graph the image of point $B(-4, 1)$ under a translation 1 unit left and 3 units up.

3. Graph the image of point $C(-2, -2)$ under a translation 5 units right and 1 unit down.

4. Graph the image of point $D(0, -1)$ under a translation 4 units right and 4 units up.

5. Graph the image of $\triangle ABC$ with vertices $A(-6, 2)$, $B(-5, 5)$ and $C(-2, -3)$ under a translation 4 units right and 2 units down.

6. Graph the image of $\triangle PQR$ with vertices $P(-5, 2)$, $Q(0, 5)$ and $R(1, 1)$ under a translation 3 units left and 7 units up.

7. Graph the image of $\triangle XYZ$ with vertices $X(-4, 1)$, $Y(-4, -6)$ and $Z(2, -5)$ under a translation 2 units right and 3 units up.

8. Graph the image of $\triangle DEF$ with vertices $D(-3, -2)$, $E(-2, 5)$ and $F(4, 2)$ under a translation 1 unit left and 4 units down.

PRACTICE ◣ LESSON 8-6

Give the coordinates of the image of each point under a reflection across the given axis.

9. $(-1, 3)$; x-axis
10. $(-3, -3)$; y-axis
11. $(5, -2)$; x-axis
12. $(0, 3)$; x-axis

13. $(2, 0)$; x-axis
14. $(-3, 1)$; y-axis
15. $(2, -6)$; x-axis
16. $(-5, -5)$; y-axis

17. $(4, -1)$; y-axis
18. $(3, 4)$; x-axis
19. $(0, -2)$; y-axis
20. $(-4, 0)$; y-axis

Use graph paper for these exercises.

21. Graph $\triangle HJK$ with vertices $H(2, 3)$, $J(6, 1)$ and $K(4, 6)$. Then graph its image under a reflection across the y-axis.

22. Graph $\triangle MPR$ with vertices $M(-3, -2)$, $P(3, 2)$ and $R(4, -4)$. Then graph its image under a reflection across the x-axis.

23. Graph square $ABCD$ with vertices $A(-4, -1)$, $B(-4, 4)$, $C(1, 4)$ and $D(1, -1)$. Then graph its image under a reflection across the y-axis.

24. Graph square $MNOP$ with vertices $M(-5, -2)$, $N(-2, -5)$, $O(1, -2)$ and $P(-2, 1)$. Then graph its image under a reflection across the x-axis.

25. Graph rectangle $EFGH$ with vertices $E(2, 4)$, $F(2, -3)$, $G(5, -3)$ and $H(5, 4)$. Then graph its image under a reflection across the y-axis.

26. Graph rectangle $WXYZ$ with vertices $W(-2, 1)$, $X(2, 5)$, $Y(4, 3)$ and $Z(0, -1)$. Then graph its image under a reflection across the x-axis.

Tell whether the dashed line is a line of symmetry.

27.

28.

29.

• •

PRACTICE ◣ **LESSON 8-1–LESSON 8-6**

Use Figure A. $\overleftrightarrow{PW} \parallel \overleftrightarrow{FN}$. **Find each measure.** (Lesson 8-1)

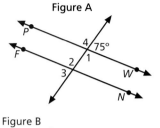

Figure A

30. $m\angle 1$ **31.** $m\angle 2$ **32.** $m\angle 3$ **33.** $m\angle 4$

34. Construct $\angle RST$ so that it measures 140°. Bisect $\angle RST$. (Lesson 8-2)

35. What is the sum of the angle measures of a pentagon? (Lesson 8-3)

36. Use Figure B. Side $\overline{AB}$ is twice the length of side $\overline{BC}$. Fold C onto the midpoint of side $\overline{AB}$. Fold D onto the midpoint of side $\overline{AB}$. Describe the figure that is formed. (Lesson 8-4)

Figure B

For Exercises 37 and 38, use grid paper. (Lessons 8-5 and 8-6)

37. Graph the image of $\triangle TRS$ with vertices $T(-3, 2)$, $R(-1, 5)$ and $S(5, -2)$ under a translation 2 units left and 5 units down.

38. Graph $\triangle DHX$ with vertices $D(-2, -5)$, $H(4, 4)$ and $X(6, -4)$. Then graph its image under a reflection across the y-axis.

Math*Works* **Career – Video Editor**

Workplace Knowhow

A video editor influences the final look of videos, cartoons, documentaries and other films. In animated films, the video editor may coordinate the writing and creation of the images, as well as check the animation and soundtrack for errors or inconsistencies. Errors can occur for many reasons. Sometimes the images do not match the soundtrack. Several artists working on the same feature may create different expressions, clothing or backgrounds. A video editor must understand mathematics, acoustics and film.

Find the errors in each animation frame.

1.

2.

3.

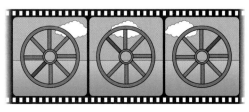

8-7 Rotations and Tessellations

Goals
- Identify and draw rotations and tessellations.
- Identify rotational symmetry.

Applications Art, Animation, Chemistry

TIME The hands of a clock rotate around the point at which they are attached in the center of the clock face.

1. What fractional part of a full turn does the minute hand make in 30 min?

2. When the minute hand has moved 90° forward, or clockwise, from 3:25, to what number does it point?

3. What degree of rotation of the minute hand will set a clock back by 15 min?

BUILD UNDERSTANDING

You have already learned about two kinds of transformations: translations and reflections. A third type of transformation is a rotation.

A **rotation** is a transformation in which a figure is *turned*, or *rotated*, about a point. The movement of the hands of a clock, the turning of the wheels of a bicycle and moving windshield wipers on a car are all examples of a rotation about a point.

To describe a rotation, you need three pieces of information.

1. the point about which the figure is rotated, called the *center of rotation*, or *turn center*

2. the amount of turn expressed as a fractional part of a whole turn, or in degrees called the *angle of rotation*

3. the direction of rotation—either *clockwise* or *counterclockwise*

Example 1

Draw the rotation image of the flag when it is turned 90° counterclockwise about a turn center, *T*.

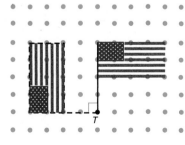

Solution

Copy the flag onto grid or dot paper, labeling point *T*. Then trace the flag and point *T* onto a sheet of paper.

Hold your pencil point on the paper at point *T*. Turn the paper one-quarter turn, or 90°.

Remove the paper and copy the image onto the grid paper.

Example 2

Find the image of △PQR with vertices P(3, 3), Q(2, 1) and R(5, 1) after a rotation of 180° clockwise about the origin.

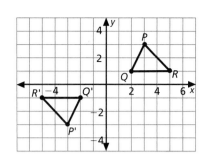

Solution

Multiply both the *x*-coordinate and the *y*-coordinate of each vertex by −1.

$$P(3, 3) \rightarrow P'(-1 \cdot 3, -1 \cdot 3) \rightarrow P'(-3, -3)$$

$$Q(2, 1) \rightarrow Q'(-1 \cdot 2, -1 \cdot 1) \rightarrow Q'(-2, -1)$$

$$R(5, 1) \rightarrow R'(-1 \cdot 5, -1 \cdot 1) \rightarrow R'(-5, -1)$$

You learned about line symmetry when studying reflections. Rotations also have a form of symmetry. In the figure shown, if you placed the point of a pencil at point *A* and rotated the parallelogram 180° clockwise or counterclockwise, it would fit exactly over its original position.

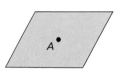

The parallelogram is said to have **rotational symmetry**. Its *order of rotational symmetry* is 2 since the parallelogram would fit over its original position 2 times in the process of a complete turn.

Problem Solving Tip

There are 360° at the center of a circle. That is why the measure of a complete turn of a rotation is given as 360°.

Example 3

Give the order of rotational symmetry for each figure.

a.

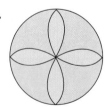

b.

c.

Solution

a. The figure fits over its original position 4 times during a complete turn, so the order of rotational symmetry is 4.

b. The figure fits over its original position 2 times during a complete turn, so the order of rotational symmetry is 2.

c. The figure fits over its original position 6 times during a complete turn, so the order of rotational symmetry is 6.

A **tessellation** is a pattern in which identical copies of one figure or a few figures fill a plane so that there are no gaps or overlaps.

In a *regular tessellation*, each shape is a regular polygon, and all the shapes are congruent. The figure shown is a regular tessellation composed of equilateral triangles.

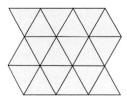

Example 4

ART An artist is designing a floor pattern using hexagons, squares and triangles so there are no gaps or overlaps between the shapes. Construct the tessellation.

Solution

Step 1 Draw a regular hexagon that will be the center of the pattern.

Step 2 On each side of the hexagon, place a square.

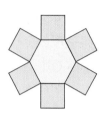

Step 3 Place an equilateral triangle between each pair of squares as shown.

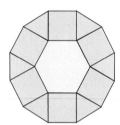

Step 4 Repeat the pattern by placing a hexagon next to each square.

◥ TRY THESE EXERCISES

1. Copy the figure onto dot or grid paper. Then draw the image of the figure when it is turned 270° clockwise about a turn center, *T*.

2. Graph the image of △*KLM* with vertices *K*(−3, 1), *L*(1, 7) and *M*(1, 3) after a rotation of 180° clockwise about the origin.

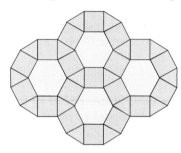

Give the order of rotational symmetry for each figure.

3.

4.

5.

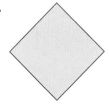

Can a regular tessellation be made from each polygon? If so, construct the tessellation.

6. equilateral triangle

7. square

8. regular octagon

◥ PRACTICE EXERCISES • For Extra Practice, see page 576.

9. Copy the figure onto dot or grid paper. Then draw the image of the figure when it is turned 90° clockwise about a turn center, *T*.

10. Graph the image of △*QRS* with vertices *Q*(2, −5), *R*(4, −1) and *S*(6, −4) after a rotation of 180° clockwise about the origin.

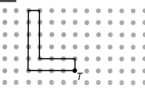

Give the order of rotational symmetry for each figure.

11.

12.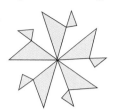

13.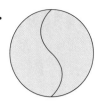

Can a tessellation be made from each polygon? If so, construct the tessellation.

14. regular pentagon   **15.** regular hexagon   **16.** regular heptagon

17. ANIMATION A computer animator is drawing a series of frames to illustrate the motion of a spinning pinwheel. Draw the next two frames if he is using a rotation of 10° clockwise.

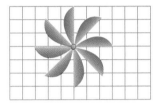

18. CHEMISTRY Igor and Madeline time a chemical reaction for a chemistry lab. If they start with the second hand on 60, through how many degrees has the tip of the second hand traveled when the reaction reaches 24 seconds?

19. Construct a tessellation consisting of irregular shapes or using different combinations of polygons.

EXTENDED PRACTICE EXERCISES

20. WRITING MATH Compare the three different types of transformations discussed in this chapter in a short paragraph.

CRITICAL THINKING Use grid paper to construct a figure having rotational symmetry of the order indicated.

21. order 3   **22.** order 5   **23.** order 6

24. CHAPTER INVESTIGATION Draw the pages of your flip book. You can plan how much each picture needs to change by dividing the total distance of movement by the number of frames. Make between 15 and 20 drawings. Assemble the book by stapling it across the top.

MIXED REVIEW EXERCISES

Draw each of the following in one-point perspective. (Lesson 4-6)

25. cone   **26.** triangular prism   **27.** rectangular prism

Identify each figure. (Lesson 4-3)

28.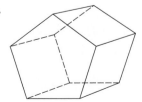

29.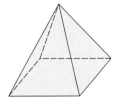

30.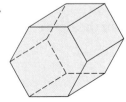

Chapter 8 Review

VOCABULARY ◤

1. A(n) __?__ of a polygon is a segment that joins two vertices and is not a side.

2. The __?__ of a line segment is the point that separates it into two line segments of equal length.

3. A move such as a translation is called a(n) __?__ of the figure.

4. A figure has __?__ if a line drawn through it divides the figure into two parts that are mirror images of each other.

5. A(n) __?__ is a line that intersects two or more lines in a plane at different points.

6. Two angles are called __?__ if the sum of their measures is 180°.

7. A(n) __?__ is a transformation in which a figure is turned about a point.

8. Opposite angles formed by intersecting lines are __?__.

9. Before a figure is transformed, it is called the __?__.

10. The __?__ is a ray that separates an angle into two congruent adjacent angles.

a. angle bisector

b. complementary angles

c. diagonal

d. image

e. line symmetry

f. midpoint

g. preimage

h. rotation

i. supplementary angles

j. transformation

k. transversal

l. vertical angles

LESSON 8-1 ◤ Angles and Transversals, p. 352

▶ Two angles are **complementary** if the sum of their measures is 90°.

▶ Two angles are **supplementary** if the sum of their measures is 180°.

▶ **Adjacent angles** share a vertex and a side, but have no interior points in common. When two lines intersect in a plane, the angles not adjacent to each other are **vertical angles**.

▶ When a **transversal** intersects a pair of parallel lines, **corresponding angles** have the same measure.

Use the figure at the right to name a pair of the following.

11. complementary angles   12. supplementary angles

13. vertical angles   14. adjacent angles

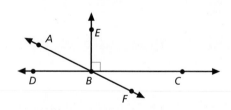

Find each measure if $m\angle ABE = 42°$.

15. $m\angle ABD$ 16. $m\angle CBF$ 17. $m\angle EBF$

In the figure, $\overleftrightarrow{AD} \parallel \overleftrightarrow{CB}$ and $m\angle OED = 150°$. Find each measure.

18. $m\angle 1$ 19. $m\angle 3$

20. $m\angle 2$ 21. $m\angle 5$

22. the complement of $\angle 1$

23. the supplement of $\angle 4$

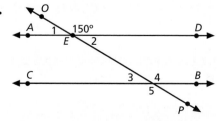

LESSON 8-2 ▨ Beginning Constructions, p. 356

▶ A **perpendicular bisector** of a line segment is a line, ray or line segment that is perpendicular to a line segment at its **midpoint**.

▶ An **angle bisector** is a ray that separates an angle into two congruent adjacent angles.

24. Using a protractor, construct ∠*ABC* whose measure is 160°. Then bisect ∠*ABC*.

25. Trace $\overline{CD}$. Then construct its perpendicular bisector.

26. Trace $\overleftrightarrow{MN}$. Then construct a line perpendicular to $\overleftrightarrow{MN}$ through point *P*.

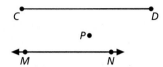

LESSON 8-3 ▨ Diagonals and Angles of Polygons, p. 362

▶ The number of **diagonals** in a polygon with *n* sides is given by the formula $\dfrac{n(n-3)}{2}$.

27. Find the number of diagonals in a hexagon.

28. Find the unknown angle measure in the figure.

29. What is the sum of the angle measures of a dodecahedron (12 sides)?

LESSON 8-4 ▨ Problem Solving Skills: Modeling Problems, p. 366

▶ A strategy for understanding some problems and arriving at its solution is to **act it out**.

30. Arrange three one-inch squares so that any two adjoining sides align exactly. Find the perimeter of each figure that can be formed this way.

31. Julie is taking a picture of the Spanish Club's five officers. The club president will always stand on the left and the vice-president will always stand on the right. How many different ways can she arrange the officers for the picture?

32. Suppose you run 10 yd forward and then 5 yd backward. How many sets will you run to reach the end of a 100-yd field?

LESSON 8-5 ▨ Translations in the Coordinate Plane, p. 370

▶ A **translation**, or slide, of a figure is a **transformation** that produces a new figure exactly like the original. The new figure is the **image**, and the original is the **preimage**.

33. Graph the image of point *M*(6, −2) under a translation 2 units left and 3 units up.

34. Graph the image of △*ABC* with vertices *A*(−2, −3), *B*(0, 5), and *C*(−1, −1) under a translation 2 units right and 3 units down.

35. Describe the translation of △*DEF* to △*D'E'F'* in the figure at the right.

36. Find the coordinates of △*DEF* after a translation 3 units to the right and 1 unit down.

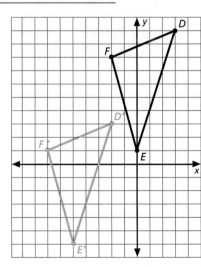

LESSON 8-6 ◼ Reflections and Line Symmetry, p. 374

▶ A **reflection** is a transformation in which a figure is flipped, or reflected, over a *line of reflection.*

▶ A figure has **line symmetry** if a line drawn through it divides the figure into two matching parts that are reflections of one another across the line.

37. Give the coordinates of the image of $(3, -2)$ under a reflection across the *x*-axis.

38. Give the coordinates of the image of $(3, -2)$ under a reflection across the *y*-axis.

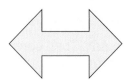

39. Trace the figure and draw all the lines of symmetry.

Find the number of lines of symmetry for the following figures.

40. rhombus

41. regular octagon

42. parallelogram

43. square

LESSON 8-7 ◼ Rotations and Tessellations, p. 380

▶ A **rotation** is a transformation in which a figure is turned, or rotated, about a point. To describe a rotation, use the *center of rotation, angle of rotation* and the *direction of rotation.*

▶ A figure has **rotational symmetry** when the image of the figure coincides with the figure after a rotation of less than 360°.

▶ A **tessellation** is a pattern in which identical copies of a figure fill a plane so that there are no gaps or overlaps.

For Exercises 44–46, use Figures A and B.

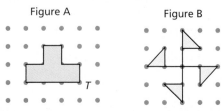

Figure A   Figure B

44. Copy Figure A onto dot or grid paper. Then draw the image of the figure when it is turned 270° clockwise about a turn center, *T*.

45. Give the order of rotational symmetry for Figure B.

46. Can a tessellation be made with Figure B? If so, construct the tessellation.

47. Find the image of $\triangle RST$ with vertices $R(5, -3)$, $S(1, 0)$, and $T(3, -4)$ after a rotation of 180° clockwise about the origin.

CHAPTER INVESTIGATION

EXTENSION Exchange your flipbook with a classmate. Discuss any geometric properties that have been used in the design, such as geometric shapes and figures, parallel and perpendicular lines and others.

Chapter 8 Assessment

Use the figure to name each pair of angles.

1. complementary   2. supplementary   3. vertical   4. adjacent

5. If $m\angle ABD = 25°$ find $m\angle DBC$.

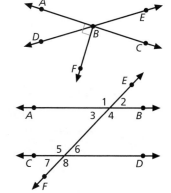

In the figure $\overleftrightarrow{AB} \parallel \overleftrightarrow{CD}$, and $m\angle 5 = 113°$. Find each measure.

6. $m\angle 1$ 　　　　　　 7. $m\angle 2$ 　　　　 8. $m\angle 3$ 　 9. $m\angle 4$

10. a complement of $\angle 6$

11. Using a protractor, construct $\angle ABC$ whose measure is $100°$.
Then bisect $\angle ABC$.

12. Trace $\overline{JK}$. Then construct its perpendicular bisector.

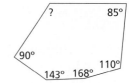

13. Using only quarters, dimes and nickels, in how many ways can you give
someone 70¢ change with 4 coins?

Refer to the polygon.

14. Find the total number of diagonals.

15. Find the sum of the angle measures.

16. How many diagonals can be drawn from one vertex?

17. Find the unknown angle measure.

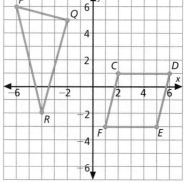

18. Graph the image of parallelogram
CDEF under a translation 2 units to the
left and 3 units up.

19. Graph the image of $\triangle PQR$ under a
reflection across the x-axis.

20. Graph the image of $\triangle PQR$ after a $90°$
turn counterclockwise about the origin.

Refer to the figures.

21. How many lines of symmetry does
the figure on the left have?

22. What is the order of rotational
symmetry for the figure on the right?

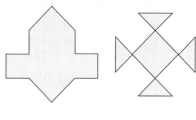

23. Copy the figure onto dot or grid paper. Then draw
the image of the figure when it is turned $180°$
counterclockwise about a turn center, *T*.

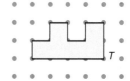

Standardized Test Practice

Part 1 | Multiple Choice

Record your answers on the answer sheet provided by your teacher or on a sheet of paper.

1. According to the graph, the greatest increase in temperature occurred between which two days? (Lesson 1-6)

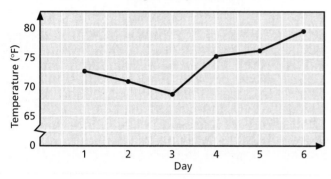

High Temperature

Ⓐ 1 and 2
Ⓑ 3 and 4
Ⓒ 4 and 5
Ⓓ 5 and 6

2. The giant stones of Stonehenge are arranged in a circle 30 m in diameter. What is the circumference of the circle? Use 3.14 for π. (Lesson 2-7)

Ⓐ 47.1 m
Ⓑ 94.2 m
Ⓒ 188.4 m
Ⓓ 706.5 m

3. The perimeter of a square is 64 in. What is the length of one side? (Lesson 3-9)

Ⓐ 4 in.
Ⓑ 8 in.
Ⓒ 16 in.
Ⓓ 32 in.

4. You and your friend spent a total of $15 for lunch. Your friend's lunch cost $3 more than yours did. How much did you spend for lunch? (Lesson 5-4)

Ⓐ $6
Ⓑ $7
Ⓒ $9
Ⓓ $12

5. Which equation represents the graph? (Lesson 5-8)

Ⓐ $x > -3$
Ⓑ $x \leq -3$
Ⓒ $x \geq -3$
Ⓓ $x < -3$

6. The regular price of a ring is $495. It is on sale at a 20% discount. What is the sale price of the ring? (Lesson 6-3)

Ⓐ $99
Ⓑ $374
Ⓒ $390
Ⓓ $396

7. Write an equation of the line in slope-intercept form. (Lesson 7-6)

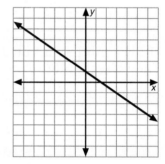

Ⓐ $y = 3x - 1$
Ⓑ $y = \frac{2}{3}x + 1$
Ⓒ $y = -\frac{3}{2}x + 1$
Ⓓ $y = -\frac{2}{3}x + 1$

8. Which angle is complementary to $\angle BCD$? (Lesson 8-1)

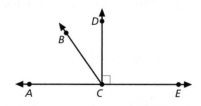

Ⓐ $\angle ACB$
Ⓑ $\angle ECD$
Ⓒ $\angle DCA$
Ⓓ $\angle BCE$

9. What is the sum of the angle measures in a nonagon? (Lesson 8-3)

Ⓐ 900°
Ⓑ 1080°
Ⓒ 1260°
Ⓓ 1440°

10. What set of coordinates shows a reflection of the point with vertices $(-4, 3)$ over the y-axis? (Lesson 8-6)

Ⓐ $(-4, -3)$
Ⓑ $(4, 3)$
Ⓒ $(-3, 4)$
Ⓓ $(-3, -4)$

Preparing for the Standardized Tests
For test-taking strategies and more
practice, see pages 587-604.

Part 2   Short Response/Grid In

**Record your answers on the answer sheet
provided by your teacher or on a sheet of paper.**

11. How many kilograms are in 743 g?
(Lesson 2-2)

12. Identify the number of faces, vertices, and
edges for the figure. (Lesson 4-2)

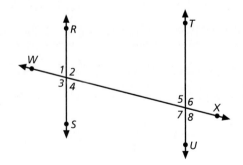

13. 157 is what percent of 2512? (Lesson 6-2)

14. On October 19, 1987, the stock market
opened at 2246.74 points and closed at
1738.42 points. What was the percent of
decrease? Round to the nearest tenth of a
percent. *(Lesson 6-7)*

15. Find the *x*-intercept of $y = -5x - 1$.
(Lesson 7-4)

**For Questions 16–18, use the figure below.
Assume $\overleftrightarrow{RS} \parallel \overleftrightarrow{TU}$. (Lesson 8-1)**

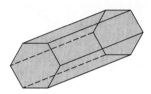

16. Identify the alternate exterior angles and their
relationship.

17. If $m\angle 2 = 112.6°$, find $m\angle 5$.

18. Describe the relationship between
$\angle 3$ and $\angle 4$.

19. Find the unknown angle measure in the
figure. (Lesson 8-3)

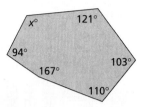

20. Find the new coordinates of $\triangle ABC$ with
vertices $A'(0, -2)$, $B'(-3, 5)$, and $C'(2, 4)$ after
a translation of 3 units to the left and 2 units
up. (Lesson 8-5)

Part 3   Extended Response

**Record your answers on a sheet of paper. Show
your work.**

21. Suppose a roller coaster climbs 208 ft higher
than its starting point making a horizontal
advance of 360 ft. When it comes down, it
makes a horizontal advance of 44 ft.
(Lesson 7-7)

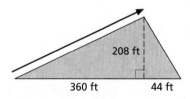

   a. How far will it travel to get to the top of
the ride?

   b. How far will it travel on the downhill
track?

   c. Compare the total horizontal advance,
vertical height, and total track length.

Test-Taking Tip

Questions 10 and 20
For some tests you will be allowed to write in your testing
booklets. Sketching a graph in the margin of your test booklet
can help you work the problem.

Polynomials

THEME: Natural Disasters

Polynomials often combine variable expressions with the properties and dimensions of two- and three-dimensional objects. The study of polynomials is a stepping stone to other areas of mathematics and science, including physics, calculus, and meteorology.

Nature can be nurturing and peaceful. But sometimes nature acts in powerful, unpredictable, and dangerous ways. Similar to the unknown variables in a polynomial, natural disasters have unknown elements and features. Predicting and planning for a natural disaster is an attempt to solve the unknown.

- **Meteorologists** (page 403) record, study, and research the atmosphere's physical characteristics, processes, and the way it effects our environment.

- **Disaster relief coordinators** (page 423) help people prepare for and recover from a natural disaster. They are prepared to travel to the scene of a natural disaster anywhere in the world.

Saffir-Simpson Scale (Hurricanes)

Category	Wind speed	Severity	Storm surge*
1	74-95 mi/h	Weak	4-5 ft
2	96-110 mi/h	Moderate	6-8 ft
3	111-130 mi/h	Strong	9-12 ft
4	131-155 mi/h	Very strong	13-18 ft
5	156 mi/h and greater	Devastating	18 ft and great

* Above normal tides.

Math
Online

mathmatters1.com/chapter_theme

Richter Scale (Earthquakes)

Magnitude	Description
2.5	Generally not felt, but recorded on seismographs*
3.5	Felt by many people
4.5	Some local damage may occur
6.0	A destructive earthquake
7.0	A major earthquake
8.0 and above	Great earthquakes

* A tool used to measure and record vibrations within the earth and of the ground.

Each single-integer increase represents 10 times more ground movement and 30 times more energy released.

Fujita Scale (Tornadoes/Wind Storms)

Rank	Wind speed	Damage	Strength
F-0	Up to 72 mi/h	Light	Weak
F-1	73-112 mi/h	Moderate	Weak
F-2	113-157 mi/h	Considerable	Strong
F-3	158-206 mi/h	Severe	Strong
F-4	207-260 mi/h	Devastating	Violent
F-5	261 mi/h and greater	Incredible	Violent

Data Activity: Earthquakes, Hurricanes, and Tornadoes

Use the three tables for Questions 1–3.

1. The difference between numbers on the Richter scale can be represented by 10^x and 30^x, where x represents the difference. Comparing a 3.0 earthquake with a 1.0 earthquake, a 3.0 has $10^{3.0-1.0}$, or 100 times more ground movement and $30^{3.0-1.0}$, or 900 times more energy released than a 1.0 earthquake. How many times more ground movement and energy does a 6.0 earthquake have than a 3.0 earthquake?

2. A certain location has winds of 135 mi/h. If this were a hurricane, what is its category? If this were a tornado, what is its rank?

3. A beach resort has an 8-ft wall for hurricane protection. If a hurricane with 128 mi/h winds hits this resort, will the wall protect it? Explain.

CHAPTER INVESTIGATION

Cloud seeding attempts to weaken the force of storms such as hurricanes, tropical storms, hailstorms, and lightning strikes. This process places millions of tiny chemical drops onto a storm cloud. This causes the liquid in the cloud to change to ice crystals.

Working Together

The eye of a tropical storm has a radius of x miles. The length of the cloud formation extends about 20 mi on both sides of the eye. The cloud's width is 8 times the radius. Use the ratio of 0.17 oz of particles to 0.62 mi² of cloud surface to find how many ounces of chemicals are needed to *seed* this storm. Use the Chapter Investigation icons to find the solution.

The skills on these two pages are ones you have already learned. Use the examples to refresh your memory and complete the exercises. For additional practice on these and more prerequisite skills, see pages 536–544.

DIVISIBILITY RULES

A factor is a number that divides another number evenly. Finding factors can be made easier when you know the rules of divisibility.

Write the word or number(s) that complete(s) each sentence.

1. Numbers that have 5 as a factor end in 0 or _____.

2. The sum of the digits of a multiple of 9 is always a multiple of _____.

3. If half of a given number is divisible by 2, the given number is divisible by _____.

4. Numbers that have 6 as a factor can be divided evenly by _____, _____, and _____.

5. Multiples of _____ have digits whose sum is always a multiple of 3.

6. _____ numbers are always divisible by 2.

LIKE TERMS

When simplifying expressions, like terms can be combined through addition and subtraction. Like terms are real numbers or expressions that have the same variable.

Examples Like terms

$17, 3\frac{1}{4}, 4.65, -7\frac{2}{3}$

$4x, 21x, -42x$

$6y^3, \frac{3}{4}y^3, -y^3$

Unlike terms

$42, -42x$

$-14y, 35x, 24z$

$5t, 7t^2, 41st$

Name the like terms in each set.

7. $35, 4\frac{1}{8}, 9t, -14, 4y$

8. $34k, 22m, -16k, \frac{2}{3}k, 9j$

9. $4x, -17x^2, 24x^2, 13x^3, \frac{5}{8}x^2$

10. $-16y, 24.75, 24k, -6\frac{3}{4}, 4x$

Simplify each expression by combining like terms.

11. $4x + 26 - 17x + 4y$

12. $38\frac{3}{4} + 17t - 30.4 - r$

13. $14.5 - 16k + -24k^2 + 28k$

14. $28s + (-24st) - 34t - 17st - 7st$

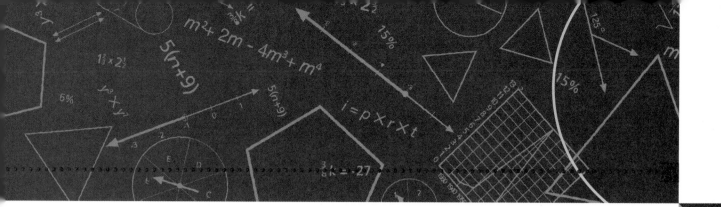

GCF — GREATEST COMMON FACTOR

When you compare the factors of one or more numbers, the greatest common factor (GCF) is the greatest factor that is a factor of every number.

Examples   The factors of 36 are 1, 2, 3, 4, 6, 9, 12, 18 and 36.
The factors of 64 are 1, 2, 4, 8, 16, 32 and 64.
The GCF of 36 and 64 is 4.

Find the factors of each number.

15. 28   **16.** 32   **17.** 48   **18.** 49

19. 225   **20.** 128   **21.** 99   **22.** 54

Find the GCF of each set of numbers.

23. 28 and 32   **24.** 48 and 128   **25.** 49 and 54

26. 225 and 100   **27.** 64, 32, and 128   **28.** 99, 54, and 36

DISTRIBUTIVE PROPERTY

You can use the distributive property to simplify an expression. Multiply the numbers inside the parentheses individually by the factor outside the parentheses.

Examples $3(4 + 7) = 3 \cdot 4 + 3 \cdot 7$ $-7(5 - 2y) = -7 \cdot 5 - (-7 \cdot 2y)$
$= 12 + 21$ $= -35 - (-14y)$
$= 33$ $= -35 + 14y$

Simplify each expression using the distributive property.

29. $9(7 + 4)$ **30.** $10(14 - 5)$ **31.** $-6(14 + 6)$

32. $25(7k + 6)$ **33.** $4w(7 - 3)$ **34.** $-6x(-7 - 4y)$

EXPONENTS

An exponent tells the number of equal factors to multiply.

Example   In 4^3, the exponent, 3, tells you to find the product of three 4s.
$4^3 = 4 \cdot 4 \cdot 4 = 64$

Simplify each expression by finding the product.

35. 5^3 **36.** 3^5 **37.** 9^2 **38.** 5^5

Write each product in exponential form.

39. $5 \cdot 5 \cdot 5 \cdot 5$ **40.** $7 \cdot 7 \cdot 7$ **41.** $x \cdot x \cdot x \cdot x \cdot x \cdot x$

Introduction to Polynomials

Goals
- Write polynomials in standard form.
- Simplify polynomials.

Applications   Finance and Natural disasters.

Algeblocks can be used to model variable expressions. The length of each side of the x-block can be assigned the value x.

By calculating the area of each Algeblock, you find that each yellow square Algeblock represents x^2 and is called the x^2-block. The area of each x-block is x, and the area of each green unit block is 1.

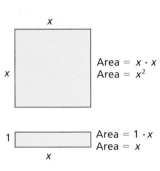

1. Set two x^2-blocks side by side. What expression is represented?

2. Model the expression $4x$ using x-blocks.

3. Combine the x^2-blocks from Question 1 with the x-blocks from Question 2. What expression do the Algeblocks now represent?

BUILD UNDERSTANDING

Each of these expressions is called a monomial.

$$4 \qquad x \qquad -3y \qquad 2y^2 \qquad 5xy \qquad -6x^2y$$

A **monomial** is an expression that is a number, a variable, or the product of a number and one or more variables.

In a monomial such as $3y$, the number 3 is called the **coefficient**. In monomials such as x, xy, and x^2y, the coefficient is 1. A monomial such as -4, which contains no variable, is a **constant**.

The *sum* or *difference* of two or more monomials is a **polynomial**. Each monomial is a **term** of the polynomial. A polynomial with two terms is called a **binomial**. A polynomial with three terms is called a **trinomial**.

Reading Math

The prefixes *mono-* and *poly-* are from Greek words meaning "one" and "many." The prefixes *bi-* and *tri-* are Latin prefixes meaning "two" and "three."

binomials: $2x + 3$ $x^2 - x^2y$ $4xy + (-5y)$
↑   ↑
term term

trinomials: $x^2 + y - 7$ $2xy^2 + (-5xy) + (-2)$
↑   ↑   ↑
term term term

Check Understanding

What are the coefficients of the trinomial?

$$2xy^2+(-5xy)+(-2)$$

Is this the same trinomial as $2xy^2 - 5xy - 2$? Explain.

You can think of a monomial as a polynomial of one term. A polynomial is written in **standard form** when its terms are arranged in order from greatest to least powers of one of the variables.

Example 1

Write each polynomial in standard form.

a. $8n^2 + 5n + 6n^3 + 9n^4$ **b.** $7y + 11 + y^2$ **c.** $5z - 2z^3 - 6 - 4z^2$

Solution

a. Order the terms from the greatest power of n to the least power of n.

$$8n^2 + 5n + 6n^3 + 9n^4 = 9n^4 + 6n^3 + 8n^2 + 5n$$

b. Order the terms from the greatest power of y to the least power of y. Write the constant last.

$$7y + 11 + y^2 = y^2 + 7y + 11$$

c. First write the polynomial as a sum of monomials with negative coefficients. Then write the polynomial in standard form.

$$5z - 2z^3 - 6 - 4z^2 = 5z + (-2z^3) + (-6) + (-4z^2)$$
$$= -2z^3 + (-4z^2) + 5z + (-6), \text{ or } -2z^3 - 4z^2 + 5z - 6$$

In a polynomial, terms that are exactly alike, or that are alike except for their numerical coefficients, are called **like terms**.

like terms: x^2y, $5x^2y$, $-4x^2y$ *unlike terms*: xy, x^2y, $2xy^2$

When a polynomial contains two or more like terms, simplify the polynomial by adding the coefficients of the like terms. This process is called *combining like terms*. A polynomial is simplified when all the like terms have been combined.

Example 2

Simplify.

a. $7x^2 + 2 + 9x^2$ **b.** $4d^2 - 6d - 3d^2 - 5d$ **c.** $-15m + 7m^2 + 9 - 3m$

Solution

a. $7x^2 + 2 + 9x^2 = 7x^2 + 9x^2 + 2$ Rearrange or collect like terms.

$\qquad\qquad\quad = (7 + 9)x^2 + 2$ Combine like terms by applying the distributive property.

$\qquad\qquad\quad = 16x^2 + 2$

b. $4d^2 - 6d - 3d^2 - 5d = 4d^2 + (-3d^2) + (-6d) + (-5d)$ Rearrange or collect like terms.

$\qquad\qquad\qquad\quad = (4 + (-3))d^2 + (-6 + (-5))d$ Combine like terms.

$\qquad\qquad\qquad\quad = 1d^2 + (-11d)$

$\qquad\qquad\qquad\quad = d^2 - 11d$

c. $-15m + 7m^2 + 9 - 3m = 7m^2 + (-15m) + (-3m) + 9$

$\qquad\qquad\qquad\quad = 7m^2 + [-15 + (-3)]m + 9$

$\qquad\qquad\qquad\quad = 7m^2 + (-18m) + 9$

$\qquad\qquad\qquad\quad = 7m^2 - 18m + 9$

Write each polynomial in standard form.

1. $5m + m^3 + 4m^2 + 2m^4$ **2.** $5d + d^2 + 14$

3. $-3x - 4x^2 + 2 + 5x^3$ **4.** $0.5y - 1.3 + 2.7y^3 + y^4$

Tell whether the terms in each pair are like or unlike terms.

5. $2s, -5s$ **6.** $7xy, 7xz^2$ **7.** $d^2t, -d^2t$ **8.** $x^2y, 2xy^2$

9. WRITING MATH Explain why x^2y and $2xy$ are not like terms.

Simplify. Be sure your answer is in standard form.

10. $4x^2 + 7x^3 - 3x^2 - 8$ **11.** $5c^2 - 4c - 2c^2 - 3c$

12. $-4y + 2y^2 + 5 + 3y$ **13.** $\frac{3}{4}n + \frac{1}{2}n^3 - \frac{3}{8} + \frac{1}{2}n^4$

14. MODELING Write the polynomial represented by the Algeblocks shown.

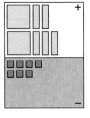

Write each polynomial in standard form.

15. $4b - 2b^2 + 7$ **16.** $n^3 + 2n^2 + 7n + 4n^4$ **17.** $5n^5 - 8n + 2n^3$

18. $4 - 4z - 5z^2 + 5z^3$ **19.** $m^2 + 2m - 4m^3$ **20.** $p^2 + 2p - 3p^3 - 9$

21. $2a^4 + 3 + 2a^2 + 3a^3 + 2a$ **22.** $\frac{1}{3}n - n^5 + \frac{4}{5}n^2 + \frac{2}{3}n^3$ **23.** $-t^3 + 4t^2 - 11t + 21$

24. $0.6y^5 - 2.4y^7 + y^4 + 3.1y$ **25.** $8s - 14s^5 + 9s^2 - s^3 + 2$ **26.** $0.1x + 0.5 + 0.7x^2$

Tell whether the terms in each pair are *like* or *unlike* terms.

27. $-6a, 9a^2$ **28.** $n^2, -3n^2$ **29.** r^3s, rs^3

30. $yz, 8yz$ **31.** $-14a^2b, 7a^3b$ **32.** $17x^3y^2, -x^3y^2$

Simplify. Be sure your answer is in standard form.

33. $-6a - 3 + 9a$ **34.** $-4b + 7 + 8b$ **35.** $-2 + 4c - 9c$

36. $n^2 + 2n - 4n^2$ **37.** $7r^2 + 4r^2 + r^2 + r$ **38.** $4x^2 + 3x^2 + x^3 - 6x$

39. $7s^2 - s + 6s^2 - s^2 - s$ **40.** $2x^2 + 3x - 2 + x^2 - 3$ **41.** $-2 + 18h^2 - 9h^3 + 4h^2$

42. $-7r + 4r^2 + 8r - 12r^2 - 12$ **43.** $\frac{1}{2}y + \frac{3}{4}y^2 + \frac{2}{3} + \frac{3}{4}y + \frac{1}{2}y^2$ **44.** $4.3p + 0.5p^2 - 8.1 - 0.9p^2 + 1.6$

45. WRITING MATH In your own words, explain like terms. Tell how like terms are used when simplifying polynomials.

46. Express the perimeter of the figure shown in simplest form.

$3x$

$2x + 5$

47. MONEY Tyron wants to buy a fruit drink. In his backpack he finds 2 quarters, 6 dimes, and 3 nickels. In his pocket he has 1 quarter, 4 dimes, and 1 nickel. Use q, d and n to represent the coins as a polynomial expression in simplest form. How much money in coins does Tyron have to buy a fruit drink?

EARTHQUAKES Use the table for Exercises 48 and 49. Choose a variable to represent the average magnitude for each of the five ranges of the Richter Scale.

Richter scale	World-wide occurrence
8 and higher	1 per year
7.0-7.9	18 per year
6.0-6.9	120 per year
5.0-5.9	800 per year
4.9 or less	9150 per year

48. Write a polynomial to represent the total magnitude of earthquakes measuring 6.0 or greater in one year.

49. Write a polynomial to represent the total magnitude of earthquakes measuring 6.9 or less in one year.

DATA FILE For Exercises 50–52, refer to the data about the maximum speed of animals on page 518. Let x represent the maximum speed of a chicken. Represent the maximum speed of the following animals in terms of x.

50. six-lined race runner

51. elk

52. jackal

53. Explain why the expression $\frac{3x}{y}$ is not a polynomial.

■ EXTENDED PRACTICE EXERCISES

Simplify each polynomial, if possible. Otherwise, write *not possible*. Terms are like only if they contain the same variables raised to the same powers. Write your answer in standard form.

54. $yz + 8yz$

55. $xz^3 + 4xz^3 - 3yz^3$

56. $15 - k - 2km - 9km$

57. $2xy + 3xy + 3x^2 - x^2 - x$

58. $r^3s - 4s + rs^2$

59. $17x^3z + 18x^3y + 2xy^3 + xy - 4xz^3$

The *degree of a polynomial in one variable* is the highest power of the variable that appears in the polynomial. What is the degree of each polynomial?

60. $x^2 + 3x - 5$

61. $5a - 4 + a^3 - 6a^2$

62. $m + 4$

63. MODELING Show or sketch the polynomial $3x^2 + x + 6$ using Algeblocks.

64. Tell whether the statement is *true* or *false*: If the degree of a polynomial in one variable is 4, then the polynomial will have four terms. Explain.

65. CHAPTER INVESTIGATION The cloud formation of the tropical storm over the Atlantic Ocean is approximately rectangular. Write a polynomial expression that represents the length of the cloud formation. Write a polynomial expression that represents the width of the cloud.

■ MIXED REVIEW EXERCISES

Complete. (Lesson 2-2)

66. 52 ft = _____ yd _____ ft

67. 24 fl oz = _____ pt _____ c

68. 105 in. = _____ ft _____ in.

69. 16 cm = _____ mm

70. 5.7 km = _____ m

71. 147 g = _____ kg

72. FINANCE The price of a share of stock at Monday's close was $57. The price dropped $5 on Tuesday, dropped $2 on Wednesday, rose $3 on Thursday, rose $2 on Friday. What was the price of a share of stock at the end of the week? (Lesson 3-1)

Add and Subtract Polynomials

Goals ■ Add and subtract polynomials

Applications Landscaping

EDUCATION When learning arithmetic, children begin by saying "2 apples added to 4 apples is 6 apples." In algebra that would be $2a + 4a = 6a$.

1. Write an expression if Ahmed has 2 apples and 3 bananas.

2. Write an expression if Jen has 3 apples and 4 bananas.

3. Add your results from Questions 1 and 2 to create a new expression.

4. Write the simplified expression to represent the total number of apples and bananas that Ahmed and Jen have together.

◢ BUILD UNDERSTANDING

Just as you can add, subtract, multiply or divide real numbers, you can perform each of these basic operations with polynomials. To add two polynomials, simplify by combining like terms.

Example 1

Simplify.

a. $3m + (2m - 6)$ **b.** $(2a + 3) + (4a - 1)$ **c.** $(x^2 + 4x - 2) + (3x^2 + 7)$

Solution

a. $3m + (2m - 6) = (3m + 2m) - 6$ Use the associative property. $3m$ and $2m$ are like terms.

$\qquad\qquad\qquad = (3 + 2)m - 6$ Use the distributive property.

$\qquad\qquad\qquad = 5m - 6$

b. $(2a + 3) + (4a - 1) = (2a + 4a) + (3 - 1)$ Group like terms.

$\qquad\qquad\qquad\qquad = (2 + 4)a + 2$ Use the distributive property.

$\qquad\qquad\qquad\qquad = 6a + 2$

c. $(x^2 + 4x - 2) + (3x^2 + 7) = (x^2 + 3x^2) + 4x + (-2 + 7)$

$\qquad\qquad\qquad\qquad\qquad = (1 + 3)x^2 + 4x + 5$

$\qquad\qquad\qquad\qquad\qquad = 4x^2 + 4x + 5$

Check Understanding

Represent the expression in Example 1, part a, using Algeblocks.

Polynomials can also be added by lining up like terms vertically.

Example 1, part c, is shown using the vertical method.

$$
\begin{array}{r}
x^2 + 4x - 2 \\
+\ 3x^2 \qquad\ +\ 7 \\
\hline
4x^2 + 4x + 5
\end{array}
$$

Subtraction of polynomials is like subtraction of real numbers. To subtract one polynomial from another, add its opposite and simplify. To find the opposite of a polynomial, write the opposite of each term of the polynomial.

Polynomial	*Opposite*
$x + 4$	$-x + (-4)$ or $-x - 4$
$4f^2 - 2f - 7$	$-4f^2 + 2f + 7$
$-3x^2 + 5x + 2$	$3x^2 + (-5x) + (-2)$ or $3x^2 - 5x - 2$

Example 2

Simplify.

a. $9y - (2y - 4)$ **b.** $(-x + 2) - (-8x - 3)$ **c.** $(5x^2 - 11) - (3x^2 - 3x + 2)$

Solution

a. $9y - (2y - 4) = 9y + (-2y + 4)$ Add the opposite of $(2y - 4)$.

$\qquad\qquad = [9y + (-2y)] + 4$ $9y$ and $-2y$ are like terms.

$\qquad\qquad = (9y - 2y) + 4$

$\qquad\qquad = (9 - 2)y + 4$ Use the distributive property.

$\qquad\qquad = 7y + 4$

b. $(-x + 2) - (-8x - 3) = (-x + 2) + (8x + 3)$ Add the opposite of $(-8x - 3)$.

$\qquad\qquad = (-x + 8x) + (2 + 3)$

$\qquad\qquad = (-1 + 8)x + 5$ Write the coefficient of $-x$ as -1.

$\qquad\qquad = 7x + 5$

c. $(5x^2 - 11) - (3x^2 - 3x + 2) = (5x^2 - 11) + (-3x^2 + 3x + (-2))$

$\qquad\qquad = [5x^2 + (-3x^2)] + 3x + [-11 + (-2)]$

$\qquad\qquad = (5 - 3)x^2 + 3x + (-11 - 2)$

$\qquad\qquad = 2x^2 + 3x - 13$

◥ TRY THESE EXERCISES

Add.

1. $4x^2 + 3x - 2$

   $+ \ x^2 \qquad + 6$

2. $2m^2 + 2m - 5$

   $+ \qquad\quad 3m + 4$

3. $5y^2 + 2y + 3$

   $+ \ -2y^2 \qquad - 4$

Simplify.

4. $4x + (3x - 5)$

5. $(4m + 2) + (5m - 3)$

6. $(2r^2 + 3r - 4) + (2r^2 + 3)$

7. $6y - (4y - 3)$

8. $(-z + 3) - (-4z - 2)$

9. $(3c^2 - 8) - (2c^2 - 2c + 6)$

10. WRITING MATH Will the sum of two binomials always result in another binomial? Explain your answer.

PRACTICE EXERCISES • For Extra Practice, see page 577.

Simplify.

11. $2y + (4y - 3)$

12. $\frac{1}{4}k + \left(\frac{2}{3}k + 4\right)$

13. $9m + (-2m - 7)$

14. $(3m + 2n) + (2m - 3n)$

15. $(2x^2 - 1) + (3x^2 + 7)$

16. $(2x - 4y) + (x + 3y)$

17. $\left(\frac{8}{9}m - \frac{7}{9}n\right) + \left(-\frac{5}{9}m - \frac{2}{9}n\right)$

18. $(3x^2 - 2x) + (x^2 - 4x)$

19. $(3y^2 - 4) + (y^2 - 5y + 3)$

20. $(5a^2 + 3) + (-7a^2 - 4)$

21. $(2x^2 - 3x + 5) + (4x^2 + 6x - 8)$

22. $(a^2 - 2a + 1) + (a^2 - 4)$

23. $\frac{1}{3}x - \left(-\frac{3}{6}x\right)$

24. $2k - (2k + 4)$

25. $6z - (2z + 7)$

26. $(7x - 5) - (5x + 3)$

27. $(10 - 2k) - (7 - 3k)$

28. $\left(z^2 - \frac{3}{4}z + 10\right) - \left(\frac{1}{2}z^2 + 5z - 6\right)$

29. $(2z - 3y) - (5z - 5y)$

30. $(7x - 2y) - (3x + 4y)$

31. $(8p^2 + 5) - (3p^2 + 2p - 9)$

32. $(2a^2 - 3a + 5) - (a^2 - 2a - 4)$

33. $(4x - 4xy + 3y) - (4x - 5xy + 3y)$

34. $(3a + 9b - 12) - (4a + 7b - 13)$

35. $(11m^3 + 2m^2 - m) - (-6m^3 + 3m^2 + 4m)$

36. $(3d^3 - 10d^2 + d + 1) + (5d^3 + 2d^2 - 3d - 6)$

Find the term that makes the statement true.

37. $-9d + \underline{\ ?\ } = 5d$

38. $3m + (-7m) + \underline{\ ?\ } = m$

39. $-x + (-7x) + \underline{\ ?\ } = -x$

40. $7h^2 + \underline{\ ?\ } + (-h^2) + 4h^2 = 4h^2$

41. VOLCANOES Tacana, an active volcano in Guatemala, has a height of $(12x + 4y)$ ft. Etna, an active volcano in Italy, has a height of $(10x + 9y + 90)$ ft. How many feet taller is Tacana than Etna?

42. LANDSCAPING A garden has a length of $7a - 3b$, and its width measures $4a - b$. Find the amount of fencing needed to enclose the garden.

TRAVEL For Exercises 43–46, use the table. Let d represent the price for a gallon of diesel gas, r for regular unleaded, and p for premium unleaded. Write a polynomial expression for the amount of sales for each of the three hours.

Gallons Purchased at Ted's Car Care

Hour	Diesel	Regular	Premium
1	102	383	289
2	99	305	260
3	97	401	326

43. hour 1   **44.** hour 2   **45.** hour 3

46. Add the polynomial expressions in Exercises 43–45.

Use the triangle and quadrilateral shown.

47. Write an expression in simplest form for the perimeter of triangle *RST*.

48. Write an expression in simplest form for the perimeter of quadrilateral *ABCD*.

49. Find the difference when the perimeter of triangle *RST* is subtracted from that of quadrilateral *ABCD*.

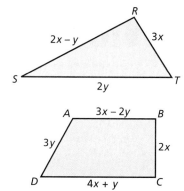

Simplify, showing all steps. Write a justification for each step.

50. $2x^2 - (3x^2 - 7x + 1)$

51. $(4p^2 - 11p + 8) + (-2p^2 + 5p + 3)$

 MODELING Find the sum of the expressions represented on each pair of sentence mats.

52.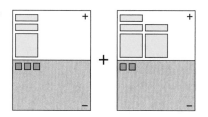

53.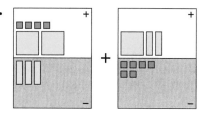

EXTENDED PRACTICE EXERCISES

Simplify.

54. $(4a^2bc - 3abc^2 + 2abc) + (8abc^2 - 5a^2bc - ab^2c)$

55. $(12abc - 3ab^2c + 4a^2bc) - (-3ab^2c - 4abc + 4a^2bc)$

Find each difference.

56. $(4n^2 - 2n + 1) - (n^2 + 6n - 5)$

57. $(n^2 + 6n - 5) - (4n^2 - 2n + 1)$

58. **CRITICAL THINKING** Compare your answers to Exercises 56 and 57. How are they related?

 59. **WRITING MATH** Analyze your answers to Exercises 56–58. Make a general statement about what happens when you reverse the order of the polynomials in a subtraction problem.

MIXED REVIEW EXERCISES

Write each answer in simplest form. (Lesson 2-2)

60. $\quad$ 4 ft $\;$ 5 in.
$\quad +$ 1 ft 10 in.

61. $\quad$ 8 lb 2 oz
$\quad -$ 3 lb 9 oz

62. $\quad$ 4 gal 3 qt
$\quad -$ 1 gal 3 qt 1 pt

Solve each equation. Check the solution. (Lessons 5-4 and 5-5)

63. $147.6 - \dfrac{k}{3.9} = 142.9$

64. $4(d + 6) = -2(d - 4)$

65. $9b + 7 - 3b = 3 - 8(b + 1)$

66. **MANUFACTURING** The diameter of a car wheel is 84.6 cm. Find about how far the car travels for one turn of the wheel. Round to the nearest tenth. (Lesson 2-7)

Review and Practice Your Skills

Write each polynomial in standard form.

1. $4t^2 - 8 + 2t$
2. $-3x + 2x^2 - 7x^3$
3. $8 + 3h^2 - 5h$
4. $a^5 + a^2 - 5a + a^3$
5. $-2m - m^3 + 3m^2 + 4$
6. $6k^5 + 1 - k^3 - k$
7. $\frac{1}{2}g^2 - \frac{2}{3}g + \frac{1}{4} + \frac{1}{3}g^4$
8. $0.4 + 1.4y^2 - 2.3y$
9. $-0.1d^3 + d^5 + 0.2d^4$

Tell whether the terms in each pair are *like* or *unlike*.

10. $4a, 2b$
11. $d, -5d$
12. $3w^2, -w^2$
13. $6z^3, 3z^6$
14. $-2g^3, 12g^3$
15. $9x^4, 2x^4$
16. $-2xy, -2xy^2$
17. $4ab, -ab^2$
18. $3m^2n, 2mn^2$

Simplify. Be sure your answer is in standard form.

19. $-2g + g$
20. $4w + 2 - w$
21. $-3q - 2q + q$
22. $b + 4b + 1$
23. $2d - 6 + 4d$
24. $-5x + 2x + 4$
25. $3y^2 - 2y - y^2$
26. $4s^2 + s - s^2 + 2s$
27. $-w^2 + 2w^3 + 2w - w^2$
28. $\frac{3}{5}p^2 - \frac{1}{5}p - \frac{1}{2}p^2 + p$
29. $-0.2x + 0.3x^2 + 1.4x$
30. $0.5r - 0.25r^2 - 0.5 + 0.75r^2$

Simplify.

31. $5d + (2 - 3d)$
32. $-x + (3x - 2)$
33. $(4 + m) + 5m$
34. $(-7h + 2h^2) + 3h$
35. $(x - 3) + (-2x - 4)$
36. $(-6a + a^2) + (2a^2 - 3)$
37. $(4x + y) + (-2x - y)$
38. $(-3m + 2n) + (-n - 3m)$
39. $(2z + 1 - z^2) + (4 - 3z^2)$
40. $(5p - 2q + 3) + (-1 + 4q + 3p)$
41. $8p - (2 + 3p)$
42. $(3g + 2g) - (-6g)$
43. $-2m - (m^2 + -3m)$
44. $6x - (3x - 4)$
45. $(-3z + 2) - (-3 + z)$
46. $(4m - 3n) - (n + 2m)$
47. $(-6y^2 + 3y) - (y - 2y^2)$
48. $(3a^2 + 2 - a) - (-8 + 2a)$
49. $(4g + 2h) - (-h + 2 - 3g)$
50. $(7y - 2y^2) - (1 + 2y^2 - y)$
51. $(-p^2 + 2p + 3) - (4p + 3p^2 - 4)$
52. $(-3k + 6 - 2k^2) - (2k + 3k^3 - 2k^2)$

Tell whether the terms in each pair are *like* or *unlike*. (Lesson 9-1)

53. $2d^2, -d^2$

54. $-y, -8y$

55. $4x^2, x^3$

56. $4ab^2, -2a^2b$

57. $3m^2n^3, 2m^2n^3$

58. $-cd^2, 2d^2$

Simplify. Be sure your answer is in standard form. (Lesson 9-1)

59. $4n - 2 + 3n$

60. $-3 - g + 3g$

61. $6b + 2b - b + 2b$

62. $-p + p^2 - 4 - 2p^2$

63. $6w^2 + w + 2w - 4w^2$

64. $4r - 2 - 3r^2 + 2r$

65. $7y^2 + 2y - 3y^3 - y$

66. $-k^2 + 3k - 2k^2 - 2k^3$

67. $3a^3 + a - 2a^3 + 3a$

68. $-5x + 2x^2 + 2x^3 - 3x^2$

69. $2.5h^2 + 3.5 + 0.5h - h^2$

70. $-0.5g - 2.5g^2 + 0.5g$

Simplify. (Lesson 9-2)

71. $-4f + (-2 + 3f)$

72. $(4m - 2n) + (-3m)$

73. $(2x + 3) + (-4x^2 - 6)$

74. $(-p + 2q) + (q^2 - q - 3p)$

75. $(4r + 3) - (6 - 2r)$

76. $(-5y) - (-2x - 3y)$

77. $(-w + 2w^2) - (3w^2 + 2w)$

78. $(2x - 3y) - (y + 3 + 4x)$

79. $(3e^3 + e - 2e^2) - (4e + 5)$

80. $(3h + 2 - k) - (4k + 3h^2 - 4)$

Math*Works*
Workplace Knowhow
Career – Meteorologist

Meteorology is the study of the atmosphere. This is applied in forecasting weather, air-pollution control, air and sea transportation, and the study of climate. Meteorologists use mathematics and formulas to measure and forecast things such as air pressure, temperature, humidity, and wind speed. Wind speed is often measured in *knots*. The following conversions are used when referring to a knot.

Use these conversions for Questions 1–4. Let *s* represent the speed in miles per hour for every knot.

1. A meteorologist reports that the wind speed atop Mt. McKinley, 20,320 ft above sea level, is 35 knots. At sea level, the wind speed is 15 knots. Find the expression for the difference in wind speed in miles per hour.

2. In a certain location a meteorologist reports that the wind speed is 20 knots. At 12,000 ft above this location the wind speed increases by 25 knots. At 24,000 ft above the initial location the wind speed decreases 15 knots from the wind speed at 12,000 ft. Find the polynomial expression to represent the wind speed at 24,000 ft in miles per hour.

9-3 Multiply Monomials

Goals
- Multiply monomials.
- Solve area problems by multiplying monomials.

Applications Engineering, Architecture

Algeblocks can be used to find the area of figures when the dimensions contain variables. To use Algeblocks, place the blocks for one factor along the horizontal axis and place the blocks for the other factor along the vertical axis of the Quadrant Mat.

1. Find the area of a rectangle that is x-block by y-block, as shown.

2. What monomial represents the light orange Algeblock?

3. Use x-blocks and y-blocks to model a rectangle with width $2x$ and length $3y$.

4. Using Algeblocks, find the area of the modeled rectangle.

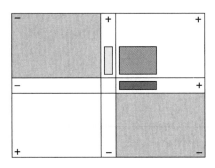

BUILD UNDERSTANDING

Recall that the order in which factors are multiplied does not affect the product.

$$a \cdot b = b \cdot a \qquad \text{(commutative property)}$$

The grouping of factors does not affect the product.

$$(a \cdot b) \cdot c = a \cdot (b \cdot c) \qquad \text{(associative property)}$$

These properties can be used to find a product of two monomials.

Example 1

Simplify.

a. $(4a)(5b)$ b. $(-3k)(2p)$ c. $(-\frac{1}{2}x)(-4y)$

Reading Math

Parentheses are often used to show multiplication.

Solution

a. $(4a)(5b) = (4)(5)(a)(b)$ Use the commutative and associative properties.

$\qquad\qquad = 20ab$

b. $(-3k)(2p) = (-3)(2)(k)(p) = -6kp$

c. $(-\frac{1}{2}x)(-4y) = (-\frac{1}{2})(-4)(x)(y) = 2xy$

Product Rule for Exponents	$a^m \cdot a^n = a^{m+n}$ To multiply two powers having the same base, add the exponents.

Example 2

Simplify.

    a. $x^2 \cdot x^4$ **b.** $(-3k)(-2k^3)$ **c.** $(ab^3)(a^3b^4)$

Solution

    a. $x^2 \cdot x^4 = x^{2+4}$
                $= x^6$

    b. $(-3k)(-2k^3) = (-3)(-2)(k \cdot k^3)$ k means k^1.
                       $= 6k^{1+3} = 6k^4$

    c. $(ab^3)(a^3b^4) = (a \cdot a^3)(b^3 \cdot b^4)$
                    $= a^{1+3} \cdot b^{3+4}$ Use the product rule for each base, a and b.
                    $= a^4b^7$ Because a and b are unlike bases, you cannot add exponents.

Power Rule for Exponents	To find a power of a monomial that is a power, multiply exponents. $$(a^m)^n = a^{m \cdot n}$$

Example 3

Simplify.

    a. $(c^2)^4$ **b.** $(y^3)^8$

Solution

    a. $(c^2)^4 = c^{2 \cdot 4}$ **b.** $(y^3)^8 = y^{3 \cdot 8}$
              $= c^8$ $= y^{24}$

Power of a Product Rule for Exponents	To find a power of a product, find the power of each factor and multiply. $$(ab^m)^n = a^n(b^{m \cdot n})$$

Example 4

Simplify.

    a. $(3z)^2$ **b.** $-(4y^2)^2$ **c.** $(-2c^2)^3$

Solution

    a. $(3z)^2 = (3^2)(z^2)$ **b.** $-(4y^2)^2 = -(4)^2(y^2)^2$ **c.** $(-2c^2)^3 = (-2)^3(c^2)^3$
              $= 9z^2$ $= -16y^{2 \cdot 2}$ $= -8c^{2 \cdot 3}$
                                      $= -16y^4$ $= -8c^6$

Simplify.

1. $(3x)(3y)$

2. $(-b)(5d)$

3. $(4g)(-2h)$

4. $(-\frac{1}{3}a)(-3b)$

5. $(-2y)(y^2)$

6. $(2p^2r^3)(5pr^3)$

7. $(2y)^3$

8. $(3z^3)^2$

9. $(-m^2)^3$

10. $(-3c^2)^2$

11. Write an expression for the area of a square with the length of a side $3x^2$.

12. **WRITING MATH** Explain how expressing the product of x^2 and x^3 differs from expressing the product of x^2 and y^3.

Simplify.

13. $(2a)(3b)$

14. $(-4m)(-3n)$

15. $(6k)(-2m)$

16. $\left(\frac{1}{3}x\right)\left(-\frac{5}{6}z\right)$

17. $(-2k)(3p)$

18. $(-6a)\left(-\frac{1}{6}f\right)$

19. $(2c)(3.5d)$

20. $(-3p)(-4.1r)$

21. $(2a)^2$

22. $(2a^2)(3a^3)$

23. $(-3y^2)(4y^3)$

24. $(-x)(x^5)$

25. $(3m^2)^3$

26. $(7y)(y^4)$

27. $(5b^3)(-2b)$

28. $(4p^2)\left(-\frac{3}{4}p\right)$

29. $(-3a^2)(-2a^3)$

30. $(5k^4)(-3k^2)$

31. $(2p^3)(-4p^5)$

32. $\left(\frac{1}{2}x\right)^2$

33. $(3y^4)^5$

34. $(-2d^3)^3$

35. $(-3x^2)^3$

36. $(9y)(3y)^2$

37. $(3x^2)(2x)^3$

38. $(2c)^2(3c)^2$

39. $(2a^2)^2$

40. $(-4b^4)^3$

41. $(-y^2)^2(-5y^3)^2$

42. $(-6h^2)^2(-7h)^3$

43. $(2a^3b^2)(6a^3b)$

44. $(-3x^3y^3)(-4xy^2)$

45. $\left(-\frac{1}{3}p^3r\right)(9p^2r^4)$

Write an expression for the area of each figure.

46.

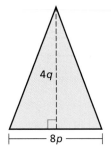

47.

```
        2m
┌──────────────────┐
│                  │ 3p
└──────────────────┘
```

48. **MODELING** What multiplication is expressed by the Algeblocks shown? Find the product.

49. Make a chart summarizing the three properties of exponents that were reviewed in this lesson. Make up one example for each property.

50. YOU MAKE THE CALL An exercise from Bernardo's math homework is shown. Is his solution correct? Explain. $(9x^2)(-2x)^2 = -36x^4$

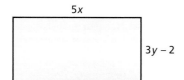

ENGINEERING Engineers have planned to install a soundproof room. They will use this diagram for the dimensions of the room.

51. Write an expression for the area of the ceiling.

52. Find the area when $x = 5$ and $y = 4$.

53. Find the area when $x = 2$ and $y = 3$.

Find each value when $a = \frac{1}{2}$ and $b = \frac{1}{3}$.

54. $(2a)(3b)$ **55.** $(3ab)(a^2)$ **56.** $(6ab)(-2ab)$

GRAPHING Use a graphing calculator to graph each polynomial function. How are the graphs different? How are they alike? By looking at a polynomial, can you predict what its graph will look like?

57. $y = 2x^2 + 2x + 1$ **58.** $y = 2x$ **59.** $y = -5 - 3x$ **60.** $y = -x^2 - 2x + 5$

▪ EXTENDED PRACTICE EXERCISES

Write an expression for the area of each figure.

61.

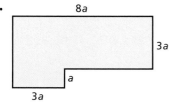

62.

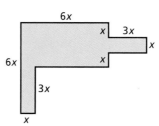

Write an expression for the volume of each figure.

63.

64.

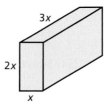

65.

66. ARCHITECTURE On buildings, architects install sheer walls of concrete to absorb the force of earthquakes. How many square units of wall space must be painted in the building shown?

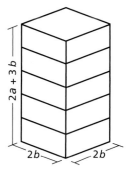

▪ MIXED REVIEW EXERCISES

Find the unit rate. (Lesson 2-6)

67. 30 mi in 40 min   **68.** 175 mi on 7 gal of gas   **69.** 480 words in 15 min

70. Karen can read 15 pages in 30 minutes. How long will it take her to read 450 pages?

Multiply a Polynomial by a Monomial

Goals ■ Solve problems by multiplying a polynomial by a monomial.

Applications Photography, Astronomy, Agriculture, Retail

A construction worker was working with a piece of wood that was 4 units longer than it was wide.

1. Draw a diagram of the piece of wood.

2. If x represents the width, how can you represent the length in terms of x?

3. Label the length and width on the diagram.

4. What formula would you use to find the wood's area?

◥ BUILD UNDERSTANDING

You can use the distributive property and the rules for exponents to find the product of a monomial and a polynomial.

Distributive Property	A factor outside parentheses can be used to multiply each term within the parentheses.

Example 1

Simplify.

a. $3x(x + 4)$ **b.** $-2x(x^2 - 3x + 11)$ **c.** $-n(n^3 + 2n)$

Solution

a. $3x(x + 4) = 3x(x) + 3x(4)$ Use the distributive property.

$= 3x^2 + 12x$ Apply the product rule for exponents.

b. $-2x(x^2 - 3x + 11) = -2x[x^2 + (-3x) + 11]$

$= -2x(x^2) + [-2x(-3x)] + [-2x(11)]$

$= -2x^3 + 6x^2 + (-22x)$

$= -2x^3 + 6x^2 - 22x$

c. $-n(n^3 + 2n) = -n(n^3) + [-n(2n)]$

$= -n^4 + (-2n^2)$

$= -n^4 - 2n^2$

Check Understanding

Represent the expression in Example 1, part a, using Algeblocks.

E x a m p l e 2

PHOTOGRAPHY For a science exhibit, Gail uses a photograph of Saturn's rings that is 12 in. wide and 16 in. long. When she adds a caption below the photograph, the total width will be $(12 + c)$ in.

a. What formula will you use to find the area of the photograph?

b. Write an expression for the total area occupied by the photograph and the caption. Then simplify the expression.

Solution

a. Use the formula for the area of a rectangle.

$A = l \cdot w$

b. $A = 16(12 + c)$ Replace l with 16 and w with $(12 + c)$.

$A = 16(12) + 16(c)$

$A = 192 + 16c$

The total area occupied by the photograph and the caption is $(192 + 16c)$ in.^{2}

◥ TRY THESE EXERCISES

Simplify.

1. $4x^2(x + 7)$

2. $2b(b^2 + 12)$

3. $3g^2(g^2 - 7)$

4. $2z(3z^2 + z - 8)$

5. $6k(2k^2 + k)$

6. $-x(3x^2 - x - 6)$

7. $-3x^2(x^2 - 1)$

8. $y(y^2 - 3y + 4)$

9. $2c^2(-10c - c + 5)$

10. $5s^3(s^2 - 10s + 12)$

11. $\frac{1}{2}x(x - 1)$

12. $3y^2\left(\frac{2}{3}y + \frac{1}{6}\right)$

 13. WRITING MATH In your own words, describe how to multiply a polynomial by a monomial.

14. PHOTOGRAPHY A photograph of the Grand Canyon was 12 in. long and 8 in. wide. After Pablo trimmed one side of the photograph, its width was $(8 - t)$ in. Write an expression for the area of the trimmed photograph. Then simplify it.

 15. YOU MAKE THE CALL Margo found the area of the rectangle shown. Her work is displayed. Is she correct? Explain.

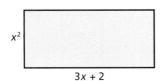

$x^2(3x + 2) = x^2(3x) + x^2(2)$

$= 3x^2 + 2x^2$

$= 5x^2$

The area is $5x^2$ units squared.

Simplify.

16. $2x(x + 3x^2)$

17. $-4(-y + 2y^2)$

18. $2a(3a + 1)$

19. $x(3x + 5)$

20. $c(c^2 + 9c)$

21. $-4(y + 2)$

22. $-3(2z - 8)$

23. $k(k^2 - 4k + 1)$

24. $-2(a^2 - 3a + 2)$

25. $-3m(4m - 7)$

26. $5d(9d^2 + 4d - 7)$

27. $3x(x^2 + 4x - 3)$

28. $10w(-3w + 2)$

29. $2(a^2 + 5a - 1)$

30. $-2m(m^2 - 3m - 4)$

31. $-5y(y^2 - 3y - 8)$

32. $2(p^2 - 5p - 1)$

33. $-t(2t^2 + 8t - 24)$

34. $-3(f^2 - 2f - 8)$

35. $4p(p^2 - 3p - 2)$

36. $-r(r^3 - 9r + 6)$

37. $a(4 - a - 5a^2)$

38. $2x(2x^2 - 4x + 3)$

39. $4a^2(-a^2 - 3a + 9)$

Write an expression for the area of each rectangle. Then simplify the expression.

40.

5x
2x − 1

41.
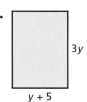
3y
y + 5

42.
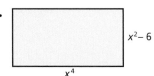
x² − 6
x⁴

43.
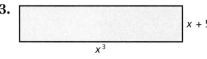
x + 5
x³

RETAIL The profit made in cents on the sale of roses is given by the expression $0.05t(3t - 25)$ where t is the number of roses sold.

44. Simplify the expression for the profit of roses.

45. Find the profit if the company sells 100 roses on a particular day.

46. Let r represent an even integer. Then $r + 2$ is the next even integer. Write an expression for the product of the two integers. Then simplify the expression.

 47. WRITING MATH Draw a rectangle to represent $y(y + 1)$ and another to represent $(y + 1)y$. Explain how they are the same and how they are different.

48. ASTRONOMY Ganymede, one of the largest moons in the solar system, has a diameter of $(w^2v)(w^5v)$ mi, where $w = 2$ and $v = 5$. Find the diameter of this moon.

49. AGRICULTURE During a tornado, a farmer's soybean crops were destroyed. According to the diagram, how many square units were lost?

House | Corn
Wheat
Barn | Hay
4x | Soybeans
2x + 6

Simplify.

50. $4(a - 3) - 3a$

51. $-2p - 2(p - 3) + 6$

52. $4(m - 1) + 3(m + 5)$

53. $\frac{1}{2}x(2x^2 + 6xy - 14y^2)$

54. $x(2x^2 - x + 2) - x(x^3 + 4x^2 - 3x)$

55. $(4x^2y - 3xy^2 + 7y^3)(4xy)$

 56. MODELING What multiplication is modeled on the Quadrant Mat? Find the product.

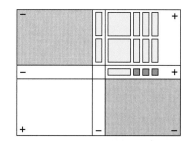

◼ EXTENDED PRACTICE EXERCISES

Use the diagram. It indicates the number of seconds it took a beam of sunlight to travel first to the earth, then to the moon.

57. Write an expression for the total time it took the light to reach the moon.

58. Suppose that light travels at a rate of c mi/sec. Write a product for the distance from the sun to the moon. Then simplify. (Hint: $d = rt$)

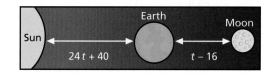

59. Use $c = 186{,}000$ mi/sec and $t = 19$ sec to approximate the distance from the sun to the moon.

Write an expression for the area of the shaded region. Then simplify the expression.

60.

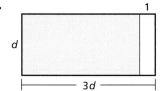

61.

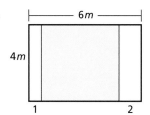

 62. CHAPTER INVESTIGATION Find the surface area of the cloud formation.

◼ MIXED REVIEW EXERCISES

Find the area of each kite to the nearest tenth. (Lesson 2-9)

63.

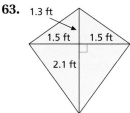

64.

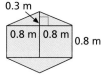

65.

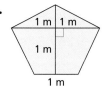

66. Find the area of a 2.5 ft by 6.3 ft rectangle. (Lesson 2-4)

67. COMMISSION Michelle sells home theater systems. She receives a base salary of $350 a week and a commission on sales. One week she sold $10,150 worth of equipment. Her combined salary and commission was $502.25. What is her commission rate? (Lesson 6-5)

68. DATA FILE Refer to the data on the busiest airports in the U.S. on page 534. What is the percent of increase of the airport with the largest jump in rank? Round to the nearest tenth of a percent. (Lesson 6-7)

Review and Practice Your Skills

PRACTICE ■ LESSON 9-3

Simplify.

1. $(4a)(-2b)$
2. $(-6y)(2z)$
3. $(5t)(3s)$
4. $(-2x)(-4y)$
5. $(q)(-3p)$
6. $(-3f)(-2g^2)$
7. $(3h)(-h)$
8. $(3c^2)(2c^3)$
9. $(2y)(-2y^3)$
10. $(4t^2)(3t)$
11. $(4r)(-r^4)$
12. $(3d^2)(-2d^2)$
13. $(-5b)^2$
14. $(-d^4)(3d^2)$
15. $(-5w^2)(2w)$
16. $(3z)(4z^2)$
17. $(-2b)^3$
18. $(-x^3)(-3x)$
19. $(2e^5)(e^2)$
20. $(3p)^2$
21. $(6a^3)(-2a^6)$
22. $(w)(2w)^2$
23. $(5s)^2(-3s)$
24. $(2p)^3(3p^2)$
25. $(5g^2)(-2g)^2$
26. $(-3a)^2(-a^3)$
27. $(4e^4)(-2e)^3$
28. $(-4y^2)^2(2y)$
29. $(2k^3)^2(3k)^2$
30. $(-3m)^2(-2m^2)^3$
31. $(2m^2n)(-3mn)^2$
32. $(-xy)^2(2xy^3)$
33. $(4c^2d^2)^2(cd^2)$
34. $(-3ab)^2(-ab)^3$
35. $(3p^4q)^2(2pq^3)^2$
36. $(-2st)^3(4s^2t)^2$

PRACTICE ■ LESSON 9-4

Simplify.

37. $4g(4g + 2)$
38. $-r(-3 - 2r)$
39. $3z(2z^2 + 4)$
40. $-2(-3m + 2m^2)$
41. $3a(-2a^2 + 3a)$
42. $-2b(7 + 4b^2)$
43. $k(4k - 8)$
44. $2p(-p^2 + 3p)$
45. $-6x(-2x - 3x^2)$
46. $2y(3 - 2y^2)$
47. $-5c(1 - 2c)$
48. $3n(-4n^2 + 2n^3)$
49. $3r(2r^2 - r + 3)$
50. $-3q(1 + q^2 - q)$
51. $-2g(-5g^2 + 4 + 3g)$
52. $-w(-4w^2 + 2w - 1)$
53. $4d(2d - d^2 + 3)$
54. $7a(10a^2 + 2 - 3a)$
55. $4e(3 + 2e^2 - 3e)$
56. $2z(-4z^2 - 3z - 2)$
57. $3y(6 - 4y^2 + y)$
58. $8s(7s + 2s^2 - 1)$
59. $-3j(3j^2 + 2j + 6)$
60. $-h(-2h^2 - 2h + 6)$
61. $-2x(-6 - 3x + 5x^2)$
62. $e(-2 + 3e^2 - 4e)$
63. $2w(3 + w - 4w^2)$
64. $3h(-h^2 - 3h + 1)$
65. $4m(7 + 2m - 10m^2)$
66. $3c^2(-c - 2c^2 + 1)$
67. $2p^2(6p^2 + 2 - 3p)$
68. $-2y^2(2y^2 - 2y - 2)$
69. $-4f(3f + 2 + 5f^2)$
70. $-d^3(2 - d + 3d^2)$
71. $6a^2(a^2 + 2a + 3)$
72. $-3x(8 - 2x^2 - 4x)$

73. Write an expression for the area of a rectangle with a length of $x^2 + 4x + 8$ and a width of $6x$. Then simplify the expression.

74. Write an expression for the area of a rectangle with a length of $3a^4 - a^3 + 3a$ and a width of $4a^2$. Then simplify the expression.

Simplify. Be sure your answer is in standard form. (Lesson 9-1)

75. $-7e + e^2 - 2e^2 + e$

76. $3z^2 + 2z^2 - z + 2z^3$

77. $4a + 2a^2 + 3 + 2a^2$

78. $6p - 2p^2 - 3p - 4p^3$

79. $-x^2 + 2x^3 - x^2 + 3x^3$

80. $5d - 2d^2 + 3d^3 - 2d^2$

Simplify. (Lesson 9-2)

81. $(4b + 3) + (-2 - 3b)$

82. $(-2m - 3 + 4n) + (6 - n + 5m)$

83. $8x - (3 + 2x)$

84. $(4p - 2q) - (1 + p - 2q)$

85. $(4x + 2x^2 - 3) - (x - 6 + 4x^2)$

86. $(2c^2 + 3d - 3c) - (6c - d + 4d^2)$

Simplify. (Lesson 9-3)

87. $(-3a)(2b)$

88. $(y^2)(-4y)$

89. $(6w^4)(3w^2)$

90. $(-2x)^3(-3x)^2$

91. $(4m^2n)^2(2mn^3)$

92. $(-pq)^4(3p^2q)^2$

Simplify. (Lesson 9-4)

93. $3d(-2 + d)$

94. $-a(4a + 3a^2)$

95. $4m^2(-2m + 3)$

96. $-2z(z - 2z^2 + 3)$

97. $3h(4 - 2h + 5h^2)$

98. $-2g^2(3g^2 + 4 - 5g)$

Mid-Chapter Quiz

Simplify. (Lessons 9-1–9-4)

1. $5x + 5 - 8x + 8$

2. $7m^2 - 4m^3 - 8m^2 + 6m + 5m^3$

3. $-4r + 7r^2 + 9r - 1$

4. $14t^3 + 5t - 3t^2 - 6t^3 + 2 - 8t$

5. $ab - 5ab + 4a^2b - 3ab^2$

6. $(8m^2 + 3m) - (5m^2 - 6m)$

7. $(4t^2 - 6t) + (t^2 - 3t)$

8. $(8x^3 - 2x) + (3x - 5) - (-4x^3 + 6x)$

9. $-6p(p^3 - 4p^2 + p)$

10. $3a^2(-a^3 - 4a + 6)$

11. $(3b)(4d)$

12. $(-5m)(-3n)$

13. $(8t)(-2t)$

14. $(-a^3)(4a^5)$

15. $(2b^3)(4b^4)$

16. $4c(c + 5c^2)$

17. $-3(-d + 6d^2)$

18. $5s^3(2s^2 - 3s)$

19. Let n represent an integer. Write an expression for the product of that integer and an integer that is 6 greater than 3 times the integer n. Then simplify the expression. (Lesson 9-4)

20. A farmer has a rectangular plot of land measuring x mi wide and $3x + 2$ mi long. He purchases another plot measuring $5x$ mi wide and $3x + 2$ mi long. Write an expression for the total area of land. (Lesson 9-4)

Factor Using Greatest Common Factor (GCF)

Goals ■ Factor polynomials using the greatest common factor.

Applications Statistics, Meterology

Two rectangles are shown. Assume that each unit square represents 1 ft². Do the rectangles appear to have the same area?

1. Find the area of each rectangle.

2. What other whole number dimensions can a rectangle with the same area have?

3. What are all the factors of 36?

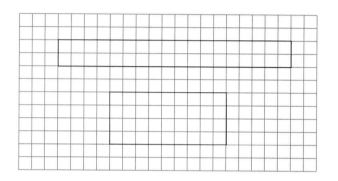

▼ BUILD UNDERSTANDING

Recall that to simplify expressions, you apply the distributive property.

$$2(3x + 4) = 2(3x) + 2(4) = 6x + 8$$

Begin with the factors, and multiply them to find the product. This process can be reversed to find what factors were multiplied to obtain the product. The reverse process is called **factoring**.

The **greatest common factor (GCF)** of two or more monomials is the *common factor* with the greatest numerical factor and variable or variables of greatest degree.

Example 1

Find the greatest common factor of the monomials.

a. $4x$ and $16xy$ **b.** $-6a^2b$ and $12ab$

Solution

a. Write the factors of $4x$. $(2^2)(x)$

Write the factors of $16xy$. $(2^4)(x)(y)$

The GCF of $4x$ and $16xy$ is $(2^2)(x)$, or $4x$.

b. Write the factors of $-6a^2b$. $(-1)(2 \cdot 3)(a^2)(b)$

Write the factors of $12ab$. $(2^2 \cdot 3)(a)(b)$

The GCF of $-6a^2b$ and $12ab$ is $(2 \cdot 3)(a)(b)$, or $6ab$.

To factor a polynomial, look for the greatest monomial factor common to all the terms. Then write the polynomial as the product of the greatest monomial factor and another polynomial.

Example 2

Factor each polynomial.

a. $3a + 6$ 　　　　　　　　　　　　　　　**b.** $4x^2y - 18x$

Solution

a. Find the GCF of each term of $3a + 6$.

　　factors of $3a$: $(3)(a)$　　　factors of 6: $(2 \cdot 3)$

The GCF is 3. Use the GCF to rewrite the polynomial.

$$3a + 6 = 3 \cdot a + 3 \cdot 2$$
$$= 3(a + 2) \qquad \text{Use the distributive property.}$$

So, $3a + 6 = 3(a + 2)$.

b. Find the GCF of each term.

　　factors of $4x^2y$: $(2^2)(x^2)(y)$　　　factors of $18x$: $(2 \cdot 3^2)(x)$

The GCF is $2x$. Use the GCF to rewrite the polynomial.

$$4x^2y - 18x = (2x)(2xy) - (2x)(9)$$
$$= 2x(2xy - 9)$$

So, $4x^2y - 18x = 2x(2xy - 9)$.

The GCF of some polynomials is an expression, not a monomial term. The process to factor such polynomials is sometimes called *chunking*. Chunking refers to the process of collecting several pieces of information and grouping them together as a single piece of information. In algebra, you are chunking when you think of a polynomial as a variable expression instead of a group of individual terms.

Example 3

Factor $n(2 + n) + 5(2 + n)$.

> ### Reading Math
>
> Read the expression in Example 3 as "*n* times the quantity two plus *n*, plus five times the quantity two plus *n*."

Solution

$n(2 + n) + 5(2 + n) = (n + 5)(2 + n)$　　　Use the distributive property to factor out the chunk $(2 + n)$.

So, $n(2 + n) + 5(2 + n) = (n + 5)(2 + n)$.

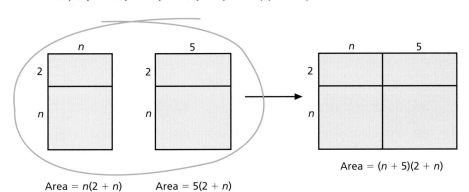

Area $= n(2 + n)$　　　Area $= 5(2 + n)$

Area $= (n + 5)(2 + n)$

$$n(2 + n) + 5(2 + n) = (n + 5)(2 + n)$$

Find the greatest common factor of the monomials.

1. $3y$ and $15yz$ **2.** $4cd^2$ and $20cd$ **3.** $24x^3$ and $8x^2$ **4.** $6x^2y$ and $10x^2y^2$

Factor each polynomial.

5. $4x + 12$ **6.** $2n + 4$ **7.** $y^2z - 9y$ **8.** $6a^2b - 15a$

Factor each expression.

9. $a(a - 3) + 2(a - 3)$ **10.** $(t + 5)t - (t + 5)7$

11. STATISTICS The mean of the four numbers a, b, c and d can be found using the formula $M = \frac{1}{4}a + \frac{1}{4}b + \frac{1}{4}c + \frac{1}{4}d$. Rewrite the formula by factoring.

12. WRITING MATH List 3 examples of how you group items in everyday life.

▼ **PRACTICE EXERCISES** • For Extra Practice, see page 578.

Find the greatest common factor of the monomials.

13. $8y$ and $4xy^2$ **14.** $3c^2$ and $18cd$

15. $5ab^2$ and $15a^2b$ **16.** $7r^2s$ and $9rs$

17. $10s^2t$ and $25s^3t^2$ **18.** $14x^2y^2$ and $42x^2y$

Match each set of monomials with their greatest common factor.

19. $2x^2$, $2x^2y$, $2x^2z$ **a.** $2ab$

20. $6ab^2$, $-4a^2b$, $12a^2b^2$ **b.** xy

21. $-2x^3y$, $-x^2y$, $-2xy^2$ **c.** x^2

22. ax^2, $2bx^2$, $3cx^2$, $5dx^2$ **d.** $2x^2$

Factor.

23. $6 - 12x$ **24.** $5x^2 - 3x$

25. $2x^2 - 6x$ **26.** $35a - 7ab$

27. $9 - 18x$ **28.** $5x^2 - 4x$

29. $4xy - y^2$ **30.** $t^2 - 2t$

31. $4n^2 - 16$ **32.** $12ab - 3b^2$

33. $35mn - 15m^2n^2$ **34.** $24x^2 - 6xy$

35. $28m^2 - 14mn$ **36.** $36ab - 25a^2b^2$ **37.** $6y^2 - 12y + 15$

38. $3x^2 + 6x^2y - 18xy$ **39.** $xy + xy^2 - y^2$ **40.** $25a^3 + 50a^2 + 5a$

41. $8ab - 16a^2b + 32ab^2$ **42.** $7xy - 14x^2y^2 + 28xyz$ **43.** $27m^2n^3 - 18mn^2 - 3m^2n$

44. $r(r + 6) + 2(r + 6)$ **45.** $3(k - 1) + k(k - 1)$ **46.** $(x - 3)x + 4(x - 3)$

47. $12(y + 7) - y(7 + y)$ **48.** $c^2(c + 1) + c(c + 1) + 5(c + 1)$ **49.** $(z - 5)z^2 - (z - 5)z$

Find the missing factor.

50. $8x^5 = (2x^2)(\underline{\quad ? \quad})$

51. $-6a^3b^2 = (2ab^2)(\underline{\quad ? \quad})$

52. $21c^2d = (3c^2)(\underline{\quad ? \quad})$

53. $12xy^2 = (3x)(\underline{\quad ? \quad})$

54. WRITING MATH Explain the order of operations to follow to simplify this expression. Simplify the expression, then factor.

$$6x^2 - x[3x - 4x(2x - 3)]$$

◤ EXTENDED PRACTICE EXERCISES

Factor each polynomial.

55. $\frac{1}{3}xyz - \frac{1}{3}xy^2z - \frac{1}{3}x^2yz^2$

56. $14f^2g - 42fg^2 - 56g^2h$

57. $4rs + 4st - 4tu + 4ru$

FARMING Silos are sometimes shaped like cylinders that store food products before they are shipped. For Exercises 58 and 59, use the formula for the surface area of a cylinder with radius r and height h.

$$SA = 2\pi rh + 2\pi r^2$$

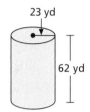

23 yd

62 yd

58. Rewrite the formula with the right side as a polynomial in factored form.

59. Find the approximate surface area for the inside of a silo with the given dimensions. Use 3.14 for π.

Write an expression in factored form for the area of each shaded region.

60.

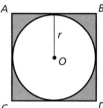

61.

◤ MIXED REVIEW EXERCISES

The data represent the number of people who mailed their tax return at 20 post offices on April 15. (Lesson 1-3)

210	120	550	220	320
290	310	420	280	360
410	570	430	290	210
560	480	390	320	180

62. Make a stem-and-leaf plot in which 2|1 represents 210 people.

63. Identify any modes.

64. Which is greater for these data, the mean or the range?

Multiply or divide. Check your answer. (Lesson 1-3)

65. $-4 \cdot 5$

66. $28 \div (-7)$

67. $12 \cdot 3$

68. $-64 \div (-4)$

Divide by a Monomial

Goals
- Divide a monomial by a monomial.
- Divide a polynomial by a monomial.

Applications Area and Nature

You know that division "undoes" multiplication.

$$-4 \cdot 9 = -36, \text{ so } -36 \div 9 = -4.$$

1. Simplify the product $3g \cdot 5h$.

2. Write a division equation that "undoes" the multiplication you performed in Question 1.

3. Write a division equation that "undoes" each of these.

   a. $2a \cdot 5a = 10a^2$

   b. $6y \cdot 4y^2 = 24y^3$

   c. $2m \cdot 7mn = 14m^2n$

▼ BUILD UNDERSTANDING

You have found the product of two monomials by multiplying the coefficients and multiplying the variables.

$$3m \cdot 2n = (3 \cdot 2)(mn) = 6mn$$

Because division is the inverse of multiplication, the process of multiplication can be reversed to obtain the related division sentence.

$$3m \cdot 2n = 6mn, \text{ so } 6mn \div 3m = 2n.$$

A different way to find the quotient $6mn \div 3m$ is shown.

$$\frac{6mn}{3m} = \frac{6}{3} \cdot \frac{mn}{m} \qquad \text{Divide both coefficients by their GCF, which is 3.}$$

$$= 2 \cdot \frac{mn}{m} \qquad \text{Divide the variable parts by their GCF, which is } m.$$

$$= 2n$$

This division illustrates the following rule for dividing monomials.

Dividing Monomials	To simplify the quotient of two monomials, find the quotient of any numerical coefficients. Then find the quotient of the variables.

Example 1

Simplify.

   a. $\dfrac{-4mn}{2m}$

   b. $\dfrac{wxz}{wz}$

Mental Math

What will be the coefficient of each quotient?

1. $\dfrac{-16ab}{8b}$ 2. $\dfrac{225pq}{25p}$ 3. $\dfrac{36st}{-6s}$

Solution

a. $\dfrac{-4mn}{2m} = \dfrac{-4}{2} \cdot \dfrac{mn}{m} = -2 \cdot n = -2n$

b. $\dfrac{wxz}{wz} = \dfrac{x}{1} = x$ Both coefficients are 1.

The quotient rule for exponents can be used to find the quotients of monomials having the same base.

Quotient Rule for Exponents	$\dfrac{a^m}{a^n} = a^{m-n}$ To divide two powers having the same base, subtract the exponent of the denominator from that of the numerator.

Example 2

Simplify.

a. $\dfrac{-16x^8}{4x^4}$

b. $\dfrac{-24x^3y^5}{-3x^2y}$

Solution

a. $\dfrac{-16x^8}{4x^4} = \left(\dfrac{-16}{4}\right)x^{8-4} = -4x^4$

b. $\dfrac{-24x^3y^5}{-3x^2y} = \left(\dfrac{-24}{-3}\right)x^{3-2}y^{5-1} = 8xy^4$

Each of the following statements is true.

$\dfrac{18 + 24 + 36}{6} = \dfrac{78}{6} = 13$ and $\dfrac{18}{6} + \dfrac{24}{6} + \dfrac{36}{6} = 3 + 4 + 6 = 13$

So you can arrive at this conclusion.

$\dfrac{18 + 24 + 36}{6} = \dfrac{18}{6} + \dfrac{24}{6} + \dfrac{36}{6}$

This example suggests the following general rule.

Dividing a Polynomial by a Monomial	$\dfrac{a + b}{c} = \dfrac{a}{c} + \dfrac{b}{c}$ when a, b and c are real numbers and c is not equal to 0.

Check Understanding

Refer to Example 2, part b. Explain why the exponent 2 in the denominator is subtracted from the exponent 3 in the numerator but not from the exponent 5.

Example 3

Simplify.

a. $\dfrac{6a + 9}{3}$

b. $\dfrac{2x^4 + 8x^3 + 12x^2}{2x^2}$

Solution

a. $\dfrac{6a + 9}{3} = \dfrac{6a}{3} + \dfrac{9}{3} = 2a + 3$ Divide each term of the polynomial by the divisor.

b. $\dfrac{2x^4 + 8x^3 + 12x^2}{2x^2} = \dfrac{2x^4}{2x^2} + \dfrac{8x^3}{2x^2} + \dfrac{12x^2}{2x^2}$

$= x^2 + 4x + 6$

Simplify.

1. $\dfrac{-6xy}{2y}$

2. $\dfrac{abc}{-ac}$

3. $\dfrac{8cd}{-4cd}$

4. $\dfrac{12wz}{3z}$

5. $\dfrac{27x^7y}{3x^2}$

6. $\dfrac{-21a^7b^8}{-3ab^5}$

7. **WRITING MATH** In Exercise 6, why is the exponent 5 in the denominator subtracted from the exponent 8 in the numerator but not from the 7?

Simplify.

8. $\dfrac{4x + 8}{2}$

9. $\dfrac{6y - 12}{3}$

10. $\dfrac{-14b + 21}{7}$

11. $\dfrac{2a^2 + 4a + 12}{2}$

12. $\dfrac{3y^4 + 6y^3 + 12y^2}{3y}$

13. $\dfrac{4x^5 - 8x^3 - 12x^2}{4x^2}$

▼ **PRACTICE EXERCISES** • For Extra Practice, see page 579.

Simplify.

14. $\dfrac{20cd}{4d}$

15. $\dfrac{18gh}{2g}$

16. $\dfrac{40ry}{10r}$

17. $\dfrac{-12ab}{2a}$

18. $\dfrac{-12pq}{3q}$

19. $\dfrac{-12st}{-4t}$

20. $\dfrac{18x^6}{-9x^3}$

21. $\dfrac{24z^6}{6z^5}$

22. $\dfrac{45a^3}{-9a}$

23. $\dfrac{y^3z}{y^2z}$

24. $\dfrac{x^3y^2z}{x^2y}$

25. $\dfrac{cd^6}{d^5}$

26. $\dfrac{4w^2x^2}{-2wx}$

27. $\dfrac{ab^5}{b^3}$

28. $\dfrac{-x^5y^5}{x^3y^4}$

29. $\dfrac{28a^8}{7a^2}$

30. $\dfrac{18b^9}{-3b}$

31. $\dfrac{gh^3}{-h^2}$

32. $\dfrac{x^3}{x^2y}$

33. $\dfrac{x^2y^2}{y^2z}$

34. $\dfrac{5z^3}{15z}$

35. $\dfrac{2a + 12}{2}$

36. $\dfrac{18y + 6}{2}$

37. $\dfrac{9a - 18b}{3}$

38. $\dfrac{-24x^2 + 16x}{8x}$

39. $\dfrac{27x^2 - 9x + 3}{3}$

40. $\dfrac{32c - 8b - 4a}{4}$

41. $\dfrac{6x^3 - 9x^2 + 3x}{3x}$

42. $\dfrac{28x^2y^2 - 21xy^2 + 14xy}{7xy}$

43. $\dfrac{9a^2b^2c^2 - 15abc^2}{3abc}$

44. **WRITING MATH** Why is the quotient of $\dfrac{12x^2}{2x^2}$ equal to 6?

45. A rectangle has an area of $36ab$ square units. The width is $4b$ units. Write an expression for the length of the rectangle.

Write an expression for the unknown dimension of each rectangle.

46. Area: $25pq$

$5p$

?

47. Area: $36abc$

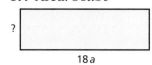

?

$18a$

48. Area: $28x^3y^2$

$4xy$

?

49. The product of $3x^2y^5$ and a certain monomial is $21x^7y^9z^2$. Find the monomial.

50. The area of the square shown is $25m^4n^8$. Write an expression for the unknown dimension.

?

51. **MODELING** Write a division sentence for the Algeblocks shown.

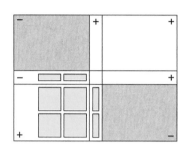

EXTENDED PRACTICE EXERCISES

Simplify.

52. $\dfrac{-18r^6s^2t^3}{-3r^2st^2}$

53. $\dfrac{24a^4b - 12a^2b^2 + 18a^3b^3}{3a^2b}$

54. $\dfrac{35c^3d^2e - 50c^4d^3e^2}{5cde}$

55. $\dfrac{8z^2xy - 12x^2zy + 16zx^2y^2 - 24z^2x^3y^3}{4zxy}$

56. **CRITICAL THINKING** Write an expression for the volume of a rectangular prism with a length of $6xz$, a width of $3y^2$ and a height of $5z$.

57. A prism with a volume equal to the one in Exercise 55 has a length of $15y$ and a width of $2z$. Write an expression for its height.

58. **CHAPTER INVESTIGATION** Complete this proportion for how much dry ice is needed to seed this tropical storm.

$$\frac{0.17}{d} = \frac{0.62}{\text{surface area of a cloud}}$$

MIXED REVIEW EXERCISES

DATA FILE For Exercises 59–62, refer to the data about champion trees on page 529. (Lesson 2-7)

59. Find the radius at 4.5 ft of the giant sequoia tree.

60. Find the radius at 4.5 ft of the sugar pine tree.

61. Find the difference in your answers for Exercises 60 and 61.

62. Theresa collected data on the height and growing rate of a tree. She found that the equation $y = 3x - 6$, where x is the age of the tree in years and y is the height of the tree in feet, describes the tree's rate of growth. Graph the equation. Then find the height of the tree when it is 7 years old. (Lesson 7-6)

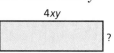

Review and Practice Your Skills

Find the greatest common factor of the monomials.

1. $3d$ and $12cd$

2. $2pq$ and $6p$

3. $10c$ and $5cd$

4. $6ab$ and $2a^2b$

5. $5m^2n$ and $15m$

6. $4xy^2$ and $12x^2y$

7. $3g^2h^2$ and $18g^2h$

8. $20fg^2$ and $4f^2g^2$

9. $3s^2t$ and $21st$

Match each set of monomials with their greatest common factor.

10. $12x^2y$, $16x^3y$, $24x^2y^2$ **a.** $3x^2y$ **b.** $4x^2y$ **c.** $4xy^2$

11. $2m^4$, $3m^2$, $4m^3$ **a.** m^2 **b.** $2m$ **c.** $2m^2$

12. $6a^2b$, $8a^2b^2$, $12a^2b$ **a.** $2a^2$ **b.** $2a^2b$ **c.** $2ab^2$

Factor.

13. $9w + 3$

14. $-6 - 12y$

15. $5d - 20d^2$

16. $3h + 12g^2h$

17. $10c^2d - 5cd^2$

18. $2m^2n - 4m^2n^2$

19. $-xz + x - x^2z$

20. $st^2 - 2s^2t + st$

21. $a^2c - ac^2 + a^2c^2$

22. $-4ag^2 + 2g^2 - 8a^2g$

23. $3xy + 12x^2y - 6y^2$

24. $20p^2 + 16p^2q - 8p^2q^2$

25. $4(n + 2) - n(n + 2)$

26. $(g - 5)4 - (g - 5)g$

27. $5(3 - z^2) + 2z(3 - z^2)$

Simplify.

28. $\dfrac{2m^2}{m}$

29. $\dfrac{24c^3}{8c}$

30. $\dfrac{-9d^2}{3d}$

31. $\dfrac{-2a^2b^2}{b}$

32. $\dfrac{18m^2n}{2n}$

33. $\dfrac{10x^2y^2}{-5xy^2}$

34. $\dfrac{p^3q^2}{p^2q}$

35. $\dfrac{50gh^2}{10h}$

36. $\dfrac{14c^2d^2}{2cd}$

37. $\dfrac{36a^2x^3}{12x^2}$

38. $\dfrac{3y^2z^3}{3yz}$

39. $\dfrac{6ab^4c}{3b^2c}$

40. $\dfrac{4d - 8}{4}$

41. $\dfrac{9m^2 - 6n}{3}$

42. $\dfrac{4q - 2q^2}{2q}$

43. $\dfrac{6t^2 - 8t}{2}$

44. $\dfrac{6y^2 + 9y}{3y}$

45. $\dfrac{15c^2 - 5c^3}{5c}$

46. $\dfrac{8x + 4w - 20}{4}$

47. $\dfrac{2x - 10x^2 + 8x^3}{2x}$

48. $\dfrac{3p^2q + 9pq^2 - 3p^2q^2}{3pq}$

49. $\dfrac{4m^2}{2m}$

50. $\dfrac{15a^2 + 6ab^2}{3b}$

51. $\dfrac{2p^3q - 16p^2q^2 + 20pq^3}{2pq}$

Simplify. (Lesson 9-2)

52. $(4 - 2a) + (-a - 6)$

53. $(3x - 2x^2) + (x - 6x^2 + 4)$

54. $(3y^2 + 2y - 3) - (6 - 2y^2)$

55. $(5a + 2 - 3b) - (-b + 4a - 1)$

Simplify. (Lesson 9-3)

56. $(-3m)(4n^2)$

57. $(5f^2)(-2f)^3$

58. $(-2yz^2)^3(-3yz)^2$

Simplify. (Lesson 9-4)

59. $5y(2y^2 - 3)$

60. $-3g(6 + 3g^2 - g)$

61. $2m^2(3m - 4 + 5m^2)$

Factor. (Lesson 9-5)

62. $8fg^2 - 14g$

63. $6xy + 3x - 9y$

64. $14mn^2 + 4m^2n - 7m^2n^2$

65. $2(c + 3) + c(c + 3)$

66. $(h^2 - 4)h + (h^2 - 4)$

67. $3w^2(2 - w) - 5(2 - w)$

Math*Works*
Workplace Knowhow

Career – Disaster Relief Coordinator

Disaster relief coordinators aid the victims of natural disasters, as well as help minimize the damage prior to a natural disaster. Their services include offering shelter and food to victims, rebuilding damaged homes and preparing for disaster situations. Disaster relief coordinators organize the building of levees in efforts to minimize flood damage. Bags are half-filled with sand. The first layer of bags is laid parallel to the flow of the river. Two sandbags are placed side-by-side, so that the width of the levee is two sandbags. The second layer is perpendicular to the flow of the river. Each layer then alternates in direction. See the single stack of sandbags shown below. The final height of the levee is equal to one-third of the levee's width.

1. Draw a model of the bottom layer of a stack of sandbags. Label the dimensions, using l for length and w for width.

2. Write a monomial expression for the height of the levee.

3. Write an expression in feet for the number of sandbags that are needed along the bottom layer of a levee that is 1 mi long.

4. Write an expression in feet for the number of sandbags that are needed along the bottom layer of a levee that is x mi long.

Problem Solving Skills: Algeblocks and Area

It can help to **use a model picture or diagram**, such as Algeblocks, to solve problems. Many careers involve creating a model to solve a problem. An automobile designer generates a computer model before manufacturing a car. An architect uses floor plans and models to determine the characteristics of a building or structure. Restaurant employees use a diagram of the tables to seat customers.

Problem Solving Strategies

Guess and check

Look for a pattern

Solve a simpler problem

Make a table, chart or list

✔ Use a picture, diagram or model

Act it out

Work backwards

Eliminate possibilities

Use an equation or formula

Problem

LANDSCAPING A square garden measures x meters by x meters. Suppose the width doubled and the length increased by 2 m.

a. Sketch or show the new garden using Algeblocks.

b. Find the area of the new garden.

Solve the Problem

a. First, represent the original garden. Next, add Algeblocks to represent the dimensions of the new garden. Then complete the area of the new garden.

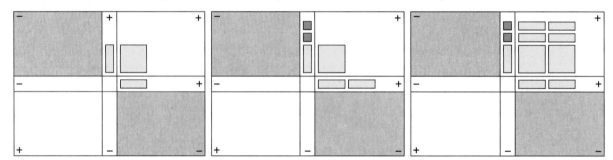

b. Total the Algeblocks to find the area of the new garden.

$$x^2 + x^2 + x + x + x + x = 2x^2 + 4x$$

The area of the new garden is $2x^2 + 4x$ m.

◥ TRY THESE EXERCISES

State the multiplication and product represented on each Quadrant Mat.

1.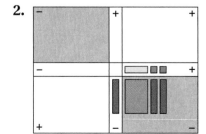

2.

3.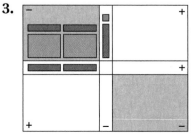

Sketch or show each product on an Algeblocks Mat.

4. $x(x + 2)$

5. $x(2y)$

6. $y(2y + 1)$

Five-step Plan

1 Read
2 Plan
3 Solve
4 Answer
5 Check

PRACTICE EXERCISES

INTERIOR DESIGN Tile needs to be ordered for a bathroom floor. The width of the bathroom is y feet. The floor is twice as long as it is wide.

7. Use Algeblocks to model the area of the bathroom.

8. How many square feet of tile must the builder order?

9. What is the area of the bathroom if the width is 1.8 ft?

10. What is the area of the bathroom if the width is $2\frac{1}{4}$ ft?

CONSTRUCTION A family will knock down two walls to expand the size of their kitchen. Currently, each side of the square room measures x meters. The room's length will double and its width will increase by 5 m.

11. Draw a diagram of the current kitchen. In a different color draw the kitchen addition.

12. What will be the dimensions of the new kitchen?

13. Use Algeblocks to model the area of the original kitchen.

14. Find the area of the original kitchen.

15. Use Algeblocks to model the area of the new kitchen.

16. Find the area of the new kitchen.

17. By how many square meters will the kitchen increase?

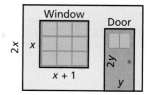

MUSIC The walls, ceiling and floor of a recording studio are carpeted to absorb sound. A window and a door separate the musicians from the recording engineers. The diagram is labeled with dimensions of the wall, window and door in feet.

18. Use Algeblocks to model the area of the window.
19. Find the area of the window.

20. Use Algeblocks to model the area of the door.
21. Find the area of the door.

22. Use Algeblocks to model the area of the wall.
23. Find the area of the wall.

24. If the entire wall will be carpeted except for the window and door, how many square feet of carpeting will be used to cover the wall?

MIXED REVIEW EXERCISES

Use the Pythagorean Theorem to solve each problem. Round to the nearest tenth. (Lesson 7-9)

25. Kenji has 60 m of string. How much farther can he let out the kite with the remaining string?

26. Suppose Kenji pulled in the string so that the kite was still 26 m above the ground, but the ground distance from Kenji to the kite was also 26 m. How much extra string would he have?

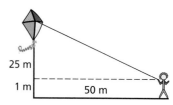

Chapter 9 Review

VOCABULARY ◥

1. The __?__ of two or more monomials is the common factor with the greatest numerical factor and variable or variables of greatest degree.

2. In a polynomial, terms that are exactly alike, or that are alike except for their numerical coefficients, are called __?__.

3. A polynomial with three terms is called a __?__.

4. In $5m$, 5 is called the __?__.

5. A __?__ is an expression that is a number, a variable, or the product of a number and one or more variables.

6. In the expression $2x^2 + 3x - 6$, 6 is called the __?__.

7. Ordering the terms of a polynomial from the greatest power of the variable to the least power is writing the polynomial in __?__.

8. The reverse of the Distributive Property is called __?__.

9. Each monomial is a __?__ of the polynomial.

10. An example of a __?__ is $2xy - 3y$.

a. binomial

b. coefficient

c. constant

d. factoring

e. Greatest Common Factor (GCF)

f. Least Common Multiple (LCM)

g. like terms

h. monomial

i. polynomial

j. standard form

k. term

i. trinomial

LESSON 9-1 ◥ Introduction to Polynomials, p. 394

▶ A **monomial** that contains no variable is called a **constant**.

▶ The *sum* or *difference* of two or more monomials is called a **polynomial**. One way to simplify a polynomial is to combine **like terms**. A polynomial is written in **standard form** when its terms are arranged in order from greatest to least powers of one of the variables.

Simplify. Be sure your answer is in standard form.

11. $9n - n$

12. $-5m + 3m^2 + m^3 + 4m^2$

13. $9x^2 + 3x - 2x^3 + 3x^2 + 7x$

14. $8q + 3 + 4q - 2$

15. $4r^2 + 11r + 5r^2 - 3r$

16. $n^2 - 3n + 10 + 3n$

17. $-5p^2 + 2p + 3p^2 - 4p$

18. $6w^2 + 8w - 9 + 7w$

19. $7t^2 + 8t^3 - 9t - 8t^2$

20. $6x + 2(2x + 7)$

21. $5n + 12n + 36$

22. $3x + 6y + 12x + 4y$

23. $x^2 + 3x + 2x + 5x^2$

24. $4a^3 + 6a + 3a^3 + 8a$

25. $3m^2 + 2m - 4m^2 + 8$

26. $j^2 - 3j^3 + 8j^2 + j - 12$

27. $5n^3 + 3n^2 - 6n^3 + 4n^2$

28. $8k^2 - 7k + 3k^2 + 7k$

LESSON 9-2 ◥ Add and Subtract Polynomials, p. 398

▶ To *add* polynomials, combine like terms.

▶ To *subtract* a polynomial, add its opposite and then simplify.

Simplify.

29. $(8h + 4) + (5 - 5h)$ **30.** $(13a^2 + 7a) - (4a^2 - a)$ **31.** $3m - (3m + 8)$

32. $(r^2 + 9) + (-4r^2 - 6r + 10)$ **33.** $(g^2 + 3g - 6) + (6g^2 - 6g)$

34. $(-2m + 10) + (5m - 3)$ **35.** $(4x^2 - 7x) + (8x + 5)$

36. $(3h^2 - h) + (h - 2h^2)$ **37.** $(-5w + 6) + (7w - 3)$

38. $(k^2 - 12) - (k^2 + 6k - 9)$ **39.** $(3u^2 - 9) - (u^2 + 21u + 2)$

40. $(9x^2 - x - 2) - (3x^2 - x - 4)$ **41.** $(y^2 + y + 1) - (y^2 - y + 1)$

42. $(3x^2 + 2x - 4) - (4x^2 + 2x - 4)$ **43.** $(z^2 - z + 3) - (2z^2 + 5z - 2)$

LESSON 9-3 ◼ Multiply Monomials, p. 404

▶ To multiply two powers having the same base, add the exponents.
$a^m \cdot a^n = a^{m+n}$

▶ To find a power of a monomial that is a power, multiply the exponents.
$(a^m)^n = a^{m \cdot n}$

▶ To find the power of a product, find the power of each factor and then multiply.
$(ab)^m = a^m b^m$

Simplify.

44. $(2a)(-4b)$ **45.** $(x^2 y^2)^3$ **46.** $-k(k^3 - 7k - 5)$

47. $(3x)(4y)$ **48.** $(-2a)(4b)$ **49.** $(-3t)(-4w)$

50. $(2y)(-3z)$ **51.** $(-2x^2)(3xy)$ **52.** $(4gh)(-2g^2 h)$

53. $(-4uv^2)(3u^2 v)$ **54.** $(-5c^2 d)(2c)$ **55.** $(j^2 k^2)(-3j^2 kl)$

56. $(-2x^2)(5x)$ **57.** $(-8a^2 b)(2b^2)$ **58.** $(6q)(7qr)$

59. $(-5x^3 yz^2)^2$ **60.** $(-2mn^2)^3$ **61.** $(3wx^3 y^2)^2$

62. $(-3x^2 y^2)^2(2xy)$ **63.** $(-5ab)(a^2 b^2)^3$ **64.** $(4m^2 n)^2(3mnj)^2$

LESSON 9-4 ◼ Multiply a Polynomial by a Monomial, p. 408

▶ The distributive property states that a factor outside parentheses can be used to multiply each term within the parentheses.

Simplify.

65. $3x^2(x + 2)$ **66.** $-4t(t^3 - t^2)$ **67.** $8y^3(3y^2 - 9y + 7)$

68. $g(2g + 5)$ **69.** $-b(9b - 6)$ **70.** $(4y + 7)y$

71. $(6y - 6)(-y^2)$ **72.** $-k^2(2k - 3)$ **73.** $c(7c^2 + 3c - 4)$

74. $7m(m^2 + 3m + 4)$ **75.** $8a(a + a^2 - 3)$ **76.** $-3t(3t^2 - 2t + 6)$

77. $3d(4d^2 - 8d - 15)$ **78.** $-4y(7y^2 - 4y + 3)$ **79.** $2m^2(5m^2 - 7m + 8)$

80. $2h(-7h^2 - 4h)$ **81.** $-\frac{1}{4}m(8m^2 + m - 7)$ **82.** $-\frac{2}{3}n^2(-9n^2 + 3n + 6)$

LESSON 9-5 ■ Factor Using Greatest Common Factors (GCF), p. 414

▶ To **factor** a polynomial, look for the greatest monomial factor common to all the terms. Then write the polynomial as the product of the greatest monomial factor and another polynomial.

Factor each polynomial.

83. $6x^2y - 24x$

84. $ab^3 - a^2b^2 + b^2$

85. $y(y + 5) - 6(y + 5)$

86. $64 - 40ab$

87. $4d^2 + 16$

88. $6r^2s - 3rs^2$

89. $15cd + 30c^2d^2$

90. $32a^2 + 24b$

91. $36xy^2 - 48x^2y$

LESSON 9-6 ■ Divide by a Monomial, p. 418

▶ To simplify the quotients of two monomials, find the quotient of any numerical coefficients then find the quotient of the variables.

▶ To divide two powers having the same base, subtract the exponent of the denominator from the opponent of the numerator.

$$\frac{a^m}{a^m} = a^{m-m}$$

▶ When a, b and c are real numbers and c is not equal to 0,

$$\frac{a + b}{c} = \frac{a}{c} + \frac{b}{c}.$$

Simply.

92. $\dfrac{24mn}{-3m}$

93. $\dfrac{-15a^3b^2}{5ab}$

94. $\dfrac{12c + 15}{3}$

95. $\dfrac{4x^4 + 6x^3 + 14x^2}{2x^2}$

LESSON 9-7 ■ Problem Solving Skills: Algeblocks and Area, p. 424

▶ To solve some problems you may need to use a picture or a model, such as Algeblocks.

Sod needs to be ordered for a backyard. The width of the backyard is y feet. The yard is three times as long as it is wide.

96. Use Algeblocks to model the area of the backyard.

97. How many square feet of sod must the homeowner order?

98. What is the area of the backyard if the width is 95.5 ft?

99. What is the width of the backyard if the length is $75\frac{3}{4}$ ft?

100. A 7-lb application should be used for every 1000 ft^2 of sod. How many pounds of fertilizer would be used on the lawn in Exercise 98? Round to the nearest tenth.

CHAPTER INVESTIGATION

EXTENSION Work in a group to draw a model of the tropical storm that will be seeded. The plane that will seed the clouds has a wingspan of 18 m. The plane will travel back and forth across the storm. Trace the shortest path of the plane so that it covers the entire area only once.

Chapter 9 Assessment

Tell whether the terms in each pair are *like* or *unlike* terms.

1. $-4x, 3x$

2. $7t^2, -2t^2$

3. $x^4s^3, 6x^3s^4$

Simplify. Write your answer in standard form.

4. $4m + (-3m) + 2m$

5. $3x^2 + (-2x^2) + 4x^2$

6. $-8s^3 + 6s + 2s^3 + 4s - 3s^2$

7. $a^2(a^3 - 3a + c)$

8. $(6z^3 + 4z - 5) + (3z^2 - z + 2)$

9. $(5p^2 + 4p - 8) - (3p^2 + 2p - 12)$

11. $(-2x)(3xy)$

12. $(-2m^2p^2)(pq)$

13. $(g^3x^2)^3$

14. $(4k^3)(-2k^4)$

15. $\dfrac{12ab}{4a}$

16. $\dfrac{-6p^2}{2p^2}$

17. $\dfrac{15t^2m^7}{-3tm^3}$

18. $3(y - 2)$

19. $(-3m)(2m - 1)$

Factor each polynomial.

21. $14p - 8$

22. $8m^2n + 24m$

23. $2rs - 12r^2s^2 + 10rst$

24. $-5a(b + 3) + a(b + 3)$

25. $6x^3(y^2 + 7) - 9x(y^2 + 7)$

26. $8xq^5 - 2xq - 6xq^3$

27. ASTRONOMY Three meteorite fragments are located at the vertices of triangle *ABC*. Write and simplify an expression for the region's perimeter.

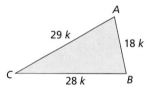

Simplify.

28. $\dfrac{-15ab}{3a}$

29. $\dfrac{-36x^3y^2z}{-9xz}$

30. $\dfrac{18q^5 - 21q - 9q^3}{3q}$

31. $\dfrac{25a^3b^2}{15ab^2}$

32. $\dfrac{16q^3 + 4q^2 - 3q}{3q}$

33. $\dfrac{-6x^8y}{3xz}$

Solve.

34. A rectangular garden has an area of $8p^3t^5$ square units. The width is $4pt$ units. Write and simplify an expression for the length.

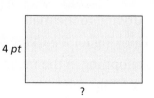

4 *pt*

?

35. The diagram shows the dimensions in feet of a doctor's office wall. The doctor wants to paint this wall. How many square feet will he paint?

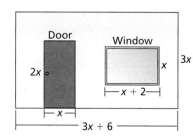

Standardized Test Practice

Part 1 Multiple Choice

Record your answers on the answer sheet provided by your teacher or on a sheet of paper.

1. What is the volume of the cylinder? Use $\pi = 3.14$. (Lesson 4-7)

   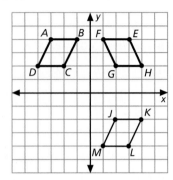

   Ⓐ 7033.6 yd^3

   Ⓑ 14,067.2 yd^3

   Ⓒ 98,470.4 yd^3

   Ⓓ 393,881.6 yd^3

2. Which open sentence best describes the graph? (Lesson 5-8)

   Ⓐ $x < -3$

   Ⓑ $x \leq -3$

   Ⓒ $x \geq -3$

   Ⓓ $x > -3$

3. The graph shows Mrs. Meyer's trip to the mall by car. What was she doing between 3:00 P.M. and 5:00 P.M.? (Lesson 7-1)

   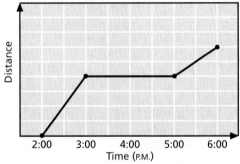

   Ⓐ She was driving to the mall.

   Ⓑ She was looking for a parking space.

   Ⓒ She was shopping in the mall.

   Ⓓ She was driving home from the mall.

4. If $\triangle EGH$ is reflected over the y-axis, what are the coordinates of H'? (Lesson 8-6)

   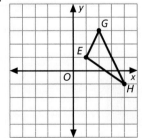

   Ⓐ $(4, 1)$

   Ⓑ $(4, -1)$

   Ⓒ $(-4, 1)$

   Ⓓ $(-4, -1)$

5. Which statement about the figures is true? (Lesson 8-7)

   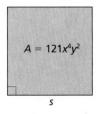

   Ⓐ Parallelogram $JKLM$ is a reflection image of $\square ABCD$.

   Ⓑ Parallelogram $EFGH$ is a translation image of $\square ABCD$.

   Ⓒ Parallelogram $JKLM$ is a translation image of $\square EFGH$.

   Ⓓ Parallelogram $JKLM$ is a translation image of $\square ABCD$.

6. Which polynominal is written in standard form? (Lesson 9-1)

   Ⓐ $6a^2 - 17 + a^3 + 9a^4$

   Ⓑ $-b - 2b^2 + 5b^3 - b^4$

   Ⓒ $10c^3 - c^2 + 7c + 20$

   Ⓓ $2d^5 + 12d - 8 - d^2$

7. Let x be an integer. What is the product of twice the integer added to three times the next consecutive integer? (Lesson 9-3)

   Ⓐ $5x$

   Ⓑ $5x + 1$

   Ⓒ $5x + 2$

   Ⓓ $5x + 3$

8. The area of a square is $121x^4y^2$. What is the length of each side? (Lesson 9-6)

   Ⓐ $11x^2y$

   Ⓑ $11x^2y^2$

   Ⓒ $30.25xy$

   Ⓓ $30.25xy^2$

Preparing for the Standardized Tests
For test-taking strategies and more
practice, see pages 587–604.

Part 2 Short Response/Grid In

**Record your answers on the answer sheet
provided by your teacher or on a sheet of paper.**

9. The scale of a drawing is 1 cm : 3 m. A room
   in the drawing is 10 cm long. What is the
   length of the actual room in meters?
   (Lesson 2-8)

Use the figure for Questions 10–12. (Lesson 4-2)

10. Identify the number of faces.

11. Identify the number of vertices.

12. Identify the number of edges.

13. What is the surface area of the rectangular
    prism in square centimeters? (Lesson 4-9)

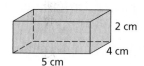

2 cm

4 cm

5 cm

14. Alicia rode her bicycle 44 km in $3\frac{2}{3}$ hr. Find
    her average rate of speed in kilometers per
    hour. Use the formula $d = rt$. (Lesson 5-6)

15. What is the amount of interest earned in one
    year on $900, if the rate of interest is 9.5% per
    year? (Lesson 6-5)

16. A ladder is propped against a window ledge
    that is 15 ft above the ground. If the base of
    the ladder is 8 ft away from the building, how
    long is the ladder in feet? (Lesson 7-7)

Test-Taking Tip

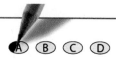

Question 13
Remember to rewrite the percent as a fraction or decimal to
compute the answer.

17. Which angles have the
    same measure as $\angle 1$?
    (Lesson 8-1)

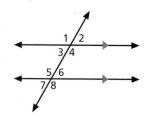

18. The measures of two sides of a triangle are
    $3x + 4y$ and $5x - y$. If the perimeter is
    $10x + 5y$, what is the measure of the third
    side? (Lesson 9-2)

19. Simplify: $-3t(2t^2 + 3t - 5)$. (Lesson 9-3)

20. A landscape architect is designing a square
    garden with an area of $144x^2$. If the lengths of
    each side are doubled, what will be the area of
    the garden? (Lesson 9-7)

Part 3 Extended Response

**Record your answers on a sheet of paper. Show
your work.**

21. Nick borrows $2500 at an annual interest rate
    of 12.9%. The term of the loan is 24 mo.
    (Lesson 6-5)

    a. Find the simple interest.

    b. What is the total amount due on the loan?

22. Refer to the graph. (Lesson 7-3)

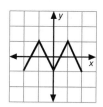

    a. Is this relation a function? Use the vertical
       line test.

    b. Explain why the vertical line test is a valid
       method to determine whether a relation is
       a function.

Probability

THEME: Genetics

Probability is the chance that a given event will occur. Some areas where probability is used are sports, weather, politics, and agriculture. You use probability when you guess at answers, play games with dice or cards, and enter a sweepstakes.

Are you short or tall? Do you have blue, green, or brown eyes? Can you roll or curl your tongue? Genetics determine all of these traits, plus more. Researchers and scientists make new discoveries in genetics each day.

- **Genetic counselors** (page 445) work with families to identify risks for inherited disease. The genetics of each family are used to calculate the probability of such defects.

- **Research geneticists** (page 463) work in sophisticated laboratories applying modern genetic technology to the study of blood, agriculture, medicine development, and police work.

Math Online

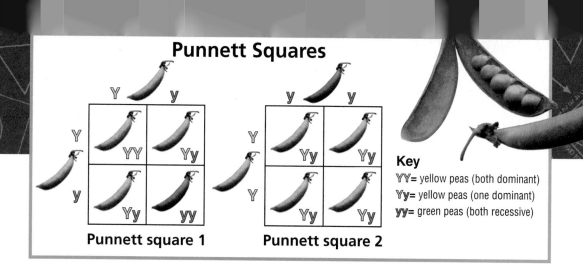

Punnett Squares

Punnett square 1

	Y	y
Y	YY	Yy
y	Yy	yy

Punnett square 2

	y	y
Y	Yy	Yy
Y	Yy	Yy

Key
YY= yellow peas (both dominant)
Yy= yellow peas (one dominant)
yy= green peas (both recessive)

Data Activity: Punnett Squares

Genes determine many physical traits in plants and animals. A single gene can have several different forms, called *alleles*. Dominant alleles express themselves more than others, and mask or hide recessive alleles. A person must inherit the same recessive allele from both parents in order for the recessive trait to be expressed.

A Punnett square uses capital letters to represent dominant alleles and lowercase letters to represent recessive alleles. Punnett square 1 shows possible outcomes of crossing a yellow pea plant with a yellow pea plant. Punnett square 2 shows possible outcomes of crossing a green pea plant with a yellow pea plant.

Use the Punnett squares for Questions 1–3.

1. How many offspring does each Punnett square show?

2. Suppose you cross two green peas. What color peas could the offspring possibly have? Why?

3. What percent of the offspring of a green pea (yy) and a yellow pea (Yy) will have green peas?

CHAPTER INVESTIGATION

Genetics determine physical qualities, hair type and facial features.

Working Together

Study the likelihood that certain traits will pass to children by finding theoretical and experimental probabilities. Assume that two parents each have one recessive allele for each trait in the chart shown. Remember that a person is equally likely to receive either allele from each parent. The Chapter Investigation icons will guide your progress.

Unattached earlobes	EE, Ee or eE
Attached earlobes	ee
Widow's peak	WW, Ww or wW
Continuous hairline	ww
Cleft chin	CC, Cc or cC
Smooth chin	cc
Dimples	DD, Dd or dD
No dimples	dd
Long eyelashes	LL, Ll or lL
Short eyelashes	ll
Large eyeballs	BB, Bb or bB
Small eyeballs	bb

The skills on these two pages are ones you have already learned. Use the examples to refresh your memory and complete the exercises. For additional practice on these and more prerequisite skills, see pages 536-544.

SIMPLIFYING FRACTIONS

Fractions are often more helpful when they are written in lowest terms. The same is true with probability. Remember to divide the numerator and denominator by the greatest common factor.

Example $\dfrac{24}{64}$ GCF = 8

$\dfrac{(24 \div 8)}{(64 \div 8)} = \dfrac{3}{8}$

Write each fraction in simplest form.

1. $\dfrac{14}{28}$ 2. $\dfrac{24}{32}$ 3. $\dfrac{48}{60}$ 4. $\dfrac{6}{18}$

5. $\dfrac{42}{56}$ 6. $\dfrac{26}{52}$ 7. $\dfrac{44}{80}$ 8. $\dfrac{48}{112}$

9. $\dfrac{60}{125}$ 10. $\dfrac{63}{288}$ 11. $\dfrac{75}{500}$ 12. $\dfrac{84}{120}$

RATIOS

The probability of an event is written as a ratio. Knowing how to write ratios is necessary for finding probability of a desired outcome.

Examples   The ratio of blue eyes to brown eyes can be written blue : brown.

$20 : 10$

Write as a fraction and reduce. $\dfrac{20}{10} = \dfrac{2}{1}$

The ratio of males with blue eyes to females with blue eyes can be written blue male : blue female.

$8 : 12$

Write as a fraction and reduce. $\dfrac{8}{12} = \dfrac{2}{3}$

Eye Color	Male	Female
hazel	14	24
blue	8	12
brown	6	4
green	1	1
gray	2	1

Use the data table to write each ratio as a fraction in simplest form.

13. hazel : green

14. brown : blue

15. green male : green female

16. blue female : hazel female

17. all females : all males

18. female blue or brown : male blue or brown

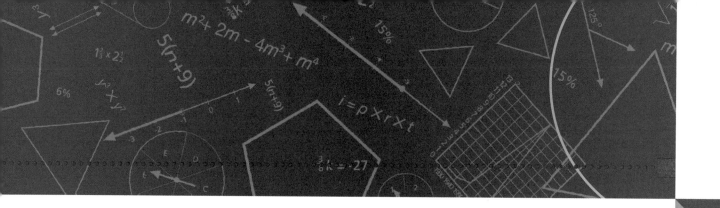

WRITING FRACTIONS AS PERCENTS

The probability of an event in everyday life is often presented as a percent. Weather forecasts use probability to describe the possibility of rain—"There's a 40% chance of a thunderstorm this afternoon."

Example   Write $\frac{3}{4}$ as a percent.

$$
\begin{array}{r}
0.75 \\
4\overline{)3.00} \\
\underline{28} \\
20 \\
\underline{20} \\
0
\end{array}
$$

$0.75 \cdot 100 = 75\%$

Write each fraction as a percent. Round to the nearest tenth of a percent.

19. $\frac{2}{5}$ **20.** $\frac{3}{8}$ **21.** $\frac{1}{3}$ **22.** $\frac{4}{10}$

23. $\frac{7}{12}$ **24.** $\frac{3}{16}$ **25.** $\frac{5}{8}$ **26.** $\frac{6}{20}$

27. $\frac{4}{9}$ **28.** $\frac{5}{7}$ **29.** $\frac{8}{15}$ **30.** $\frac{12}{25}$

MULTIPLYING FRACTIONS

Multiplication can be used to find the probability of two or more events happening simultaneously.

Examples $\frac{2}{5} \cdot \frac{25}{48} =$ $\frac{1}{500,000} \cdot 875,000 = \frac{875,000}{500,000}$

$\frac{1\cancel{2}}{1\cancel{5}} \cdot \frac{\cancel{25}^5}{\cancel{48}_{24}} = \frac{5}{24}$ $= \frac{875}{500}$

$$= 1\frac{3}{4}$$

Simplify.

31. $\frac{4}{5} \cdot \frac{10}{13}$ **32.** $\frac{7}{8} \cdot \frac{9}{14}$ **33.** $\frac{6}{7} \cdot \frac{15}{18}$ **34.** $\frac{9}{10} \cdot \frac{24}{35}$

35. $\frac{25}{28} \cdot \frac{14}{15}$ **36.** $\frac{42}{49} \cdot \frac{7}{20}$ **37.** $\frac{21}{35} \cdot \frac{7}{9}$ **38.** $\frac{45}{56} \cdot \frac{35}{36}$

39. $250 \cdot \frac{1}{48}$ **40.** $1000 \cdot \frac{1}{250}$ **41.** $3600 \cdot \frac{1}{360}$ **42.** $5000 \cdot \frac{1}{100}$

43. $6400 \cdot \frac{1}{8000}$ **44.** $10,000 \cdot \frac{1}{12,000}$ **45.** $\$1,000,000 \cdot \frac{1}{20,000,000}$

46. $\$500,000 \cdot \frac{1}{15,000}$ **47.** $\$26,000,000 \cdot \frac{1}{50,000}$ **48.** $\$125,000 \cdot \frac{1}{500,000}$

10-1 Introduction to Probability

Goals ■ Find the probability of an event.

Applications Weather, Health

Work with a partner. Use a number cube with faces labeled 1 through 6.

1. Toss a number cube 20 times. How many times does each face land up?

2. Predict how many times 5 will land face up if you toss the number cube 40 times.

3. Toss a number cube 40 times and record the number of times 5 lands face up.

4. Compare this number to your prediction. How close was your prediction?

◣ BUILD UNDERSTANDING

An **event** is an outcome or a combination of outcomes. The event of tossing a six-sided number cube can have six possible results, or **outcomes**. The outcomes (1, 2, 3, 4, 5, and 6) are **equally likely** to happen.

The **probability** of an event is a numerical measure of chance that is from 0 to 1. Probability is a ratio written as a percent, a fraction, or a decimal that compares favorable outcomes to possible outcomes. A **favorable outcome** is the particular outcome whose likelihood you are calculating. A favorable outcome is called the *desired outcome*.

Probability of an Event	$P(E) = \dfrac{\text{number of favorable outcomes}}{\text{number of possible outcomes}}$

Example 1

Find each probability. Use the spinner.

a. $P(2)$ **b.** $P(\text{blue})$ **c.** $P(10)$ **d.** $P(\text{blue, red, or green})$

Solution

a. The favorable outcome is the one section labeled 2.

$$P(2) = \frac{1}{6} \quad \text{one favorable outcome} \atop \text{six possible outcomes}$$

b. Three of the six sections are blue.

$$P(\text{blue}) = \frac{3}{6} = \frac{1}{2}$$

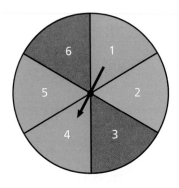

c. None of the six sections is 10 so there are no favorable outcomes.

$P(10) = \dfrac{0}{6} = 0$ The probability of any *impossible event* is 0.

d. All the outcomes are favorable.

$P(\text{blue, red, or green}) = \dfrac{6}{6} = 1$ The probability of a *certain event* is 1.

Check Understanding

In Example 1 part c, explain what is meant by an impossible event.

In Example 1 part d, how do you know that the event is certain to occur?

Example 2

A bag contains 3 purple, 4 green, 4 blue, and 4 white marbles. Find the following. Give your answers as percents rounded to the nearest tenth.

a. $P(\text{green or white})$ **b.** $P(\text{not purple})$

Solution

a. Find the total number of marbles.

$3 + 4 + 4 + 4 = 15$

Add to find the number of *green or white* marbles.

$4 + 4 = 8$

$P(\text{green or white}) = \dfrac{8}{15}$

$8 \div 15 \approx 0.533 = 53.3\%$

b. Subtract to find the number of *not purple* marbles.

$15 - 3 = 12$

$P(\text{not purple}) = \dfrac{12}{15}$

$12 \div 15 = 0.8 = 80\%$

The **odds of an event** is also a numerical measure of chance. This ratio compares favorable outcomes to unfavorable outcomes and is often written as *a* to *b* or *a* : *b*. The formula below is used for computing the *odds in favor of an event*.

Odds in Favor of an Event	odds in favor of $E = \dfrac{\text{number of favorable outcomes}}{\text{number of unfavorable outcomes}}$

Example 3

Find the odds in favor of each event by using the spinner.

a. B   **b.** green   **c.** not A

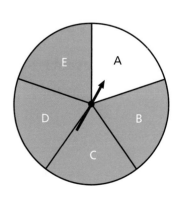

Solution

There are 5 possible outcomes.

a. The odds in favor of B $= \dfrac{1 \text{ B section}}{4 \text{ non-B sections}} = \dfrac{1}{4}$ or 1 to 4.

b. Two sections are green. The odds in favor of green $= \dfrac{2}{3}$ or 2 to 3.

c. Four sections are not A. The odds in favor of not A $= \dfrac{4}{1}$ or 4 to 1.

Find each probability using the spinner. Give your answers as fractions and percents.

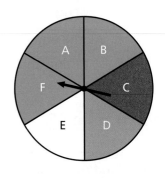

1. *P*(blue)

2. *P*(not blue)

3. *P*(C or D)

4. *P*(Z)

5. *P*(red, green, or blue)

6. *P*(A, B, C, D, E, or F)

Use the spinner above to find the odds in favor of each event.

7. C or D

8. green

One of the cards is picked without looking. Find each probability.

9. *P*(O)

10. *P*(C, O, or E)

11. *P*(B)

12. *P*(T)

13. WRITING MATH Describe the probability of randomly choosing two positive integers whose product is even. How would the answer change for the sum of two such integers?

14. GENETICS Use the Punnett squares on page 435. What is the probability that the offspring for each Punnett square is a green pea (yy)?

Find each probability using the spinner. Give your answers as fractions and percents.

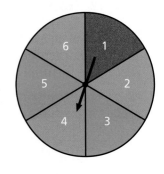

15. *P*(blue)

16. *P*(4 or 5)

17. *P*(even number)

18. *P*(not green)

19. *P*(number divisible by 2)

20. *P*(10)

21. *P*(not red)

22. *P*(6)

23. *P*(number > 0)

24. *P*(red)

Use the spinner above to find the odds in favor of each event.

25. 2 or 6

26. red

27. green or 5

28. 2, 3 or 4

One card is drawn randomly. Find each probability.

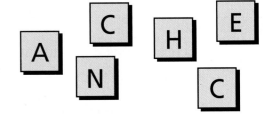

29. *P*(vowel)

30. *P*(C)

31. *P*(N or H)

32. *P*(consonant)

33. *P*(consonant or vowel)

34. *P*(S)

35. HEALTH Odds are often used with medical treatments. For example, the odds in favor of recovery using a new medicine are 1 to 4. Find the probability of recovery using the medicine.

A number cube with faces labeled from 1 through 6 is tossed. Find each probability.

36. $P(6)$ **37.** $P(2)$ **38.** P(even number)

39. P(prime number)   **40.** P(number > 10)   **41.** P(2, 3, 5 or 6)

42. P(composite number)   **43.** P(odd number)   **44.** P(1 or 4)

45. WRITING MATH Use the spinner from Exercises 15–28. In your own words, tell if it is possible to not spin a green in 100 spins.

46. DATA FILE Refer to the data on the world's tallest buildings on page 521. Suppose the names of the ten tallest buildings are put into a hat and one name is drawn randomly. What is the probability that the building drawn is 100 or more stories tall?

EXTENDED PRACTICE EXERCISES

What is the probability of selecting the shaded regions in the figures? (Hint: Subtract the area of the small figure from the area of the large figure.)

47.

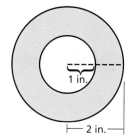

1 in.

2 in.

48.

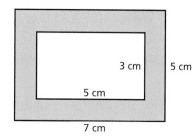

3 cm

5 cm

5 cm

7 cm

CRITICAL THINKING If the probability of an event occurring is $\frac{3}{5}$, then the event has a 60% chance of occurring. A certain event can be described as "100% sure." For Exercises 49–51, guess the percent chance of each event.

49. A person chosen at random from a crowded street will be a woman.

50. The sun will rise in the west tomorrow.

51. Water will freeze when its temperature is lowered to $-10°$F.

52. WEATHER A meteorologist says that there is a 30% chance of snow tomorrow. What is the probability that it will not snow tomorrow? Write the probability as a percent.

MIXED REVIEW EXERCISES

Find the slope of each line. (Lesson 7-5)

53. $y = 7x - 12$ **54.** $2y = -x + 5$

55. $y = 56 - \frac{3}{4}x$ **56.** $x + 3y = 12$

57. On a coordinate plane, graph the ordered pairs $(-1, -2)$, and $(4, 3)$. Draw a line through the points. Make a table showing three additional points that lie on the same line. Write the equation for the line. (Lesson 7-4)

Experimental Probability

Goals ■ Find experimental probability.

Applications Entertainment, Sports, Food service, Industry

Work with a partner. Use a number cube with faces labeled 1 through 6.

1. Use the probability formula to find the probability of rolling a 5 on the first toss.

2. Now toss the number cube 60 times. Keep a tally of the results of your experiment in a table.

3. Examine the results of your experiment. How many times did the face labeled 5 actually appear?

4. Compare your results with your classmates.

◣ BUILD UNDERSTANDING

The probability of an event based on the results of an experiment is called the **experimental probability**.

Example 1

The table shows the outcomes from an experiment in which a number cube was tossed 50 times.

Number on cube	1	2	3	4	5	6
Number of outcomes	7	9	10	2	11	11

Find each experimental probability.

a. $P(6)$

b. $P(3)$

c. P(even number)

d. $P(4 \text{ or } 5)$

Solution

Add the number of outcomes to find the total tosses. The cube was tossed a total of 50 times.

a. $P(6) = \dfrac{11}{50}$

b. $P(3) = \dfrac{10}{50} = \dfrac{1}{5}$

c. $P(\text{even number}) = \dfrac{22}{50} = \dfrac{11}{25}$

d. $P(4 \text{ or } 5) = \dfrac{13}{50}$

> **Check Understanding**
>
> Write the solution for each part of Example 1 as a percent.

Experimental probabilities are often used when all outcomes are not equally likely. For instance, when you toss a paper cup, it could land up, down or on its side. So there are three outcomes, but they are not equally likely. You would expect that the cup would most often land on its side.

Example 2

A paper cup is tossed 50 times. The results are shown in the table. Find the experimental probability of the cup landing on its side.

Solution

Count the tally marks. The cup landed on its side 36 times.

$$P(\text{side}) = \frac{36}{50} = \frac{18}{25}$$

Outcome	Tally
up	|||
down	⊬⊬
side	⊬⊬⊬⊬⊬ ⊬⊬⊬⊬|

A **simulation** acts out an event so that you can find the outcomes. Tossing a coin or number cube or spinning a spinner are ways you can simulate a situation. Each time you simulate the problem, you complete one **trial**.

Example 3

ENTERTAINMENT Dena is a contestant on a TV game show. She has to answer 10 true-false questions. To win a prize, she must answer at least 6 of the questions correctly. Predict the probability of Dena winning a prize.

Solution

Each question has two possible outcomes: either Dena answers correctly or incorrectly. So the game can be simulated by tossing 10 coins.

Let heads represent a correct answer. Let tails represent an incorrect answer. Toss all 10 coins at once. Possible outcomes are shown.

	Number of heads (correct answers)	Number of tails (incorrect answers)
1st toss	4	6
2nd toss	7	3
3rd toss	5	5
4th toss	6	4
5th toss	7	3

> **Check Understanding**
>
> Why is a coin toss appropriate for simulating a true-false test?
>
> What simulation would be appropriate for a multiple-choice test with 6 choices? with 4 choices?

In five tosses, heads showed on six or more coins three times.

$$P(\text{winning}) = \frac{3}{5} \quad \frac{\text{number of favorable outcomes}}{\text{number of possible outcomes}}$$

$$3 \div 5 = 0.60$$

So you predict that Dena has a 60% chance of winning a prize.

You can use a random sample to find the experimental probability of an event. Using this probability, conclusions can be drawn about the population sample.

To obtain a random sample, you can simulate the experiment. Consult a random number list or use a calculator or computer with a random number generator. Most graphing calculators will generate random numbers between 0 and 1.

Example 4

In a bottle factory, 1000 bottles are selected at random. Of these, 3 are defective. What is the probability of a bottle being defective?

Solution

$$P(\text{defective}) = \frac{3}{1000} \quad \frac{\text{number defective in sample}}{\text{total number in sample}}$$

The probability of a bottle being defective is $\frac{3}{1000}$, or 0.3%.

◥ TRY THESE EXERCISES

A number cube is tossed 20 times. The outcomes are recorded in the table. Find each experimental probability.

Number on cube	1 2 3 4 5 6
Number of outcomes	3 5 2 5 4 1

1. $P(1)$　　　　　**2.** $P(2)$　　　　**3.** $P(3)$

4. $P(4)$　　　　　**5.** $P(5)$　　　　**6.** $P(6)$

7. $P(\text{even number})$　**8.** $P(2 \text{ or } 4)$　**9.** $P(\text{odd number})$

Jessie chose a marble from a bag 80 times. With each selection, she recorded the color and returned the marble to the bag. Use the results in the table to find each experimental probability.

10. $P(\text{striped})$

11. $P(\text{blue})$

12. $P(\text{red})$

13. $P(\text{striped or blue})$

Outcome	Tally	Total
striped	卌 卌 卌 卌 卌	25
blue	卌 卌 卌 卌 卌 卌 卌	35
red	卌 卌 卌 卌	20

14. **SPORTS** A basketball team won 62 out of 120 games. What is the experimental probability that the team will win their next game?

15. **WRITING MATH** Jontay guesses the answers to 15 questions on a true-false test. Explain how to determine the experimental probability of Jontay getting 12 or more answers correct.

◥ PRACTICE EXERCISES • For Extra Practice, see page 580.

FOOD SERVICE The table shows the number of muffins served in the cafeteria one morning. Find the experimental probability of each event.

Blueberry	Cranberry	Bran	Cinnamon	Apple	Other
22	6	10	8	4	10

16. $P(\text{blueberry})$　　**17.** $P(\text{not blueberry})$　　**18.** $P(\text{cranberry})$

19. $P(\text{not bran})$　　**20.** $P(\text{bran})$　　**21.** $P(\text{apple or cinnamon})$

22. $P(\text{cinnamon})$　　**23.** $P(\text{not cranberry})$　　**24.** $P(\text{not apple or other})$

25. INDUSTRY On a production line, 18 defective bolts were found among 2000 randomly selected bolts. What is the probability of a bolt being defective?

ENGINEERING A quality control engineer at Everglow Bulbs tested 400 light bulbs and found 6 of them to be defective.

26. What is the experimental probability that an Everglow light bulb will be defective?

27. In a shipment of 75,000 light bulbs, how many are likely to be defective?

28. DATA FILE Refer to the data on lower-back injuries on page 530. Suppose 875 lower-back-injury patients are polled randomly and asked how they injured their back. How many will respond that their injury was caused by bending?

29. GENETICS Use the Punnett squares on page 435. Toss two coins 100 times to simulate the cross in Punnett square 1. How close is the experiment to the probabilities in the Punnett square? Design an experiment to simulate the other squares.

◣ EXTENDED PRACTICE EXERCISES

Make an unfair "coin" by taping a nickel and a penny together. Be sure your coin has one head and one tail.

30. Flip the coin 10 times. Use your results to predict how many tails you will get if you flip the coin 50 times. Then flip the coin 40 more times for a total of 50 flips. Was the prediction you made after 10 flips accurate for 50 flips?

31. CRITICAL THINKING If you predict the number of tails in 500 flips using your results for 50 flips, do you think your prediction will be more reliable than predicting from 10 flips? Explain.

32. Make four cards with an X on one side. Place the cards face down and tell your partner that each card has an X or 0 on it. Have your partner try to guess each symbol as you concentrate on each card.

33. How many cards should your partner guess correctly by chance?

34. How many times should your partner guess correctly to be convinced that he or she can read minds?

◣ MIXED REVIEW EXERCISES

Use the figure for Exercises 35–38. (Lesson 8-1)

35. What is the measure of ∠AXF?

36. Are $\overleftrightarrow{AD}$ and $\overleftrightarrow{EB}$ parallel, perpendicular, or neither?

37. Name three angles that have the same angle measure as ∠AXF?

38. If ∠AXB has an angle measure of 2x, what is the measure of ∠BXC?

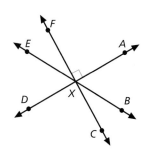

Review and Practice Your Skills

Find each probability using Spinner 1. Give your answers as fractions and percents.

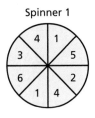
Spinner 1

1. $P(6)$ **2.** $P(4)$ **3.** P(even number)

4. P(1 or 2)   **5.** $P(7)$ **6.** P(number less than 4)

Use Spinner 1 to find the odds in favor of each event.

7. 5   **8.** 5 or 6   **9.** odd number   **10.** number less than 3

Find each probability using Spinner 2. Give your answers as fractions and percents. Round percents to the nearest tenth of a percent.

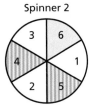
Spinner 2

11. P(shaded)   **12.** P(striped and even)

13. P(white and odd)   **14.** P(white)

15. P(striped or odd)   **16.** P(white or even)

Use Spinner 2 to find the odds in favor of each event.

17. even   **18.** striped   **19.** shaded   **20.** striped and odd

The letters A, B, C, D, E, and F are written on six cards. One card is drawn randomly. Find each probability.

21. $P(D)$ **22.** P(A or B)   **23.** P(vowel)   **24.** $P(G)$

The table shows the outcomes from an experiment in which a number cube was tossed 50 times.

Number on cube	1	2	3	4	5	6
Number of outcomes	6	8	9	6	11	10

Find each experimental probability.

25. $P(3)$ **26.** P(1 or 2)   **27.** P(even number)   **28.** P(odd number)

A spinner is divided into seven nonequal parts, with a letter of the alphabet in each part. The spinner is spun 50 times. This table shows the outcomes.

Letter	A	B	C	D	E	F	G
Number of outcomes	6	4	15	5	10	3	7

Find each experimental probability.

29. $P(E)$ **30.** $P(M)$ **31.** P(A or B)   **32.** P(A or G)

33. $P(F)$ **34.** $P(D)$ **35.** P(E, F or G)   **36.** P(consonant)

Find each probability using the spinner. Give your answers as fractions and percents. Round percents to the nearest tenth of a percent. (Lesson 10-1)

37. P(R or P)　　**38.** P(white and Q)　　**39.** P(R, P or Q)

40. P(white)　　**41.** P(shaded or Q)　　**42.** P(striped and P)

Use the spinner to find the odds in favor of each event. (Lesson 10-1)

43. striped　　**44.** white, not P　　**45.** shaded and P　　**46.** shaded or P

A spinner is divided into eight nonequal parts, with a letter of the alphabet in each part. The spinner is spun 100 times. This table shows the outcomes.

Letter	A	B	C	D	E	F	G	H
Number of outcomes	6	18	12	14	15	5	20	10

Find each experimental probability. (Lesson 10-2)

47. P(D)　　**48.** P(B or D)　　**49.** P(K)　　**50.** P(A, B or C)

51. P(A)　　**52.** P(C or H)　　**53.** P(vowel)　　**54.** P(consonant)

MathWorks Career – Genetic Counselor
Workplace Knowhow

Genetic counselors work with families to identify the risks for inherited disorders and birth defects. They analyze inheritance patterns and genetic traits of their clients. Genetic counselors take part in medical education and patient support groups.

Use the Punnett squares shown to find the probability of each event.

	D	d
D	DD	Dd
d	Dd	dd

Punnet Square A

	D	d
D	DD	Dd
D	DD	Dd

Punnet Square B

Key: DD -will definitely get the disease
　　Dd -might develop the disease
　　dd -no chance of developing
　　　 the disease

1. Punnett square A: a child will *not* develop this disease

2. Punnett square B: a child could develop the disease

3. Punnett square B: a child will definitely have the disease

4. Punnett square B: a child will *not* develop the disease

5. Punnett square A: a child might develop this disease

10-3 Sample Spaces and Tree Diagrams

Goals ■ Use a tree diagram to find possible outcomes in a sample space.

Applications Recreation, Literature, Genetics

Suppose that your teacher surprises you with a true-false quiz. You haven't done the reading yet, so you have to guess at the answers.

1. List the possible answers if the quiz has only one question.

2. Now suppose that the quiz has just two questions. List the possible answers.

3. What if the quiz has just three questions? List the possible answers. How can you make sure you are accounting for all the possibilities?

▼ BUILD UNDERSTANDING

To find a probability, it is important that you account for all outcomes. The **sample space** is the set of all possible outcomes in a probability experiment. When you toss a coin twice, there are four possible outcomes in the sample space. You can show the outcomes in a list or a table.

{HH, TT, HT, TH}

First toss	H	T	H	T
Second toss	H	T	T	H

A **tree diagram** is another way to show possible outcomes in a probability experiment. You can use a tree diagram to picture a sample space and then count the outcomes.

Example 1

A penny and a nickel are tossed. Make a tree diagram of the sample space. How many possible outcomes are there?

Solution

When each coin is tossed there are two possibilities.

    penny → heads or tails

    nickel → heads or tails

The tree diagram displays the sample space. There are four possible outcomes.

Penny	Nickel	Outcomes
H	H	HH
	T	HT
T	H	TH
	T	TT

Example 2

Three coins are tossed. Find *P*(exactly 2 tails).

Solution

Make a tree diagram to picture the sample space.

Coin 1	Coin 2	Coin 3	Outcomes
	H	H	HHH
		T	HHT
H	T	H	HTH
		T	HTT
	H	H	THH
		T	THT
T	T	H	TTH
		T	TTT

There are 8 possible outcomes.

Only 3 of these 8 outcomes contain exactly 2 tails.

   HTT, THT, TTH

So P(exactly 2 tails) $= \dfrac{3}{8}$.

Check Understanding

Explain the difference between an event and a sample space.

◥ TRY THESE EXERCISES

List each sample space.

1. tossing a coin

2. spinning the spinner shown

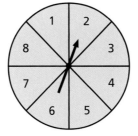

Make a tree diagram to find the number of possible outcomes in each sample space.

3. tossing four coins

4. tossing a coin and tossing a number cube

Suppose that you toss a coin and spin the spinner above. Find each probability.

5. *P*(tail and 5)

6. *P*(head and even number)

7. GENETICS Use Punnett square 1 on page 435. In how many ways could the offspring have yellow peas with one recessive allele (Yy)?

8. WRITING MATH Suppose you will guess at the answers to a true-false test. Tell what happens to your grade as the number of true-false questions increases. Is guessing ever a good strategy? Why or why not?

List each sample space.

9. picking a card from those shown

10. spinning Spinner 1

Use a tree diagram to find the number of possible outcomes in each sample space.

11. tossing a penny, a nickel and a dime

12. picking a card from those shown and tossing a coin

13. tossing a coin and spinning Spinner 1

14. tossing a nickel, tossing a number cube and spinning Spinner 1

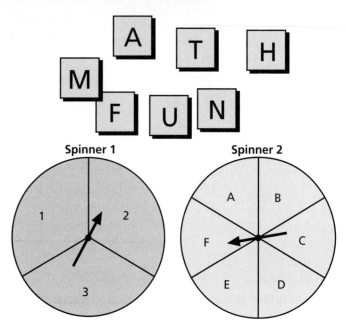

Spinner 1   Spinner 2

Suppose you spin both Spinner 1 and Spinner 2. Find each probability.

15. *P*(3 and E)

16. *P*(2 and consonant)

17. *P*(odd number and vowel)

18. *P*(prime number and C)

For Exercises 19–25, use these three spinners.

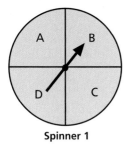

Spinner 1

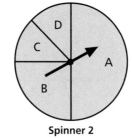

Spinner 2

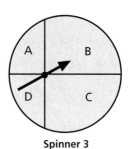

Spinner 3

List the sample space for each spinner.

19. Spinner 1

20. Spinner 2

21. Spinner 3

Find *P*(A) for each spinner.

22. Spinner 1

23. Spinner 2

24. Spinner 3

25. **CHECK YOUR WORK** Compare your answers to Exercises 22 and 24. How are the spinners alike? How are they different? Explain.

26. **LITERATURE** In Greek mythology, Jason enters a maze to find the Golden Fleece. Use a tree diagram to find how many different courses he can select.

27. What is Jason's probability of success?

Jason

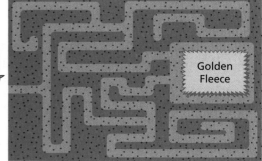

Golden Fleece

RECREATION The Highlands Recreation Center has the following rules for coding their membership cards.

>   *Rule 1* Letters and numbers may be used only once in each card.

>   *Rule 2* A membership code must be two numbers followed by two letters.

Using the given rules, list the possible codes for membership cards that can be made from the letters and numbers given for each.

28. 4, 9, A, J, R   **29.** 1, 6, 7, 5, P, N   **30.** 2, 8, U, I, O, E

31. Choose two numbers and two letters to create your own membership codes. Make a tree diagram to show all possible outcomes in this sample space.

32. WRITING MATH Write a definition of a sample space.

33. DATA FILE Refer to the data on maximum heart rate on page 531. Suppose three teens will be a sample set. If the teens are 15, 16 and 17, what are the possible maximum heart rates that could be recorded?

◤ EXTENDED PRACTICE EXERCISES

An arrangement of items in a particular order is called a *permutation*. Each time you pick an item for a position in a permutation, there will be one less item to pick from for the next position.

To find the number of permutations of 4 things, calculate the product $4 \cdot 3 \cdot 2 \cdot 1$. This product is written as 4!, and is called 4 *factorial*. The number of permutations of n different items is $n(n-1)(n-2)\ldots(2)(1)$. This is written as $n!$, which is read n factorial.

34. How many different "words" can be made with the letters *a, b, c, d* and *e* if all the letters are used? (The words do not have to make sense in any language.)

35. Find 3!   **36.** Find 8!   **37.** Find 6!

38. CALCULATOR Find 11! using the factorial function in the Math Menu on your graphing calculator.

39. CRITICAL THINKING What is 0 factorial? Verify your answer on your calculator. Explain your results.

40. CHAPTER INVESTIGATION Assume that each parent has one recessive allele for each trait. Flip a coin to find whether the other allele is dominant or recessive. Make a Punnett square for each trait, using the genetic makeup of each parent.

◤ MIXED REVIEW EXERCISES

Simplify. (Lesson 5-5)

41. $5c + (-15) + 8d - 12c$ **42.** $14x^2 - 24x + 7y - (-5x)$ **43.** $7j + 5(3 + 2jk) - jk$

Solve. (Lesson 5-4)

44. $27 - 11g + (-16) = 0$ **45.** $21 - (2x + 12) = 17$ **46.** $\frac{2}{3}(3x - 12) + 9 = -x$

47. Find the value of x if twice the sum of -4 and $3x$ equals 27. (Lesson 5-4)

48. How many feet are in a mile? How many yards are in a mile? (Basic Math Skills)

10-4 Counting Principle

Goals ■ Use the counting principle to find the number of outcomes.

Applications Sports, Retail, Travel

Work with a partner.

Suppose that you and a friend are making sandwiches for lunch. You can use either chicken or roast beef. For bread your choices are pita, wheat or pumpernickel.

1. How many different kinds of sandwiches can you make? (Use only one type of deli meat and one type of bread for each sandwich.) Make a tree diagram to solve the problem.

2. Name other ways you think you can find the number of different sandwiches besides using a tree diagram.

◤ BUILD UNDERSTANDING

As the number of choices in an event increases, it becomes less convenient to count the possible outcomes on a tree diagram. Another method to find the number of outcomes is the **counting principle**.

Counting Principle	Find the number of possible outcomes of an event by multiplying the number of outcomes at each stage of the event.

Example 1

RETAIL Benita needs a computer system for her home office. She must choose which computer, monitor and printer to buy. The table shows the possibilities from which she can choose.

Brand of computer : Plum, Ideal, Sunflower, AA Tech

Size of monitor : 14", 17", 21"

Type of printer : inkjet, laser

From how many different configurations can she choose?

Solution

There are three stages involved in choosing a configuration: select a computer, select a monitor and select a printer.

Multiply the number of choices at each stage.

computer · monitor · printer = number of configurations possible

$$4 \quad \cdot \quad 3 \quad \cdot \quad 2 \quad = 24$$

Benita can choose from 24 possible configurations.

Check Understanding

What would be an advantage of using a tree diagram in Example 1?

Example 2

In a game, a player tosses a number cube and chooses one of 26 alphabet cards. Using the counting principle, find P(prime number, T or Q).

Solution

First find the number of possible outcomes.

number cube · alphabet cards = number of possible outcomes

6 · 26 = 156

Then find the number of favorable outcomes.

There are 3 prime numbers: 2, 3, and 5.

There are 2 letters: T and Q.

prime numbers · letters = number of favorable outcomes

3 · 2 = 6

P(prime number, T or Q) $= \dfrac{6}{156} = \dfrac{1}{26}$

Math: Who, Where, When

The earliest book dealing with the theory of probability was *Art of Conjecturing* (1713) by Jacques Bernoulli.

▼ TRY THESE EXERCISES

Use the counting principle to find the number of possible outcomes.

1. Choose a stereo system from 2 receivers, 3 CD players and 4 cassette decks.

2. Make a sandwich using wheat, rye, or pumpernickel bread and peanut butter or cheese as a filling.

3. Toss a six-sided number cube 3 times.

4. Toss a coin four times.

Use the counting principle to find each probability.

5. A number cube is tossed 3 times. Find P(4, 4, 4).

6. A coin is tossed 5 times. Find P(all heads or all tails).

7. A coin is tossed 10 times. Find P(10 tails)

8. A number cube is tossed 4 times. Find P(1, 1, 2, 2).

9. **WRITING MATH** Describe how to determine different ways you can order soup from a restaurant that serves 5 kinds of soup in 3 different sizes, with and without salt.

10. **DATA FILE** Refer to the data on the 2000 Summer Olympics on page 533. Suppose that three randomly selected winners of the United States 2000 Olympics were arranged in order, one gold winner, one silver winner, and one bronze winner. In how many ways can they be arranged?

Use the counting principle to find the number of possible outcomes.

11. Choose a snack from 3 desserts and 4 kinds of herbal tea.

12. Choose breakfast from 4 cereals, 5 juices, and 2 kinds of toast.

13. Choose a pasta from spaghetti, rigatoni or linguini and a sauce from plain, meat, mushroom, garden vegetable, or herb.

14. Choose pants from denim, khaki, or plaid and a shirt from white, tan, red, blue, black, or yellow.

15. Toss a coin 3 times.

16. Toss a six-sided number cube 2 times.

17. Toss a coin 6 times.

18. Toss a six-sided number cube 10 times.

Use the counting principle to find the probability.

19. A coin is tossed 6 times. Find P(all heads).

20. A coin is tossed 3 times and a six-sided number cube is tossed twice. Find P(all tails and both numbers even).

21. A six-sided number cube is tossed 5 times. Find P(1, 2, 3, 4, 5).

22. A six-sided number cube is tossed 10 times. Find P(all even numbers).

Use the counting principle to answer the following questions.

23. How many different three-letter "words" can be made using any of the letters A through Z? (Assume that a letter may be repeated.)

24. How many different three-letter "words" can be made using any of the letters A through Z if each letter can only be used once and the first letter is a vowel?

25. A lock company uses 3 numbers for its combination locks. If the numbers range from 0 to 99, how many different combinations can be used if a number may be repeated?

26. What if each number can only be used once in each combination?

27. **WRITING MATH** Write a problem that has 24 outcomes.

28. How many different four-digit numbers can be made using only the digits 1, 2 and 3? (Assume that a digit may be repeated.)

29. **RETAIL** A type of telephone is available with or without an answering machine. Color selections are white, red, black, blue and yellow. Each phone has a programmable memory for 3, 6, 9, 12 or 20 telephone numbers. How many different types of telephone are available for sale?

30. **TRAVEL** Suppose that a car has room to seat 4 passengers. In how many ways can 4 people be arranged in the 4 seats if they all are able to drive?

31. In how many ways can 4 people be arranged in the 4 seats if only one of them is able to drive?

SPORTS Three students are going to run a relay race. The track coach has to decide which student will run in each position (first, second and third).

32. In how many ways can the coach assign a student to the first position?

33. Once a student has been assigned to the first position, in how many ways can the coach assign a student to the second position?

34. Once students have been assigned to the first and second positions, in how many ways can the coach assign a student to the third position?

35. In how many ways can the coach assign the 3 students to the 3 positions?

EXTENDED PRACTICE EXERCISES

GENETICS Children can have straight(SS), wavy(Ss) or curly(ss) hair. If two parents have wavy hair, then the probability of their children having straight hair is $\frac{1}{4}$, wavy hair is $\frac{1}{2}$ and curly hair is $\frac{1}{4}$. Find the following probabilities.

36. two children, the first with straight hair and the second with curly hair

37. two children, one with straight hair and one with curly hair

38. two children, neither with curly hair

39. three children, all with wavy hair

40. three children, the first with straight hair, the second with wavy hair and the third with curly hair

41. If there are four children with curly hair, what is the probability of the fifth child having curly hair?

42. CHAPTER INVESTIGATION Complete the Punnett squares for each trait. Calculate the theoretical probability for each trait, and make a table of the data like the one shown. Make a sketch of the face with the traits with the greatest probability.

Trait	Probability
Attached earlobes	
Continuous hairline	
Smooth chin	
No dimples	
Short eyelashes	
Small eyeballs	

MIXED REVIEW EXERCISES

Draw a representation of each figure. (Lesson 4-1)

43. $\overleftrightarrow{NP}$ **44.** $\overrightarrow{LK}$ **45.** plane G **46.** $\angle ABC$

47. MODELING Suppose you have 5 straws of the same length that you can use to represent line segments. What polygons can you make using some or all of the straws? Name each polygon and tell the number of vertices in each. (Lesson 4-2)

48. Name a figure that is equilateral but not equiangular. (Lesson 4-2)

Review and Practice Your Skills

Spinner 1

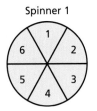

List each sample space.

1. tossing a coin   2. spinning Spinner 1   3. spinning Spinner 2

Use a tree diagram to find the number of possible outcomes in each sample space.

4. tossing three coins

5. tossing a coin and spinning Spinner 1

Spinner 2

6. tossing a coin and spinning Spinner 2

7. tossing two coins and spinning Spinner 2

8. spinning Spinner 1 and spinning Spinner 2

Suppose you spin both Spinner 1 and Spinner 2. Find each probability.

9. P(3 and A)

10. P(even number and C)

11. P(2 and A or B)

12. P(odd number and C)

13. P(1 or 2 and C)

14. P(even number and C or B)

Use the counting principle to find the number of possible outcomes.

15. Choose one rug, one paint color and one wallpaper from these choices:
    4 different rugs, 3 different paints, 2 different wallpapers.

16. Choose a meal from 6 salads, 10 sandwiches and 2 drinks. The meal must
    have 1 salad, 1 sandwich and 1 drink.

17. Choose a code number with three one-digit numbers and one letter. The
    numbers can each be any of the digits 0–9. The letter can be any of the
    choices A–Z.

18. A spinner is divided into five equal parts with the letters A, B, C, D and E.
    Choose a four-letter password by spinning four times.

19. Choose a four-digit password by rolling a 1–6 number cube four times.

Use the counting principle to find the probability.

20. In Exercise 17, what is the probability of getting 206B?

21. In Exercise 17, what is the probability of getting an odd number and a vowel?

22. In Exercise 18, what is the probability of getting the same letter four times?

23. In Exercise 18, what is the probability of getting all vowels?

24. In Exercise 19, what is the probability of getting four odd digits?

Find the probability. (Lesson 10-4)

25. A person chooses a two-digit number by rolling a 1-6 number cube twice. What is the
    probability of getting a number greater than 30?

Find each probability using the spinner. (Lesson 10-1)

26. P(even number) **27.** P(prime number) **28.** P(striped or odd)

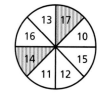

The table shows the outcomes when a number cube was tossed 60 times.

Number on cube	1	2	3	4	5	6
Number of outcomes	8	12	10	9	9	12

Find each experimental probability. (Lesson 10-2)

29. $P(3)$ **30.** $P(5)$ **31.** $P(2 \text{ or } 4)$ **32.** P(odd number)

Use a tree diagram to find the number of possible outcomes in each sample space. (Lesson 10-3)

33. tossing a 1–6 number cube and a dime   **34.** tossing two 1–6 number cubes

Use the counting principle to find the number of possible outcomes. (Lesson 10-4)

35. Toss a coin eight times.   **36.** Choose a letter A–Z and a whole number greater than 0 and less than 5.

Mid-Chapter Quiz

You are to choose a letter at random from the word MISSISSIPPI. Find each probability. (Lesson 10-1)

1. $P(M)$ **2.** $P(I)$ **3.** $P(S)$ **4.** $P(I \text{ or } P)$

5. P(vowel)   **6.** P(consonant)   **7.** $P(Q \text{ or } R)$ **8.** P(letter following E in the alphabet)

A spinner is spun 80 times with the results shown. Find each experimental probability. (Lesson 10-2)

Spinner number	1	2	3	4	5
Number of outcomes	13	15	16	21	15

9. $P(2)$ **10.** $P(4)$ **11.** $P(6)$ **12.** $P(3 \text{ or } 5)$

Use a tree diagram to find the number of possible outcomes for the sample space. (Lesson 10-3)

13. Picking a letter from the word WORK and spinning a spinner with two equal divisions labeled 1 and 3.

14. Picking a vowel and tossing a coin.

Use the counting principle to find the number of outcomes. (Lesson 10-4)

15. Rolling a number cube 3 times.

16. Choosing a sweatshirt from 4 colors and 5 styles.

17. Rolling a number cube and then tossing a coin twice.

18. Find P(odd number and all heads) for Exercise 17.

10-5 Independent and Dependent Events

Goals ■ Find the probabilities of independent and dependent events.

Applications Sports, Safety, Construction, Literature

A box contains one red, one white and one green chalk. Suppose that a chalk is selected from the box and replaced, and then a chalk is selected again.

1. Make a tree diagram to show the sample space.

2. Use the tree diagram to find P(red, then white).

Suppose that a chalk is selected from the box but is not replaced. Then another chalk is selected.

3. Make a tree diagram to show the sample space.

4. Use the tree diagram to find P(red, then white).

◤ BUILD UNDERSTANDING

Two events are **independent** when the result of the second event does not depend on the result of the first event.

If two events are independent, the probability that both events will occur is the product of their individual probabilities.

Independent Events	If two events X and Y are independent, then $$P(X \text{ and } Y) = P(X) \cdot P(Y)$$

Example 1

A bag contains 3 red apples and 2 yellow apples. An apple is taken from the bag and replaced. Then an apple is taken from the bag again. Find P(red, then red).

Solution

There are 5 apples in the bag, and 3 of these are red. Because the first apple is replaced, the second selection is independent of the first. There are 3 red apples in the bag when the second apple is chosen.

$$P(\text{red, then red}) = P(\text{red}) \cdot P(\text{red})$$
$$= \frac{3}{5} \cdot \frac{3}{5}$$
$$= \frac{9}{25}$$

Two events are **dependent** if the result of the first event affects the result of the second event. The probability that both events will occur is the product of the probability of X and the probability of Y, given that X has occurred.

Dependent Events	If two events X and Y are dependent and Y follows X, $$P(X \text{ and } Y) = P(X) \cdot P(Y \text{ after } X \text{ has occurred})$$

Example 2

A bag contains 3 red apples and 2 yellow apples. An apple is taken from the bag and is not replaced. Then another apple is taken from the bag.

Find P(red, then red).

Solution

Because the first apple is not replaced, the second event is dependent on the first. On the first selection there are 5 apples in the bag, 3 of which are red.

$$P(\text{red}) = \frac{3}{5}$$

On the next selection, there are only 4 apples in the bag. Assuming that a red apple was removed, only 2 of the apples in the bag are red.

$$P(\text{red after red}) = \frac{2}{4} = \frac{1}{2}$$

Multiply the two probabilities.

$$P(\text{red, then red}) = \frac{3}{5} \cdot \frac{1}{2} = \frac{3}{10}$$

◤ TRY THESE EXERCISES

A silverware drawer contains 10 forks, 16 teaspoons and 4 steak knives. A utensil is taken from the drawer and replaced. Then a utensil is taken from the drawer again.

Find each probability.

1. P(fork, then teaspoon)   2. P(steak knife, then fork)

3. P(fork, then fork)   4. P(teaspoon, then teaspoon)

5. **CONSTRUCTION** Jenna is laying bricks for a fireplace. A box contains 2 red bricks, 3 white bricks and 5 gray bricks. Find P(white, then gray) if she selects two bricks from the box without replacement.

6. A change purse contains 2 quarters, 3 dimes and 5 pennies. Find P(quarter, then dime) if the coins are selected without replacement.

7. **WRITING MATH** Explain the difference between independent and dependent events. Include an example of each type of event.

8. **DATA FILE** Refer to the data on the utilization of selected media on page 523. Suppose a household was polled randomly in 2000 and asked if they owned cable television and a VCR. What is the probability of the household owning both?

A game is played by spinning the spinner then randomly choosing a card. After each turn, the card is replaced. Find the probability of each event.

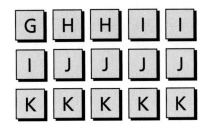
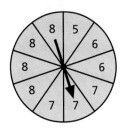

9. $P(5 \text{ then } G)$

10. $P(6 \text{ then } G)$

11. $P(5 \text{ then } K)$

12. $P(8 \text{ then } J)$

13. $P(7 \text{ then } I)$

14. $P(7 \text{ then } H)$

15. $P(6 \text{ then } J)$

16. $P(6 \text{ then } K)$

17. $P(5 \text{ then } H)$

18. $P(7 \text{ then } J)$

19. $P(8 \text{ then } G)$

20. $P(5 \text{ then } I)$

21. Are the events in Exercises 9–20 dependent or independent?

State whether the pairs of events are independent or dependent.

22. TJ will play in the basketball game today. He will score more than 10 points.

23. You got an A on the last math test. You will get an A on a history test.

24. It will storm on Saturday. You will go swimming Saturday at the beach.

Ten cards numbered 1 to 10 are in a box. A card is taken from the box and not replaced. Then a second card is chosen. Find the probability of each event.

25. $P(4 \text{ then } 4)$

26. $P(5 \text{ then odd})$

27. $P(6 \text{ then prime})$

28. $P(\text{prime then prime})$

29. $P(1 \text{ then } 2)$

30. $P(\text{even then } 9)$

31. WRITING MATH Explain how the probabilities change if the cards are replaced.

SPORTS A bucket contains 8 orange golf tees and 6 blue golf tees. Po Sin randomly selects 2 tees.

32. What is the probability that he selects no orange tees?

33. What is the probability that he selects one blue tee and one orange tee?

34. What is the probability that he selects two orange tees?

Two of the cards shown are chosen at random without replacing the first card before choosing the second. What is the probability of each event?

35. $P(\text{both vowels})$

36. $P(\text{vowel then consonant})$

37. $P(\text{both consonants})$

38. How do the answers for Exercises 35–37 change if the second card is chosen after the first card is replaced?

LITERATURE In a writing contest, 6 nonfiction stories and 4 fiction stories are chosen as finalists. Prizes are awarded randomly. The judges put the titles of the winning stories into a box and select 3 winners at a time. Find each probability.

39. P(3 fiction)

40. P(nonfiction, then fiction, then nonfiction)

41. P(3 nonfiction)

42. P(fiction, then nonfiction, then fiction)

43. Are the events in Exercises 39–42 dependent or independent?

Julita passes through 3 traffic lights on her way to school. The probability that each light is green is 0.4 and the probability that each light is red is 0.6.

44. What is the probability of Julita getting three green lights?

45. What is the probability of Julita getting two green lights and then a red light?

46. What is the probability of getting two red lights and then one green light?

47. **SAFETY** Marcus comes to a street with a stop sign. The probability of a car coming from the left is $\frac{2}{3}$ and from the right is $\frac{1}{2}$. What is the probability that he can safely cross the highway because there are no cars approaching?

48. What is the probability that Marcus can safely turn right?

49. Suppose the probability of a car coming straight at Marcus is $\frac{1}{3}$. What is the probability of him safely turning left?

■ EXTENDED PRACTICE EXERCISES

CRITICAL THINKING A wallet contains three $20-bills, one $10-bill and two $1-bills. Use this information to give an example for each Exercise.

50. two independent events with probability of $\frac{1}{6}$

51. two dependent events with probability of $\frac{1}{10}$

52. two dependent events with probability of 0

53. two independent events with probability of 0

54. **GENETICS** Two families, the Beckers and the Smiths, each have two children. At least one of the Smith children is a boy, and the oldest Becker child is a boy. Given this information, why are the probabilities of each family having two boys different?

■ MIXED REVIEW EXERCISES

Solve. (Lessons 2-3, 2-4 and 4-7)

55. The volume of a cylinder is 9.42 cm³. What is its height if its diameter is 1 cm? Use 3.14 for π. (Lesson 4-7)

56. The perimeter of a trapezoid is 48 m. Its two bases measure 17 m and 18 m. What is the sum or measure of its two remaining sides? (Lesson 2-3)

57. The area of a square is 148 ft². Find the length of one side of the square to the nearest tenth. (Lesson 2-4)

58. The volume of a rectangular prism is 225 cm³. If its height and width are 7.5 cm, what is its length? (Lesson 4-7)

Problem Solving Skills: Make Predictions

Recall from Chapter 1 that a random sample is a group chosen from a population so that each member has an equal chance of being chosen. The responses of a random sample can be used to make predictions and generalizations about the entire population.

When results from a sample population are available, you can use the fraction or decimal associated with a particular response to predict the total population's results. By first predicting or estimating, you can then determine if your final answer is reasonable.

Problem Solving Strategies
✔ Guess and check
Look for a pattern
Solve a simpler problem
Make a table, chart or list
Use a picture, diagram or model
Act it out
Work backwards
Eliminate possibilities
Use an equation or formula

Problem

POLITICAL SCIENCE Channel 3 conducts a poll before the mayoral election. The city has 250,000 registered voters. The polltakers ask a random sample of 225 voters which candidate they prefer. Results are shown. Predict how many will vote for Brimmage. Check your prediction.

Candidate	Voters planning to vote
Miselli	35
Nguyen	80
Brimmage	110

Solve the Problem

The 110 votes Brimmage received in the poll is about one-half of the sample. So Brimmage is expected to win about one-half of the election votes.
$250{,}000 \div 2 = 125{,}000$

First, to confirm your prediction, use the poll results to find $P(\text{Brimmage})$.

$$P(\text{Brimmage}) = \frac{110}{225} = \frac{22}{45} \approx 0.489$$

Second, find how many votes Brimmage might receive in the election.

$$P(\text{Brimmage}) \cdot \text{number of voters} = \text{predicted number of votes}$$

$$0.489 \quad \cdot \quad 250{,}000 \quad = \quad 122{,}222$$

Brimmage can expect to receive about 122,222 votes in the election. The prediction of 125,000 votes is a good prediction.

TRY THESE EXERCISES

INDUSTRY A quality control inspector estimates that 1 out of every 6 aluminum cans is not sealed properly.

1. Throw a number cube 10 times to create random sample data. Predict the number of defective cans in 1,000,000 cans.

2. Find the probability of finding a defective can.

3. Check your prediction. How many defective cans can be expected in 1,000,000 cans?

Five-step Plan

1 Read
2 Plan
3 Solve
4 Answer
5 Check

ENTERTAINMENT A cable company asks a random sample of 500 of its 50,000 subscribers their favorite kind of programming. Of those polled, 150 chose comedies, 80 chose sports, 225 chose movies and 45 chose wildlife shows.

Based on the random sample, determine the following.

4. the most popular program

5. the least popular program

6. the number who would choose sports

7. the number who would choose movies

Find each probability.

8. P(wildlife shows)

9. P(movies)

10. P(comedies)

11. P(sports)

Check your predictions from Exercises 4–7. If all of the subscribers were polled, how many would you expect to select the following?

12. the most popular program

13. the least popular program

14. the number who chose sports

15. the number who chose movies

16. **YOU MAKE THE CALL** Myisha thinks that 45,000 people will select sports in the survey above. Is she correct? Explain.

POLITICAL SCIENCE Of the 750,000 registered voters in Centerville, 1000 are polled about their preferences in an upcoming election. The results are shown.

17. Based on the random sample, estimate how many of the 750,000 voters will be undecided.

18. Find P(undecided votes).

19. Check your prediction from Exercise 17. How many voters can be expected to be undecided?

Bendel	370
Morales	325
Greenberg	155
Undecided	150

20. **GENETICS** Based on the data shown, predict the number of students in your class with each blood type.

21. Find the probability of each blood type among students in your class.

22. Check your prediction from Exercise 20. How many students in your class should have each blood type?

23. **WRITING MATH** Explain which would provide results closest to the data in the table, polling your class or polling your school. Why?

Blood type	Percent
A Positive	34
A Negative	6
B Positive	9
B Negative	2
AB Positive	4
AB Negative	1
O Positive	38
O Negative	6

◤ **MIXED REVIEW EXERCISES**

Simplify. (Lessons 9-2 and 9-4)

24. $(2h + -4k) + (k - 3h)$

25. $(7x + 3y + (-6)) - (x - 3y + 11)$

26. $-6(a - 2b) - 4(2b - a)$

27. $4(g^3 + 4g^2k + (-6k^2))(-5k)$

28. Michal is thinking of a number. If he raises this number to the third power, then multiplies by -3 and adds 17 to the result, he gets -358. What is Michal's number? (Lesson 9-7)

Review and Practice Your Skills

Ten cards with the even numbers 2 through 20 are in a box. A card is taken from the box and then replaced. A second card is chosen. Find the probability of each event.

1. P(8 then 4)
2. P(2 then even)
3. P(6 then odd)
4. P(two numbers < 10)
5. P(two numbers > 10)
6. P(two even numbers)
7. P(two equal numbers)
8. P(two multiples of 10)
9. P(two multiples of 4)

Ten cards with the odd numbers 3 through 21 are in a box. A card is taken from the box and NOT replaced. A second card is chosen. Find the probability of each event.

10. P(3 then 21)
11. P(3 then odd)
12. P(prime then odd)
13. P(11 then prime)
14. P(two numbers < 10)
15. P(two numbers > 10)
16. P(two odd numbers)
17. P(two prime numbers)
18. P(two equal numbers)

From a total group of 50,000 people, a random sample of 1,000 people is chosen. The people in the sample are asked what color car they like best. Of those polled, 220 chose red, 360 chose white, 285 chose black and 135 chose some other color.

Based on the sample, *estimate* the number in the total group who would choose each of the following.

19. a red car
20. a white car
21. a black car
22. a red or a white car
23. a red or a black car
24. the most popular color
25. a car that isn't red
26. a car that isn't black
27. the least popular color

Write each probability as a decimal. Round to the nearest hundredth.

28. P(a red car)
29. P(a white car)
30. P(a black car)
31. P(a red or a white car)
32. P(a red or a black car)
33. P(the most popular color)
34. P(a car that isn't red)
35. P(a car that isn't black)
36. P(the least popular color)

Use the probabilities to compute the number in the total group who would choose each of the following.

37. a red car
38. a car that isn't black
39. a black car
40. a red or a white car
41. a red or a black car
42. the most popular color

From a total of 30,000 people, a random sample of 500 people is chosen. Of those polled, 167 want a new library, 205 don't want a new library and 128 are undecided. Compute the number in the total group who would do the following.

43. vote for a new library
44. vote against a new library
45. have not decided

Find each probability using Spinner 1. (Lesson 10-1)

Spinner 1

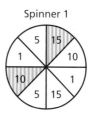

46. P(multiple of 5)

47. P(white or odd)

48. P(not striped)

A spinner has six nonequal parts, with a letter of the alphabet in each part. The spinner is spun 100 times. The table shows the outcomes. Find each experimental probability. (Lesson 10-2)

U	10
V	32
W	18
X	25
Y	4
Z	11

49. P(W)   **50.** P(X or Y)   **51.** P(T)

Find the number of possible outcomes. (Lesson 10-3)

52. tossing two coins and a 1–6 number cube   **53.** tossing four coins

Find the probability. (Lesson 10-4)

54. An outfit is picked at random from 8 skirts, 3 sweaters, and 2 jackets. One skirt, one sweater, and one jacket are red. What is the probability of choosing the all-red outfit?

Math*Works* Career – Research Geneticist
Workplace Knowhow

Research geneticists study the branch of biology that deals with heredity and genetic variation. They apply modern technology to agriculture, medicine and police work. Blood types are an important genetic factor. Humans can have four possible blood types: A, B, O or AB. In different parts of the world, the fraction of persons with blood types A, B, O and AB differs.

Use the table for Exercises 1–4.

1. If a researcher tests a random sample of 500 Chinese citizens, how many can be expected to have type O blood?

2. If the blood of 1000 people is tested, which country has the greatest probability of testing a person with type A blood? Why?

3. If the blood of 150,000 Bolivian Indians is tested, how many of each blood type might you expect to find?

4. If the blood of 1000 Austrians is tested, which blood type would you expect to find in 66 people?

Frequencies of ABO Blood Groups in Some Human Populations

Population	Blood Group			
	O	A	B	AB
Armenians	0.289	0.499	0.132	0.080
Austrians	0.427	0.391	0.115	0.066
Bolivian Indians	0.931	0.053	0.016	0.001
Chinese	0.439	0.270	0.233	0.058
Danes	0.423	0.434	0.101	0.042
Eskimos	0.472	0.452	0.059	0.017

10-7

Expected Value and Fair Games

Goals
- Find expected value.
- Use expected value to determine fairness of games.

Applications Finance, Sports, Entertainment, Construction

A sweepstakes winner is selected randomly, and the probability of winning depends on the overall number of entries. Based on past sweepstakes, a company expects 201,000,000 entries. Prizes and approximate probabilities of winning are shown.

Prize		Probability of winning
First	$5,000,000	1 out of 201,000,000
Second	150,000	1 out of 201,000,000
Third	100,000	1 out of 201,000,000
Fourth	25,000	1 out of 100,500,000
Fifth	10,000	1 out of 50,250,000
Sixth	5,000	1 out of 25,125,000

1. Is the chance of winning this sweepstakes worth the price of a postage stamp?

2. Which factor is the greatest influence on your decision to enter a sweepstakes, the payoff or the probability of winning?

3. If the prize money is distributed equally among all the entries, would you enter the sweepstakes?

◥ BUILD UNDERSTANDING

Expected value is the amount you can expect to win or lose in situations in which the winners are determined randomly. To find the expected value of a sample space, separate the sample space into a number of events where no two events have any common outcomes.

To find expected value (E) use the event's probability and the event's *payoff*. The expected value is the sum of the products of probabilities multiplied by their payoffs.

Expected Value	$E = [P(A) \cdot \textbf{payoff for } A] + [P(B) \cdot \textbf{payoff for } B]$ for outcomes A and B.

The number of events equals the number of products in the formula. For two events there are two products in the formula. For three events, there are three.

Example 1

The student council sells stickers to raise money. The Fan Club company states the probability of making $3000 is 0.8 and the probability of losing $1000 is 0.2. Fan Club claims that on average, schools make over $2000. Is the claim accurate?

Solution

The sample space is separated into two events: making $3000 and losing $1000. Use a negative sign for an amount lost or spent.

Expected Value	=	Probability of Event	·	Payoff of Event	+	Probability of Event	·	Payoff of Event
E	=	0.8	·	$3000	+	0.2	·	−$1000

$$E = \$2400 + (-\$200) = \$2200$$

The claim is accurate since the expected value is more than $2000.

Example 2

Find the expected value of the sweepstakes described in the lesson introduction. Round to the nearest ten thousandth.

Solution

There are 6 events in the sample space, so there must be 6 terms in the formula.

$$E = \left(\frac{1}{201,000,000} \cdot \$5,000,000\right) + \left(\frac{1}{201,000,000} \cdot \$150,000\right) +$$

$$\left(\frac{1}{201,000,000} \cdot \$100,000\right) + \left(\frac{1}{100,500,000} \cdot \$25,000\right)+$$

$$\left(\frac{1}{50,250,000} \cdot \$10,000\right) + \left(\frac{1}{25,125,000} \cdot \$5000\right)$$

$$E \approx 0.0249 + 0.0007 + 0.0005 + 0.0002 + 0.0002 + 0.0002$$
$$\approx \$0.0267$$

The expected value is $0.03 per entry.

The expected value of a **fair game** equals 0. For a game to be considered fair, the chances of winning are equal for all participants.

Example 3

Find the expected value to determine whether the game described is fair.

Two children play a game with 3 colored cards. A bag contains 1 red, 1 blue, and 1 yellow. Each child pulls a card from the bag. The child with the red card receives 2 points. Otherwise, that child loses 1 point.

Solution

$$E = \left(\frac{1}{3} \cdot 2\right) + \left[\frac{2}{3} \cdot (-1)\right] \qquad E = (P(\text{red}) \cdot 2) + (P(\text{not Red}) \cdot (-1))$$

$$E = \frac{2}{3} - \frac{2}{3} = 0$$

This game is fair since the expected value is 0.

1. A sample space has four equally likely outcomes. The payoffs for the outcomes are 1, 2, 3 and 4. What is the expected value of the sample space?

2. A sample space has ten equally likely outcomes. The payoffs for five of the outcomes is $2. The payoffs for five of the outcomes is $5. What is the expected value of the sample space?

3. A charity raffles off a $20,000 car by selling 5000 tickets for $2 per ticket. What is the expected value?

4. **WRITING MATH** Use an example to explain how the expected value can determine the fairness of a game.

5. Tristann and Keenan flip a coin 3 times. If one or two coins land tails up, then Tristann scores 3 points. Otherwise, Tristann loses 1 point. What is the expected value of the game?

6. Is the game described in Exercise 5 fair?

PRACTICE EXERCISES • For Extra Practice, see page 582.

7. A sample space has six equally likely outcomes. The payoffs for the outcomes are 1, 2, 3, 4, 5 and 6. What is the expected value of the sample space?

8. A sample space has five equally likely outcomes. The payoffs for four of the outcomes are 2. The expected value of the sample space is 4. What is the payoff for the fifth outcome?

9. Suppose a student has a $\frac{1}{5}$ probability of being tardy. The teacher assigns 10 min of detention for each tardy but subtracts 2 min from the detention for each time the student is not tardy. What is the average number of minutes of detention for the student?

10. The table shows the results of a telephone company's survey to determine the number of times the telephone rings before it is answered. What is the average number of rings per phone call?

Number of rings	Experimental probability
1	0.1
2	0.05
3	0.25
4	0.35
5	0.13
6	0.08
7	0.03
8	0.01

Suppose a charity raffles off 1000 $1 tickets for a $600 television.

11. What is the expected value for the purchase of one ticket?

12. Is $1 a fair price to pay for a ticket? Explain.

13. **CONSTRUCTION** A contractor bids on a construction project. There is a 0.7 probability of making $200,000 profit and a 0.3 probability of losing $250,000. What is the expected value?

14. **SPORTS** The World Series is the championship game between the National and American Leagues in baseball. The champion is the first to win four games. If two teams are evenly matched, the probability of the series ending in 4 games is $\frac{1}{8}$, 5 games $\frac{1}{4}$, 6 games $\frac{5}{16}$ and 7 games $\frac{5}{16}$. What is the expected number of games?

15. The table shows the results of a survey conducted to find out how many times a person changes jobs after age 25. Find the average number of job changes.

16. FINANCE An insurance company insures a $14,000 car. The company determines that the probability of an accident causing a total loss is 0.05. A loss of half the cost of the car is 0.1. A loss of a quarter of the cost of the car is 0.2. What yearly premium payment should the company charge to break even?

Number of job changes	Experimental probability
0	0.01
1	0.02
2	0.04
3	0.08
4	0.11
5	0.15
6	0.25
7	0.2
8	0.09
9	0.05

ENTERTAINMENT In the carnival game "Go Fish," participants "fish" for different plastic fish that workers put on hooks from behind a curtain. The fish are selected at random from a bucket containing 15 red fish, 10 silver fish and 5 gold fish. The prizes are $0.05 for a red fish, $0.10 for a silver fish and $0.15 cents for a gold fish.

17. What is the expected value for the game?

18. What is a fair price to play the game?

EXTENDED PRACTICE EXERCISES

19. Use the dart board shown. Four points are scored for the inner circle, 3 points for the next region, 2 points for the next region and 1 point for the outer region. What is the expected value for each dart thrown?

TECHNOLOGY Conduct an experiment using a calculator or computer that can generate random integers between 1 and 20.

20. Give yourself 2 points for each multiple of 5, 1 point for each multiple of 7 and −1 point for 11. Find the expected value of this game.

21. Is it a fair game?

22. Generate 10 random integers between 1 and 20. Record your results.

23. Did the results of your experiment indicate that this game is fair? Explain.

24. CHAPTER INVESTIGATION Assume that one allele for the offspring is recessive. Flip a coin to find whether the other allele is dominant or recessive. Sketch a face with the traits determined by experimental probability.

MIXED REVIEW EXERCISES

Trace each angle. Then do construction indicated. (Lesson 8-2)

25. copy

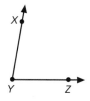

26. bisect

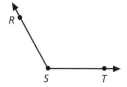

27. bisect

28. Suppose ∠J and ∠K are supplementary angles and ∠J and ∠L are complementary angles. If ∠J measures 30°, what is m∠K and m∠L? (Lesson 8-1)

Chapter 10 Review

VOCABULARY

Choose the word from the list that best completes each statement.

1. The expected value of a __?__ is 0.

2. The __?__ of an event is between 0 and 1 and is usually written as a fraction in lowest terms.

3. A(n) __?__ acts out an event so that you can find the outcomes.

4. The set of all outcomes is called the __?__.

5. A(n) __?__ is an outcome or a combination of outcomes.

6. Probability based on the results of a series of trials is called __?__.

7. A(n) __?__ shows the total number of possible outcomes.

8. If the result of the first event affects the result of the second event the two events are __?__.

9. The result of an experiment can also be called a(n) __?__.

10. To find the number of outcomes by multiplying the number of outcomes at each stage in the event is called the __?__.

a.	counting principle
b.	dependent
c.	event
d.	expected value
e.	experimental probability
f.	fair game
g.	independent
h.	outcome
i.	probability
j.	sample space
k.	simulation
l.	tree diagram

LESSON 10-1 ■ Introduction to Probability, p. 436

▶ A **favorable outcome** is the desired outcome.

▶ $P(E) = \dfrac{\text{number of favorable outcomes}}{\text{number of possible outcomes}}$

▶ Odds in favor of $E = \dfrac{\text{number of favorable outcomes}}{\text{number of unfavorable outcomes}}$

A number cube, with faces labeled from 1 through 6, is tossed.

11. Find $P(\text{number} < 5)$.

12. Find $P(\text{prime number})$.

13. The odds in favor of an even number.

14. The odds in favor of an odd number.

Find each probability using the spinner. Give your answers as fractions and percents.

15. $P(8)$

16. $P(\text{red})$

17. $P(\text{even})$

18. $P(\text{prime})$

19. $P(\text{greater than 5})$

20. $P(\text{less than two})$

21. $P(\text{not yellow})$

22. $P(\text{not red})$

LESSON 10-2 ◢ Experimental Probability, p. 440

▶ The probability of an event based on the results of an experiment is called the **experimental probability**.

▶ Each time you **simulate** a problem, you complete one **trial**.

A number cube is tossed 20 times. The outcomes are recorded in this table.

23. Find $P(1)$.

24. Find P(even number).

25. Find P(3 or 4)

Number on cube	1	2	3	4	5	6
Number of outcomes	3	5	2	5	2	3

A spinner divided into four equal sections is spun 20 times. The outcomes are recorded in this table.

26. Find P(red).

27. Find P(blue or green).

28. Find P(not yellow).

Color on Spinner	red	blue	green	yellow
Number of Outcomes	12	2	4	2

LESSON 10-3 ◢ Sample Spaces and Tree Diagrams, p. 446

▶ You can make a **tree diagram** to picture a **sample space** and then count the outcomes.

Two coins are tossed.

29. Draw a tree diagram to show the sample space.

30. How many possible outcomes are there?

LESSON 10-4 ◢ Counting Principle, p. 450

▶ Find the number of possible outcomes of an event by multiplying the number of outcomes at each stage of the event.

Use the counting principle to find the number of possible outcomes.

31. tossing a coin four times

32. choosing a snack from 2 fruits and 3 kinds of herb tea

33. sweatshirts in four sizes and four colors

34. rolling a number cube twice

35. tossing two coins and rolling a number cube

36. choosing a car with two or four doors, a four-or six-cylinder engine, and a choice of six exterior colors

Use the counting principle to find each probability.

37. Three coins are tossed. Find P(H, H, T).

38. An 8-sided die is rolled three times. Find P(7, 7, 7).

39. A coin is tossed and a card is drawn from a standard deck of cards. Find P(H, any heart).

40. A number cube is tossed three times. Find P(3, 3, 3)

LESSON 10-5 ◣ Independent and Dependent Events, p. 456

▶ If two events X and Y are independent, then $P(X \text{ and } Y) = P(X) \cdot P(Y)$

▶ If two events X and Y are dependent and Y follows X, then
$P(X \text{ and } Y) = P(X) \cdot P(Y \text{ after } X \text{ has occurred})$

A bag contains 3 red, 2 white, and 5 green marbles. One is taken from the bag, replaced, and another is taken. Find the probability of each event.

41. P(red, then green)

42. P(white, then white)

43. P(red, then white)

44. P(green, then red)

A change purse contains 3 quarters, 2 dimes, and 5 nickels. Two coins are selected from the purse without replacement. Find the probability of each event.

45. P(dime, then nickel)

46. P(quarter, then dime)

47. P(nickel, then dime)

48. P(nickel, then quarter)

LESSON 10-6 ◣ Problem Solving Skills: Make Predictions, p. 460

▶ You can use information from a random sample of a population to make predictions about the entire population.

▶ Some probability experiments are easier to understand if a simpler experiment is conducted to simulate the more difficult one.

A newspaper company asks a random sample of 300 of its 30,000 subscribers their favorite time of day to receive their newspapers. Of those polled, 150 chose morning, 40 chose noon, and 110 chose evening. If all of the subscribers were polled how many would you expect to select the following?

49. the most favorite time of day

50. the number who chose evening

Find each probability.

51. P(noon)

52. P(morning)

LESSON 10-7 ◣ Expected Value and Fair Games, p. 464

▶ **Expected value** is $E = [P(A) \cdot \text{payoff for } A] + [P(B) \cdot \text{payoff for } B]$

53. Mark's karate club is thinking of selling magazines to raise money. The magazine company states the probability of making $5000 is 0.7, and the probability of losing $2000 is 0.3. The company claims that on average clubs make over $3000. Is the claim accurate?

54. Two children play a game with 5 cards numbered 1–5. Each child pulls a card from the bag. The child with the winning number receives 4 points. The other child loses 1 point. The number 5 wins over the numbers 1, 2, 3, and 4. Determine whether the game is fair. Explain your answer.

CHAPTER INVESTIGATION

EXTENSION Compare the two face sketches you made. How are they alike? How are they different? For which traits did the theoretical probability match the experimental probability? For which traits did the probabilities not match?

Chapter 10 Assessment

Find each probability. Use this spinner.

1. $P(A)$

2. $P(E \text{ or } D)$

3. $P(Z)$

4. $P(\text{white})$

5. $P(A, B, C, D, E, \text{ or } F)$

Use the spinner above to find the odds in favor or each event.

6. C or D

7. Green

8. Blue or A

9. A, B, or D

Use a tree diagram to find the number of possible outcomes in each sample space.

10. tossing a nickel and then tossing a number cube

11. spinning a spinner like the one shown and then tossing a quarter.

Use the counting principle to find the number of possible outcomes.

12. tossing a coin three times

13. choosing a stereo configuration from 2 receivers, 3 CD players, and 5 speakers

14. choosing a dinner consisting of either beef, lamb, fish, or chicken; with either peas or carrots; and with either milk, juice, or iced tea.

15. a coin is tossed six times. Find $P(\text{all tails})$.

16. A bag contains 1 red marble, 3 blue marbles, and 1 orange marble. Shanice takes one marble from the bag, replaces it, and takes another. Find $P(\text{blue, then orange})$.

A number cube is tossed 30 times. The outcomes are recorded in this table.

Number on cube	1	2	3	4	5	6
Number of outcomes	2	4	7	3	8	6

17. Find $P(5)$.

18. Find $P(\text{even number})$.

19. Find $P(2 \text{ or } 6)$.

20. Suppose a charity raffles off 2000 tickets for a $800 go-cart for $2 per ticket. What is the expected value?

21. Carmen and Nicholas are tossing a pair of coins. Carmen earns two points when one coin lands heads up and the other lands tails up. Nicholas earns two points when both coins land heads up or tails up. Is this a fair game? Explain.

Standardized Test Practice

Part 1 Multiple Choice

Record your answers on the answer sheet provided by your teacher or on a sheet of paper.

1. A gasoline company wants to survey its customers. From a list of their credit card holders, they send a questionnaire to every 20th person. What type of sampling are they using? (Lesson 1-1)
   (A) clustered
   (B) convenience
   (C) random
   (D) systematic

2. Complete: 716 mg = ___?___ kg. (Lesson 2-1)
   (A) 7.16 kg
   (B) 0.716 kg
   (C) 0.0716 kg
   (D) 0.000716 kg

3. What is the perimeter of a rectangle with a length of 24.3 mi and a width of 8.1 mi? (Lesson 2-3)
   (A) 3 mi
   (B) 16.2 mi
   (C) 64.8 mi
   (D) 196.83 mi

4. A truckload of dirt is dumped in front of a home. The pile of dirt is cone-shaped with a height of 7 ft and a diameter of 15 ft. Find the volume of the dirt to the nearest hundredth. Use 3.14 for π. (Lesson 4-8)
   (A) 412.33 ft^3 (B) 769.30 ft^3
   (C) 1236.38 ft^3 (D) 1648.50 ft^3

5. What is the surface area of the rectangular prism? (Lesson 4-9)
   (A) 38 cm^2
   (B) 40 cm^2
   (C) 76 cm^2
   (D) 96 cm^2

6. Solve: $4(k - 9) = 24$. (Lesson 5-4)
   (A) -5 (B) 8
   (C) 11   (D) 15

7. If the area of a square is 225 m^2, what is the length of one side? (Lesson 5-6)
   (A) 3.87 m   (B) 15 m
   (C) 25 m   (D) 56.25 m

8. Compare: $3 \times (-14)$ ■ $-13 \times (-3)$. (Lesson 5-9)
   (A) >   (B) <
   (C) =   (D) ≈

9. What percent of 900 is 270? (Lesson 6-1)
   (A) 20%   (B) 30%
   (C) 50%   (D) 130%

10. Janelle was shopping for a coat. She found one that she liked for $100. The coat was on sale for $79. What was the percent discount? (Lesson 6-3)
    (A) 15%   (B) 18%
    (C) 21%   (D) 79%

11. Complete: ∠5 and ___?___ are alternate interior angles. (Lesson 8-1)

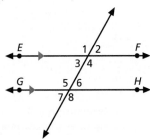

   (A) ∠1   (B) ∠4
   (C) ∠7   (D) ∠8

Test-Taking Tip

Question 5
One method to solve multiple choice questions with equations or inequalities is to substitute each answer choice for the variable in the equation.

Preparing for the Standardized Tests
For test-taking strategies and more
practice, see pages 587-604.

Part 2 Short Response/Grid In

**Record your answers on the answer sheet
provided by your teacher or on a sheet of paper.**

12. A map scale is 1 in. : 10 mi. What is the actual
number of miles between two cities if the
distance on the map is 7.5 in.? (Lesson 2-8)

13. What is the area of the figure? Use 3.14 for π.
(Lesson 2-9)

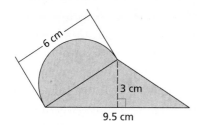

14. A garden has 6 rows of equal numbers of
pepper plants. Each row is 40 ft long. The
plants are placed 5 ft apart starting at the
corner of the garden. How many pepper
plants are in the garden? (Lesson 3-4)

15. What is the volume of the
right circular cone? Use 3.14
for π. (Lesson 4-8)

16. What is the domain of the relation {(−1, 3),
(4, 6), (−4, 8), (6, −4)}? (Lesson 7-3)

17. What are coordinates of the image of the
point (3, −3) under a reflection across
the *y*-axis? (Lesson 8-6)

18. In a box of 85 plugs, 15 are defective. What is
the probability of choosing a defective plug?
(Lesson 10-1)

19. A single die is rolled. What are the odds that a
prime number was *not* rolled? (Lesson 10-1)

20. How many outcomes are possible if two
number cubes are rolled and their results are
added? (Lesson 10-4)

21. Raffle tickets numbered 1 through 30 are
placed in a box. Tickets for a second raffle
numbered 21 through 48 are placed in
another box. One ticket is randomly drawn
from each box. Find the probability that both
tickets are even. (Lesson 10-5)

Part 3 Extended Response

**Record your answers on a sheet of paper.
Show your work.**

22. Shalimar wants to make a circular enclosure
for her pet rabbits. Use 3.14 for π. (Lesson 2-7)
   a. If she uses 25 ft of fencing, what is the
   radius of the circular enclosure? Round to
   the nearest tenth.
   b. What is the area of the enclosure? Round
   to the nearest square foot.

23. Wallpaper border comes in 5-yd rolls for
$22.95. Katy purchases enough border for a
12 ft-by-11 ft bedroom. (Lesson 3-4)
   a. What is the perimeter of the room?
   b. How many rolls of border will cover the
   perimeter of the room?
   c. How much does this cost?

24. Carrie and Jim want to order a pizza from the
Pizza Palace. (Lesson 10-4)

Pizza Palace		
Crust	**Toppings**	**Size**
thin	pepperoni	individual
homestyle	sausage	small
thick	mushroom	medium
deep dish	green pepper	large

   a. How many different one-topping pizzas
   can be ordered?
   b. They decide to order a large one-topping
   pizza. How many choices are possible?
   c. If a one-topping pizza is chosen at random,
   what is the probability that it will be a
   large deep dish pizza with mushrooms?

Reasoning

THEME: Mysteries

Reasoning is an important part of geometry, statistics, and algebra. Reasoning is used to determine if an answer is reasonable, to solve mathematical puzzles, and to write geometric proofs.

Reasoning plays an important part in many careers, including detective work, forensics, scientific research, and psychology. Fans of science fiction, mysteries and riddles exercise their reasoning skills to separate illusion from reality and to discover answers that aren't obvious.

- **Private investigators** (page 487) solve mysteries to catch criminals, find the sources of illegal activity, or discover evidence connected with crimes. Some specialize in one field, such as finance or missing persons. Others do executive protection and bodyguard work.

- **Cryptographers** (page 505) determine the contents of a message that is written in an unknown code. Mathematical patterns and matrices are often the keys to decoding a secret message.

Math Online

mathmatters1.com/chapter_theme

Great Fictional Detectives
(in alphabetical order)

Fictional Detectives	Authors Who Created Them
Uncle Abner	Melville Post
Lew Archer	Ross Macdonald
Father Brown	G. K. Chesterton
Albert Campion	Margery Allingham
Charlie Chan	Earl Derr Biggers
C. Auguste Dupin	Edgar Allen Poe
Dr. Gideon Fell	John Dickson Carr
Sherlock Holmes	A. Conan Doyle
Inspector Maigret	Georges Simenon
Philip Marlowe	Raymond Chandler
Miss Marple	Agatha Christie
Perry Mason	Erle Stanley Gardner
Hercule Poirot	Agatha Christie
Ellery Queen	Ellery Queen
Sam Spade	Dashiell Hammett
Lord Peter Wimsey	Dorothy L. Sayers
Nero Wolfe	Rex Stout

Data Activity: Mystery Writers

Two mystery writers named Frederic Dannay and Manfred B. Lee wrote as a team. The two men were cousins and they published their work under the pen name Ellery Queen.

Use the table to determine if the statement is true. Write *true,* *false,* **or** *cannot tell.*

1. Agatha Christie created more than one fictional detective.

2. A. Conan Doyle only created one fictional detective.

3. G. K. Chesterton created Lord Peter Wimsey.

4. Ellery Queen is the only character written by an author using a pen name.

5. There are only three female authors on the list.

CHAPTER INVESTIGATION

Mystery and suspense novels are among the best-selling books of all time. Characters use their reasoning skills and logic to solve mysteries and apprehend criminals. Many mystery and suspense fans enjoy using their own skills and experience to predict how the mystery is solved before the end of the book.

Working Together

Read a mystery or suspense story, and write about the types of reasoning used by the character. Decide whether you think the author included the most logical method. Use the Chapter Investigation icons to guide your progress.

11 Are You Ready?

Refresh Your Math Skills for Chapter 11

The skills on these two pages are ones you have already learned. Use the examples to refresh your memory and complete the exercises. For additional practice on these and more prerequisite skills, see pages 536-544.

WORKING WITH SETS

To solve problems involving reasoning, it's helpful to be able to sort items into groups and know when the groups overlap. Set notation makes this process more simple.

Examples S = {red, green, blue, black, brown}
R = {white, yellow, orange, red, purple}

The symbol $\cup$ represents a union. The union of two or more sets is a combination of all items in all sets. $S \cup R$ = {red, green, blue, black, brown, white, yellow, orange, purple}

The symbol $\cap$ represents an intersection. The intersection of two or more sets is the elements that occur in all sets. $S \cap R$ = {red}

Use sets R and S above and set T listed below to find each union and intersection.

T = {red, orange, yellow, green, blue, purple}

1. $S \cup T$ **2.** $S \cap T$ **3.** $R \cup T$ **4.** $R \cap T$

5. $R \cap (S \cup T)$ **6.** $R \cap S \cap T$ **7.** $R \cup S \cup T$ **8.** $T \cap (R \cup S)$

PROBLEM SOLVING STRATEGIES—PATTERNS

Finding a pattern can help you solve reasoning problems.

Examples   Find the pattern in each sequence of numbers. Use the pattern to find the next three numbers.

3, 11, 19, 27, 35, . . . The difference between each consecutive pair of numbers in the sequence is 8. The next three numbers are 43, 51, and 59.

0, 2, 4, 8, 16, . . . If you multiply a number in the sequence by 2, you get the next number. The next three numbers are 32, 64, and 128.

Find the pattern and use it to find the next three numbers in each series.

9. 1, 2, 4, 7, 11, . . .   **10.** 1, 3, 9, 27, 81, . . .

11. 1, 1, 2, 3, 5, . . .   **12.** 0, 1, 4, 9, 16, . . .

13. 81, 76, 91, 86, 101, . . .   **14.** 1, 5, 13, 29, 61, 125, . . .

PROBLEM SOLVING STRATEGIES—DRAW A PICTURE

Pictures help you visualize a problem, which can help you find the solution.

Solve each problem by drawing a picture.

15. Michelle takes her motor boat out on the lake. She went 0.95 km north, 0.87 km east, 1.1 km north, and 0.57 km east. If she takes a straight path back to the dock, how far will she go and in which direction?

16. A swimming pool of width w, and length $5w$, has a deck of width $0.25w$ surrounding the pool. The area of the pool and the deck combined is 500 m^2. What is the width of the pool?

PROBLEM SOLVING STRATEGIES—MAKE A TABLE

Tables are helpful to use when solving logic problems. Use the table to organize the information given.

Find which animal each person owns and the street where they live.

17. Vasha, Diane, Chad and Leya live on First, Second, Third and Fourth Streets. Vasha and Diane live on odd-numbered streets and one of them owns a dog. Chad is allergic to horses, but owns a cat. Leya lives on Second Street and owns an animal with 4 legs. The person who lives on First Street owns a bird. Diane is a friend of the person who owns a bird. The person who lives on First Street keeps an animal in a cage.

	First	Second	Third	Fourth	Cat	Dog	Bird	Horse
Vasha								
Diane								
Chad								
Leya								

PROBLEM SOLVING STRATEGIES—GUESS AND CHECK

Some problems only give sketchy details. Your best plan of action may be to guess an answer then check to see if it is correct. If it isn't, revise your guess and check again. Each time you make a revision, you should get closer to the correct answer.

18. In a 500-m race, Chachi finished before Shea but after Michael. Shea didn't finish last. Mika finished before Michael. Taylor was the middle runner at the finish, coming in right before Tyrique. In what place did Monica finish?

19. Colin has found three odd numbers whose sum is 51. The difference between any pair of these numbers is no greater than 8. Only one of the three numbers is a composite number. What are the three numbers?

11-1 Optical Illusions

Goals ▪ Use optical illusions to make statements.
          ▪ Determine the truth value of statements.

Applications Architecture, Construction, Art, Design

Draw the 5-unit by 13-unit rectangle shown on grid paper.

1. Cut out the rectangle. Cut it into pieces along the lines.

2. Rearrange the pieces to form a square. Tape the pieces together.

3. What do you think the relationship is between the area of the original rectangle and the area of the square?

4. Count square units to compute the area of the original rectangle. Now count square units to compute the area of the square. What can you conclude about the relationship between the area of the original rectangle and the square?

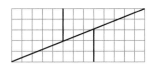

◤ BUILD UNDERSTANDING

In an **optical illusion**, the human eye perceives, or pictures, something that is not true. In this illustration of railroad tracks, the rails appear to become closer together and the distances between the railroad ties seem to grow shorter. The trees along the sides of the railroad tracks appear to become smaller. None of these perceptions is true. The picture is an optical illusion.

You can make a statement about what appears to be true in an optical illusion. A **statement** is a sentence that describes a particular relationship. Statements can be tested to determine if they are true or false.

Example 1

Make a statement about the lengths of $\overline{AB}$ and $\overline{AC}$. Determine if the statement is true or false.

Solution

Line segment AC crosses more slant lines than line segment AB.

   Statement $\overline{AC}$ is longer than $\overline{AB}$.

Measure both segments with a ruler. You will find that they have the same length. Therefore, the statement is false.

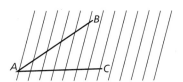

Example 2

Make two statements about the lengths and directions of $\overline{DE}$ and $\overline{FG}$ in the figure. Check to see if each statement is true or false.

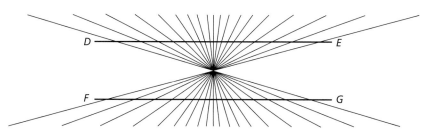

Solution

Notice that $\overline{FG}$ crosses the same number of lines as $\overline{DE}$. However, $\overline{FG}$ extends farther out past the lines. Also, $\overline{FG}$ and $\overline{DE}$ seem to bend away from each other in the middle. From these observations, you can make the following statements.

Statement 1 $\overline{FG}$ is longer than $\overline{DE}$.

Statement 2 $\overline{FG}$ and $\overline{DE}$ are not parallel to each other.

Place a ruler next to $\overline{DE}$, then next to $\overline{FG}$. You will find that they are both straight line segments and have the same length. So, Statement 1 is false.

Now measure the distance between the left ends of the segments, the right ends of the segments and the middles of the segments. All three distances are the same, so the line segments are parallel. Therefore, Statement 2 is also false.

▼ TRY THESE EXERCISES

Refer to the figure.

1. Write a statement about the lengths of $\overline{EF}$ and $\overline{GH}$.

2. Check to see if your statement is true or false.

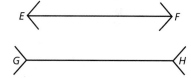

Refer to the figure.

3. Write a statement about $\overline{AB}$, $\overline{AC}$ and $\overline{BC}$.

4. Check to see if your statement is true or false.

5. **ARCHITECTURE** To make a building look taller, an architect might use tall, narrow windows. To make a room look wider, an interior designer might use horizontal stripes. Describe a building that uses geometry and optical illusions to enhance a particular feature.

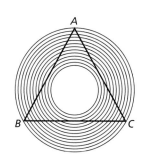

Refer to the figures shown. How many blocks does each contain?

6.

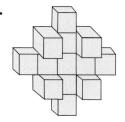

7.

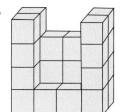

8. Write a statement about lines l and m. Check to see if it is true.

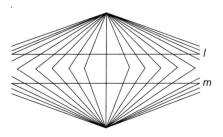

9. Write a statement about $\overline{AB}$ and $\overline{BC}$ Check if it is true.

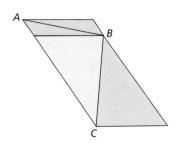

10. Write a statement about $\overline{DE}$ and $\overline{FG}$. Check to see if it is true.

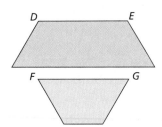

11. Write a statement about the dots. Check if it is true.

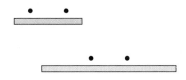

 WRITING MATH Each optical illusion can be viewed two different ways. Write a description of what you see in each figure. Justify each statement.

12.

13.

14. DATA FILE Refer to the data on the tallest buildings on page 521. Locate photographs of as many of these buildings as you can. Examine the geometric shapes visible on the exterior of each building. Do any of them have characteristics that might be misleading? Explain the features of each building and describe your visual perception of each structure.

Examine the two-dimensional figures shown. Could you make a three-dimensional model with sticks, straws or other objects? Justify your answer.

15.

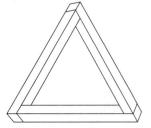

16.

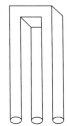

17.

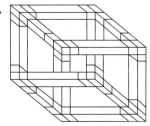

18. **MODELING** Refer to the figure shown. How many blocks does it contain? Make a model using unit blocks to check your answer.

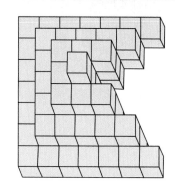

19. **WRITING MATH** Choose one of the optical illusions in this lesson. Write a paragraph to describe the illusion and explain what in the picture tricks the eye.

Study the two drawings. Describe at least one unusual thing in each.

20.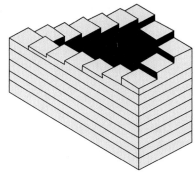

21.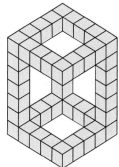

■ EXTENDED PRACTICE EXERCISES

PAPER FOLDING To make a *Möbius strip*, cut a 2-in. by 11-in. strip from a piece paper. Put one twist in the paper and tape the ends together.

22. Draw a line down the length of the strip. Describe what you notice. Is this an optical illusion?

23. Predict what would happen if you cut along the line drawn. Then cut along the line and see if your prediction is correct. If not, describe what did happen.

24. What difference do you think putting two twists in the paper before taping would make?

25. **CHAPTER INVESTIGATION** Read a mystery or suspense story. Choose a story from the table on page 477 or a different story your teacher approves.

■ MIXED REVIEW EXERCISES

Solve.

26. Find the coordinates of the vertices of △*ABC* after rotating it 180° about the origin. (Lesson 8-7)

27. Find the coordinates of △*ABC* after reflecting it over the *x*-axis. (Lesson 8-6)

28. Does the triangle have a line of symmetry? If so, write the equation of the line. (Lesson 8-6)

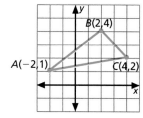

11-2 Inductive Reasoning

Goals ■ Use inductive reasoning to make and test conjectures.

Applications Detective work, Science, Number theory, Recreation

Use a sheet of grid paper and a straightedge to complete the following steps.

Step 1 Draw a rectangle that is 4 units by 5 units. Be sure to draw the sides of the rectangle on the grid lines of the grid paper.

Step 2 Start in the lower left-hand corner of the rectangle and draw a diagonal line that bisects the squares of the grid paper.

Step 3 When a side of the rectangle is met, "bounce" off the side and draw another diagonal line that bisects the grid squares.

Step 4 Repeat the process of bisecting squares and "bouncing" off the sides of the rectangle until you reach a corner of the rectangle.

Step 5 Count the number of "bounces" including the original corner and the final corner.

Step 6 Compare the result to the dimensions of your rectangle.

1. Draw a rectangle that is 3 units by 5 units. Complete Steps 2–6.

2. Compare your results with a classmate. What conclusion can you draw regarding the number of "bounces" and the dimensions of the rectangle?

3. Draw another rectangle. Is your conclusion true for this rectangle as well?

▼ BUILD UNDERSTANDING

People in many walks of life—detectives, scientists, advertisers, politicians, mathematicians and others—pay special attention to things that happen repeatedly.

They hope that by examining events, they will discover a pattern or rule that applies to all such events. That is, they hope to arrive at a conclusion by using *inductive reasoning*.

Inductive reasoning is the process of reaching a conclusion based on a set of specific examples. The conclusion is called a **conjecture**. The set of examples can be gathered by experience, observation or experiments.

Math: Who, Where, When

In 1742, mathematician Christian Goldbach made the following conjecture.

Every even whole number greater than 2 can be written as the sum of two prime numbers.

Examples: $8 = 3 + 5$ $\qquad$ $20 = 7 + 13$

$\qquad\qquad$ $50 = 13 + 37$ $\qquad$ $100 = 11 + 89$

This conjecture has never been proven, but no one has ever found an example that disproves it.

Example 1

CALCULATOR Use inductive reasoning to make a conjecture about the ones digit of 3^{16}.

Solution

Use a calculator to find several successive powers of 3. Look for a pattern in the ones digits.

$3^1 = \qquad 3$ $\qquad 3^2 = \qquad 9$ $\qquad 3^3 = \qquad 27$ $\qquad 3^4 = \qquad 81$

$3^5 = \quad 243$ $\qquad 3^6 = \quad 729$ $\qquad 3^7 = \quad 2187$ $\qquad 3^8 = \quad 6561$

$3^9 = 19{,}683$ $\qquad 3^{10} = 59{,}049$ $\qquad 3^{11} = 177{,}147$ $\qquad 3^{12} = 531{,}441$

In these examples, the ones digits repeat the pattern 3, 9, 7, 1.

Next, make a conjecture based on the pattern.

When the exponent of 3 is a multiple of 4, the ones digit is 1.

Finally, make a specific prediction based on the conjecture and test it.

Since 16 is a multiple of 4, the ones digit for 3^{16} is 1.

In fact, $3^{16} = 43{,}046{,}721$, a number whose ones digit is 1.

A conjecture may be true, but it is not *necessarily* true. To prove that a conjecture is false, you need to find just one case, called a **counterexample**, where the conjecture does not hold true.

Example 2

Jamar observed that he could express fractions with two-digit numerators and denominators in lowest terms.

$$\frac{16}{64} = \frac{1}{4} \qquad\qquad \frac{19}{95} = \frac{1}{5} \qquad\qquad \frac{26}{65} = \frac{2}{5}$$

Check Understanding

If a rule or pattern holds for several examples, will it hold for all examples?

Based on these three examples, he made the following conjecture.

Whenever the ones digit in the numerator is the same as the tens digit in the denominator, they can be canceled to get an equivalent fraction.

Is Jamar's conjecture always true?

Solution

If possible, find a case in which Jamar's conjecture is false.

Using Jamar's rule: $\qquad\qquad$ Using division:

$$\frac{12}{24} \rightarrow \frac{1}{4} \qquad\qquad \frac{12}{24} = \frac{12 \div 12}{24 \div 12} = \frac{1}{2}$$

Since $\frac{1}{4} \neq \frac{1}{2}$, this is a counterexample. Jamar's conjecture is false.

Example 3

CALCULATOR A piece of paper one-thousandth of an inch thick is torn into two pieces, and the two pieces are stacked on top of each other. If these two pieces are repeatedly torn and stacked, how many tears will it take to get a pile that is more than 1 mi high?

Problem Solving Tip

Most graphing calculators have a "reenter" key that can be useful in editing an expression.

Some number patterns can be explored by using the "previous answer" feature of a graphing utility.

Solution

5280 ft = 1 mi

12 in. = 1 ft

$12 \cdot 5280 = 63{,}360$ in. = 1 mi

Enter 0.001 into a calculator and press ENTER. Then press $\times$ 2. Repeatedly press ENTER until the value 67,108.864 is attained. Counting the number of Enters gives 26 tears.

TRY THESE EXERCISES

Use inductive reasoning to find the ones digit of the following.

1. 4^6 2. 4^7 3. 2^9 4. 58^7

5. Merola found that the sum for each of three pairs of numbers was a two-digit number whose digits are the same.

   $16 + 61 = 77 \qquad 17 + 71 = 88 \qquad 18 + 81 = 99$

   She made this conjecture. Take any two-digit number and reverse the digits. The sum of that number and the original number is a two-digit number whose digits are the same. Is Merola's conjecture always true?

6. Examine the following sequence of numbers. Describe a pattern or rule for the sequence and give the next four numbers.

   $1, 4, 2, 5, 3, 6, 4 \ldots$

7. **RECREATION** Next week, Paul's mother is bringing oranges to his soccer game. Looking at the roster she sees that there are 12 players on the team and she figures that each player will eat about 2 oranges. So she decides to bring 24 oranges to the game. Did Paul's mother use inductive reasoning to make her decision? Explain why or why not.

PRACTICE EXERCISES • For Extra Practice, see page 583.

Use inductive reasoning to find the ones digit of the following.

8. 3^{10} 9. 2^{10}

10. any power of 6   11. any odd-numbered power of 9

12. Choose an even number. Add 20. Multiply the sum by 2. Divide the product by 4. Subtract 10 from the quotient. Multiply the difference by 2. What number results? Repeat this experiment with a different starting number. What number results? Make a conjecture.

Use the given examples to complete each conjecture.

13. $3 \cdot 5 = 15$ $7 \cdot 9 = 63$ $13 \cdot 5 = 65$ $23 \cdot 27 = 621$
The product of two odd numbers is an ___ number.

14. $8 \cdot 12 = 96$ $12 \cdot 16 = 192$ $36 \cdot 74 = 2664$
The product of two even numbers is an ___ number.

15. $2 + 8 = 10$ $14 + 16 = 30$ $6 + 62 = 68$ $46 + 76 = 122$
The sum of two even numbers is an ___ number.

16. MYSTERY Private Investigator Harvey Gumshoe has been staking out the apartment of the infamous burglar Sticky Fingers for the past few weeks. He observes that every day at 12:30 P.M., Fingers leaves his apartment and walks down the street to the cafe for lunch. Based on his observations, Gumshoe conjectures that on Wednesday of next week at 12:30 P.M., Sticky Fingers will leave his apartment and go to lunch. Is this an example of inductive reasoning? Explain why or why not.

17. NUMBER THEORY What is $9,999,999,999 \cdot 5,555,555,555$? Since the value is too large for most calculators, try to find a pattern by investigating smaller values. For example, start with $5 \cdot 9$, then $55 \cdot 99$, and continue increasing the number of fives and nines until you see a pattern.

18. WRITING MATH Write about a real-life situation in which you may use inductive reasoning.

■ EXTENDED PRACTICE EXERCISES

19. YOU MAKE THE CALL Nicki looks at the numerical pattern 4, 11, 19, 28 . . . and concludes that the next three numbers in the sequence are 37, 47, 58. Is she correct? If not, what mistake did Nicki make, and what are the next three numbers in the sequence?

CRITICAL THINKING Test each conjecture by finding at least five examples that support it or one counterexample that disproves it. Explain what is wrong with each false conjecture, and change it to make it true if possible.

20. The difference between any two whole numbers is always positive.

21. The sum of consecutive odd whole numbers starting with 1 is always a perfect square. For example, $1 + 3 + 5 = 9$, and 9 is a perfect square.

22. The sum of n consecutive positive integers starting with 1 equals $\frac{n(n + 1)}{2}$.
For example, $1 + 2 + 3 + 4 + \ldots + 99 + 100 = \frac{100(100 + 1)}{2} = 5050$.

■ MIXED REVIEW EXERCISES

Determine if each figure has rotational symmetry less than 360°. Describe the least rotation needed. (Lesson 8-7)

23.    **24.**   **25.**    **26.**

Review and Practice Your Skills

Write a statement about each figure. Check to see if your statement is true or false.

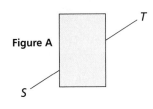

Figure A

Figure B

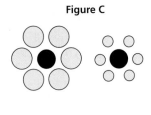

Figure C

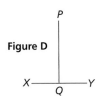

Figure D

Figure E

Figure F

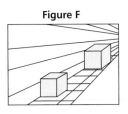

1. line segment *ST* in Figure A

2. line segments *AB* and *CD* in Figure B

3. the inner black circles in Figure C

4. line segments *XY* and *PQ* in Figure D

5. the white triangle in Figure E

6. the two cubes in Figure F

Use inductive reasoning to find the ones digit of the following.

7. 2^{22} 8. 3^{25} 9. 4^{16} 10. 8^{11}

11. any even numbered power of 9   12. any power of 5

13. the square of a number that ends in 10   14. the square of a number that ends in 25

15. the cube of a number that ends in 3   16. the cube of a number that ends in 6

Use the given examples to complete each conjecture.

17. $3 \cdot 6 = 18$ $5 \cdot 10 = 50$ $1 \cdot 12 = 12$ $7 \cdot 40 = 280$ $11 \cdot 26 = 3146$
The product of an odd number and an even number is an (odd/even) number.

18. $3^2 = 9$ $3^3 = 27$ $5^2 = 25$ $5^3 = 125$ $7^2 = 49$ $7^3 = 343$
An odd number to any odd power is an (odd/even) number.

19. $(-1)^3 = -1$ $(-1)^5 = -1$ $(-1)^7 = -1$ $(-3)^3 = -27$ $(-3)^5 = -243$
A negative number to any power is a (positive/negative) number.

20. factors of 4: 1, 2, 4   factors of 6: 1, 2, 3, 6   factors of 8: 1, 2, 4, 8
Even numbers greater than 2 have (more/less) than two factors.

Write a statement about each figure. Check to see if your statement is true or false. (Lesson 11-1)

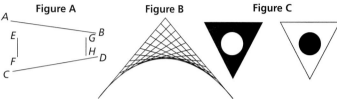

Figure A   Figure B   Figure C

21. line segments *EF* and *GH* in Figure A

22. the curved line in Figure B

23. the two circles in Figure C

Use inductive reasoning to find the ones digit of the following.

24. 7^{10} **25.** any even numbered power of 4   **26.** any power of 11

Use the given examples to complete each conjecture. (Lesson 11-2)

27. $6 - 1 = 5$ $8 - 5 = 3$ $12 - 5 = 7$ $20 - 3 = 17$ $26 - 15 = 11$
An odd number subtracted from an even number is an (odd/even) number.

28. $2^2 = 4$ $2^3 = 8$ $4^2 = 8$ $4^3 = 64$ $6^2 = 36$ $6^3 = 216$
An even number to any power is an (odd/even) number.

MathWorks — Career – Private Investigator
Workplace Knowhow

Private investigators gather evidence, trace debtors, and research facts. They conduct computer database searches, interview witnesses, and monitor people's activities. Attorneys, businesses, financial corporations, retail stores, and the general public employ private investigators. A private investigator is gathering clues about four separate burglaries committed in Centertown. She discovers a white glove on the floor at the scene of each burglary.

Decide if each conclusion is *valid* or *invalid*, given the evidence.

1. The thief, or someone who is helping the thief, is the person who leaves the white glove on the floor.

2. The thief always wears white gloves while committing burglaries.

3. It is possible that the thief wears a white glove while committing burglaries.

4. Everyone who wears white gloves is a thief.

5. A white glove will be found at the scene of the next burglary committed in Centertown.

6. Suppose that the private investigator finds a white glove at the scene of ten other burglaries. Do these added instances prove that a white glove will be found at the scene of the next burglary?

7. Determine if there are any other conclusions that might be warranted, given the facts.

11-3 Deductive Reasoning

Goals
- ■ Write conditional statements and identify them as true or false.
- ■ Identify valid and invalid deductive arguments.

Applications Advertising, Detective work, Earth science, Literature

Study the map of Illinois shown and read these four statements. Decide which of the statements are true and which are false.

1. If you live in Chicago, then you live in Illinois.
2. If you live in Illinois, then you live in Chicago.
3. If you do not live in Chicago, then you do not live in Illinois.
4. If you do not live in Illinois, then you do not live in Chicago.

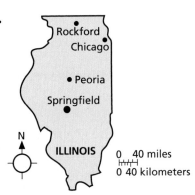

◤ BUILD UNDERSTANDING

In everyday life, we often make statements that involve "if" and "then." Such statements are called *if-then statements*, or *conditional statements*.

A conditional statement has two parts, a **hypothesis** and a **conclusion**. A hypothesis is a statement containing information that leads to a conclusion.

> *hypothesis*   *conclusion*
>
> If it rains, then it is cloudy.

Chicago, Illinois

Example 1

EARTH SCIENCE Write two conditional statements for the phrases.

> The temperature is 0°C.
> Water freezes.

Solution

Make one statement the hypothesis and the other the conclusion.

> *Statement 1* If the temperature is 0 °C, then water freezes.
>
> *Statement 2* If water freezes, then the temperature is 0 °C.

Some conditional statements are true and others are false. A *counterexample* shows that a conditional statement is false by satisfying the hypothesis but not the conclusion.

Example 2

Is the conditional statement true or false? If false, give a counterexample.

a. If a number is divisible by 10, then it is divisible by 5.

b. If a number is divisible by 5, then it is divisible by 10.

Solution

a. True. Certainly, any number ending in 0 is divisible by 5.

b. False. 25 is a counterexample because 25 is divisible by 5 but not by 10.

Deductive reasoning is the process of using facts, rules, definitions, or properties to reach a valid conclusion. It allows you to say that a conclusion is true for a given case. The following is an example of deductive reasoning.

Conditional statement If I do my homework by 6:00 P.M., then I may go to a movie.

Given information I did my homework by 6:00 P.M.

Conclusion Therefore, I may go to a movie.

In a **valid argument**, a new statement is obtained from the original statement and some given information through the process of deductive reasoning. If there is an error in the reasoning, then the argument is said to be *invalid*.

Example 3

Is this argument valid or invalid? Use a picture to help you decide.

Conditional statement If an animal is a cat, then it loves tuna.

Given information Fido loves tuna.

Conclusion Therefore, Fido is a cat.

Solution

Draw a shaded region for cats inside the region for tuna lovers. The argument is not valid because Fido might be a dog that loves tuna.

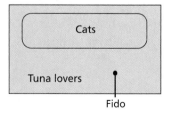

◥ TRY THESE EXERCISES

Write two conditional statements for the phrases.

1. There is no gasoline in the tank.
The car will not run.

2. Jodiann makes this free throw.
We win the game.

State whether the following conditional statements are true or false. If false, give a counterexample.

3. If a woman lives in the U.S. capitol, then she lives in Washington, D.C.

4. If two numbers are even, then their sum is an odd number.

5. If a shape is a square, then it has 4 sides.

Is this argument valid or invalid? Use a picture to help you decide.

6. If a flower is red, then it is pretty.
This rose is red.
Therefore, this rose is pretty.

7. If the battery is dead, then the car will not start.
The car did not start.
Therefore, the battery is dead.

8. **MYSTERY** If Selina's fingerprints are found on the door, then she is the thief.
Selina's fingerprints were not found on the door.
Therefore, Selina is not the thief.

PRACTICE EXERCISES • For Extra Practice, see page 583.

State whether the following conditional statements are true or false. If false, give a counterexample.

9. If two numbers are odd, then their product is odd.

10. If a number is divisible by 6, then it is divisible by 12.

11. If a shape has 4 sides, then it is a square.

Is this argument valid or invalid? Use a picture to help you decide.

12. If two numbers are even, then their sum is even.
The sum of 7 and 9 is even.
Therefore, 7 and 9 are even.

13. If an animal is a bird, then it can fly.
An ostrich is a bird.
Therefore, an ostrich can fly.

14. If a fruit is an orange, then it is a citrus fruit.
This fruit is not an orange.
Therefore, this fruit is not a citrus fruit.

Ostrich

Write each statement as an if-then statement.

15. All even numbers are divisible by 2.

16. All squares are quadrilaterals.

17. All whales are mammals.

18. All rectangles have 4 right angles.

Is the argument valid or invalid? Use a picture to help you decide.

19. All giraffes have long necks.
Titus has a long neck.
Therefore, Titus is a giraffe.

20. All multiples of 100 are multiples of 10.
500 is a multiple of 100.
Therefore, 500 is a multiple of 10.

Sometimes conditional statements can be strung together. Write a new if-then statement from the following conditional statements.

21. If a number is a whole number, then it is an integer. If a number is an integer, then it is a rational number.

22. If people live in Mexico City, then they live in Mexico. If people live in Mexico, then they live in North America.

Write a valid logical argument for each picture.

23.

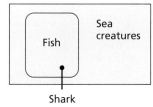

24.

Multiples of 2
Multiples of 4
16

25. ADVERTISING Advertisements are often misleading. An advertisement might show a young person at a party while a voice says "Sizzling Soda." The advertiser wants you to think that drinking Sizzling Soda leads to popularity. Think of some examples of misleading statements used by advertisers. Explain the error in reasoning that the advertiser hopes consumers will make.

26. WRITING MATH Suppose that a friend of yours is having difficulty understanding the difference between inductive and deductive reasoning. Write your friend a brief note explaining the differences between the two kinds of thinking. Include an example of each type of thought process.

◼ EXTENDED PRACTICE EXERCISES

27. CRITICAL THINKING Determine whether the following statement is always true. If it is not, provide a counterexample. *If the mathematical operation ∗ is defined for all numbers a and b as a ∗ b = a + 2b, then the operation ∗ is communative.*

LITERATURE The following statements are from Lewis Carroll's book *Symbolic Logic*. Draw a conclusion and write a valid argument for each.

28. All well-fed canaries sing loud. No canary is melancholy if it sings loud.

29. All puddings are nice. No nice things are wholesome. This dish is a pudding.

30. CHAPTER INVESTIGATION Write a paragraph describing the reasoning the characters use to solve the mystery.

◼ MIXED REVIEW EXERCISES

Simplify. (Lesson 9-3)
$f(x) = 4x^3 \quad g(x) = 2x^{-2} \quad h(x) = 6x$

31. $f(x)[g(x)]$

32. $g(x)[h(x)]$

33. $f(x)[h(x)]$

34. $f(x)[g(x)][h(x)]$

35. Find the probability of drawing two aces from a standard deck of cards without replacement. (Lesson 10-5)

Venn Diagrams

Goals
- Draw Venn diagrams.
- Use Venn diagrams to solve problems.

Applications   Statistics, Sports, Community service, Transportation

There are 10 basketball players and 18 baseball players.

1. Are there necessarily 28 different players? Explain.

2. Complete the table to illustrate your example.

Sport	Basketball	Baseball	Both sports	Either sport
Number of players	■	■	■	■

3. Discuss the difference between "both" and "either."

▼ BUILD UNDERSTANDING

Some problems can be solved by using a diagram. In a **Venn diagram**, sets of data are represented by overlapping circles. These circles show relationships among the sets so that all possible combinations have a distinct area in the diagram. The size or area of the circles has no significance to the number of data in a set. The diagram shows the relationship between basketball players and baseball players.

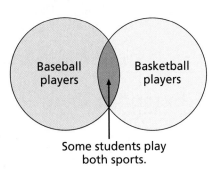

Some students play both sports.

Example 1

COMMUNITY SERVICE Use the Venn diagram.

a. How many volunteers painted the school?

b. How many painted the school but did not clean the park?

c. How many cleaned the park and painted the school?

d. How many different volunteers painted the school or cleaned the park?

Community Service Volunteers

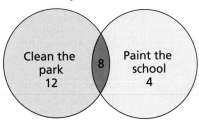

Solution

a. Adding 4 + 8, there were 12 volunteers who helped paint the school.

b. There were 4 volunteers who painted the school but did not clean the park.

c. There were 8 volunteers who both cleaned the park and painted the school.

d. There were 4 + 12 + 8 = 24 different volunteers.

To represent the entire population discussed, the overlapping circles are commonly placed within a rectangle. This rectangle represents the *universal set.*

Example 2

A poll of 100 students finds the data shown.

a. How many students play soccer or basketball?

b. How many of the 100 students polled play no sport?

Number of students	Sport
66	Basketball
50	Soccer
48	Hockey
22	Soccer and basketball
30	Hockey and basketball
28	Soccer and hockey
12	All three sports

Solution

Make a Venn diagram like the one shown below at the left. Put 12 in the region marked A because there are 12 students who play all three sports.

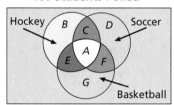

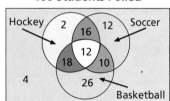

Since 28 students play hockey and soccer (regions A and C together) and region A has 12 people, region C has 28 − 12 = 16 people. Find the value for each other letter. The completed diagram is shown above at the right.

a. 16 + 12 + 12 + 10 + 18 + 26 = 94 students play soccer or basketball.

b. 100 − (2 + 16 + 12 + 12 + 18 + 10 + 26) = 4 students play no sport. Indicate this by placing a 4 in the lower left corner of the universal set.

◤ TRY THESE EXERCISES

Use the Venn diagram shown.

1. How many students own a dog?

2. How many students own a dog but not a cat?

3. How many students own both a dog and a cat?

4. How many different students own a dog or a cat?

FASHION Tamara surveyed 25 students at lunch. Of the students surveyed, 18 wore sneakers, 12 wore jeans and 9 wore both sneakers and jeans.

5. Make a Venn diagram for the survey results.

6. How many students did not wear either sneakers or jeans?

7. How many different students wore sneakers or jeans?

8. How many students wore sneakers but do not wear jeans?

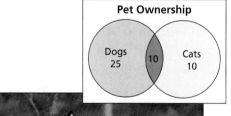

Use the Venn diagram shown.

9. How many students are in chorus?

10. How many students are in chorus but not band?

11. How many students are in both chorus and band?

12. How many different students are in chorus or band?

Students in Chorus or Band

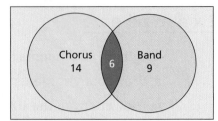

STATISTICS For his statistics class, Takoda researched people's preferences for a new store opening in the mall. He surveyed 120 people at the mall. Of the people surveyed, 62 would like a new bookstore, 48 would like a new music store, and 23 would like both new stores.

13. Make a Venn diagram of the results of Takoda's survey.

14. How many people do not want either store to be included in the mall?

15. How many people want a new music store but not a new bookstore?

RECREATION In a class of 30 students, 20 are going on vacations during the summer, 15 will work, and 3 will do neither activity.

16. Make a Venn diagram of the results of the survey.

17. How many students are planning on both working and going on a vacation?

18. How many different students are planning on working or going on a vacation?

ENTERTAINMENT Use the Venn diagram shown. List the value in the region for each group.

19. Teenagers who like mystery movies.

20. Teenagers who like comedy and action movies.

21. Teenagers who like mystery and comedy movies.

22. Teenagers who do not like comedy movies.

23. Teenagers who don't like mystery, comedy, or action movies.

Favorite Movies of 36 Teenagers

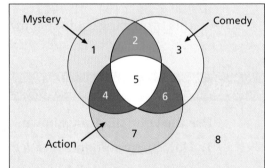

RECREATION A state park conducted a survey of 250 people to determine their favorite camping activities. The results are shown in the Venn diagram.

24. How many like to swim and hike but not canoe?

25. How many different people like to swim or hike?

26. How many like to swim or hike but not canoe?

27. How many participate in all three activities?

28. How many people surveyed do not like to do any of the activities listed?

29. **WRITING MATH** Write a Venn diagram problem of your own. Exchange your problem with a classmate and solve each other's problem.

Survey of 250 Campers

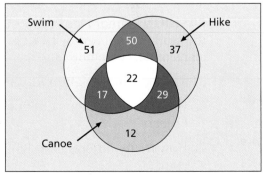

30. DATA FILE Refer to the data about trips by purpose on page 534. The data refer to the primary purpose of travel, but many people travel for more than one reason. For example, someone on a business trip may also play golf or go swimming, so that there is some overlap between business and leisure. Make a possible Venn diagram of the information given in the table to illustrate some of the different combinations of vacation purposes.

SPORTS At a school sports banquet, a survey was conducted concerning what sports each person played. The results showed that of those surveyed, 18 students play a fall sport, 16 play a winter sport, and 21 play a spring sport. It was also noted that 5 play a fall and a winter sport, 8 play a fall and a spring sport, and 7 play a winter and a spring sport. Only 3 students play a fall, a winter and a spring sport.

31. Make a Venn diagram for the survey.

32. How many students play only in fall? in winter? in spring?

33. How many students were surveyed at the banquet?

34. How many students play a fall or winter sport but not a spring sport?

◣ EXTENDED PRACTICE EXERCISES

TRANSPORTATION A survey of 200 commuters produced the following data: 118 use the expressway, 98 wear sunglasses, 84 listen to the radio, 40 listen to the radio and wear sunglasses, 58 listen to the radio and use the expressway, 62 use the expressway and wear sunglasses, 24 do all three.

35. How many people wear glasses but do not use the expressway or listen to the radio?

36. How many people only use the expressway?

37. How many people listen to the radio or use the expressway?

38. How many people do none of these things?

39. Make a Venn diagram for the survey.

◣ MIXED REVIEW EXERCISES

Make tree diagrams to determine the sample space for each. (Lesson 10-3)

40. the number of unique outfits made from 3 pairs of shorts, 4 T-shirts, and either tennis shoes or sandals

41. the number of two-topping pizzas made from sausage, pepperoni, green peppers, onions, bacon, and meatballs

42. POLITICAL SCIENCE People were randomly called and asked who they would vote for in a local election. What is the experimental probability that Forrest will win the election with the following results. (Lesson 10-2)

Conrad	35,137
Forrest	46,250
Undecided	18,613

Review and Practice Your Skills

State whether each conditional statement is true or false. If false, give a counterexample.

1. If two numbers are odd, then the sum of the numbers is odd.

2. If a number has exactly two factors, then the number is prime.

3. If a shape is a square, then it has four right angles.

4. If a number is divisible by 5, then it is divisible by 25.

5. If a shape has four equal sides, then it is a square.

Is each argument valid or invalid? If it is invalid, give a counterexample.

6. If two numbers are negative, then their product is positive.
   The numbers −3 and −8 are negative numbers.
   Therefore, the product of −3 and −8 is a positive number.

7. If two figures are congruent, then they have the same area.
   Triangles *ABC* and *PRQ* have the same area.
   Therefore, triangles *ABC* and *PRQ* are congruent.

8. All multiples of 6 are divisible by 3.
   24 is divisible by 3.
   Therefore, 24 is a multiple of 6.

9. All the chess club members are under 20.
   Devora is a member of the chess club.
   Therefore, Devora is younger than 20.

Write each of the following as an if-then statement.

10. All butterflies are insects.

11. All even numbers are divisible by 2.

12. All pentagons have five sides.

13. All powers of 6 end in the digit 6.

Fifty people were surveyed about what kinds of exercise they do. The results are shown in the Venn diagram. Use it to answer the questions.

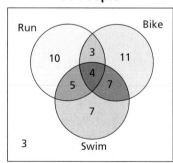

50 People

14. How many people run?

15. How many people swim?

16. How many people bicycle?

17. How many do all three?

18. How many run or swim?

19. How many run and swim?

20. How many don't do any of the three kinds of exercise?

TRANSPORTATION In a group of 70 people, 28 own a car, 48 own a bicycle, and 4 people own neither item.

21. Make a Venn diagram of the data.

22. How many of the people own either a car or a bike?

23. How many of the people own both a car and a bike?

Write a statement about each figure. Check to see if your statement is true or false. (Lesson 11-1)

Figure A

24. line segment *AB* in Figure A   **25.** the diagonal lines in Figure B

Use the given examples to complete the conjecture. (Lesson 11-2)

26. $(-1)^2 = 1$ $(-1)^4 = 1$ $(-2)^2 = 4$ $(-2)^4 = 16$

A negative number to an even power is a (positive/negative) number.

Figure B

Is each argument valid or invalid? If it is invalid, give a counterexample. (Lesson 11-3)

27. All the runners finished the race.
Kenji was a runner in this race.
Therefore, Kenji finished the race.

28. All cats have four legs.
Alice, the cow, has four legs.
Therefore, Alice is a cat.

A survey was taken of 30 people in their early twenties. Of the people surveyed, 13 were in school, 15 worked and 6 did neither. (Lesson 11-4)

29. Make a Venn diagram.

30. How many of the people are either in school or working?

31. How many of the people are both working and in school?

Mid-Chapter Quiz

Define the following. (Lessons 11-1, 11-2, and 11-3)

 1. optical illusion   **2.** statement   **3.** conjecture   **4.** counterexample

 5. Use inductive reasoning to find the ones digit of 2^{40}. (Lesson 11-2)

Test each conjecture to see if it is always true. (Lesson 11-3)

 6. The square of a number ending in 5 always ends in 25.

 7. A number is divisible by 6 if the sum of the digits is divisible by 2 and 3.

 8. Pick any 2-digit number. Multiply it by 111. Multiply the product by 91. The product is a 6-digit number which repeats the 2-digit number three times.

Is each conditional statement true or false? If false, give a counterexample. (Lesson 11-3)

 9. If the square of a number is divided by the number, the quotient is 1.

10. If a number is divisible by 12, then the number is also divisible by 6.

Is the argument valid or invalid? (Lesson 11-3)

11. Dogs are animals. Animals need food. So, dogs need food.

Use Logical Reasoning

Goals
- Make a table to solve non-routine problems.
- Solve non-routine problems involving logic.

Applications   Private investigation, Logic puzzles, Sports

Pablo has five hats. Three are red and two are blue. Orlando, Jaimie, and Calvin are seated in a single-file row. Pablo places a hat on each of their heads, discarding the two extra hats.

Jaimie and Calvin cannot tell which color hat they wear, even though they can see the color of the hats on the other players. Orlando then states correctly which color hat he wears.

1. What are the possible scenarios for the three hats placed on the heads?

2. How can Orlando tell which color hat he wears when the others cannot?

3. If Calvin saw two red hats could he state the color of his hat with any certainty?

4. If Calvin saw two blue hats could he state the color of his hat?

5. How can the hats be arranged so that both Calvin and Jaimie know the color of their hats?

◣ BUILD UNDERSTANDING

The **process of elimination** is a logical reasoning tool that can be used to solve many non-routine problems. Using the process of elimination involves looking at all of the possible answers to a problem and eliminating the ones that are not feasible. For example, on a multiple choice test, it is often helpful to test all of the choices in order to eliminate the ones that are obviously incorrect.

Example 1

Make a conclusion from the following statements.

Levanna either likes algebra or history.

Levanna does not like working with variables.

Solution

Since algebra involves working with variables, you can deduce that Levanna likes history.

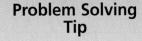

Problem Solving Tip

To solve simple reasoning problems, you can perform one or two eliminations mentally.

Example 1 can be solved with a single elimination. However, more complicated reasoning problems that contain a lot of information can be too confusing to solve mentally.

In order to solve such complex logic problems, it can help to make a table. Using the process of elimination, you can complete the table to gain information.

Example 2

Dexter, Maria and Phil each play baseball, badminton and tennis, but not necessarily in that order. Each person plays only one sport, and no two people play the same sport. Dexter does not play a sport that requires a ball. Phil does not play baseball.

Which sport does each play?

Solution

Step 1 Make a table listing each person and each sport.

	Baseball	Badminton	Tennis
Dexter			
Maria			
Phil			

Step 2 Since Dexter does not play a sport that requires a ball, place an **X** in the boxes for baseball and tennis. Place a ✔ in the box for badminton. Since Phil does not play baseball, place an **X** in the box for baseball.

	Baseball	Badminton	Tennis
Dexter	X	✔	X
Maria			
Phil	X		

Step 3 Since each person plays only one sport, place an **X** in the badminton boxes for Maria and Phil.

	Baseball	Badminton	Tennis
Dexter	X	✔	X
Maria		X	
Phil	X	X	

Step 4 There is only one choice of sport for Phil. Place a ✔ in the box for tennis for Phil.

	Baseball	Badminton	Tennis
Dexter	X	✔	X
Maria		X	
Phil	X	X	✔

Step 5 By the process of elimination, Maria must be the baseball player. So Dexter plays badminton, Maria plays baseball, and Phil plays tennis.

◥ TRY THESE EXERCISES

Draw a conclusion from the following statements.

1. Amy either likes ice cream or yogurt.
   Amy does not like yogurt.

2. I either have Math third period or seventh period.
   I have English seventh period.

3. Brett will either go surfing or golfing while on vacation.
   Brett is afraid of the water.

Make a table to solve the following problem.

4. The Whites, Browns and Greens all live in houses and own cars that are either white, brown or green. None of the families lives in a house or has a car that is the same color as the name. The Browns' car is the same color as the Greens' house. What color house and car does each family own?

PRACTICE EXERCISES • For Extra Practice, see page 584.

Make a conclusion from the given information without using a table.

5. Either Billy or Naomi earned a score of 100% on the math test.
   Billy scored 90% on the math test.

6. Three friends took a spelling test and are trying to figure out who did the best. Liang did better than Kyle, who did better than Jeter.

7. Either Mary, Terry or Gerry is not married.
   Terry attended Mary's wedding.
   Gerry is celebrating her first wedding anniversary.

8. Two friends, Marcellus and Amielo, are deciding who Ginny likes.
   Ginny said she likes someone who isn't blonde.
   Amielo has dark brown hair, and Marcellus has blonde hair.

Make a table to solve Exercises 9–15.

9. Three dogs are named Tillie, Spot and Tippie. Two are long-haired dogs, a collie and a Pekingese. One is a short-haired basset hound. Tippie does not have short hair. The collie lives next door to Tillie. Tippie lives next door to the Pekingese. Tillie is not a basset hound. Which is the Pekingese?

10. **ENTERTAINMENT** Three singers appear at a concert. One is named Jones, one is named Salazar and one is named Friedman. One sings only jazz, one sings only opera and one sings only folk music. Jones dislikes the opera singer, but the folk singer is a friend of Jones. Salazar performs before the opera singer in the concert. Who is the folk singer?

11. My three cousins, Sue, DeeDee and Betty, just got married, but I have forgotten each husband's name. I remember that my cousins married Tom, Bill and Harry. I heard Betty say that she hoped Tom didn't argue with her husband. I also remember that DeeDee married Tom's brother and that Harry married DeeDee's sister. Who married whom?

12. **MYSTERY** Hector is a private investigator called to the scene of a crime. There are four suspects: the butler, the cook, the chauffeur, and the maid. All four suspects are wearing identical uniforms and refuse to tell Hector their jobs. Hector is sure that the butler committed the crime. The four suspects are named Alexandra, Beatrice, Cecil, and Delphine. Hector discovers that Delphine cannot drive. Cecil is the only person who knows where the cheese is when Hector asks for a snack. Neither Alexandra nor Delphine can tell Hector where the mops are kept. Who is the butler?

13. Armand, Belinda, Colette and Dimitri are four artists. One is a potter, one is a painter, one is a violinist and one is a writer. Armand and Colette saw the violinist perform. Belinda and Colette have modeled for the painter. The writer wrote a story about Dimitri and plans to write a story about Armand. Armand does not know the painter. Who is the writer?

14. Carol, Tawnee, and Esther go to the same high school. The girls' fathers, Mr. Smith, Mr. Jones and Mr. King, teach at the high school. Neither Carol nor Tawnee is named Smith. Esther and Tawnee are in Mr. King's class. Mr. Jones teaches Carol and Esther. None of the girls has her father for a teacher. Who is Carol's father?

15. **SPORTS** Amir, Ely and Zeke each play either football, basketball or baseball. The basketball player is taller than the football player. The football player is taller than the baseball player. Ely is 5 in. taller than Zeke, and 2 in. taller than Amir. Match each person with his sport.

16. **WRITING MATH** Write your own logical reasoning problem. Be as creative as possible. Exchange problems with a classmate and solve.

EXTENDED PRACTICE EXERCISES

17. **CRITICAL THINKING** The softball team has six players: Loraine, Hisa, Mila, Dana, Edna and Queshia. Loraine had more hits than Mila, who had more hits than Dana. Edna had fewer hits than Loraine and Mila but more than Dana. Hisa had the same number of hits as Loraine but fewer than Queshia. Arrange the players by the number of hits.

18. **DATA FILE** Refer to the data on the 2000 Olympics on page 533. Carlos won one of his country's 11 medals of a certain color. Michael won one of his country's 13 medals of a certain color. Carlos and Michael are not from the same country. Michael's country won the same number of gold medals as silver medals. Carlos is not British. Which country is Carlos from?

19. **CHAPTER INVESTIGATION** Were the story's conclusions drawn logically? Use the vocabulary in the chapter to explain the logic.

MIXED REVIEW EXERCISES

Find the probability of each event. (Lessons 10-1 and 10-5)

20. P(red)

21. P(green or orange)

22. P(red, then blue)

23. P(orange or green, then blue)

Solve.

24. About how many times would you expect to spin green or orange in 50 spins? (Lesson 10-6)

25. Lian wins if she spins either blue, green, or orange. Otherwise Julie wins. Is this a fair game? Explain. (Lesson 10-7)

Problem Solving Skills: Reasonable Answers

A common mathematical skill is looking at a situation and seeing if the numbers "make sense," that is, checking to see if a proposed answer is reasonable. For example, if you are planning a long trip, you might approximate the amount of money you will need and then go over a possible list of expenses to see if your amount seems reasonably sufficient.

One strategy, **eliminate possibilities**, is often used to determine whether or not answers are reasonable. When using this strategy, account for all the possibilities, then organize them so that you are able to dismiss some as impossible.

Problem Solving Strategies

Guess and check

Look for a pattern

Solve a simpler problem

Make a table, chart or list

Use a picture, diagram or model

Act it out

Work backwards

✔ Eliminate possibilities

Use an equation or formula

Problem

Suppose that one car travels at a speed of 30 mi/h and another car travels at 45 mi/h. How long will it take the faster car to catch the slower car if the slower car has a 90 mi head start? Select the most reasonable answer.

A. 1 h   B. 2 h   C. 4 h   D. 6 h

Solve the Problem

D. The fact that it takes the faster car 2 h just to make up the head start distance makes the first two choices unreasonable, so they should be eliminated. In 4 h, the car with the head start has gone 210 mi and the car in pursuit has gone only 180 mi. In 6 h, they both have traveled 270 mi.

◤ TRY THESE EXERCISES

1. A car is speeding at 70 mi/h when picked up on a state trooper's radar. If the speeder has a 1-mi head start, how fast must the state trooper go to catch the speeder within the state line, which is 5 mi away from the trooper? Select the most reasonable answer.

   A. 70 mi/h   B. 80 mi/h

   C. 90 mi/h   D. 100 mi/h

2. At Mike's you can buy a burger for $2.00 and a drink for $1.00. At Bob's you can buy a burger for $1.50 and a drink for $1.50. Paige says that if she buys a burger at Bob's and a drink at Mike's she can save $0.50 on the burger and $0.50 on the drink, for a savings of $1.00. Tanita says that since she spent $2.50 and would have spent $3.00 at either place, then she only saved $0.50. Whose answer is more reasonable? Explain.

Five-step Plan
1 Read
2 Plan
3 Solve
4 Answer
5 Check

3. Ten years ago a father was three times as old as his daughter. The sum of their ages today is 80. How old is the father? Select the most reasonable answer.

    A. 25   B. 35   C. 55   D. 60

4. A dog is on a 20-ft leash attached to a stake. If the dog walks in a circle at the end of the leash, about how far can he walk before returning to where he started? Select the most reasonable answer.

    A. 9 ft   B. 62 ft   C. 20 ft   D. 126 ft

5. A 22 yr-old man has a 2 yr-old son, so he is $\frac{1}{11}$ the age of his father.

When the father is 24 yr-old, the boy will be 4-yr old, or $\frac{1}{6}$ his father's age.

When the father is 25 yr-old, the boy will be 5-yr old, or $\frac{1}{5}$ his father's age.

When the father is 30 yr-old, the boy will be 10-yr old, or $\frac{1}{3}$ his father's age.

When the father is 40 yr-old, the boy will be 20-yr old, or $\frac{1}{2}$ his father's age.

Assumption The son will eventually be the same age as his father. Is this a reasonable assumption? Explain your answer.

6. A turtle challenged a rabbit to a 2-mi race on the condition that the turtle had a 1-mi head start. The turtle stated that when the rabbit had run 1 mi, he would be $\frac{1}{2}$ mi ahead. When the rabbit had run another $\frac{1}{2}$ mi he would be $\frac{1}{4}$ mi ahead. When the rabbit had run another $\frac{1}{4}$ mi, the turtle would be $\frac{1}{8}$ mi ahead. Hence, the turtle would always be ahead and would win the race. Is this a reasonable answer? Explain.

7. A man told his son that he was leaving him nothing in his will. But when the will was read, the man had left the son a piece of land. The son was happy to learn that the triangular piece of land had sides of 250 ft, 300 ft, and 600 ft. Is it reasonable to say he received something? Explain your answer.

MIXED REVIEW EXERCISES

Factor. (Lesson 9-5)

    8. $42r + 56r^2$ **9.** $16gh^2 - 28g^2h$ **10.** $12x^2 + 18xy - 36y^3$

    11. $x(t + 7) - 4(t + 7)$ **12.** $28rt - 32r + 16r^2t^2$ **13.** $14m - 49mt^2$

Simplify. (Lesson 9-6)

    14. $\frac{42bc}{7c}$ **15.** $\frac{27y^7}{-3y^3}$ **16.** $\frac{r^2s^3t^4}{st^2}$ **17.** $3d - \frac{12cd}{3d}$

    18. $-jk^3 + 3j^2 - \frac{2j}{j}$ **19.** $f^4 + 16f^5h^2 - \left(\frac{20f^4}{-4f^3}\right)$ **20.** $15v^3 + 10v^2 - \frac{5v}{5v^2}$

Review and Practice Your Skills

Draw a conclusion from the given information without using a table.

1. Shantia will learn either water skiing or snow skiing.
   Shantia does not like cold weather.

2. Bart has either math or history during fourth period.
   Today in third period Bart learned how to graph an equation.

3. Julia is older than Mia. Sharon is younger than Julia.

4. Two neighbors, Marcia and Ted, are a lawyer and an engineer.
   Last Wednesday Marcia spent all day preparing a legal presentation.

5. Two friends, Kanya and Bonita, play the flute and the clarinet.
   Kanya is married to the flute player's brother.

Make a table to solve each problem.

6. Joe, Paula, and Mark each have a car. The cars are red, black, and white. The person with the red car is Joe's sister-in-law. Mark thinks white cars always look dirty. What color is each person's car?

7. Harriet, Fred and Wendy like to read mystery books, biographies and fiction. Each person has a favorite subject. Wendy bought a biography of Georgia O'Keefe for Harriet. Fred does not like mystery books. What is each person's favorite subject?

Select the most reasonable answer for each problem.

8. Two cities are 275 mi apart. Two cars leave the cities at the same time. One is going 50 mi/h; the other is going 60 mi/h. When do the cars meet?

   A. 1 h 30 min   B. 2 h 30 min   C. 4 h 35 min   D. 5 h 30 min

9. Walnuts cost $2.75 lb and peanuts are $1.50 lb. How many pounds of peanuts should be mixed with 4 lb of walnuts to get a mixture worth $2.00/lb?

   A. 2 lb   B. 4 lb   C. 5 lb   D. 6 lb

10. Six years from now a woman will be four times as old as her son. The product of their ages today is 52. How old is the son right now?

    A. 2 yr   B. 6 yr   C. 8 yr   D. 10 yr

11. The number 64 is both a square and a cube. What is the next greatest whole number that has this property?

    A. 1   B. 512   C. 729   D. 1331

12. Two sides of a triangle are 20 in. and 34 in. Which of these could be the length of the third side of the triangle?

    A. 10 in.   B. 34 in.   C. 54 in.   D. 58 in.

13. Use inductive reasoning to find the ones digit of 7^{12}. (Lesson 11-2)

14. Is the argument below valid or invalid? (Lesson 11-3)

If a figure is a square, then it is also a parallelogram.
The quadrilateral *ABCD* is a parallelogram.
Therefore, *ABCD* is a square.

MARKET RESEARCH In a survey of 100 neighbors, 47 want a new school, 50 want a new library and 12 want neither. (Lesson 11-4)

15. Make a Venn diagram.

16. How many people want both a school and a library?

17. Denzel has two sisters, Marcy and Ruth. The oldest child is not a boy. Ruth was born 2 years after Denzel. Find the youngest child by making a table. (Lesson 11-5)

18. Select the most reasonable answer. Kentare has 15 dimes and quarters worth $2.55. How many dimes does he have? (Lesson 11-6)

A. 3 dimes   B. 5 dimes   C. 8 dimes   D. 12 dime

MathWorks Career – Cryptographer
Workplace Knowhow

Secret codes are important to the military, banks, the U.S. postal service, and other businesses and organizations. A cryptographer studies codes and code breaking. The process of putting a message in coded form is called encoding. Taking it from coded form back to its original form is called decoding or deciphering. A cryptographer studies patterns and mathematics to decode a message.

A shift code replaces each letter of the alphabet with a letter that is a certain number of places away. For example, in a +2 shift code, the letter L represents the letter N, because N is two letters beyond L in the alphabet.

Refer to the messages written in code. Each message contains the word "math."

1. What is the shift code in Message 1?

2. Decipher Message 1.

3. What is the shift code in Message 2?

4. Decipher Message 2.

5. Which code is more difficult to decipher? Explain.

Message 1

JXQEFPZEXIIBKDFKDXKA CRK.

Message 2

15 3 22 10   20 7 3 14 14 1

6 17 7 21   15 3 22 22 7 20.

Non-Routine Problem Solving

Goals ■ Solve non-routine problems involving multiple steps.
■ Solve non-routine problems involving reasoning processes.

Applications Landscaping, Detective work, Number sense

Work in groups of four students.

1. Three students sit in a triangle facing each other.

2. The fourth student prepares three pieces of tape, marking one or none of them with an **X**. This student then places one piece of tape on the forehead of each student in the triangle.

3. By looking at the other students, each student decides whether the tape on his or her own forehead is marked with an **X**.

4. How can each player determine whether his or her tape is blank or marked with an **X**?

▼ BUILD UNDERSTANDING

Seemingly difficult problems can often be solved if you try looking at them with a different point of view. Experimentation, or **trial and error**, can be useful in helping you solve non-routine problems.

Example 1

COOKING You need exactly 4 qt of water for a recipe. You have plenty of water available, but you only have 3-qt and 5-qt measuring cups. How can you measure exactly 4 qt of liquid by using only the two measuring cups that you have on hand?

Solution

There are many possible solutions to this problem, one of which is illustrated in the diagram. In this solution, container A holds 3 qt and container B holds 5 qt.

> **Problem Solving Tip**
>
> One useful strategy that may help you figure out a non-routine problem is using pictures to think of and visualize possible solutions.

Pour 3 qt from container A into container B.

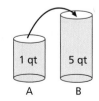

Fill container A again. From it, fill container B. Now container A has 1 qt and container B has 5 qt.

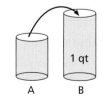

Empty container B. Place the 1 qt from container A into B.

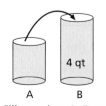

Fill container A. Empty the 3 qt from container A into container B. Now container B has 4 qt.

Example 2

Divide the figure shown into two pieces, each having the same size and shape.

Solution

Suppose that your initial approach is to try to solve the problem by making only one cut. The two figures that result probably will not have the same shape, such as the figure shown at the left.

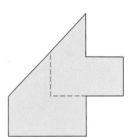

Now make two cuts to solve the problem. The figure at the right shows that two cuts will indeed give two figures that have the same size and shape.

◥ TRY THESE EXERCISES

1. A girl put a caterpillar in a glass jar at 8:00 P.M. when she went to bed. Every hour the caterpillar climbed up the wall of the jar 2 in. and slid back 1 in. If the jar is 6 in. high, at what time did the caterpillar reach the top of the jar without sliding back?

2. If it takes 1 min to make a cut, how long will it take to cut a 10-ft log into ten equal parts?

3. What is the greatest number of pieces into which you could cut a round pizza with four straight cuts of a knife?

4. A pair of sneakers and the laces cost $15, and the sneakers cost $10 more than the laces. How much do the laces cost?

5. How many triangles are there in the figure?

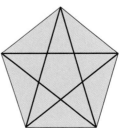

6. There are eight marbles that are exactly alike in size and shape, but one is heavier than the others. Using a balance scale, how can you find the heaviest marble in two weighings?

7. **NUMBER SENSE** How can you score exactly 100 points with 6 darts if they all hit the target?

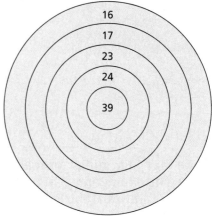

8. Suppose you have a 5-gal, an 11-gal and a 13-gal container. How could you take 24 gal of water and divide it equally into three parts using only these containers?

9. **LANDSCAPING** A landscaper wants to build a fence around a garden that is 50 ft on each side. If the fencing comes in 10-ft sections, how many posts will the landscaper need?

10. Trace the four shapes shown. Cut them out and arrange the pieces to form an equilateral triangle. Then rearrange them to form a square.

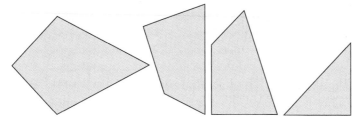

For Exercises 11–17, suppose you have a 3-in. cube painted green.

11. How many cuts would it take to cut the cube into 1-in. cubes?

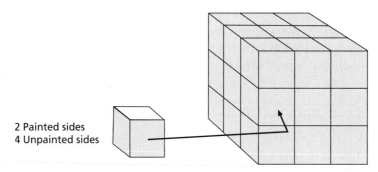

2 Painted sides
4 Unpainted sides

12. How many cubes would you have?

13. How many cubes would have no painted sides?

14. How many cubes would have exactly one painted side?

15. How many cubes would have exactly two painted sides?

16. How many cubes would have exactly three painted sides?

17. How many cubes would have exactly four painted sides?

18. Copy the figure shown. Add three straight lines to get nine, non-overlapping triangles.

19. WRITING MATH How does drawing a picture help you solve many non-routine problems? Describe some of the advantages of this technique.

20. Two mothers and three daughters went to see a ball game. There were only three tickets, yet each one had a ticket. How could this be?

21. ERROR ALERT Miguel reasoned that since he needed to take the elevator from the first floor to the eighth floor, he would be traveling 8 stories. Is he correct? If not, what mistake did he make?

■ EXTENDED PRACTICE EXERCISES

22. MYSTERY Suppose you are solving a crime of missing money. The only two people who could have committed the crime are the butler or the maid. The maid admits that she saw the money while cleaning but put it under the desk blotter for safe keeping. The butler admits that he saw the money and placed it inside the book that was on the desk for safe keeping. He remembers placing the money between pages 99 and 100. Solve the mystery. Your textbook may help you.

23. How many different size squares can be drawn in a 4 by 4 grid of dots?

24. A man has to take a wolf, a goat and some cabbage across a river. His rowboat has enough room for the man plus either the wolf or the goat or the cabbage. If he takes the cabbage with him, the wolf will eat the goat. If he takes the wolf, the goat will eat the cabbage. Only when the man is present is the goat safe from the wolf and the cabbage safe from the goat. How does he get them across? *Hint*: He makes the round trip three times before making his final crossing.

■ MIXED REVIEW EXERCISES

Name the type of graph that is most appropriate for each situation.
(Lessons 1-3 through 1-8)

25. to compare changes over time

26. to compare women's and men's 12-min-run times for a particular year

27. to see if two sets of data correlate

28. to show how a quantity is divided up

Chapter 11 Review

VOCABULARY

Choose the word from the list that best completes each statement.

1. A(n) __?__ shows that a conjecture is false.

2. A(n) __?__ shows the relationships between two or three different classes of things.

3. A picture that the eye sees as true that is actually not true is called a(n) __?__.

4. The __?__ is the *Then* part of a conditional statement.

5. The __?__ is the *If* part of a conditional statement.

6. The process of reaching a conclusion based on a set of specific examples is called __?__.

7. The __?__ can be represented by a rectangle in a Venn diagram.

8. Suppose you are given a conditional statement and an additional piece of information upon which you can form a logical argument. This process is called __?__.

9. An *if-then* statement is also known as a(n) __?__ statement.

10. The process of experimenting to find a solution is known as the __?__ method.

a.	concept map
b.	conclusion
c.	conditional
d.	counterexample
e.	deductive reasoning
f.	hypothesis
g.	inductive reasoning
h.	optical illusion
i.	scientific theorem
j.	trial and error
k.	universal set
l.	Venn diagram

LESSON 11-1 ◣ Optical Illusions, p. 478

▶ A **statement** is a sentence that describes a particular relationship. You can often test a statement to find out whether it is true or false.

Write a statement about each figure. Then check your statement.

11.

12.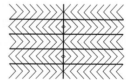

13. Write a description of what you see in the figure. Justify your answer.

14. What is the relationship between the lengths of pencil L and pencil R?

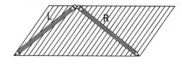

LESSON 11-2 ◣ Inductive Reasoning, p. 482

▶ **Inductive reasoning** is the process of reaching a conclusion based on a set of specific examples. The conclusion is called a **conjecture**. A conjecture may be true, but it is not necessarily true. You need to find just one **counterexample** to prove it false.

15. Pick a number. Double it. Add 6. Add the original number. Divide by 3. Add 4. Subtract the original number. What do you get? Repeat the process several times using a different original number. Then make a conjecture about the process.

16. Each figure below is a regular polygon with a side extended at each vertex. Measure the angles formed by these extended sides and find the sum of the angles for each figure. Make a conjecture about these angles. Is your conjecture always true?

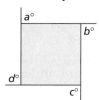 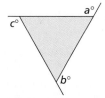 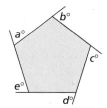

LESSON 11-3 ◣ Deductive Reasoning, p. 488

▶ A conditional statement has two parts, a hypothesis and a conclusion. A **hypothesis** is a statement containing information that leads to a **conclusion**.

Write two conditional statements that can be made from these two sentences.

17. It has been snowing. The streets are slippery.

18. An even number is divisible by 2. The number 5280 is even.

Determine whether each argument is valid or invalid.

19. If today is Monday, then I will play tennis.
Today is Monday.
Therefore, I will play tennis.

20. If it rains on me, then I will get wet.
It did not rain on me.
Therefore, I did not get wet.

LESSON 11-4 ◣ Venn Diagrams, p. 492

▶ In a **Venn diagram**, a collection is represented by a circular region inside a rectangle. The rectangle is called the *universal set*.

A survey of 250 people was conducted to determine their preferences in car accessories. The results are shown in the Venn diagram.

21. How many people prefer front-wheel drive?

22. How many people prefer white-wall tires, but not front-wheel drive?

23. How many people prefer bucket seats but not white-wall tires?

24. How many people prefer front-wheel drive, white-wall tires, and bucket seats?

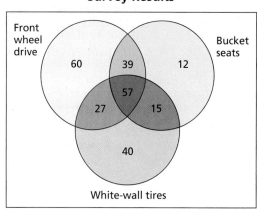

Survey Results

25. At the supermarket, 83 customers bought wheat bread, 83 bought white bread, and 20 bought rye bread. Of those who bought exactly two kinds of bread, 6 bought rye and wheat, 10 bought white and rye, and 12 bought white and wheat. Four customers bought all three types of bread. Draw a Venn diagram to find how many customers bought only rye bread.

LESSON 11-5 ◣ Use Logical Reasoning, p. 498

▶ The *process of elimination* is a logical tool that can be used to solve many non-routine problems.

26. Three students named Ashley, Brandon, and Conchita receive three different grades on a test. One received an A, one received a B and one received a C. No student received a grade matching the first letter of the student's name. Ashley did not receive a B. What did each receive?

27. Rick, Carlos, and Lee are neighbors. Their hobbies are painting, fixing cars, and gardening. Their occupations are doctor, teacher, and lawyer. The gardener and the teacher both graduated from the same college. Both the lawyer and Rick have poodles, as does the painter. The doctor bandaged the painter's broken thumb. Carlos and the lawyer have lived next door to each other for five years. Lee beat both Carlos and the gardener in tennis. What is each person's hobby and occupation?

LESSON 11-6 ◣ Problem Solving Skills: Reasonable Answers, p. 502

▶ One strategy, **eliminate possibilities**, is often used to determine if answers are reasonable.

28. Twenty years ago, Rhoda was four times as old as Li. Now Rhoda is only twice as old as Li is now. How old is Li? Select the most reasonable answer.

    A. 60 years   B. 55 years   C. 30 years   D. 15 years

29. What value of n makes the sum $n + 25$ equal -18?

    A. -43 B. -7 C. 7   D. 43

30. Solve the equation $y = -8(-4)(-2)(-1)$.

    A. -64 B. -32 C. 32   D. 64

LESSON 11-7 ◣ Non-Routine Problem Solving, p. 506

▶ Experimentation, or trial and error, can be useful when solving non-routine problems.

31. X is an even number, and it is greater than the product of 4 and 8, but less than the quotient of 100 and 2. The sum of its digits is 6. What number is X?

CHAPTER INVESTIGATION

EXTENSION Rewrite the ending to the story you read. Use a different type of reasoning to reach a conclusion. Will the story reach the same conclusion?

Present your results to the class. Include a synopsis of the story, the type of reasoning you used, the way you applied that type of reasoning to the story, and the conclusion you reached.

Chapter 11 Assessment

1. Make a statement about line segments *MN* and *ST*. Then check to see whether your statement is true or false.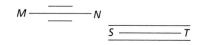

2. Choose any single-digit number. Multiply it by 9. Then multiply by 12,345,679. What number do you get? Repeat the process with several numbers. Then make a conjecture about the results of the process.

3. Experiment with several different sets of numbers and make a conjecture about the product of an even number, an odd number and an even number.

4. Write two conditional statements, using the following two sentences.

   Something is made of wood. It will burn.

Is each statement true or false? If false, give a counterexample.

5. If a man is a bachelor, then he is unmarried.

6. If a number is divisible by 7, then it is divisible by 14.

Determine whether the following arguments are valid or invalid.

7. If today is Friday, then I will go to school.
   I went to school today.
   Therefore, today is Friday.

8. If a horse wins the race, then it runs fast.
   Lightning won the race.
   Therefore, Lightning ran fast.

A survey of 77 employees in an office was conducted to determine their preferences of vending machine snacks. The results are shown in the Venn diagram.

9. How many employees wanted either raisins or granola bars?

10. How many employees wanted peanuts, raisins, and granola bars?

11. How many employees wanted peanuts but not granola bars or raisins?

12. Kerani, Renaldo, and Kathy each have different jobs. One is a writer, one is a bus driver, and one is a chef. Kerani cannot drive. Kathy cannot cook. Renaldo does not know the chef. Kathy is the writer's sister. What job does each person have?

Snack Preferences

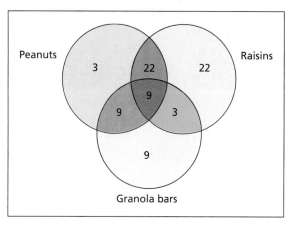

13. Montez has to either drive or walk to the library. Montez does not have his driver's license. What conclusion can be made?

14. Draw a 3 cm-by-5 cm rectangle. Then divide it into 4 squares by drawing only 3 lines.

Standardized Test Practice

Part 1   Multiple Choice

Record your answers on the answer sheet provided by your teacher or on a sheet of paper.

1. The test scores for five students are listed below. What is the mean of the scores? (Lesson 1-2)

$$67, 86, 97, 79, 92$$

Ⓐ 30   Ⓑ 84.2   Ⓒ 86   Ⓓ 421

2. Find the area of the triangle. (Lesson 2-4)

Ⓐ 28.94 m²

Ⓑ 40 m²

Ⓒ 80 m²

Ⓓ 100 m²

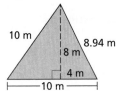

3. Which statement is true? (Lesson 3-1)

Ⓐ $-20(-4) - 20(4) > 0$

Ⓑ $-20(-4) + 20(4) < 0$

Ⓒ $-20(-4) - 20(4) = 0$

Ⓓ $-20(-4) + 20(4) = 0$

4. What is the approximate volume of the cylinder? Use 3.14 for π. (Lesson 4-7)

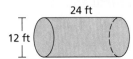

Ⓐ 288 ft³   Ⓑ 904.32 ft³

Ⓒ 2712.96 ft³   Ⓓ 10,851.84 ft³

5. What is the solution of $7(x + 4) = 21$? (Lesson 5-4)

Ⓐ 21   Ⓑ 3.5

Ⓒ 7   Ⓓ 10

6. A DVD player costs $285. The sales tax rate is 7.8%. How much does it cost to purchase this DVD player? (Lesson 6-4)

Ⓐ $262.77   Ⓑ $285.78

Ⓒ $292.80   Ⓓ $307.23

7. What is the slope of the line in the graph? (Lesson 7-5)

Ⓐ 2

Ⓑ $\frac{1}{2}$

Ⓒ $-\frac{1}{2}$

Ⓓ -2

8. What is the measure of $\angle ABC$? (Lesson 8-1)

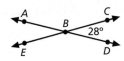

Ⓐ 28°   Ⓑ 76°

Ⓒ 152°   Ⓓ 332°

9. Simplify: $(3x^2 + 5x + 9) + (4x^2 + 6x + 2)$. (Lesson 9-2)

Ⓐ $7x^2 + 11x + 11$ Ⓑ $12x^2 + 30x + 18$

Ⓒ $7x^4 + 11x^2 + 11$ Ⓓ $12x^4 + 30x^2 + 18$

10. How many ways are possible of choosing a one-digit number followed by a letter of the alphabet? (Lesson 10-4)

Ⓐ 26 ways   Ⓑ 36 ways

Ⓒ 234 ways   Ⓓ 260 ways

11. You guess the answers on a two-question multiple choice quiz with four choices: a, b, c, and d. What is the probability that you will get both correct? (Lesson 10-5)

Ⓐ $\frac{1}{4}$ Ⓑ $\frac{1}{8}$

Ⓒ $\frac{1}{16}$ Ⓓ $\frac{1}{32}$

Test-Taking Tip

If you are allowed to write in your test booklet, underline key words, do calculations, sketch diagrams, and cross out answer choices as you eliminate them. Do not make any marks on the answer sheet except your answers.

Part 2 | Short Response/Grid In

Record your answers on the answer sheet provided by your teacher or on a sheet of paper.

12. What is the exponent of n in the simplified form of $n^5 \cdot n^4 \div n^6$? (Lesson 3-8)

13. What is the volume of the prism in cubic centimeters? (Lesson 4-7)

5 cm
4 cm
10 cm

14. A radio is on sale for $48, which is $\frac{2}{3}$ of its regular price. Use an equation to find the regular price of the radio. (Lesson 5-1)

15. What is the simple interest on a loan of $1500 at a rate of 4% for 3 years? (Lesson 6-5)

16. Graph rectangle $ABCD$ if diagonal AC has endpoints $A(-3, 4)$ and $C(4, -2)$. (Lesson 7-2)

17. How many lines of symmetry does the figure have? (Lesson 8-6)

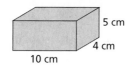

18. Simplify: $\dfrac{4n^4 + 8n^3 - 12n^2}{2n^2}$. (Lesson 9-6)

19. Each time a baby is born, the chance for either a boy or a girl is one-half. Find the probability that a family of four children has four girls. (Lesson 10-5)

20. In a recent governor's campaign, the newscasters reported that a straw poll indicated that the candidate was expected to get 54% of the vote. If 25,000 people vote, how many votes would the candidate expect? (Lesson 10-6)

21. Determine whether the following conditional statement is true or false. If false, give a counterexample.
All parallelograms have acute angles. (Lesson 11-3)

22. Li withdrew some money from her bank account. She spent one third of the money for gasoline. Then she spent half of what was left for a haircut. She bought lunch for $6.55. When she got home, she had $13.45 left. How much did she withdraw from the bank? (Lesson 11-6)

Part 3 | Extended Response

Record your answers on a sheet of paper. Show your work.

23. At a carnival, you can either flip a coin (choosing heads or tails) or roll a die (choosing a specific number 1 through 6) to win a prize. (Lesson 10-1)

   a. What is the probability of winning the prize if you choose to flip a coin?

   b. What is the probability of winning the prize if you choose to roll a die?

   c. Which method would you pick? Explain your answer.

24. Study the figure. (Lesson 11-7)

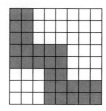

   a. What is the area of the figure?

   b. Show two ways that the figure can be divided into two regions that have that same area.

Student Handbook

Data File

Animals

Maximum Speed of Animals

Animal	Speed (mi/h)	Animal	Speed (mi/h)
Cheetah	70	Reindeer	32
Pronghorn antelope	61	Giraffe	32
Wildebeest	50	White-tailed deer	30
Lion	50	Wart hog	30
Thomson's gazelle	50	Grizzly bear	30
Quarter horse	47.5	Cat (domestic)	30
Elk	45	Man	27.89
Cape hunting dog	45	Elephant	25
Coyote	43	Black mamba snake	20
Gray fox	42	Six-lined race runner	18
Hyena	40	Squirrel	12
Zebra	40	Pig (domestic)	11
Greyhound	39.35	Chicken	9
Whippet	35.5	Spider (Tegenearia atrica)	1.17
Rabbit (domestic)	35	Giant tortoise	0.17
Mule deer	35	Three-toed sloth	0.15
Jackal	35	Garden snail	0.03

Dinosaurs

Name	Age (million years ago)	Height (feet)	Length (feet)	Weight (metric tons)	Fossil Sites
Diplodocus	145	30	89	10-11	North America
Brachiosaurus	150	39	74	80	Colorado, U.S., and Tanzania, Africa
Apatosaurus	145	23	69	30	North America
Tenontosaurus	110	7	15-21	1	North America
Tyrannosaurus	70	20	46	7	North America
Megalosaurus	145	9.5	23-26	2	England, France
Triceratops	70	13	30	5-6	North America

Gestation, Longevity, and Incubation of Animals

Longevity figures were supplied by Ronald T. Reuther. They refer to animals in captivity; the potential life span of animals is rarely attained in nature. Maximun longevity figures are from the Biology Data Book. Figures on gestation and incubation are averages based on estimates by leading authorities.

Animal	Gestation (days)	Average Longevity (years)	Maximum Longevity (yr, mo)
Baboon	187	20	35-7
Bear: Black	219	18	36-10
Grizzly	225	25	----
Polar	240	20	34-8
Beaver	122	5	20-6
Buffalo (American)	278	15	----
Bactrian camel	406	12	29-5
Cat (domestic)	63	12	28
Chimpanzee	231	20	44-6
Chipmunk	31	6	8
Cow	284	15	30
Deer (white-tailed)	201	8	17-6
Dog (domestic)	61	12	20
Elephant (African)	---	35	60
Elephant (Asian)	645	40	70
Elk	250	15	26-6
Fox (red)	52	7	14
Giraffe	425	10	33-7
Goat (domestic)	151	8	18
Gorilla	257	20	39-4
Guinea pig	68	4	7-6
Hippopotamus	238	25	----
Horse	330	20	46
Kangaroo	42	7	----
Leopard	98	12	19-4

Animal	Gestation (days)	Average Longevity (years)	Maximum Longevity (yr, mo)
Lion	100	15	25-1
Monkey (rhesus)	164	15	----
Moose	240	12	----
Mouse (meadow)	21	3	----
Mouse (dom. white)	19	3	3-6
Opossum (American)	14-17	1	----
Pig (domestic)	112	10	27
Puma	90	12	19
Rabbit (domestic)	31	5	13
Rhinoceros (black)	450	15	----
Rhinoceros (white)	----	20	----
Sea lion (California)	350	12	28
Sheep (domestic)	154	12	20
Squirrel (gray)	44	10	----
Tiger	105	16	26-3
Wolf (maned)	63	5	----
Zebra (Grant's)	365	15	----

Incubation Time (days)

Chicken	21
Duck	30
Goose	30
Pigeon	18
Turkey	26

Architecture

Golden Gate Bridge

The 10 Longest Bridge Spans in the World

Bridge	Location	Length of Main Span		Year Completed
		feet	meters	
Akashi Kaikyo	Japan	6570	2003	1998
Izmit Bay	Turkey	5538	1688	2003
Storebaelt	Denmark	5328	1624	1998
Humber	Britain	4626	1410	1981
Jiangyin Yangtze	China	4544	1385	1999
Tsing Ma	Hong Kong	4518	1377	1997
Verrazano-Narrows	Lower New York Bay	4260	1298	1964
Golden Gate	San Francisco Bay	4200	1280	1937
High Coast	Vasternorrland	3969	1210	1997
Mackinac Straits	Michigan	3800	1158	1957

Pyramids

Bent Pyramid in Egypt

Great Pyramid of Cheops in Egypt

Inca Pyramid in Peru

Pyramid of the Sun in Mexico

Pyramid of the Sun in Peru

Step Pyramid of Djoser in Egypt

10 20 30 40 50 60 70 80 90 100 110 120 130 140 150

Height (meters)

Great Pyramid of Cheops

New York, NY

The 10 Tallest Buildings in the World

Building	Location	Stories	Height of Building		Year Completed
			feet	meters	
Taipei 101	Taipei, Taiwan	101	1667	508	2004
Petronas Tower I	Kuala, Lumpur, Malaysia	88	1483	452	1998
Petronas Tower II	Kuala, Lumpur, Malaysia	88	1483	452	1998
Sears Tower	Chicago, IL, U.S.	110	1450	442	1974
Jin Mao Building	Shanghai, China	88	1381	421	1999
Two International Finance Center	Hong Kong, China	88	1352	412	2003
CITIC Plaza	Guangzhou, China	80	1283	391	1996
Shun Hing Square	Shenzhen, China	69	1260	384	1996
Empire State Building	New York, NY, U.S.	102	1250	381	1931
Central Plaza	Hong Kong, China	78	1227	374	1992

Statistics on 10 Popular U.S. Rollercoasters

Rollercoaster	Park	Location	Year Built	Height (feet)	Length (feet)	Top Speed (mi/h)	Average Time
Beast	Paramount's Kings Island	Cincinnati, OH	1979	135	7400	65	3 min 40 s
Texas Giant	Six Flags Over Texas	Arlington, TX	1990	143	4920	62	2 min
Steel Phantom	Kennywood	West Mifflin, PA	1991	225	3000	83	2 min 15 s
Alpengiest	Busch Gardens Williamsburg	Williamsburg, VA	1997	195	3828	67	3 min 10 s
Superman The Escape Tower of Terror	Six Flags Magic Mountain	Los Angeles, CA	1997	415	1235	100	28 s
Steel Force	Dorney Park	Allentown, PA	1998	200	5600	75	3 min
Mamba	Worlds of Fun	Kanas City, MO	1998	200	5600	75	3 min
Millenium Force	Cedar Point	Sandusky, OH	2000	310	6595	93	2 min 45 s
Titan	Six Flags over Texas	Arlington, TX	2001	245	5312	85	3 min 30 s
Top Thrill Dragster	Cedar Point	Sandusky, OH	2003	420	2800	120	10 s

Statue of Liberty Facts

Facts	English	Metric
Height from base to torch	151 ft 1 in.	46.50 m
Length of hand	16 ft 5 in.	5.0 m
Index finger	8 ft	2.44 m
Size of fingernail	13 in. × 10 in.	0.33 m × 0.25 m
Distance across the eye	2 ft 6 in.	0.76 m
Length of nose	4 ft 6 in.	1.48 m
Right arm length	42 ft	12.80 m
Width of mouth	3 ft	0.91 m
Tablet, length	23 ft 7 in.	7.19 m
Tablet, width	13 ft 7 in.	4.14 m

Arts & Entertainment

25 Top Films

Movie	Year Released	Total Gross (millions)
Titanic	1997	$601
Star Wars	1977	$461
E.T.	1982	$435
Star Wars: Episode I – The Phantom Menace	1999	$431
Spiderman	2002	$404
Lord of the Rings: The Return of the King	2003	$371
The Passion of the Christ	2004	$367
Jurassic Park	1993	$357
Lord of the Rings: The Two Towers	2002	$340
Finding Nemo	2003	$339
Forrest Gump	1994	$330
The Lion King	1994	$328
Harry Potter and the Sorcerer's Stone	2001	$317
Lord of the Rings: The Fellowship of the Ring	2001	$314
Star Wars: Episode II – Attack of the Clones	2002	$310
Star Wars: Episode VI – Return of the Jedi	1983	$309
Independence Day	1996	$306
Pirates of the Caribbean: The Curse of the Black Pearl	2003	$305
The Sixth Sense	1999	$294
Star Wars: Episode V – The Empire Strikes Back	1980	$290
Home Alone	1990	$286
The Matrix Reloaded	2003	$281
Shrek	2001	$268
Harry Potter and the Chamber of Secrets	2002	$262
Jaws	1975	$260

Longest Broadway Runs

Title	Number of Performances
Cats	7485
Phantom of the Opera	6749
Les Miserables	6680
Chorus Line	6137
Oh! Calcutta!	5959
Miss Saigon	4092
Beauty and the Beast	4076
42nd Street	3486
Grease	3388
Rent*	3304
Fiddler on the Roof	3242
Life with Father	3224
Tobacco Road	3182
Chicago (1996 Revival)*	2875
Hello Dolly	2844
My Fair Lady	2717
The Lion King*	2667
Cabaret	2378
Annie	2377
Man of La Mancha	2328
Abie's Irish Rose	2327
Oklahoma!	2212
Smokey Joe's Café	2037
Pippin	1944
South Pacific	1925

*Still running

Utilization of Selected Media

Item or Service	Percent of U.S. Households				
	1970	1980	1990	1995	2000
Telephone service	87.0	93.0	93.3	93.9	94.4
Radio	98.6	99.0	99.0	99.0	99.0
Television	97.1	97.9	98.2	98.3	98.2
Cable television	12.6	19.9	56.4	63.4	68.0
VCR	(N/A)	1.1	68.6	81.0	85.1

Top-Rated TV Shows

Season	Program	Rating*	TV-Owning Households (thousands)
1981-82	Dallas	28.4	81,500
1982-83	60 Minutes	25.5	83,300
1983-84	Dallas	25.7	83,800
1984-85	Dynasty	25.0	84,900
1985-86	Bill Cosby Show	33.8	85,900
1986-87	Bill Cosby Show	34.9	87,400
1987-88	Bill Cosby Show	27.8	88,600
1988-89	Roseanne	25.5	90,400
1989-90	Roseanne	23.4	92,100
1990-91	Cheers	21.6	93,100
1991-92	60 Minutes	21.7	92,100
1992-93	60 Minutes	21.6	93,100
1993-94	Home Improvement	21.9	94,200
1994-95	Seinfeld	20.5	95,400
1995-96	E.R.	22.0	95,900
1996-97	E.R.	21.2	97,000
1997-98	Seinfeld	22.0	98,000
1998-99	E.R.	17.8	99,400
1999-00	Who Wants to Be a Millionaire	18.6	100,800
2000-01	Survivor II	17.4	102,200
2001-02	Friends	15.3	105,500
2002-03	CSI	16.1	106,700

*Rating is percent of TV-owning households tuned in to program.
Data prior to 1988-89 exclude Alaska and Hawaii.

Astronomy & Space Science

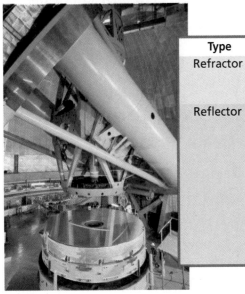

Hale Telescope

Largest Telescopes of the World

Type	Location	Size of Lens*
Refractor	Yerkes Observatory, Wisconsin	102 cm
	Lick Observatory, California	91 cm
	Paris Observatory, Meudon, France	84 cm
Reflector	W.M. Keck Telescope, Mauna Kea, Hawaii	10 m **
	Special Astrophysical Observatory, Zelenchukskaya, USSR	6 m
	Hale Telescope, Mount Palomar, California	5 m
	Cerro Tololo Inter-America Observatory, Chile	4 m
	Kitt Peak National Observatory, Arizona	4 m
	Mount Stromlo Observatory, Australia	3.8 m
	European Southern Observatory, Chile	3.6 m
	Lick Observatory, California	3 m
	McDonald Observatory, Texas	2.7 m
	Hale Observatory, Mt. Wilson, California	2.5 m

*For refractor telescopes, size is diameter of objective lens;
for reflector telescopes, size is diameter of mirror.
**multiple mirrors

Total and Partial Eclipses of the Moon 1993-2010

Date	Time* of Mideclipse (GMT)	Length of Totality (minutes)	Length of Eclipse (hours:minutes)	Date	Time* of Mideclipse (GMT)	Length of Totality (minutes)	Length of Eclipse (hours:minutes)
1993 June 4	13:02	96	3:38	2003 May 16	3:41	52	3:14
1993 Nov. 29	6:26	46	3:30	2003 Nov. 9	1:20	22	3:30
1994 May 25	3:32	Partial	1:44	2004 May 4	20:32	76	3:22
1995 April 15	12:19	Partial	1:12	2004 Oct. 28	3:05	80	3:38
1996 April 4	0:11	86	3:36	2005 Oct. 17	12:04	Partial	0:56
1996 Sept. 27	2:55	70	3:22	2006 Sept. 7	18:52	Partial	1:30
1997 March 24	4:41	Partial	3:22	2007 March 3	23:22	74	3:40
1997 Sept. 16	18:47	62	3:16	2007 Aug. 28	10:38	90	3:32
1999 July 28	11:34	Partial	2:22	2008 Feb. 21	3:27	50	3:24
2000 Jan. 21	4:45	76	3:22	2008 Aug. 16	21:11	Partial	3:08
2000 July 16	13:57	106	3:56	2009 Dec. 31	19:24	Partial	1:00
2001 Jan. 9	20:22	60	3:16	2010 June 26	11:40	Partial	2:42
2001 July 5	14:57	Partial	2:38	2010 Dec. 21	8:18	72	3:28

*Times are Greenwich Mean Time. Subtract 5 hours for Eastern Standard Time, 6 hours for Central Standard Time, 7 hours for Mountain Standard Time, and 8 hours for Pacific Standard Time. From your time zone, lunar eclipses that occur between sunset and sunrise will be visible, and those at midnight will be best placed.

Surface Temperature of the Planets

Planet	Degrees Celsius	Degrees Fahrenheit
Mercury	−173° to 430°C	−279 to 806°F
Venus	472°C	882°F
Earth	−50° to 50°C	−60° to 120°F
Mars	−140° to 20°C	−220° to 68°F
Jupiter	−110°C (temp at cloud tops)	−116°F (temp at cloud tops)
Saturn	−180°C (temp at cloud tops)	−292°F (temp at cloud tops)
Uranus	−221°C (temp above cloud tops)	−366°F (temp above cloud tops)
Neptune	−216°C (temp at cloud tops)	−357°F (temp at cloud tops)
Pluto	−230°C	−382°F

The Sun

Basic Planetary Data

Planet	Mean Distance from Sun (millions of kilometers)	Mean Distance from Sun (millions of miles)	Period of Revolution	Rotation Period	Equatorial Diameter (kilometers)	Satellites	Rings
Mercury	57.9	36.0	88 days	59 days	4,880	0	0
Venus	108.2	67.24	224.7 days	243 days	12,100	0	0
Earth	149.6	92.9	365.2 days	23 h 56 min 4 s	12,756	1	0
Mars	227.9	141.71	687 days	24 h 37 min	6,794	2	0
Jupiter	778.3	483.88	11.86 days	9 h 55 min 30 s	142,800	16	1
Saturn	1,427	887.14	29.46 yr	10 h 40 min 24 s	120,660	19	1,000(?)
Uranus	2,870	1,783.98	84 yr	17 h 14 min	51,810	15	11
Neptune	4,497	2,796.46	165 yr	16 h 3 min	49,528	8	4
Pluto	5,900	3,666	248 yr	6 days 9 h 21 min	2,300	1	?

Earth Science

Data File

Size and Depth of the Oceans

Ocean	Size (square miles)	Greatest Depth (feet)
Pacific	63,800,000	36,161
Atlantic	31,800,000	30,249
Indian	28,900,000	24,441
Arctic	5,400,000	17,881

Main Elements That Make Up the Earth's Crust

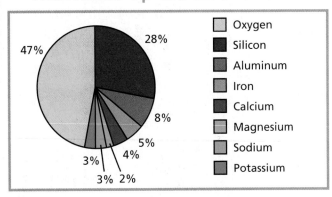

- 47% Oxygen
- 28% Silicon
- 8% Aluminum
- 5% Iron
- 4% Calcium
- 3% Magnesium
- 3% Sodium
- 2% Potassium

Mt. Everest

Highest and Lowest Continental Altitudes

Continent	Highest Point	Elevation (feet)	Lowest Point	Feet Below Sea Level
Asia	Mount Everest, Nepal-Tibet	29,028	Dead Sea, Israel-Jordan	1312
South America	Mount Aconcagua, Argentina	22,834	Valdes Peninsula, Argentina	131
North America	Mount McKinley, Alaska	20,320	Death Valley, California	282
Africa	Kilimanjaro, Tanzania	19,340	Lake Assai, Dijouti	512
Europe	Mount El'brus, Russia	18,510	Caspian Sea, Russia	92
Antarctica	Vinson Massif	16,864	Unknown	...
Australia	Mount Kosciusko, New South Wales	7,310	Lake Eyre, South Australia	52

Record Holders From Earth Science

Traditionally, the study of early life by way of its remains—which are called fossils—has been the province of earth scientists, rather than biologists. This is because fossils were the first known way to group rocks of the same age together, since it was assumed that at a given time in the past certain life-forms existed and later became extinct. Earth itself is thought to be about 4.5 billion years old.

By Size		
This group of record holders includes a couple of ancient life-forms but concentrates more on those features of Earth that are studied by earth scientists.		
Record	**Record Holder**	**Size or Distance**
Longest dinosaur	*Seismosaurus*, who lived about 150 million years ago in what is now New Mexico	100 - 120 ft long
Largest island	Greenland, also known as Kalaallit Nunnaat	About 840,000 mi^2
Largest ocean	Pacific	64,186,300 mi^2
Deepest part of ocean	*Challenger* Deep in the Marianas Trench in Pacific Ocean	35,640 ft, or 6.85 mi
Greatest tide	Bay of Fundy between Maine and New Brunswick	47.5 ft between high and low tides
Largest geyser	Steamboat Geyser in Yellowstone Park	Shoots mud and rocks about 1,000 ft in air
Largest glacier	Lambert Glacier in Antarctica (upper section known as Mellor Glacier)	At least 250 mi long
Largest known cave	Mammoth Cave, which is connected to Flint Ridge Cave system	Total mapped passageway of over 300 mi
Largest canyon on land	Grand Canyon of the Colorado, in northern Arizona	Over 217 mi long, from 4 to 13 mi wide, as much as 5,300 ft deep
Highest volcano	Cerro Aconagua in the Argentine Andes	22,834 ft high
Most abundant mineral	Magnesium silicate perovskite	About 2/3 of planet, it forms Earth's mantle

By Age		
Record	**Record Holder**	**Age in Years**
Oldest rocks	Zircons from Australia	4.4 billion
Oldest fossils	Single-celled algae or bacteria from Australia	3.5 billion
Oldest slime molds (bacteria)	Slime molds - also called slime bacteria - that gathered together to form multicellular bodies for reproduction	2.5 billion
Oldest petroleum	Oil from northern Australia	1.4 billion
Oldest land animal	Millipede? Known only from its burrows	488 million
Oldest fish	*Sacabambasis*, found in Bolivia	470 million
Oldest land plant	Moss or algae	425 million
Oldest insect	Bristletail (relative of modern silverfish)	390 million
Oldest reptile	"Lizzie the Lizard," found in Scotland	340 million
Oldest bird	*Protoavis*, fossils found near Post, Texas	225 million
Oldest dinosaur	Unnamed dinosaur the size of ostrich	225 million

Environment

Annual Emissions and Fuel Consumption for an "Average" Passenger Car

Pollutant	Problem	Amount Emitted
Hydrocarbons	Urban ozone (smog) and air toxic	2.9 g/mi
Carbon monoxide	Poisonous gas	22 g/mi
Nitrogen oxides	Urban ozone (smog) and acid rain	1.5 g/mi
Carbon dioxide	Global warming	0.8 lb/mi
Gasoline	Imported oil	0.04 gal/mi

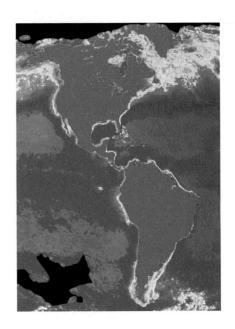

Gasoline Retail Prices, National Average
(including taxes)

Year	Unleaded Regular (dollars)	Unleaded Premium (dollars)
1996	1.23	1.41
1997	1.23	1.42
1998	1.06	1.25
1999	1.17	1.36
2000	1.51	1.69
2001	1.46	1.66
2002	1.36	1.58
2003 (Jan–June)	1.60	1.78

Earth's Water Supply

	Surface Area (square miles)	Volume (cubic miles)	Percent of Total
Salt water			
The oceans	139,500,000	317,000,000	97.2
Inland seas and saline lakes	270,000	25,000	0.008
Fresh water			
Freshwater lakes	330,000	30,000	0.009
All rivers (average level)	--	300	0.0001
Antarctic Icecap	6,000,000	6,300,000	1.9
Arctic Icecap and glaciers	900,000	680,000	0.21
Water in the atmosphere	197,000,000	3,100	0.001
Ground water within half a mile from surface	--	1,000,000	0.31
Deep-lying ground water	--	1,000,000	0.31
Total (rounded)	--	326,000,000	100.00

Selected National Champion Trees

Tree type	Girth* at 4.5 feet (inches)	Height (feet)	Crown Spread (feet)	Location
Bluegum Eucalyptus	586	141	126	Petrolia
California-Laurel	546	108	118	Grass Valley, CA
Coast Douglas-Fir	505	281	71	Olympic National Forest, WA
Coast Redwood	950	321	80	Crescent City, CA
Common Baldcypress	644	83	85	Cat Island, LA
Giant Sequoia	1024	261	108	Sequoia National Park, CA
Port-Orford-Cedar	451	219	39	Siskiyou National Forest, OR
Sitka Spruce	707	191	96	Olympic National Forest, WA
Sugar Pine	442	232	29	Dorrington, CA
Western Red Cedar	761	159	45	Olympic National Forest, WA

*Girth-The circumference or distance around something.

Contents of Garbage Cans

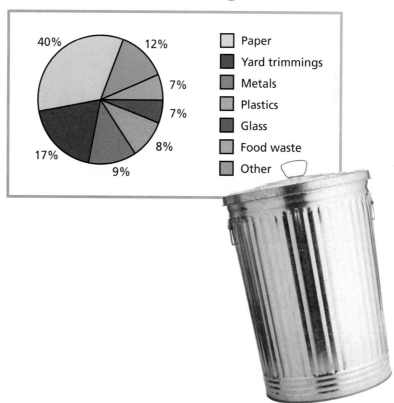

- 40% — Paper
- 12%
- 7%
- 7%
- 8%
- 9%
- 17%

Paper
Yard trimmings
Metals
Plastics
Glass
Food waste
Other

Health & Fitness

Data File

Calories Spent in Activities Per Minute

Activity	Calories spent per minute according to weight				
	110 lb	125 lb	150 lb	175 lb	200 lb
Aerobic dance (vigorous)	6.8	7.8	9.3	10.9	12.4
Basketball (vigorous, full court)	10.7	12.1	14.6	17.0	19.4
Bicycling: 13 mi/h	5.0	5.6	6.8	7.9	9.0
Bicycling: 19 mi/h	8.4	9.5	11.4	13.3	15.2
Canoeing (flat water, moderate pace)	5.0	5.6	6.8	7.9	9.0
Cross-country skiing: 8 mi/h	11.4	13.0	15.6	18.2	20.8
Golf (carrying clubs)	5.0	5.6	6.8	7.9	9.0
Handball	8.6	9.8	11.7	13.7	15.6
Horseback riding (trot)	5.7	6.5	7.8	9.1	10.4
Rowing (vigorous)	10.7	12.1	14.6	17.0	19.4
Running: 5 mi/h	6.7	7.6	9.2	10.7	12.2
Running: 7.5 mi/h	10.3	11.8	14.1	16.4	18.8
Running: 10 mi/h	12.5	14.3	17.1	20.0	22.9
Soccer (vigorous)	10.7	12.1	14.6	17.0	19.4
Studying	1.2	1.4	1.7	1.9	2.2
Swimming: 20 yd/min	3.5	4.0	4.8	5.6	6.4
Swimming: 45 yd/min	6.4	7.3	8.7	10.2	11.6
Tennis (beginner)	3.5	4.0	4.8	5.6	6.4
Walking (brisk pace): 3.5 mi/h	3.9	4.4	5.2	6.1	7.0

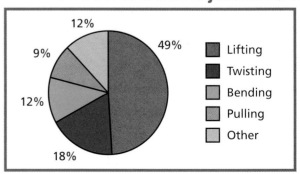

Most Common Reasons for Lower Back Injuries

- 49% Lifting
- 18% Twisting
- 12% Bending
- 12% Pulling
- 9%
- 12% Other

Years of Expected Life at Birth

Year of Birth	Male	Female
1920	53.6	54.6
1930	58.1	61.6
1940	60.8	65.2
1950	65.6	71.1
1960	66.6	73.1
1970	67.1	74.7
1980	70.0	77.5
1990	71.8	78.8
1991	72.0	78.9
1992	72.1	78.9
1993	72.1	78.9
1994	72.4	79.0
1995	72.6	78.9
1996	73.1	79.1
1997	73.6	79.4
1998	73.8	79.5
1999	73.9	79.4
2000	74.1	79.5
2001	74.4	79.8

Data File

Predicted Maximum Heart Rate

Age	Maximum Heart Rate
14	206
15	205
16	204
17	203
18	202
19	201
20	200
21	199
22	198
23	197

Sports

Ten Most Popular Boys' and Girls' Athletics Programs

Boys' Athletic Programs			
Sport	Schools	Sport	Boys
Basketball	17,333	Football	1,023,142
Track and Field (outdoor)	15,195	Basketball	540,874
Baseball	14,988	Track and Field (outdoor)	498,027
Football	13,642	Baseball	453,792
Golf	13,120	Soccer	345,156
Cross Country	12,574	Wrestling	239,845
Soccer	10,103	Cross Country	191,833
Wrestling	9,543	Golf	162,805
Tennis	9,411	Tennis	144,844
Swimming and Diving	5,588	Swimming and Diving	94,612

Girls' Athletic Programs			
Sport	Schools	Sport	Girls
Basketball	17,028	Basketball	457,165
Track and Field (outdoor)	15,032	Track and Field (outdoor)	415,602
Volleyball	14,244	Volleyball	396,682
Softball (fast pitch)	14,007	Softball (fast pitch)	357,912
Cross Country	12,083	Soccer	301,450
Tennis	9,332	Cross Country	163,360
Soccer	9,299	Tennis	162,810
Golf	8,106	Swimming and Diving	141,468
Swimming and Diving	6,120	Competitive Spirit Squads	111,191
Competitive Spirit Squads	4,644	Golf	62,159

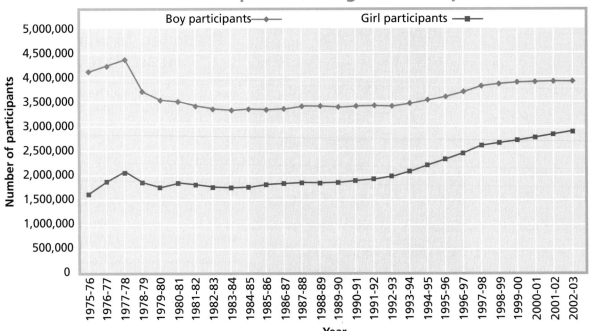

Participation in High School Sports

Summer Olympic Games

Year	Date	City	Number of Countries	Number of Events	Number of Competitors
1896	4/6-4/15	Athens	14	43	245
1900	5/14-10/28	Paris	19	86	1078
1904	7/1-11/23	St. Louis	13	89	889
1908	4/27-10/31	London	22	107	2035
1912	5/5-7/27	Stockholm	28	102	2437
1916*					
1920	4/20-9/12	Antwerp	29	152	2807
1924	5/4-7/27	Paris	44	126	2972
1928	5/17-8/12	Amsterdam	46	109	2884
1932	7/30-8/14	Los Angeles	37	117	1333
1936	8/1-9/16	Berlin	49	129	3936
1940*					
1944*					
1948	7/29-8/14	London	59	136	4092
1952	7/19-8/3	Helsinki	69	149	5429
1956	11/22-12/8	Melbourne	67	145	3178
1960	8/25-9/11	Rome	83	150	5313
1964	10/10-10/24	Tokyo	93	163	5133
1968	10/12-10/27	Mexico	112	172	5498
1972	8/26-9/11	Munich	121	195	7121
1976	7/17-8/1	Montreal	92	198	5043
1980	7/19-8/3	Moscow	80	203	5283
1984	7/28-8/12	Los Angeles	140	221	8802
1988	9/17-10/2	Seoul	159	237	8473
1992	7/25-8/9	Barcelona	169	257	9368
1996	7/19-8/4	Atlanta	197	271	10,320
2000	9/16-10/1	Sydney	199	300	10,651
2004	8/13-8/29	Athens	201	301	10,550

* Not celebrated

Top 10 Standings for 2000 Summer Olympics

National medal standings are not recognized by the
International Olympic Committee (IOC). Point totals are based
on 3 points for a gold medal, 2 for a silver, and 1 for a bronze.

Rank	Country	Gold	Silver	Bronze	Total	Points
1	United States	39	25	33	97	200
2	Russian Federation	32	28	28	88	180
3	China	28	16	15	59	131
4	Australia	16	25	17	58	115
5	Germany	14	17	26	57	102
6	France	13	14	11	38	78
7	Italy	13	8	13	34	68
8	The Netherlands	12	9	4	25	58
9	Cuba	11	11	7	29	62
10	Great Britain	11	10	7	28	60

Travel & Transportation

Busiest Airports
(Passengers Enplaned, in thousands)

Airport	1992 Total	1992 Rank	2002 Total	2002 Rank
Atlanta (Hartsfield Intl), GA	19,705	3	34,927	1
Chicago (O'Hare Intl), IL	28,948	1	27,029	2
Dallas/Ft. Worth Intl, TX	24,671	2	22,759	3
Los Angeles Intl, CA	18,395	4	19,397	4
Denver Intl, CO	13,595	6	16,077	5
Phoenix Sky Harbor Intl, AZ	10,787	7	15,952	6
Las Vegas (McCarran Intl), NV	9,347	12	15,581	7
Minneapolis-St.Paul Intl, MN	10,055	11	14,639	8
Detroit (Wayne County), MI	10,425	10	13,939	9
Houston International, TX	8,358	21	13,343	10
Seattle-Tacoma Intl, WA	8,572	19	12,222	11
Orlando Intl, FL	8,765	16	12,133	12
St. Louis (Lambert-St. Louis Muni), MO	10,436	9	11,748	13
San Francisco Intl, CA	14,208	5	11,253	14
Newark, NJ	10,479	8	11,138	15
Charlotte (Douglas Muni), NC	8,239	22	10,147	16
New York (LaGuardia), NY	9,252	14	9,889	17
Cincinnatti, OH	4,916	27	9,882	18
Philadelphia, PA	6,968	23	9,788	19
Boston (Logan Intl), MA	9,320	13	9,174	20

Los Angeles (LAX)

Trips by Purpose

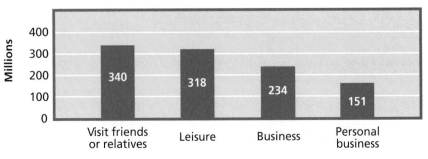

Millions

Visit friends or relatives	340
Leisure	318
Business	234
Personal business	151

Main purpose

Number of Cars and Bikes in Different Countries

Country	Cars (millions)	Bikes (millions)
Australia	7.1	6.8
China	1.2	300.0
India	1.5	45.0
Japan	30.7	60.0
Mexico	4.8	12.0
The Netherlands	4.9	11.0
United States	139.0	103.0

St. Louis, Missouri

Climate of 15 Selected U.S. Cities

Location	Average Monthly Temperature (°F)				Precipitation		Snowfall
	January	April	July	October	Average Annual (inches)	Number of Days	Average Annual (inches)
Atlanta, Georgia	41.9	61.8	78.6	62.2	48.61	115	1.9
Chicago, Illinois	21.4	48.8	73.0	53.5	33.34	127	40.3
Denver, Colorado	29.5	47.4	73.4	51.9	15.31	88	59.8
Detroit, Michigan	23.4	47.3	71.9	51.9	30.97	133	40.4
Fairbanks, Alaska	-12.7	30.2	61.5	25.1	10.37	106	67.5
Honolulu, Hawaii	72.6	75.7	80.1	79.5	23.47	100	0.0
Houston, Texas	51.4	68.7	83.1	69.7	44.76	105	0.4
Miami, Florida	67.1	75.3	82.5	77.9	57.55	129	0.0
New York, New York	31.8	51.9	76.4	57.5	42.82	119	26.1
Oklahoma City, Oklahoma	35.9	60.2	82.1	62.3	30.89	82	9.0
Pittsburgh, Pennsylvania	26.7	50.1	72.0	52.5	36.30	154	44.6
Portland, Oregon	38.9	50.4	67.7	54.3	37.39	154	6.8
Sacramento, California	45.3	58.2	75.6	63.9	17.10	58	0.1
St. Louis, Missouri	28.8	56.1	78.9	57.9	33.91	111	19.8
Washington, D.C.	35.2	56.7	78.9	59.3	39.00	112	17.0

Prerequisite Skills

❶ Place Value and Order

Example 1

Write 2,345,678.9123 in words.

Solution

The place value chart shows the value of each digit. The value of each place is ten times the place to the right.

millions	hundred thousands	ten thousands	thousands	hundreds	tens	ones	.	tenths	hundredths	thousandths	ten thousandths
2	3	4	5	6	7	8	.	9	1	2	3

The number shown is *two million, three hundred forty-five thousand, six hundred seventy-eight and nine thousand one hundred twenty-three ten-thousandths.*

Example 2

Use < or > to make this sentence true.　　　　6 ■ 2

Solution

Remember, < means "less than" and > means "greater than." So, 6 > 2.

▊ EXTRA PRACTICE EXERCISES

Write each number in words.

1. 3647　　　　　　**2.** 6,004,300.002　　　　　　**3.** 0.9001

Write each of the following as a number.

4. two million, one hundred fifty thousand, four hundred seventeen

5. five thousand, one hundred twenty and five hundred two thousandths

6. nine million, ninety thousand, nine hundred and ninety-nine ten-thousandths

Use < or > to make each sentence true.

7. 9 ■ 8　　　　　　**8.** 164 ■ 246　　　　　　**9.** 63,475 ■ 6,435

10. 52 ■ 50　　　　　**11.** 5.39 ■ 9.02　　　　　**12.** 43.94 ■ 53.69

❷ Add and Subtract Whole Numbers and Decimals

To add or subtract whole numbers and decimals, write the digits so the place values line up. Add from right to left, renaming when necessary. When adding or subtracting decimals, be sure to place the decimal point directly below the aligned decimals in the problem.

Example 1

Add 0.058, 25.39, 6346, and 1.57. The answer is called the *sum*.

Solution

```
   0.058
  25.39
6346.
+  1.57
────────
6373.018
```

The zero to the left of the decimal point is used to show there are no ones.

The decimal point is at the end of whole numbers.

Add from right to left.

Example 2

Subtract 6.37 from 27. The smaller number is subtracted from the larger number. The answer is called the *difference*.

Solution

```
 27.00
− 6.37
──────
 20.63
```

Add zeros if that helps you complete the subtraction.

▧ EXTRA PRACTICE EXERCISES

Add or subtract.

1. $23.146 - 17.215$ **2.** $46.48 - 6.57$ **3.** $52 - 1.95$

4. $0.86 + 0.75$ **5.** $83 - 82.743$ **6.** $9.45 + 13.2$

7. $14.5 - 9.684$ **8.** $913.03 - 79$ **9.** $0.8523 - 0.794$

10. $30,000 - 1237.64$ **11.** $45.3 + 160.09$ **12.** $6000 - 4362$

13. $462.09 + 32.7$ **14.** $78,850 - 56,176.8$ **15.** $3160.915 + 1920.03$

16. $17,347.85 - 12,516.90$ **17.** $107,285 - 61,500.25$

18. $6.4 + 54.2 + 938.05 + 3.7 + 47.3$ **19.** $1765.36 + 1587.50 + 1400$

20. $51,876.36 + 48,156.95 + 1417.86$ **21.** $76.2 + 80 + 56 + 9.321$

22. $567.1 + 6 + 13.452 + 100$ **23.** $5.93 + 0.06 + 96.021 + 0.0378$

24. $51.5 + 87 + 68.2$ **25.** $6532.03 + 861.006 + 3170.95$

③ Multiply Whole Numbers and Decimals

To multiply whole numbers, find each partial product and then add.

When multiplying decimals, locate the decimal point in the product so that there are as many decimal places in the product as the total number of decimal places in the factors.

Example 1

Multiply 2.6394 by 3000.

Solution

$$\begin{array}{r} 2.6394 \\ \times\ \ 3000 \\ \hline 7918.2000 \end{array} \text{ or } 7918.2$$

Zeros after the decimal point can be dropped because they are not significant digits.

Example 2

Multiply 3.92 by 0.023.

Solution

$$\begin{array}{r} 3.92 \\ \times\ 0.023 \\ \hline 1176 \\ +\ 7840 \\ \hline 0.09016 \end{array}$$

2 decimal places
+ 3 decimal places

5 decimal places

The zero is added before the nine so that the product will have five decimal places.

◢ EXTRA PRACTICE EXERCISES

Multiply.

1. 36×45

2. 500×30

3. $17,000 \times 230$

4. 6.2×8

5. 950×1.6

6. 3.652×20

7. 179×83

8. 257×320

9. 8560×275

10. 467×0.3

11. 2.63×183

12. 0.758×321.8

13. 49.3×1.6

14. 6.859×7.9

15. 794.4×321.8

16. 0.08×4

17. 0.062×0.5

18. 0.0135×0.003

19. 21.6×3.1

20. 8.76×0.005

21. 5.521×3.642

22. 5.749×3.008

23. 8.09×0.18

24. $89,946 \times 2.85$

25. 6.31×908

26. 391.05×25

27. $35,021 \times 76.34$

❹ Divide Whole Numbers and Decimals

Dividing whole numbers and decimals involves a repetitive process of estimating a quotient, multiplying and subtracting.

$$
\begin{array}{r}
34 \leftarrow \text{quotient} \\
\text{divisor} \rightarrow 7\overline{)239} \leftarrow \text{dividend} \\
21\downarrow \quad \leftarrow 3 \times 7 \\
\overline{29} \quad \leftarrow \text{Subtract. Bring down the 9.} \\
28 \quad \leftarrow 4 \times 7 \\
\overline{1} \quad \leftarrow \text{remainder}
\end{array}
$$

Example 1

Find: 283.86 ÷ 5.7

Solution

When dividing decimals, move the decimal point in the divisor to the right until it is a whole number. Move the decimal point the same number of places in the dividend. Then place the decimal point in the answer directly above the new location of the decimal point in the dividend.

$$
5.7\overline{)283.8.6} \quad \rightarrow \quad
\begin{array}{r}
49.8 \\
57\overline{)2838.6} \\
228 \\
\overline{558} \\
513 \\
\overline{45\,6} \\
45\,6 \\
\overline{0}
\end{array}
$$

If answers do not have a remainder of 0, you can add 0s after the last digit of the dividend and continue dividing.

EXTRA PRACTICE EXERCISES

Divide.

1. 72 ÷ 6
2. 6000 ÷ 20
3. 26,568 ÷ 8
4. 5.6 ÷ 7
5. 120 ÷ 0.4
6. 936 ÷ 12
7. 3.28 ÷ 4
8. 0.1960 ÷ 5
9. 1968 ÷ 0.08
10. 16 ÷ 0.04
11. 1525 ÷ 0.05
12. 109.94 ÷ 0.23
13. 0.6 ÷ 24
14. 7.924 ÷ 0.28
15. 32.6417 ÷ 9.1
16. 24 ÷ 0.6
17. 1784.75 ÷ 29.5
18. 0.01998 ÷ 0.37
19. 7.8 ÷ 0.3
20. 12,000 ÷ 0.04
21. 820.94 ÷ 0.02
22. 89,946 ÷ 28.5
23. 15 ÷ 0.75
24. 7.56 ÷ 2.25
25. 0.19176 ÷ 68
26. 0.168 ÷ 0.48
27. 5.1 ÷ 0.006
28. 55,673 ÷ 0.05
29. 84.536 ÷ 4
30. 261.18 ÷ 10
31. 134,554 ÷ 0.14
32. 90,294 ÷ 7.85
33. 59,368 ÷ 47.3
34. 11,633.5 ÷ 439
35. 28.098 ÷ 14
36. 16.309 ÷ 0.09
37. 55.26 ÷ 1.8
38. 8276 ÷ 0.627
39. 10,693 ÷ 92.8
40. 48.8 ÷ 1.6
41. 27,268 ÷ 34
42. 546.702 ÷ 0.078

⑤ Multiply and Divide Fractions

To multiply fractions, multiply the numerators and then multiply the denominators. Write the answer in simplest form.

Example 1

Multiply $\frac{2}{5}$ and $\frac{7}{8}$.

Solution

$$\frac{2}{5} \times \frac{7}{8} = \frac{2 \times 7}{5 \times 8} = \frac{14}{40} = \frac{7}{20}$$

To divide by a fraction, multiply by the reciprocal of that fraction. To find the reciprocal of a fraction, invert the fraction (turn it upside down). The product of a fraction and its reciprocal is 1. Since $\frac{2}{3} \times \frac{3}{2} = \frac{6}{6}$ or 1, $\frac{2}{3}$ and $\frac{3}{2}$ are reciprocals of each other.

Example 2

Divide $1\frac{1}{5}$ by $\frac{2}{3}$.

Solution

$$1\frac{1}{5} \div \frac{2}{3} = \frac{6}{5} \div \frac{2}{3} = \frac{6}{5} \times \frac{3}{2} = \frac{6 \times 3}{5 \times 2} = \frac{18}{10}, \text{ or } 1\frac{4}{5}$$

■ EXTRA PRACTICE EXERCISES

Multiply or divide. Write each answer in simplest form.

1. $\frac{2}{3} \div \frac{5}{6}$

2. $\frac{3}{5} \times \frac{10}{12}$

3. $\frac{5}{8} \div \frac{1}{4}$

4. $\frac{1}{2} \times \frac{2}{3}$

5. $\frac{2}{3} \times \frac{1}{2}$

6. $\frac{3}{4} \times \frac{5}{8}$

7. $\frac{1}{2} \div \frac{2}{3}$

8. $\frac{2}{3} \div \frac{1}{2}$

9. $\frac{3}{4} \div \frac{5}{8}$

10. $2\frac{2}{3} \div 1\frac{3}{5}$

11. $1\frac{1}{5} \times 2\frac{1}{4}$

12. $3\frac{1}{3} \times 1\frac{1}{10}$

13. $5\frac{2}{5} \div 2\frac{4}{7}$

14. $2\frac{4}{7} \div 5\frac{2}{5}$

15. $2\frac{4}{7} \times 5\frac{2}{5}$

16. $1\frac{7}{8} \div 1\frac{7}{8}$

17. $\frac{3}{4} \times \frac{2}{3} \times 1\frac{5}{8} \times 2\frac{2}{3}$

18. $7\frac{1}{2} \div 2\frac{1}{4}$

19. $6\frac{2}{3} \times 4\frac{1}{2} \times 5\frac{3}{8}$

20. $11\frac{5}{9} \times 6\frac{1}{12}$

21. $\frac{25}{42} \div \frac{5}{21}$

22. $\frac{13}{18} \div \frac{8}{9}$

23. $\frac{3}{8} \times \frac{11}{12} \times \frac{16}{33}$

24. $\frac{51}{56} \div \frac{17}{24}$

❻ Add and Subtract Fractions

To add and subtract fractions, you need to find a common denominator and then add or subtract, renaming as necessary.

Example 1

Add $\frac{3}{4}$ and $\frac{5}{6}$.

Solution

$$
\begin{array}{rcl}
\frac{3}{4} & = \frac{3}{4} \times \frac{3}{3} = & \frac{9}{12} \\
+\frac{5}{6} & = \frac{5}{6} \times \frac{2}{2} = & +\frac{10}{12} \\
\hline
& & \frac{19}{12}
\end{array}
$$

Add the numerators and use the common denominator.

Then simplify. $\frac{19}{12} = 1\frac{7}{12}$

Example 2

Subtract $1\frac{3}{5}$ from $5\frac{1}{2}$.

Solution

$$
\begin{array}{rcl}
5\frac{1}{2} = & 5\frac{5}{10} = & 4\frac{15}{10} \\
-1\frac{3}{5} = & -1\frac{6}{10} = & -1\frac{6}{10} \\
\hline
& & 3\frac{9}{10}
\end{array}
$$

You cannot subtract $\frac{6}{10}$ from $\frac{5}{10}$, so rename again.

◼ EXTRA PRACTICE EXERCISES

Add or subtract.

1. $\frac{1}{5} + \frac{1}{10}$

2. $\frac{2}{3} + \frac{1}{3}$

3. $\frac{5}{8} + \frac{3}{4}$

4. $\frac{6}{7} - \frac{2}{7}$

5. $\frac{3}{4} - \frac{1}{3}$

6. $\frac{5}{8} - \frac{1}{4}$

7. $2\frac{1}{2} + 3\frac{1}{2}$

8. $6\frac{5}{8} + 3\frac{7}{8}$

9. $3\frac{2}{3} + 4\frac{1}{2}$

10. $2\frac{3}{4} - 1\frac{1}{4}$

11. $5\frac{1}{8} - 3\frac{7}{8}$

12. $1\frac{1}{3} - \frac{2}{3}$

13. $6\frac{1}{2} + 5\frac{7}{9}$

14. $9\frac{2}{5} - 1\frac{1}{8}$

15. $7\frac{2}{3} + 6\frac{1}{5}$

16. $8\frac{1}{10} - 5\frac{2}{3}$

17. $6\frac{1}{2} - 5\frac{3}{5}$

18. $10\frac{5}{8} - 9\frac{3}{4}$

19. $1\frac{1}{5} + 2\frac{1}{3} + 5\frac{1}{4}$

20. $9\frac{2}{3} + 4\frac{3}{5} + 6\frac{1}{2}$

21. $10\frac{7}{8} + 3\frac{3}{4} + 6\frac{1}{2} + 2\frac{5}{8}$

❼ Fractions, Decimals and Percents

Percent means per hundred. Therefore, 35% means 35 out of 100. Percents can be written as equivalent decimals and fractions.

$35\% = 0.35$ Move the decimal point two places to the left.

$35\% = \dfrac{35}{100}$ Write the fraction with a denominator of 100.

$= \dfrac{7}{20}$ Then simplify.

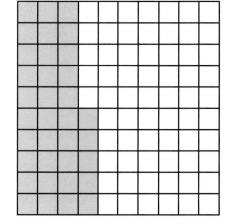

Example 1

Write $\dfrac{3}{8}$ as a decimal and as a percent.

Solution

$\dfrac{3}{8} = 0.375$ Divide to change a fraction to a decimal.

$0.375 = 37.5\%$ To change a decimal to a percent move the decimal point two places to the right and insert the percent symbol.

Percents greater than 100% represent whole numbers or mixed numbers.

$200\% = 2$ or 2.00 $350\% = 3.5$ or $3\dfrac{1}{2}$

◼ EXTRA PRACTICE EXERCISES

Write each fraction or mixed number as a decimal and as a percent.

1. $\dfrac{1}{2}$ 2. $\dfrac{1}{4}$ 3. $\dfrac{3}{4}$

4. $\dfrac{9}{10}$ 5. $\dfrac{3}{10}$ 6. $\dfrac{1}{25}$

7. $3\dfrac{7}{8}$ 8. $1\dfrac{1}{5}$ 9. $\dfrac{13}{25}$

Write each decimal or mixed number as a fraction and as a percent.

10. 0.63   11. 0.15   12. 0.4

13. 2.35   14. 10.125   15. 0.625

16. 0.05   17. 0.125   18. 0.3125

Write each percent as a decimal and as a fraction or mixed number.

19. 10%   20. 12%   21. 100%

22. 150%   23. 160%   24. 75%

25. 8%   26. 87.5%   27. 0.35%

⑧ Multiply and Divide by Powers of Ten

To multiply a number by a power of 10, move the decimal point to the right. To multiply by 100 means to multiply by 10 two times. Each multiplication by 10 moves the decimal point one place to the right.

To divide a number by a power of 10, move the decimal point to the left. To divide by 1000 means to divide by 10 three times. Each division by 10 moves the decimal point one place to the right.

Example 1

Multiply 21 by 10,000.

Solution

$21 \times 10{,}000 = 210{,}000$ The decimal point moves four places to the right.

Example 2

Find $145 \div 500$.

Solution

$$145 \div 500 = 145 \div 5 \div 100$$
$$= 29 \div 100$$
$$= 0.29$$ The decimal point moves two places to the left.

◣ EXTRA PRACTICE EXERCISES

Multiply or divide.

1. 15×100
2. $96 \times 10{,}000$
3. $1296 \div 100$
4. $9687.03 \div 1000$
5. $36 \times 20{,}000$
6. $7500 \div 3000$
7. 9×30
8. 94×6000
9. $561 \div 30$
10. $1505 \div 500$
11. $71 \times 90{,}000$
12. $9 \times 120{,}000$
13. $3159 \div 10{,}000$
14. $1{,}000{,}000 \times 0.79$
15. $601 \times 30{,}000$
16. $75 \div 300$
17. 4000×12
18. $14 \times 7{,}000{,}000$
19. $49{,}000 \div 7000$
20. $980 \div 10{,}000$
21. $216 \div 2000$
22. $108{,}000 \div 900$
23. $72 \times 10{,}000{,}000$
24. $953.16 \div 10{,}000$
25. $1472 \div 8000$
26. $490{,}000 \div 700$
27. $80 \times 90{,}000$
28. $8001 \div 90$
29. 50×6000
30. $950{,}000 \div 50{,}000$
31. $81{,}000 \times 5$
32. $1458 \times 30{,}000$
33. $452.3 \div 10$
34. $986{,}856.008 \div 10{,}000$
35. $316 \times 70{,}000$
36. $60 \div 1200$

⑨ Round and Order Decimals

To round a number, follow these rules:

1. Underline the digit in the specified place. This is the place digit. The digit to the immediate right of the place digit is the test digit.

2. If the test digit is 5 or larger, add 1 to the place digit and substitute zeros for all digits to its right.

3. If the test digit is less than 5, substitute zeros for it and all digits to the right.

Example 1

Round 4826 to the nearest hundred.

Solution

4<u>8</u>26   Underline the place digit.

4800   Since the test digit is 2 and 2 is less than 5, substitute zeros for 2 and all digits to the right.

To place decimals in ascending order, write them in order from least to greatest.

Example 2

Place in ascending order: 0.34, 0.33, 0.39.

Solution

Compare the first decimal place, then compare the second decimal place.

0.33 (least), 0.34, 0.39 (greatest)

◼ EXTRA PRACTICE EXERCISES

Round each number to the place indicated.

1. 367 to the nearest ten
2. 961 to the nearest ten
3. 7200 to the nearest thousand
4. 3070 to the nearest hundred
5. 41,440 to the nearest hundred
6. 34,254 to the nearest thousand
7. 208,395 to the nearest thousand
8. 654,837 to the nearest ten thousand

Write the decimals in ascending order.

9. 0.29, 0.82, 0.35
10. 1.8, 1.4, 1.5
11. 0.567, 0.579, 0.505, 0.542
12. 0.54, 0.45, 4.5, 5.4
13. 0.0802, 0.0822, 0.00222
14. 6.204, 6.206, 6.205, 6.203
15. 88.2, 88.1, 8.80, 8.82
16. 0.007, 7.0, 0.7, 0.07

Extra Practice

Chapter 1

Extra Practice 1–1 • Collect and Display Data • pages 6–9

GOVERNMENT You need to determine if local taxpayers would support building a new football field. Name the advantages and disadvantages of choosing a sample in each of the following ways.

1. Ask a randomly chosen sample of twenty people at a town council meeting.

2. Ask every tenth person who enters a local grocery store.

3. Ask the six people who are sitting near you at a football game.

NUTRITION You need to determine the school's most popular lunch menu. Name the advantages and disadvantages of choosing a sample in each of the following ways.

4. Ask everyone standing in line for lunch.

5. Ask everyone at your bus stop in the morning.

6. Ask everyone whose student number ends with a 3.

FOOD SERVICE You need to find if students think seniors should be able to leave the school for lunch. Name the advantages and disadvantages of choosing a sample in each of the following ways.

7. Ask everyone who comes into the library after school.

8. Ask every fifth person who enters the basketball game one evening.

9. Ask everyone in your homeroom.

Extra Practice 1–2 • Measures of Central Tendency • pages 10–13

SPORTS Here are each swim team-member's competition times in seconds for one race: 12.9   14.0   13.5   12.2   13.6   12.2   13.8   11.8

1. Find the mean, median and mode.

2. Find the range.

Tell which measure of central tendency would best represent each of the following.

3. The average number of cars parked at the park-and-ride lot every weekday.

4. The favorite pizza topping.

5. The usual cost of a house in your town.

6. The average number of students in each history class in the high school.

EDUCATION The table shows Jason's test scores in science class for the first semester.

Test	Grade
Test #1	58
Test #2	90
Mid-term	88
Test #3	93
Test #4	84
Final	85

7. Find the mean, median, and mode of the test scores.

8. Which measure of central tendency best describes Jason's average grade?

EDUCATION These data represent test scores on a math test.

90	94	67	78	73	64	83	93
80	69	79	76	69	91	82	75
62	94	97	89	67	78	78	83
94	66	77	95	89	81	88	86

1. How many scores are represented?

2. Organize the data into a stem-and-leaf plot.

3. How many students scored 89? 75? 67? 94?

WEATHER These data represent temperatures (°F) Brad recorded at 3:00 P.M. in February.

28	26	45	42	55	26	31
32	42	40	18	28	29	37
31	38	52	40	26	34	44
34	27	45	43	29	38	36

4. How many temperatures did Brad record?

5. Organize the data into a stem-and-leaf plot.

6. Write a description of the data. Note the greatest and least temperature and any outliers, clusters, and gaps.

Use the pictograph.

1. Which sport is most popular?

2. Which sport is least popular?

3. How many students prefer football?

4. How many more students prefer basketball than tennis?

5. How many more students prefer baseball than golf?

6. Make a frequency table of the data

Sports Students Prefer

Basketball	⚽ ⚽ ⚽ ⚽ ⚽ ⚽
Baseball	⚽ ⚽ ⚽ ⚽
Football	⚽ ⚽ ⚽ ⚽ ⚽
Soccer	⚽ ⚽ ⚽ ⚽ ⚽ ⚽ ⚽ ⚽
Tennis	⚽ ⚽ ⚽
Golf	⚽ ⚾

Key: ⚽ represents 10 students

At Cedar High School, the French Club has 60 members, the Key Club has 120 members, the Yearbook Committee has 45 members. Model Congress has 30 members, the Music Club has 15 members and Student Action has 75 members.

7. Construct a pictograph for the data. Choose a symbol to represent 15 students.

8. Which group has the greatest number of members?

9. Which group has the least number of members?

10. How many more members are on the Student Action committee than on the Yearbook Committee?

Soccer Games Won

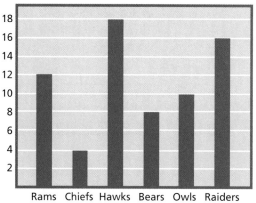

Use the bar graph.

1. How many teams are represented on the graph?

2. Which team won the most games?

3. Which team won the fewest games?

4. How many more games did the Rams win than the Bears?

5. How many more games did the Raiders win than the Chiefs?

6. If the Raiders played 23 games and there were no ties, how many games did they lose?

7. If the Owls played 25 games and there was 1 tie, how many games did they lose?

8. If the Chiefs played 2 more games than the Rams won and there were no ties, how many games did the Chiefs lose?

9. If the Bears played twice as many games as the Owls won and there were no ties, how many games did the Bears lose?

10. The Raiders won twice as many games as which other team?

The line graph shows the changes in class size for the Ridge High School Senior Class from 1998 to 2004.

11. What was the class size in 2000?

12. What year had the largest senior class?

13. When did the largest increase in class size occur?

14. How did the class size change from 2000 to 2001?

15. Draw a line graph to show the growth of a tomato plant weekly.

Size of Ridge High School Senior Class

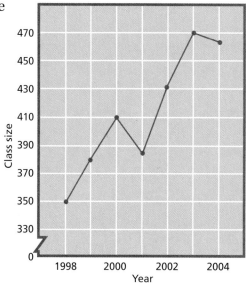

Weekly Growth of a Tomato Plant

Week	Height (in.)
1	0
2	2
3	6
4	10
5	15
6	20
7	24
8	30

16. Draw a line graph to show the number of beach passes sold during the summer.

Beach Passes Sold by Week

Week	1	2	3	4	5	6	7	8	9	10	11	12
Passes sold	30	60	80	100	130	60	120	140	150	160	100	70

A survey asked students the number of hours of sleep they average per night during the school year, then compared the data to the student's number of sick days.

Number of sick days	2	3	0	1	5	6	1	0	3	5	6	5	9	5	3	2	6
Hours of sleep	8	8	9	9	6	6	8.5	8.5	7	7.5	6.5	6.5	6	7	8.5	9	9.5

1. Draw a scatter plot and line of best fit for the data.

2. Use the line of best fit to predict the number of sick days of a student who slept five hours.

3. Is there a positive or negative correlation between the data?

4. Which data point lies father from the line of best fit? What could account for this piece of data?

Use the box-and-whisker plot.

Radial Tire Prices (sets of 4)

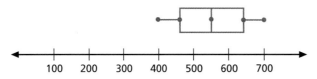

1. What are the greatest and least prices?

2. What is the range of the prices?

3. What is the median price?

4. What is the range of the middle 50%?

5. In which intervals are the prices most closely clustered?

Gather data from your classmates about the number of cousins on their mother's side of the family. Also gather data about the number of cousins on their father's side of the family.

6. Make a box-and-whisker plot that shows both sets of data.

7. What is the range for each set of data?

8. What is the median number of cousins for each set of data?

Chapter 2

Extra Practice 2–1 • Units of Measure • pages 52–55

Which unit of measure gives a more precise measurement?

1. centimeter or meter   **2.** pound or ton   **3.** milligram or centigram

Name the tool or tools that would best measure each object.

4. time taken to run a 5K race   **5.** amount of milk to put in a cake recipe

6. weight of a bunch of bananas   **7.** height of a tree

Choose the appropriate unit to estimate the measure.

8. sugar in a cake recipe: gallon, liter, cup

9. diameter of a dime: millimeter, feet, gram

Choose the best estimate for each measure.

10. capacity of an automobile's gas tank: 18 c, 18 lb, or 18 gal

11. distance from the front of the room to the back of the classroom: 25 mi, 25 mm, or 25 ft

12. length of a couch: 3 cm, 3 m, or 3 in.

Extra Practice 2–2 • Work with Measurements • pages 56–59

1. 2,000 g = ■ kg   **2.** 6 L = ■ mL   **3.** 194 g = ■ kg

4. 4 km = ■ m   **5.** 2,933 mL = ■ L   **6.** 276 mm = ■ m

7. 204 cm = ■ m   **8.** 872 mL = ■ L   **9.** 9 km = ■ m

10. 5,340 mm = ■ km   **11.** 2,500 kg = ■ g   **12.** 18 ft = ■ yd

13. 36 fl oz = ■ pt ■ fl oz   **14.** $2\frac{1}{2}$ T = ■ lb   **15.** 29 ft = ■ yd ■ ft

16. 7 qt = ■ gal ■ qt   **17.** 85 in. = ■ ft ■ in.   **18.** 2 lb 7 oz = ■ oz

19. 52 oz = ■ lb ■ oz   **20.** 7 yd = ■ ft   **21.** 24 fl oz = ■ pt ■ c

22. $2\frac{3}{4}$ lb = ■ oz   **23.** 755 m = ■ cm   **24.** 48 oz = ■ lb

Complete. Write each answer in simplest form.

25.   7 ft 5 in.
    + 4 ft 8 in.

26.   6 gal 2 qt
    − 4 gal 3 qt

27.   6 yd
    − 2 yd 1 ft

28.   7 ft 8 in.
    − 4 ft 10 in.

29.   4 yd 4 ft
    + 6 yd 3 ft

30.   3 c 6 fl oz
    ×   3

Complete.

31. 2.4 L ÷ 6 = __?__ mL   **32.** 3 m + 48 cm = __?__ cm

33. 1.7 kg − 1.36 kg = __?__ g   **34.** 460 mL + 150 mL = __?__ L

Find the perimeter of each figure drawn on the grid.

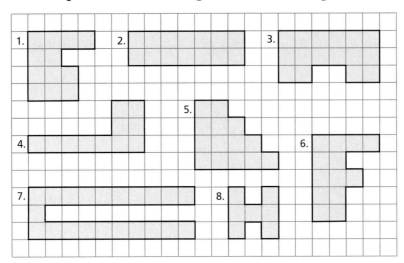

Find the perimeter of each figure.

9.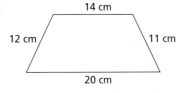
14 cm
12 cm   11 cm
20 cm

10.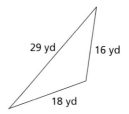
29 yd   16 yd
18 yd

11.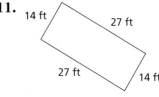
14 ft   27 ft
27 ft   14 ft

12. Find the perimeter of a football field if the length is 120 yd and the width is 53 yd.

13. Find the perimeter of a square piece of carpet if the sides measure 1.8 m.

Find the area of each figure.

1. rectangle: 0.3 cm by 0.8 cm

2. parallelogram: $b = 21$ m, $h = 30$ m

3. square: $\frac{3}{4}$ -yd sides

4. triangle: $h = 23.5$ km, $b = 5.2$ km

5. triangle: $h = 405$ mi, $b = 122$ mi

6. parallelogram: $b = 18$ in., $h = 30$ in.

7. square: 19-mm sides

8. rectangle: $1\frac{1}{2}$ ft by 2 ft

Find the area of each figure.

9.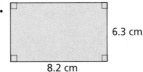
6.3 cm
8.2 cm

10.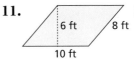
12 yd
12 yd

11.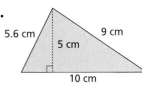
6 ft   8 ft
10 ft

12.
5.6 cm   9 cm
5 cm
10 cm

550 | **Extra Practice**

Write each ratio two other ways.

1. 3 to 7   **2.** 20 : 35   **3.** 1 : 3   **4.** $\dfrac{4}{5}$

5. $\dfrac{6}{9}$ **6.** 5 : 8   **7.** $\dfrac{4}{11}$ **8.** 5 to 1

Write each ratio as a fraction in lowest terms.

9. 3 cm : 2 cm   **10.** 5 min to 25 min   **11.** 2.4 cm to 32 mm

12. 18 qt to 4 gal   **13.** 120 min to 2 h   **14.** \$2.35 : \$5.55

Write three equivalent ratios for each given ratio.

15. $\dfrac{3}{7}$ **16.** 12 to 20   **17.** 5 : 6   **18.** 14 : 22

19. 8 to 3   **20.** $\dfrac{4}{11}$ **21.** $\dfrac{9}{10}$ **22.** 7 : 9

Are the ratios equivalent? Write *yes* or *no*.

23. 3 to 9, 6 to 18   **24.** 5 : 2, 13 : 6   **25.** $\dfrac{12}{22}, \dfrac{16}{26}$ **26.** $\dfrac{15}{20}, \dfrac{6}{8}$

27. $\dfrac{7}{28}, \dfrac{2}{8}$ **28.** 1 : 2, 100 : 200   **29.** 20 : 31, 30 : 41   **30.** $\dfrac{3}{4}, \dfrac{12}{16}$

31. A bakery is having a 50%-off bread sale. They sell 33 out of 44 white loaves and 18 out of 24 wheat loaves. Write the ratio for white loaves sold and the ratio for wheat loaves sold. Which ratio is greater?

Find the area. Use 3.14 or $\dfrac{22}{7}$ for π, as appropriate. Round to the nearest tenth.

1.
42.7 m

2.
63 ft

3.
$19\frac{1}{2}$ in.

4. Find the circumference of the circles in Exercises 1–3.

Find the area of a circle with the given dimensions. Round to the nearest tenth. Use 3.14 for π.

5. $d = 14$ mm   **6.** $r = 2.2$ cm

7. $d = 10.5$ ft   **8.** $r = 2.3$ yd

9. Find the circumference of each circle in Exercises 5–8. Round to the nearest tenth. Use 3.14 for π.

10. The diameter of the top of a soup can is 6.5 cm. Find the circumference and the area of the top of the can. Round to the nearest tenth. Use 3.14 for π.

Extra Practice

Tell whether each statement is a proportion. Write *yes* or *no*.

1. $\frac{3}{4} \overset{?}{=} \frac{4}{5}$

2. $\frac{3}{5} \overset{?}{=} \frac{6}{10}$

3. $2 : 8 \overset{?}{=} 1 : 4$

4. $\frac{6}{9} \overset{?}{=} \frac{16}{24}$

5. $11 : 15 \overset{?}{=} 33 : 37$

6. $\frac{8}{32} \overset{?}{=} \frac{3}{12}$

Use mental math to solve each proportion.

7. $\frac{1}{3} = \frac{?}{12}$

8. $\frac{4}{10} = \frac{?}{5}$

9. $? : 9 = 8 : 18$

10. $\frac{?}{20} = \frac{30}{40}$

11. $? : 14 = 6 : 7$

12. $6 : 9 = 4 : ?$

13. $\frac{?}{8} = \frac{5}{10}$

14. $4 : 1 = ? : 3$

15. $\frac{9}{5} = \frac{27}{?}$

Find the actual or drawing length.

16. Scale, 1 cm : 8 m

Drawing length, 5 cm

Actual length, ?

17. Scale, 1 in. : 3 yd

Drawing length, ?

Actual length, 60 yd

18. Scale, 1 cm: 5 km

Drawing length, ?

Actual length, 30 km

19. A Garden Club uses 50 lb of potting soil to fill 15 pots. How many pots could be filled with 10 lb of potting soil?

Find the area of each figure. Round to the nearest tenth. Use 3.14 for π.

1.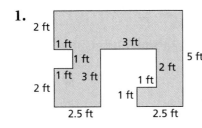

2.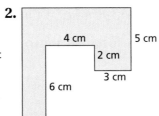

3.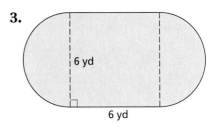

4. Find the area of the flower garden shown.

5. One bag of mulch covers 12 square yards. How many bags are needed for the flower garden shown?

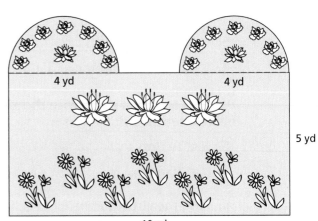

Estimate the area of each object using the units of measure given.

6. a stove top in meters

7. a book cover for your math book in inches.

8. a piece of computer paper in centimeters

9. your desk top in feet

Chapter 3

Use a number line to add or subtract.

1. $9 + 2$ **2.** $7 + (-3)$ **3.** $-3 - (-4)$ **4.** $6 - 4$

Find each absolute value.

5. $|-6|$ **6.** $|-2.9|$ **7.** $|65|$ **8.** $\left|\dfrac{2}{3}\right|$

Add or subtract using the rules for addition and subtraction.

9. $4 + (-7)$ **10.** $3 + 9$ **11.** $-4 + (-9)$

12. $-11 - 9$ **13.** $-7 - (-7)$ **14.** $4 + (-4)$

15. $-3 + 3$ **16.** $5 - (-8)$ **17.** $-14 + (-2)$

18. $11 - (-5)$ **19.** $-3 - 8$ **20.** $4 - (-13)$

Simplify.

21. $8 + 7 + (-7)$ **22.** $3 + (-5) + 9$

23. $-8 + (-5) + (-7)$ **24.** $-\dfrac{1}{3} + \dfrac{2}{9} + \left(-\dfrac{7}{9}\right)$

25. $-5 + (-3) + (-7) + (-1)$ **26.** $6 + (-6) + 8 + (-14)$

Without subtracting, tell whether the answer will be positive or negative.

27. $87 - (-7)$ **28.** $-13 - (-5)$ **29.** $-8 - (-15)$ **30.** $-13 - 29$

31. $-53 - (-41)$ **32.** $66 - 14$ **33.** $12 - 16$ **34.** $-9 - 10$

Multiply or divide.

1. $7 \cdot 2$ **2.** $7 \cdot (-3)$ **3.** $-36 \div (-4)$ **4.** $64 \div 2$

5. $-6 \cdot (-10)$ **6.** $-7 \cdot 9$ **7.** $-65 \div 5$ **8.** $\dfrac{24}{(-3)}$

9. $14 \div (-7)$ **10.** $-27 \div 3$ **11.** $-24 \div (-8)$ **12.** $-11 \cdot -9$

13. $-7 \cdot (-7)$ **14.** $4 \div (-4)$ **15.** $\dfrac{3}{-3}$ **16.** $7 \cdot (-11)$

17. $-14 \div (-2)$ **18.** $-11 \cdot (-5)$ **19.** $3 \cdot (-5)$ **20.** $2 \cdot (-4)$

Replace each ■ with <, >, or =.

21. $-3 \cdot (-2)$ ■ $3 \cdot 2$ **22.** $5 \div (-5)$ ■ $-5 \div 5$ **23.** $4 \cdot (-6)$ ■ $-1 \cdot (-5)$

24. $12 \div (-6)$ ■ $-16 \div (-2)$ **25.** $-36 \div (-9)$ ■ $-\dfrac{40}{4}$ **26.** $6 \cdot (-7)$ ■ $-8 \cdot 5$

27. $-4 \cdot (-3)$ ■ $-4 \cdot 3$ **28.** $-15 \div 3$ ■ $-1 \cdot (-5)$ **29.** $-\dfrac{300}{3}$ ■ $2 \cdot 2$

30. $-5 \div 1$ ■ $-5 \div (-1)$ **31.** $-20 \cdot (-3)$ ■ $3 \cdot 2$ **32.** $30 \div 6$ ■ $-30 \div (-6)$

33. $-16 \div 2$ ■ $-16 \div (-2)$ **34.** $-9 \cdot (-3)$ ■ $9 \cdot (-3)$ **35.** $45 \div 5$ ■ $45 \div (-5)$

Simplify. Use the order of operations.

1. $2 \cdot (-9) \div (6 + 3)$

2. $5 + (-4 \cdot 7) - 3$

3. $-4 \cdot (-3 + 2)^2 + 4$

4. $-\frac{1}{3}(5 - 2) + (-6)$

5. $-2 \cdot (-4) + (-16) \div 4$

6. $5^2 \div (-5) + 6$

7. $(-10 - 8) \cdot 9 - 8$

8. $8^2 \div (-2) \div (-2)$

9. $-24 \div (-8) - 6 \cdot 2 + 6$

10. $[8 \div (-4)]^2 + 4 \cdot (-4)$

11. $6 + (-2 \cdot 3)^2 \div 6$

12. $-2 \cdot (2 + 3)^2 + 10$

13. $-16 \div (3 + 1) \cdot (-2)$

14. $-4 + (-6) - (-24) \div 2$

15. $(10^2 \div 5) + 8 \div 4$

16. $(-12 - 2) \div 5 + (8 - 3)$

17. $(4^2 + 3) - (-4) \div (-2)$

18. $-28 \div 7 - 4 \cdot 3 + (-6)$

Use the symbols +, −, ·, or ÷ to make each number sentence true.

19. $5 \blacksquare 8 \blacksquare 5 = 18$

20. $1 \blacksquare 8 \blacksquare 3 = 11$

21. $4 \blacksquare 1 \blacksquare 12 = 48$

22. $4 \blacksquare 8 \blacksquare 2 = 10$

23. $3 \blacksquare 8 \blacksquare 6 = 4$

24. $16 \blacksquare 6 \blacksquare 16 = 6$

Write *true* or *false* for each. For any that are false, insert parentheses to make them true.

25. $4 \cdot 5 + 5 = 40$

26. $2 \cdot 8 - 3 = 13$

27. $8 - 6 \div 2 = 1$

28. $\frac{1}{3} + \frac{1}{3} \cdot 3 = \frac{4}{3}$

29. $3 + 12 - 6 \div 3 = 5$

30. $2^3 - 4 \cdot 6 \div 2 = -4$

To which sets of numbers do the following belong?

1. 12.275

2. -3

3. $\frac{5}{9}$

4. $-4.1234\ldots$

5. 7

6. -125

7. $-\frac{3}{4}$

8. 1.26

Complete. Name the property you used.

9. $6(4 + \blacksquare) = (6 \cdot 4) + (6 \cdot 3)$

10. $2(9 - 5) = (\blacksquare \cdot 9) - (\blacksquare \cdot 5)$

11. $(13 + 8) + 52 = 13 + (8 + \blacksquare)$

12. $(7 \cdot \blacksquare) \cdot 2 = 7 \cdot (5 \cdot 2)$

13. $9 \cdot 3 = 3 \cdot \blacksquare$

14. $\blacksquare + 16 = 16 + 13$

15. $(10 \cdot 9) \cdot \blacksquare = 10 \cdot (9 \cdot 6)$

16. $5 \cdot \blacksquare = 3 \cdot 5$

17. $(\blacksquare + 6) + 15 = 4 + (6 + 15)$

18. $\blacksquare(5 + 8) = (7 \cdot 5) + (7 \cdot 8)$

19. $(\blacksquare \cdot 5) - (\blacksquare \cdot 3) = 9(5 - 3)$

20. $6 + 3 = \blacksquare + 6$

Simplify using mental math. Name the properties you used.

21. $16 + 32 + 14$

22. $\frac{1}{6} \cdot \frac{1}{3} \cdot 6$

23. $33 \cdot 12$

24. $25 \cdot 7 \cdot 4$

25. $0.46 + 7.18 + 0.54$

26. $18 + 34 + 6 + 2$

State whether the following sets are closed under the given operation.

27. integers, multiplication

28. whole numbers, subtraction

Write a variable expression. Let *n* represent "a number."

1. five more than a number

2. four less than a number

3. negative one added to a number times two

4. eight divided by a twice a number

5. a number minus nine

6. four less than six times a number

Evaluate each expression. Let $a = 5$, $b = -2$, and $c = 3$.

7. $7c$ 8. $(-2)a$ 9. $a + 9$ 10. $4b - 8$

11. $4c \div b$ 12. $a - (-3)$ 13. $3a - (-b)$ 14. $c^2 + a$

Write an expression to describe each situation.

15. seven more than three times h horses

16. f fishes decreased by two

17. ten more than half of y yards

18. nine dollars shared equally by x people

19. a temperature of 80° increased by t degrees

20. a distance d times 50 increased by 18

Write in exponential form.

1. $3 \cdot 3 \cdot 3 \cdot 3$ 2. $5 \cdot 5 \cdot 5 \cdot 5 \cdot 5 \cdot 5$ 3. $\dfrac{1}{2} \cdot \dfrac{1}{2} \cdot \dfrac{1}{2} \cdot \dfrac{1}{2} \cdot \dfrac{1}{2}$

4. $0.09 \cdot 0.09 \cdot 0.09$ 5. $\dfrac{1}{4 \cdot 4 \cdot 4 \cdot 4}$ 6. $2 \cdot 2 \cdot 2 \cdot 2 \cdot 2 \cdot 2 \cdot 2 \cdot 2$

Write in standard form.

7. $\left(\dfrac{1}{2}\right)^0$ 8. 6^3 9. 21^1 10. 5^{-2}

11. 10^4 12. $(0.4)^2$ 13. 4^{-3} 14. 2^5

15. $8.28 \cdot 10^4$ 16. $7 \cdot 10^{-3}$ 17. $3 \cdot 10^6$ 18. $4.1269 \cdot 10^5$

19. $4.1 \cdot 10^{-2}$ 20. $6.790012 \cdot 10^8$ 21. $3.0492 \cdot 10^{-1}$ 22. $9.9999 \cdot 10^3$

Write in scientific notation.

23. 300,000   24. 6,000,000,000   25. 15,890,000

26. 7123   27. 0.002345   28. 0.0000009

29. 0.32   30. 900,100   31. 1,678,000,000

32. 0.1579   33. 0.000237   34. 0.0025

Extra Practice

Use the product rule to multiply.

1. $4^4 \cdot 4^8$
2. $x^{12} \cdot x^3$
3. $3^4 \cdot 3^2$
4. $h^0 \cdot h^6$
5. $5^7 \cdot 5^7$
6. $d^9 \cdot d^8$
7. $7^5 \cdot 7^2$
8. $y^4 \cdot y^8$
9. $x^3 \cdot x^7$
10. $6^0 \cdot 6^8$
11. $m^1 \cdot m^9$
12. $8^4 \cdot 8^2$

Use the quotient rule to divide.

13. $4^5 \div 4^4$
14. $x^{10} \div x^0$
15. $10^9 \div 10^4$
16. $m^7 \div m^2$
17. $x^{10} \div x^5$
18. $3^4 \div 3^2$
19. $x^{15} \div x^8$
20. $p^{20} \div p^{11}$
21. $h^8 \div h^1$
22. $5^8 \div 5^1$
23. $9^{14} \div 9^9$
24. $12^6 \div 12^6$

Use the power rule.

25. $(7^4)^5$
26. $(x^{10})^6$
27. $(y^1)^4$
28. $(9^5)^5$
29. $(x^4)^3$
30. $(4^4)^6$
31. $(10^9)^3$
32. $(t^{14})^0$
33. $(6^7)^7$
34. $(x^3)^{12}$
35. $(x^6)^4$
36. $(5^9)^3$

Use the laws of exponents.

37. $8^{30} \cdot 8^{15}$
38. $w^{10} \div w^5$
39. $(7^6)^2$
40. $(10^3)^{12}$
41. $x^9 \div x^1$
42. $k^6 \div k^6$
43. $h^{20} \div h^{10}$
44. $m^3 \div m^0$
45. $h^5 \div h^2$
46. $4^{16} \cdot 4^0$
47. $(14^8)^8$
48. $6^{23} \cdot 6^4$

Find the value of each variable.

49. $4^8 \cdot 4^a = 4^{15}$
50. $6^b \div 6^5 = 6^5$
51. $(7^8)^c = 7^{24}$
52. $9^{17} \div 9^9 = 9^d$
53. $11^e \cdot 11^7 = 11^7$
54. $5^{15} \div 5^f = 5^5$
55. $(3^g)^2 = 3^{16}$
56. $8^6 \div 8^h = 8^3$

Find each square.

1. 4^2
2. 13^2
3. $\left(\dfrac{2}{5}\right)^2$
4. $(0.005)^2$
5. $(-\sqrt{42})^2$
6. $\left(-\dfrac{2}{3}\right)^2$
7. $(-5.6)^2$
8. 22^2
9. 14^2
10. $(-15)^2$
11. $\left(\dfrac{1}{7}\right)^2$
12. $(0.15)^2$

Find each square root.

13. $-\sqrt{289}$
14. $\sqrt{\dfrac{1}{400}}$
15. $-\sqrt{0.81}$
16. $\sqrt{\dfrac{49}{16}}$
17. $\sqrt{\dfrac{4}{121}}$
18. $-\sqrt{10{,}000}$
19. $-\sqrt{0.0009}$
20. $\sqrt{\dfrac{225}{900}}$
21. $-\sqrt{\dfrac{64}{625}}$
22. $-\sqrt{0.0225}$
23. $\sqrt{\dfrac{1}{9}}$
24. $-\sqrt{\dfrac{1}{529}}$

CALCULATOR Use a calculator to find each square root. Round to the nearest thousandth.

25. $\sqrt{127}$ **26.** $\sqrt{311}$ **27.** $\sqrt{888}$ **28.** $\sqrt{633}$

29. $\sqrt{952}$ **30.** $\sqrt{477}$ **31.** $\sqrt{268}$ **32.** $\sqrt{912}$

Chapter 4

Extra Practice 4–1 • Language of Geometry • pages 156–159

Draw each geometric figure. Then write a symbol for each figure.

1. point *R*   **2.** ray *GR*   **3.** angle *HRD*   **4.** line segment *WX*

5. line *JK*   **6.** angle *C*   **7.** plane *EFG*   **8.** line *RD*

Use symbols to complete the following.

9. Name the line four ways.

10. Name two rays with *J* as an endpoint

Draw a figure to illustrate each of the following.

11. Points *M, N,* and *O* are noncollinear.   **12.** Line *p* intersects plane $\mathcal{H}$ at point *Q*.

13. Points *R, S, T,* and *W* are noncoplanar.   **14.** Planes $\mathcal{T}$ and $\mathcal{U}$ intersect at line *x*.

Find the measure of each angle.

15. $\angle FOG$ **16.** $\angle BOG$

17. $\angle DOG$ **18.** $\angle EOG$

19. $\angle AOC$ **20.** $\angle AOB$

21. $\angle AOF$ **22.** $\angle AOE$

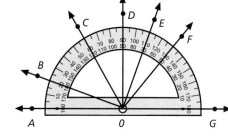

Use a protractor to draw an angle of the given measure.

23. 60°   **24.** 145°   **25.** 29°   **26.**
90°

27. 30°   **28.** 68°   **29.** 175°   **30.** 112°

Extra Practice 4–2 • Polygons and Polyhedra • pages 160–163

If each figure is a polygon, identify it, and tell if it is regular. If the figure is not a polygon, explain why.

1.   **2.**   **3.**   **4.**

Determine whether each statement is *true* or *false*.

5. Every rectangle is a parallelogram.   **6.** Equilateral polygons are always equiangular.

7. A polygon always has more than 3 vertices.   **8.** Every octagon is equiangular.

Classify each triangle.

9.
120°

10.

11.

12.

Identify the number of faces, vertices and edges for each figure.

13.

14.

15.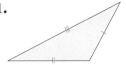

16.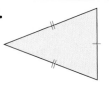

Extra Practice 4–3 • Visualize and Name Solids • pages 166–169

Identify each three-dimensional figure.

1. It has one base that is a square. The other faces are triangles.

2. It has two bases that are identical triangles. The other faces are parallelograms.

3. It has four faces that are equilateral triangles.

4. Its base is a circle and it has one vertex.

Identify the three dimensional figure that is formed by each net.

5.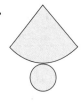

6.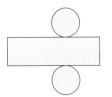

7.

8.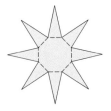

Name a three-dimensional figure that has the same shape as each object.

9. tennis ball

10. construction barrel

11. 2 by 4 wood beam

12. A-frame tent

13. drink can

14. marble

Extra Practice 4–5 • Isometric Drawings • pages 174–177

Use the figure to name the following.

1. all of the parallel lines

2. all of the perpendicular lines

Show the following on one drawing.

3. Draw $\overleftrightarrow{MN}$.

4. Draw line ℓ parallel to $\overleftrightarrow{MN}$.

5. Draw $\overleftrightarrow{MT}$ perpendicular to $\overleftrightarrow{MN}$ at M.

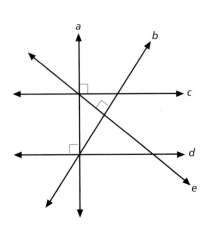

On isometric dot paper, draw the following.

6. cube: $s = 8$ yd

7. rectangular prism: $l = 3$ ft, $w = 4$ ft, $h = 5$ ft

8. cube: $s = 4.5$ cm

9. rectangular prism: 2 units by 4 units by 4 units

Make an isometric drawing of each figure.

10.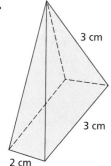

3 cm

3 cm

2 cm

11.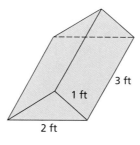

3 ft

1 ft

2 ft

12.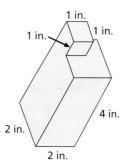

1 in.

1 in.

1 in.

4 in.

2 in.

2 in.

Extra Practice 4–6 • Perspective and Orthogonal Drawings • pages 178–181

Trace each figure. Locate and label the vanishing point and the horizon line.

1.

2.

3.

Draw each of the following in one-point perspective.

4. square pyramid

5. triangular prism

6. cone

Make an orthogonal drawing of each stack of cubes, showing the top, front and side views.

7.

8.

9.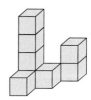

Extra Practice 4–7 • Volume of Prisms and Cylinders • pages 184–187

Find the volume of each rectangular prism.

1. $l = 3$ cm, $w = 5.1$ cm, $h = 6$ cm

2. $l = 0.01$ m, $w = 0.7$ m, $h = 0.8$ m

3. $l = 20$ in., $w = 16$ in., $h = 30$ in.

4. $l = 12$ ft, $w = 13$ ft, $h = 15$ ft

Find the volume of each cylinder.

5. $r = 240$ cm, $h = 100$ cm

6. $d = 30.1$ m, $h = 14.5$ m

7. $r = 0.05$ km, $h = 0.06$ km

8. $d = 5$ ft, $h = 18$ ft

Find the volume.

9.

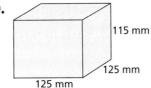
115 mm
125 mm
125 mm

10.

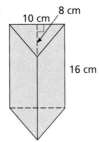
8 cm
10 cm
16 cm

11.

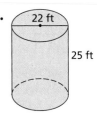
22 ft
25 ft

Which has the greater volume?

12.

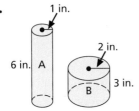
1 in.
6 in. A
2 in.
B
3 in.

13.

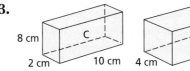
8 cm
C
2 cm
10 cm
5 cm
D
4 cm
6 cm

Extra Practice 4–8 • Volume of Pyramids and Cones • pages 188–191

Find the volume of each figure. If necessary, round to the nearest tenth.

1.

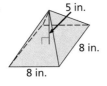
5 in.
8 in.
8 in.

2.

18 m
7 m

3.

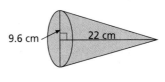
9.6 cm
22 cm

4.

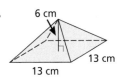
6 cm
13 cm
13 cm

5. Find the volume of a cone 5.4 ft in diameter and 6.9 ft high.

6. Find the volume of a pyramid that is 12.4 in. high and has a rectangular base measuring 7 in. by 4 in.

7. Rock salt is stored in a conical pile. Find the volume of the pile if the height is 6.3 m and the radius is 12.6 m.

8. A tent is in the form of a pyramid with a base that is 6 ft by 4 ft. The height is 4.5 ft. Find the volume.

9. A pyramid has volume of 2600 ft³. Its base area is 130 ft². What is its height?

10. A cone has a volume of 37,500 cm³. Its height is 53 cm. What is its radius?

Extra Practice 4–9 • Surface Area of Prisms and Cylinders • pages 194–197

For Exercises 1–12, find the surface area of each figure. Use 3.14 for π. Round to the nearest whole number.

1.

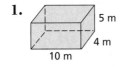
5 m
4 m
10 m

2.

7 ft
15 ft

3.

2 cm
7 cm

4.

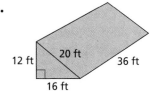
12 ft
20 ft
36 ft
16 ft

5. rectangular prism: 8 m by 15 m by 22 m
6. cylinder: $r = 16$ ft, $h = 24$ ft
7. cylinder: $d = 4$ km, $h = 9$ km
8. cube: $s = 1.8$ cm
9. rectangular prism: 10 m by 10 m by 312 m
10. cylinder: $d = 12$ in., $h = 3$ in.
11. rectangular prism: 5.6 ft by 3.2 ft by 9.1 ft
12. cylinder: $r = 1.2$ m, $h = 1.2$ m

13. A toy block is to be dipped in paint. The dimensions of the block are 2 cm by 5 cm by 12 cm. How many square centimeters of the block will be painted?

14. A gift box is cylindrical in shape. It has a radius of 2.5 in. and a height of 12 in. How many square inches of wrapping paper does the box require?

Chapter 5

Extra Practice 5–1 • Introduction to Equations • pages 208–211

Tell whether each equation is *true, false,* or an *open sentence.*

1. $6 = 10 - x$
2. $-8 = -1 - 7$
3. $35 \div (-7) = -5$
4. $3(5 + 7) = 38 - (3 - 5)$
5. $\frac{-4 + 12}{2} = 9 - (3 + 2)$
6. $6 + (8 - 5) = 4 - x + 3$

Which value is a solution of the equation?

7. $c - 6 = -2; 4, 8$
8. $7 + d = 7; 0, 1$
9. $k + 5 = -1; -4, -6$
10. $1 - n = 5; -4, -6$
11. $-8 = 8 + h; 0, -16$
12. $9 = g - 3; 6, 12$
13. $3x = 3; 0, -1, 1$
14. $6y - 3 = -9; -3, 1, -1$
15. $16 = 3n; 4\frac{1}{3}, 5\frac{1}{3}, \frac{3}{16}$
16. $17 = 8 - 2p; 5.5, 4.5, -4.5$
17. $\frac{c}{3} = 18; 6, 36, 54$
18. $4 + \frac{1}{3}k = 9; 5, 9, 15$

Use mental math to solve each equation.

19. $x + 5 = 8$
20. $c - 5 = 4$
21. $3 = 8 - k$
22. $7 + h = 0$
23. $-5n = 20$
24. $-7 = 35 \div d$
25. $-8 + p = -3$
26. $7 = 14 + y$
27. $c - 4 = -1$
28. $x + 8 = 3$
29. $c - 3 = 5$
30. $8 = 10 - k$

31. Tanya earned $28 more than Keisha. Tanya earned $136. Use the equation $136 - k = 28$ and these values for k: 8, 108, 154. Find k, the number of dollars Keisha earned.

Extra Practice 5–2 • Add or Subtract to Solve Equations • pages 212–215

Solve each equation. Check the solution.

1. $x - 5 = -4$
2. $c + 8 = -3$
3. $10 = f - 10$
4. $4 + g = -6$
5. $x - 3.2 = 4.7$
6. $0 = y + 6$
7. $\frac{5}{6} + n = -1$
8. $-4.7 = n - 2.5$
9. $-7 = p - 7$
10. $t + 1\frac{3}{8} = 3\frac{3}{4}$
11. $9 = c - 9$
12. $8 + k = -12$

13. $b - 4.5 = 3.5$ **14.** $0 = d + 11$ **15.** $\frac{2}{3} + n = -2$

16. $-6.3 = x - 4.6$ **17.** $-2.4 = y - 2.4$ **18.** $t + 1\frac{5}{6} = 3$

19. $n - 8 = 7$ **20.** $k - 3 = 6$ **21.** $c + 8 = 8$

22. $m + 8 = -5$ **23.** $-6 + a = 4$ **24.** $3.8 = n - 0.9$

25. $d - \frac{3}{8} = \frac{15}{8}$ **26.** $-9.2 = 9.3 + x$ **27.** $15 = -9 + r$

Solve each problem by writing an equation.

28. Selena owns seventeen CDs. This is 24 less than the number her cousin owns. How many CDs does her cousin own?

29. Marcus reads 185 pages of a novel. He reads 65 more pages than Christian has read. How many pages has Christian read?

Extra Practice 5–3 • Multiply or Divide to Solve Equations • pages 218–221

Solve each equation. Check the solution.

1. $-7x = -28$ **2.** $4.2x = -29.4$ **3.** $\frac{d}{3} = -3$

4. $23 = -4.6x$ **5.** $-7 = \frac{k}{12}$ **6.** $\frac{y}{-6} = -9$

7. $16.8 = 5.6x$ **8.** $13 = \frac{x}{-5}$ **9.** $-15 = \frac{m}{7}$

10. $372 = 4x$ **11.** $\frac{d}{-8} = -6$ **12.** $26 = \frac{x}{13}$

13. $-15y = 5$ **14.** $-15 = \frac{x}{-3.2}$ **15.** $8n = 8$

16. $36p = 27$ **17.** $\frac{w}{9} = -3.1$ **18.** $\frac{n}{-4} = \frac{5}{8}$

19. $1 = \frac{d}{7}$ **20.** $2.9x = 9.28$ **21.** $-n = 7$

22. $-n = -12$ **23.** $-1.7x = 5.61$ **24.** $1 = \frac{t}{-4}$

25. $\frac{n}{-5} = \frac{7}{15}$ **26.** $\frac{c}{8} = -12$ **27.** $-9x = -9$

28. $20y = 15$ **29.** $-20p = 15$ **30.** $-8 = \frac{k}{10}$

What keystrokes will you enter in your calculator to solve each equation?

31. $\frac{x}{14.2} = 220.4$ **32.** $0.035d = 1.4$ **33.** $-36y = 24$

Write an equation for each sentence.

34. The product of 6 and n is negative twelve.

35. The quotient of four and $-y$ is eight.

36. Negative three times x equals nine.

37. The quotient of nine and r is one.

38. A rectangular lot with an area of 1312.5 ft^2 has a length of 75 ft. What is the width? Write and solve an equation.

Solve each equation. Check the solution.

1. $\dfrac{x}{-6} + 5 = 4$

2. $\dfrac{3}{4} = 7x - 6$

3. $9 + 4d = -3$

4. $12 = -5 + \dfrac{x}{3}$

5. $-7 + 3k = 8$

6. $6 = \dfrac{y}{-4} + 3$

7. $6x - 5 = 19$

8. $5x + 4 = -11$

9. $-2m - 3 = -15$

10. $\dfrac{x}{2} + 5 = 4$

11. $3d + 4 = 7$

12. $-3x + 22 = 1$

13. $\dfrac{y}{-2} + 5 = 3$

14. $4 + \dfrac{x}{8} = 0$

15. $4 = 4n + 8$

16. $27 = \dfrac{p}{-3} + 34$

17. $3(w - 5) = 3$

18. $\dfrac{n}{4} - 11 = -2$

19. $\dfrac{3}{4} + 4d = 2$

20. $15 = 6(x + 1)$

21. $-5(n - 7) = 15$

22. $3n - 2.6 = 8.8$

23. $-2x - 1.7 = 1.5$

24. $14 = -6 + \dfrac{t}{3}$

25. A car service charges $39.50 per hour plus a basic fee of $50. The Hansons paid $247.50 for car service. For how many hours did they use the service?

26. Carol is 16 years old. Her age is 12 years greater than Pat's age divided by 3. Let P = Pat's age. Write and solve an equation relating Pat's age to Carol's.

Simplify.

1. $6a + 4b + 2a$

2. $-3x + 5x + 3x^2$

3. $-3y + y$

4. $7 + 7y + 8$

5. $-4p - 4p$

6. $-m + m$

7. $ab + bc + 3ac$

8. $8d + 16d$

9. $5a + 7a$

10. $7a + 4b - 2a$

11. $y + 3y$

12. $4k + 8k - 5$

13. $4 + 7p - 6p$

14. $x + 3x + 5x$

15. $3n - n + 3$

16. $9(c + 1) + 2(c + 3)$

17. $8j + k - 7j$

18. $-(4y + 4) + 3y - 6$

19. $9a - 7b - 4a + 9b$

20. $3d + 7d^2 + 12d + 7e$

21. $-5p + (-5p) + 5pq$

Solve each equation. Check the solution.

22. $9d + 10 = 2d - 11$

23. $4x - 2 = x + 7$

24. $8y + 5 = 7y + 6$

25. $9 + 8b = b + 23$

26. $-8 + 3n = n + 2$

27. $2y - 3 = 6y + 5$

28. $3y + 26 = -2y - 4$

29. $\dfrac{1}{2}y = 3y + 5\dfrac{1}{2}$

30. $5x = 15(x - 4) + 5(x + 1)$

Write an equation to represent each sentence.

31. Seven equals a number multiplied by two more than negative four.

32. Three times a number added to eight is twice a number minus 12.

Extra Practice

Solve.

1. A formula for the perimeter of a rectangle is $P = 2l + 2w$, where P = perimeter, l = length, and w = width. A rectangle has a length of 17 cm and a perimeter of 60 cm. Find the width.

2. Nilsa jogged $6\frac{3}{4}$ mi at an average rate of $4\frac{1}{2}$ mi/h. For how long did she jog?

3. A formula for the volume of a cylinder is $V = Bh$, where B = area of the base and h = height. The volume of a cylinder is 2720 in.^{3} The height is 17 in. Find the area of the base in square inches.

4. A formula for the interest earned on an investment is $I = prt$, where p = principal or the amount of money invested, r = rate of interest, and t = time in years that the money is invested. The interest on an investment is $84, the rate is 8%, and the amount of time is 3 years. Find the principal.

5. Al has a part-time job selling plants at a nursery. He earns a salary of $75 per week plus $2 for every ceramic planter he sells. His earnings can be computed using the formula $E = 75 + 2p$, where E = earnings and p = number of planters he sells. One week he earned $113. Find the number of planters he sold.

Solve each formula for the indicated variable.

6. $d = rt$, for t

7. $I = prt$, for r

8. $V = lwh$, for h

9. $P = 5s$, for s

10. $C = 2\pi r$, for r

11. $V = \pi r^2 h$, for h

12. $A = d_1 d_2$, for d_1

13. $A = \frac{1}{2}h(b_1 + b_2)$, for h

Graph each open sentence on a number line.

1. $x > 2$

2. $n < -5$

3. $b + 3 = 4$

4. $n \geq 0$

5. $d \leq 2$

6. $a - 2 = 2$

7. $y \geq -4$

8. $-4 = f - 5$

9. $0 > x$

10. $3 \leq y$

11. $5 > n$

12. $-4 < a$

13. $2s + 5 = -11$

14. $7h + 3 = -4$

15. $y \leq 3.5$

16. $t - 8 = -3$

Write three solutions of each inequality.

17. $x < 5$

18. $d \geq 4$

19. $y \leq -2$

20. $g > -3$

21. $m \leq -1.5$

22. $t > -2.4$

23. $d \leq 1.5$

24. $n \leq \frac{-3}{8}$

Write an inequality to describe each situation.

25. Lunch costs more than $4.95.

26. Delia's age is less than or equal to 17 years.

27. Every suit in the store is priced under $99.

28. The tickets cost at least $12.

Write an open sentence for each graph.

29.

30.

31.

32.

Extra Practice 5–9 • Solve Inequalities • pages 246–249

Solve and graph each inequality.

1. $x - 5 < -2$

2. $4c \geq -12$

3. $x + 7 \leq 5$

4. $3d > 0$

5. $\dfrac{c}{-4} \geq 1$

6. $2 < \dfrac{2}{3}y$

7. $14 + k > 19$

8. $15 \leq -3y$

9. $\dfrac{1}{6}r > -1$

10. $-5d < 0$

11. $-y < 4$

12. $t - (-4) < 1$

13. $7v + 2 > -5$

14. $3 + 8g < 27$

15. $9m + 4 \geq -32$

16. $4\left(c + \dfrac{1}{2}\right) < 2$

17. $\dfrac{5}{6}x \geq -5$

18. $-3(y + 2) \leq 3(2y - 3)$

19. $9a - 4a \leq -15$

20. $-2 \leq 3y - y + 4$

21. $17x - 1 > 67$

Write and solve an inequality.

22. Gerry needs at least \$187 for a vacation. She earns \$5.50 per hour at her job. Determine the least number of hours she must work.

23. Multiply a number by -3. Then subtract 8 from the product. The result is at most 22. Describe the number.

24. Russel needs at least 240 points on 3 tests to earn a B. On his first two tests he scored 74 and 79. What score must he get on the third test to receive a B.

Chapter 6

Extra Practice 6–1 • Percents and Proportions • pages 260–263

Write and solve a proportion.

1. 10% of 50 is what number?

2. 40% of what number if 64?

3. 3 is what percent of 20?

4. 85% of 50 is what number?

5. 44% of 125 is what number?

6. 42.5 is what percent of 50?

7. 16.5 is what percent of 55?

8. $66\dfrac{2}{3}\%$ of 30 is what number?

9. 1 is what percent of 40?

10. What percent of 40 is 10?

11. What percent of 88 is 11?

12. 14 is what percent of 20?

13. Janet received 28 out of 50 votes for class president. What percent of the vote did she receive?

14. Vince received 75% of the votes for class treasurer. What fraction of the vote did he receive?

Write and solve an equation.

1. What percent is 20 is 4?

2. 75% of 32 is what number?

3. 81 is what percent of 90?

4. 50% of 120 is what number?

5. 30% of 110 is what number?

6. What percent of 135 is 54?

7. $12\frac{1}{2}$% of what number if 5?

8. 720 is what percent of 900?

9. $33\frac{1}{3}$% of 75 is what number?

10. 20% of 25 is what number?

11. 4.5 is what percent of 300?

12. What percent of 130 is 13?

13. What percent of 85 is 51?

14. 18% of 12 is what number?

15. 15% of 80 is what number?

16. 42 is what percent of 56?

17. 90% of what number is 270?

18. 40% of what number if 64?

19. What percent of 125 is 45?

20. 24 is what percent of 24?

21. Tom wants to make a 25% down payment on his class ring that cost $185. What will his down payment be?

22. The regular price of a tennis racquet is $160. It is now on sale for 15% off its regular price. What is the amount of the savings?

The table shows Helen's monthly expenses. She earns $2000/mo.

Rent	$625
Car payment	$250
Insurance	$65
Phone	$25
Electricity	$50
Heat	$100
Food	$120

23. What percent of Helen's monthly income is spent on rent?

24. What percent of Helen's monthly income does she spend on phone, electricity and heat?

25. Helen spends $250/mo on entertainment. After paying all her bills, how much money can Helen put in savings each month?

Find the discount and the sale price. Round answers to the nearest cent.

1. Regular price: $79.95
   Discount: 15%

2. Regular price: $249.50
   Discount: 20%

3. Regular price: $525
   Discount: 8%

4. Regular price: $18.29
   Discount: 5%

5. Regular price: $995
   Discount: 12%

6. Regular price: $129.95
   Discount: 25%

7. Regular price: $43.95
   Discount: 33%

8. Regular price: $1090
   Discount: 16%

9. Regular price: $87.50
   Discount: 4%

Find the percent of discount.

10. Regular price: $632.79
    Sale price: $442.95

11. Regular price: $125
    Sale price: $111.25

12. Regular price: $59.19
    Sale price: $46.17

13. Regular price: $10
    Sale price: $8

14. Regular price: $0.75
    Sale price: $0.25

15. Regular price: $600
    Sale price: $400

Extra Practice

Find the income tax and net pay.

1. Income: $630/wk
   Income tax rate: 5%

2. Income: $16,930/yr
   Income tax rate: 8%

3. Income: $550/wk
   Income tax rate: 5.5%

4. Income: $890/wk
   Income tax rate: 15%

5. Income: $100,000/yr
   Income tax rate: 35%

6. Income: $70,000/yr
   Income tax rate: 15%

7. Income: $85,000/yr
   Income tax rate: 12.5%

8. Income: $43,000/yr
   Income tax rate: 8.5%

Find the amount of the sales tax and total cost of each item.

9. Shoes: $50
   Sales tax rate: 6%

10. Glasses: $250
    Sales tax rate: 7%

11. Umbrella: $25
    Sales tax rate: 5.5%

12. Boat: $15,000
    Sales tax rate: 7%

13. T-shirt: $22
    Sales tax rate: 5%

14. Phone: $29.99
    Sales tax rate: 7.5%

15. Coat: $78
    Sales tax rate: 8.75%

16. Car: $25,134
    Sales tax rate: 6.25%

Find the property tax paid by each owner.

17. Home value: $82,500
    Property tax rate: $2\frac{1}{2}$%

18. Home value: $60,250
    Property tax rate: 2%

19. Home value: $42,500
    Property tax rate: 3%

20. Home value: $100,500
    Property tax rate: $3\frac{1}{4}$%

21. Home value: $150,000
    Property tax rate: $3\frac{1}{2}$%

22. Home value: $250,000
    Property tax rate: 4%

23. Home value: $205,000
    Property tax rate: 2%

24. Home value: $403,000
    Property tax rate: 3.5%

Find the sales tax rate.

25. Price: $15.99
    Total cost: $17.27

26. Price: $10
    Total cost: $11

27. Price: $62
    Total cost: $65.41

28. Price: $823
    Total cost: $880.61

29. Price: $60
    Total cost: $66

30. Price: $19.99
    Total cost: $21.29

31. Price: $29.95
    Total cost: $31.30

32. Price: $54
    Total cost: $58.86

Find the interest and the amount due.

1. $350
2. $825
3. $75
4. $2200
5. $560
6. $6500
7. $11,400
8. $915
9. $400
10. $1268

Principal	Annual rate	Time	Interest	Amount due
$350	6%	2 years	■	■
$825	8%	6 months	■	■
$75	3%	12 months	■	■
$2200	5.5%	4 years	■	■
$560	9%	36 months	■	■
$6500	11%	9 years	■	■
$11,400	7.5%	3 years	■	■
$915	4%	60 months	■	■
$400	3.5%	24 months	■	■
$1268	7%	3 years	■	■

Find the rate of interest.

11. Principal: $250
    Interest: $30
    Time: 18 mo

12. Principal: $1230
    Interest: $332.10
    Time: 3 yr

13. Principal: $25,500
    Interest: $15,300
    Time: 5 yr

14. Vicki purchased a certificate of deposit for $5000. The certificate pays 5.2% simple interest per year. How much money will Vicki have in her account after 3 years?

15. Kevin borrows $15,000 for 6 years at 9.5% simple interest per year. How much does he repay the bank after 6 years?

Find the commission rate.

1. Total sale: $4500
   Commission: $270

2. Total sale: $235
   Commission: $4.70

3. Total sale: $8125
   Commission: $446.88

4. Total sale: $135,000
   Commission: $13,500

5. Total sale: $200,000
   Commission: $14,000

6. Total sale: $880
   Commission: $26.40

Find the total income.

7. Base salary: $490/wk
   Total sales: $700/wk
   Commission rate: 5%

8. Base salary: $49,000/yr
   Total sales: $50,000/yr
   Commission rate: 2%

9. Base salary: $750/wk
   Total sales: $3000/wk
   Commission rate: 6%

10. Base salary: $250/wk
    Total sales: $900/wk
    Commission rate: $3\frac{1}{2}$%

11. Base salary: $50,000/yr
    Total sales: $100,000/yr
    Commission rate: 4%

12. Base salary: $150/wk
    Total sales: $10,000/wk
    Commission rate: 7%

Find the percent of increase.

1. Original price: $10
New price: $13

2. Original rent: $550
New rent: $621.50

3. Original salary: $600
New salary: $630

4. Original dues: $0.75
New dues: $2.25

5. Original rent: $400
New rent: $424

6. Original number: 26
New number: 104

7. Original price: $9.99
New price: $12.49

8. Original price: $12.50
New price: $16.25

Find the percent of decrease.

9. Original taxes: $800
New taxes: $768

10. Original number: 20
New number: 5

11. Original price: $280
New price: $238

12. Original rent: $600
New rent: $579

13. Original price: $18
New price: $14.40

14. Original fare: $95
New fare: $57

15. Original salary: $500
New salary: $350

16. Original rent: $120
New rent: $48

17. The regular price of a baseball glove is $70. The sale price is $56. What is the percent of decrease?

18. Bill bought an antique rocking chair for $90. He sold it for $135. What was the percent of increase?

Chapter 7

Refer to the figure. Give the coordinates of the point or points.

1. point *A*

2. point *B*

3. two points on the *y*-axis

4. two points on the *x*-axis

5. three points with the same *x*-coordinate

6. three points with the same *y*-coordinate

7. a point in the second quadrant

8. a point in the fourth quadrant

9. a point whose coordinates are equal

10. a point whose coordinates are opposites

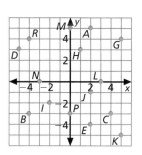

Graph each point on a coordinate plane.

11. $T(4, 2)$ **12.** $R(6, -2)$ **13.** $S(-4, 0)$ **14.** $U(0, 7)$ **15.** $N(-3, 3)$

16. $P(-3, -3)$ **17.** $K(5, 0)$ **18.** $Z(0, -1)$ **19.** $Y(3, 5)$ **20.** $O(0, 0)$

21. Sketch a rectangle on a coordinate plane with $A(4, -3)$ and $B(-4, 3)$ as endpoints of a diagonal.

22. Sketch a circle on a coordinate plane with center $C(-3, 2)$ and radius 3 units.

23. Which pair of points is closer together, $P(-4, -3)$ and $R(4, -3)$ or $G(3, 2)$ and $H(3, -5)$?

Extra Practice 7–3 • Relations and Functions • pages 314–317

Does the mapping show that the relation is a function?

1. 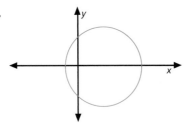   **2.** 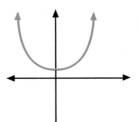   **3.**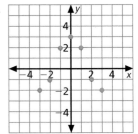

Use the vertical line test to determine if each relation is a function.

4.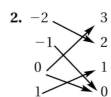

5.

6.

State the domain and range of each relation.

7. $\{(5, 2), (4, 3), (3, 4), (2, 5)\}$

8. $\{(0.4, 3.9), (1.5, 3.7), (2.4, 3.5), (1.5, 0.8)\}$

9. $\left\{\left(2, \frac{1}{2}\right), \left(-1, \frac{2}{3}\right), \left(3, \frac{2}{3}\right)\right\}$

10. $\{(4, 4), (5, 5), (-4, -4), (5, 4), (-5, -5)\}$

11. State whether each relation in Exercises 7–10 is a function.

Extra Practice 7–4 • Linear Graphs • pages 318–321

Find three solutions for each equation. Write the solutions in a table.

1. $y = x + 2$ **2.** $y = x - 2$ **3.** $y = 2x + 1$ **4.** $y = 2x - 1$

5. $y = -x + 3$ **6.** $y = -x - 1$ **7.** $y = -2x + 3$ **8.** $y = -2x - 1$

Make a table of three solutions for each equation. Then graph the equation.

9. $y = 3x$ **10.** $y = -3x + 1$ **11.** $y = \frac{1}{2}x + 1$ **12.** $y = \frac{1}{3}x - 2$

Extra Practice

Find the *x*-intercept and *y*-intercept of each equation.

13. $y = 2x + 8$ **14.** $y = -x + \dfrac{1}{2}$ **15.** $y = 3x + 2$ **16.** $y = -\dfrac{2}{3}x - 2$

17. Lois saves at the rate of $5 per week. The amount she saves is a function of the number of weeks in which she saves. Write an equation that represents this function. Then graph the function.

18. Kurt goes to the gym 4 times per week. The number of times he goes to the gym is a function of the number of weeks. Write an equation that represents this function. Then graph the function.

Extra Practice 7–5 • Slope of a Line • pages 324–327

Find the slope of each line.

1. line *AB*   **2.** line *CD*

3. line *RS*   **4.** line *TV*

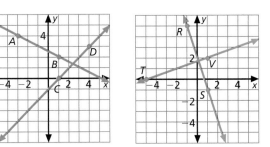

Find the slope of the line that passes through each pair of points.

5. $J(5, 6)$ and $K(6, 8)$ **6.** $N(3, 4)$ and $P(5, 10)$ **7.** $Q(4, 2)$ and $R(8, 5)$

8. $S(1, 2)$ and $T(2, -2)$ **9.** $V(5, 3)$ and $W(12, -4)$ **10.** $A(-4, -5)$ and $C(8, 2)$

11. $D(-3, 4)$ and $F(-5, 6)$ **12.** $G(1, 9)$ and $H(-8, -5)$ **13.** $E(-1, -2)$ and $B(-5, -4)$

Write an equation of each line.

14. 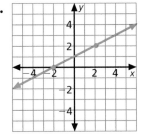   **15.** 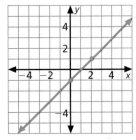   **16.**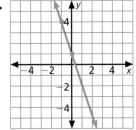

Find each slope. Convert miles to feet.

17. a hill that decreases 230 ft in 2 mi   **18.** a hill that increases 550 ft in 4.2 mi

On a coordinate plane graph the point (2, 2). Then graph the lines through (2, 2) with each slope.

19. $\dfrac{1}{2}$ **20.** $-\dfrac{1}{2}$ **21.** $\dfrac{2}{5}$ **22.** $-\dfrac{5}{2}$

Extra Practice 7–6 • Slope-Intercept Form of a Line • pages 328–331

Find the slope and *y*-intercept of each line. Then graph the line.

1. $y = 5x + 2$ **2.** $y = -\dfrac{1}{2}x - 2$ **3.** $y = -2x + 1$ **4.** $y = -x - 3$

5. $y = -\dfrac{1}{3}x + 3$ **6.** $y = -2x$ **7.** $y = 3$ **8.** $y = -\dfrac{2}{3}x - 3$

Name the slope and *y*-intercept of each line. Write an equation of the line.

9.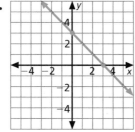

10.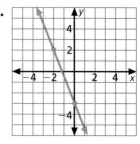

11.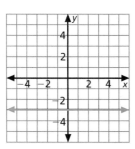

12.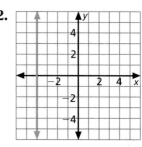

13.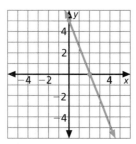

14.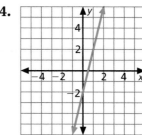

15. A phone company charges an installation fee of $65 for a phone line. Thereafter, a customer is charged $25/mo. This is represented by the equation $y = 25x + 65$, where x is the number of months a customer keeps that phone line, and y is the total amount paid. How much will a customer pay after 12 months?

Extra Practice 7–7 • Distance and the Pythagorean Theorem • pages 334–337

Graph each set of points on a coordinate plane. Then use the Pythagorean Theorem to find the distance between each pair of points.

1. $J(1, 6)$ and $K(3, 7)$
2. $N(-3, 1)$ and $P(2, 0)$
3. $Q(5, 1)$ and $R(3, 5)$
4. $S(1, 1)$ and $T(2, 2)$
5. $V(0, 0)$ and $W(2, -4)$
6. $A(4, -5)$ and $C(8, -2)$
7. $D(-3, 4)$ and $F(-5, 6)$
8. $G(1, 9)$ and $H(-8, -5)$
9. $E(-1, -2)$ and $B(-5, -4)$

Use the distance formula to find the distance between each pair of points.

10. $J(5, 6)$ and $K(6, 8)$
11. $N(3, 4)$ and $P(5, 10)$
12. $Q(4, 2)$ and $R(8, 5)$
13. $S(1, 2)$ and $T(2, -2)$
14. $V(5, 3)$ and $W(12, -4)$
15. $A(-4, -5)$ and $C(8, 2)$
16. $D(-4, 4)$ and $F(-2, 6)$
17. $G(1, 3)$ and $H(-5, -5)$
18. $E(-8, -2)$ and $B(-8, -4)$

19. A 17 ft ladder is placed 8 ft from the base of a building. How high can the ladder reach?

20. A wire at the top of a tree is attached to a hook in the ground 25 ft from the base of the tree. If the wire is 30 ft long, how tall is the tree?

Determine if the ordered pair is a solution.

1. $(-4, -12)$
$y = -\frac{1}{4}x - 11$

2. $(-1, -6)$
$y = -6x$

3. $\left(\frac{1}{2}, -3\right)$
$y = -3$

4. $(-2.5, 2)$
$y = -2x - 3$

5. $(4, -2)$

6. $(-4, -2)$

7. $(3, 7)$

8. $(3, 4)$

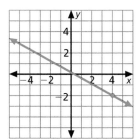 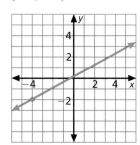 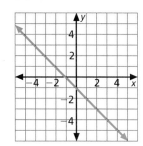 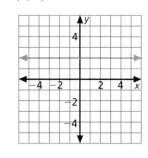

9. $(6, 2)$

10. $(-1, -1)$

11. $(4, 0)$

12. $(3, -5)$

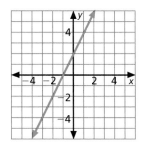 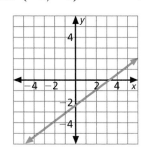 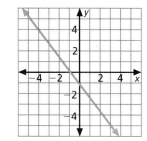 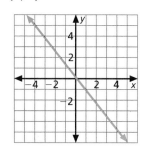

Chapter 8

Use the figure to name the following.

1. a pair of adjacent angles

2. a pair of complementary angles

3. a pair of supplementary angles

4. a pair of vertical angles

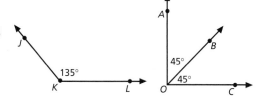

Use the figure to find the measure of each angle.

5. $\angle 2$

6. $\angle 3$

7. $\angle 4$

In the figure, $\overleftrightarrow{ST} \parallel \overleftrightarrow{XY}$. Find the measure of each angle.

8. $\angle 1$

9. $\angle 2$

10. $\angle 3$

11. $\angle 4$

12. $\angle 5$

13. $\angle 6$

14. $\angle 7$

15. $\angle 8$

16. $\angle 9$

17. $\angle 10$

18. $\angle 11$

19. $\angle 12$

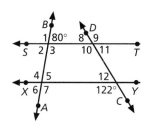

Using a protractor, construct each angle. Then bisect the angle.

1. 25°   **2.** 95°   **3.** 170°   **4.** 48°   **5.** 120°

Trace each line segment. Then construct the perpendicular bisector.

6.

7.

8.

9.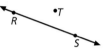

Trace each line segment. Then construct a line perpendicular to $\overleftrightarrow{RS}$ from point T.

10.

11.

12.

13.

Find the number of diagonals in each polygon. Then find the sum of the angle measures for each.

1.   **2.**   **3.**   **4.**

Find the unknown angle measure in each figure.

5.

6.

7.

8.

9.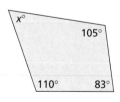

10.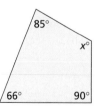

11.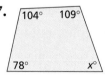

12.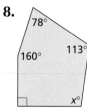

Copy each set of figures on a coordinate plane. Then graph the image of each figure under the given translation.

1. 3 units right and 1 unit up

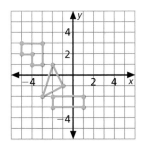

2. 3 units left and 1 unit down

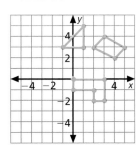

3. 2 units right and 2 units down

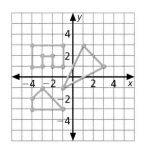

4. On a coordinate plane, graph pentagon *RSTVW* with vertices *R*(0, 1), *S*(3, 0), *T*(3, −3), *V*(−1, −4), and *W*(−2, −1). Then graph its image under a translation 2 units left and 3 units up.

5. On a coordinate plane, graph trapezoid *ABCD* with vertices *A*(−3, 1), *B*(−1, 2), *C*(−1, −2) and *D*(−3, −2). Then graph its image under a translation 5 units right and 3 units up.

Graph the point on a coordinate plane. Then graph the image of the point under a reflection across the given axis.

1. (3, 2); *x*-axis

2. (−3, 5); *x*-axis

3. (1, −4); *x*-axis

4. (−3, −3); *x*-axis

5. (4, 2); *y*-axis

6. (3, −5); *y*-axis

Graph the image of each figure under a reflection across the *y*-axis.

7.

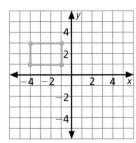

8.

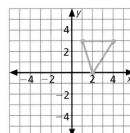

9.

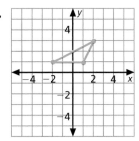

10. Graph the image of each figure in Exercises 10–12 under a reflection across the *x*-axis.

Tell whether the dashed line is a line of symmetry.

11.

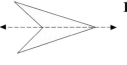

12.

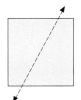

13.

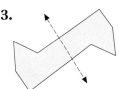

14.

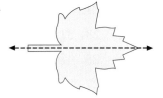

A figure has been rotated about a turn center, *T*. Describe each angle and direction of rotation.

1. 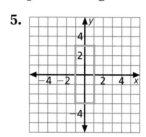   2.   3. 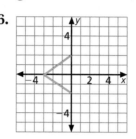   4.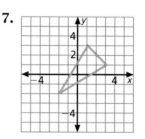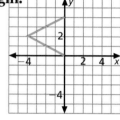

Graph the image of each figure after a 90° turn clockwise about the origin.

5.   6.   7.   8.

Chapter 9

Write each polynomial in standard form.

1. $10x^2 + 3x + 7x^5$

2. $0.8t^2 - 8 + 1.2t^3 + t^4$

3. $5y + 4y^2 + 3y^3$

4. $7x^4 + 3 + 5x + 2x^2$

5. $m + 3m^3 - 4m^2 + 11m^5$

6. $6d + 8 - 5d^4 + 4d^2 - d^8$

7. $3y + 2y^4 + 5y^3 + y^2$

8. $5z^2 + 8 + z$

9. $-4a + 3a^2 - 6a^3 + 1$

10. $3d - 4d^2 + 9$

11. $4q + 2q^2 - 3q^3 - 2$

12. $-6n + 7 + 3n^2$

Tell whether the terms in each pair are like or unlike terms.

13. $-6c, 8c^2$

14. $7d^7, 7$

15. $5kj, -6kj$

16. $-4cd^2, -6c^2d$

17. $-8hk, -16kh$

18. $9g^3, g^3$

Simplify. Be sure your answer is in standard form.

19. $-4x - 5 + 6x$

20. $f^2 + 4f - 3f^2$

21. $-5g + 8 + 9g$

22. $-3 + 6n - 8n$

23. $3k^2 + 2k^2 + k + k^2$

24. $12y^2 + 4y^2 + 8y^3 - 9y^3$

25. $4b^2 - b + 7b^2 - b^2 - 2b$

26. $2z^2 + 3z^3 + 5z^2 - 2z^3$

27. $8c^3 + 3c - c^2 - 3c^3$

28. $-5x + 3x^2 + 7x - 14x^2$

29. $-3d + 8d^2 + 7d - 11d^2 - 6$

30. $-4 + 13a^2 - 6a^3 + 5a^2$

Simplify.

1. $3a + (5a - 6)$

2. $4y + (6y + 5)$

3. $6c - (5c + 7)$

4. $8c + (-3c - 9)$

5. $(3n^2 - 2) + (4n^2 + 8)$

6. $(5y + 3) - (8y - 4)$

7. $(4x^2 + 2) + (-9x^2 - 4)$

8. $(3z^2 - 4z + 2) + (2z^2 + 7z - 6)$

9. $(b^2 - 3b + 2) + (b^2 - 3)$

10. $(5k) - (-3k)$

11. $3y - (3y + 5)$

12. $8e - (3e + 6)$

13. $(6x - 3) - (4x + 5)$

14. $(8 - 3y) - (4 - 5y)$

15. $(a^2 - 4a + 9) - (a^2 + 6a - 5)$

16. $(6q^2 + 4) - (3q^2 + 4q - 7)$

17. $(3b^2 - 4b + 6) - (b^2 - 2b - 3)$

18. $(d^2 + 3d - 6) + (d^2 - 8d - 9)$

19. $(5a^2 + 3) - (4a^2 - 2a - 6)$

20. $12r^3 + 3r^2 - r) - (-8r^3 + 2r^2 + 5r)$

21. The sixth grade class collected $3a + 4b$ cans and the seventh grade class collected $9a - 3b$ cans. Write an expression for the total cans collected.

Simplify.

1. $(3x)(4y)$

2. $(-2a)(-5b)$

3. $(6y)(-6z)$

4. $(8s)(-4t)$

5. $(-8c)\left(\dfrac{1}{4} b\right)$

6. $(-3a)(-7b)$

7. $(-4q)(3.6p)$

8. $(3x)(5.4y)$

9. $(4y^3)^2$

10. $(3z)^2$

11. $(6a^2)(-3a^3)$

12. $(9c)(c^5)$

13. $(-4e^3)(-3e^2)$

14. $(7d^2)(-5d)$

15. $(-8p^4)(6p^3)$

16. $(6f^3)(-4f^4)$

17. $(2c^2)^3$

18. $(5b)^2$

19. $(-4x^3)^2$

20. $(-3y^3)^3$

21. $(6y^2)^3$

22. $(2x^3)(-5x^5)$

23. $(8z)(-7z^4)$

24. $(-9r^6)(6r^2)$

Write an expression for the area of each figure. Then simplify the expression.

25.

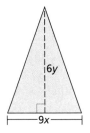

26.

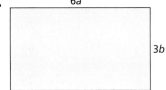

27.

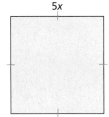

Simplify.

1. $3a(a + 2a^2)$

2. $-5(-x + 3x^2)$

3. $-2(y^2 - 6y)$

4. $8c(-4c + 3)$

5. $4b(2b + 3)$

6. $6z(5 - 3z)$

7. $-4d(2d^2 - d - 3)$

8. $3(2y^2 + y - 4)$

9. $-5(g^2 - 3g - 6)$

10. $-2(t^2 - 4t - 9)$

11. $-j(j^2 - 5j + 4)$

12. $l(3 - l - 2l^2)$

13. $2m(m^2 - 3m - 4)$

14. $-n(2n^3 - n^2 + 3n)$

15. $3p^2(-2p^2 - p + 6)$

16. $3q(2q^2 - 4q - 5)$

17. $w(3w^2 - w + 5) - w(w^3 + 5w^2 - 6w)$

18. Write an expression for the area of the rectangle. Then simplify the expression.

19. Let n represent an odd integer. Then $n + 2$ is the next odd integer. Write an expression for the product of the two integers.

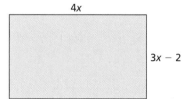

4x

3x − 2

Find the greatest common factor of the monomials.

1. $6a$ and $3ab^2$

2. $4x^2$ and $12xy$

3. $7cd^2$ and $21c^2d$

4. $8p^2q$ and $9pq$

5. $16m^2n$ and $20m^3n^2$

6. $12s^2t^2$ and $36s^2t$

Match each set of monomials with their greatest common factor.

7. $3a^2, 3a^2d, 3a^2d$ **a.** $4ac$

8. $4ac, 4acd, 8a^2c$ **b.** $3a^2c$

9. $15ac, -12a^2c, 6ac^2$ **c.** $3a^2$

10. $-4a^3c, -a^2c^2, -4ac^2$ **d.** $3ac$

11. $ac^2, 5bc^2, 4c^2e, 3c^2d$ **e.** ac

12. $-9a^2c^3, 12a^2c^2, -15a^3c$ **f.** c^2

Factor each polynomial.

13. $24x - 6xy$

14. $5 - 15a$

15. $6b^2 - 5b$

16. $18cd - 9d^2$

17. $20rs - 16r^2s^2$

18. $32d^2 - 8de$

19. $4m^2 + 8m^2n - 24mn$

20. $st + st^2 - t^2$

21. $3xy - 6x^2y + 12xy^2$

22. $9cd - 18c^2d^2 - 27cde$

23. $4x(-2 + x) + 7x(-2 + x)$

24. $(x - 5)x^3 - (x - 5)x$

25. $r^2(r + 6) + r(r + 6) + 3(r + 6)$

26. $t^3(t - 1) - t^2(t - 1) + t(t - 1) + 2(t - 1)$

Simplify.

1. $\dfrac{24ab}{6a}$

2. $\dfrac{-12cd}{4d}$

3. $\dfrac{50st}{25s}$

4. $\dfrac{-18xy}{3x}$

5. $\dfrac{-32mn}{4n}$

6. $\dfrac{-15rs}{-5s}$

7. $\dfrac{a^3b}{a^2c}$

8. $\dfrac{c^3de^2}{c^2e}$

9. $\dfrac{gh^5}{h^4}$

10. $\dfrac{6p^3q^3}{-3pq}$

11. $\dfrac{st^6}{t^3}$

12. $\dfrac{-a^4b^4}{a^2b^3}$

13. $\dfrac{3a+6}{3}$

14. $\dfrac{24b+6}{2}$

15. $\dfrac{6y-12z}{3}$

16. $\dfrac{9x-15}{3}$

17. $\dfrac{-18c+27}{9}$

18. $\dfrac{12c-18d}{3}$

19. $\dfrac{24p-6q-2r}{2}$

20. $\dfrac{8a^3-12a^2+4a}{4a}$

21. $\dfrac{27b^2c^2-18bc^2+36bc}{9bc}$

22. $\dfrac{8x^2y^2z^2-12xyz^2}{4xyz}$

23. A rectangle has an area of $36a^3b^2$ and length of $9a^2b$. Write an expression for the width.

24. The product of $4x^3y^2$ and a certain monomial is $24x^6y^4z^9$. Find the missing factor.

Chapter 10

Find each probability. Give your answer as a fraction and a percent. Use the spinner.

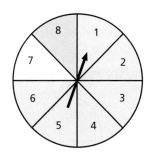

1. P(purple)

2. P(2 or 3)

3. P(odd number)

4. P(even number)

5. P(not green)

6. P(9)

7. P(not white)

8. P(8)

9. P(number ≥ 1)

Use the spinner to find the odds in favor of each event.

10. 3 or 4

11. green

One of the cards is drawn without looking. Find each probability.

12. P(consonant)

13. P(E)

14. P(T or N)

15. P(S)

Marcus tossed a paper cup 60 times. With each toss he recorded whether the cup landed up, down, or on its side. Use the results in the table to find each experimental probability.

Outcome	Tally	Total
Up	𝍲𝍲𝍲𝍲	20
Down	𝍲𝍲𝍲 |	16
Side	𝍲𝍲𝍲𝍲 ||||	24

1. P(up)
2. P(up or side)
3. P(side)
4. P(not up)
5. P(down)
6. P(up or down)

The table shows the number of pets belonging to one group of third grade students. Find each experimental probability.

Dog	Cat	Guinea pig	Bird	Hamster	Other
18	12	8	3	9	10

7. P(dog)
8. P(guinea pig)
9. P(cat)
10. P(not cat)
11. P(bird or hamster)
12. P(not dog)

13. Sasha tossed a baseball card 30 times. It landed faceup 12 times. Find the experimental probability that a tossed baseball card will not land faceup.

For Exercises 1–9, use the cards, Spinner 1, and Spinner 2. List each sample space.

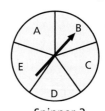

Spinner 1

Spinner 2

1. picking a card from those shown
2. spinning Spinner 1

Use a tree diagram to find the number of possible outcomes in each sample space.

3. picking a card from those shown above and tossing a coin

4. tossing a coin and spinning Spinner 1
5. tossing a coin and spinning Spinner 2

Suppose you spin both Spinner 1 and Spinner 2. Find each probability.

6. P(2 and B)
7. P(4 and vowel)
8. P(even number and A)
9. P(odd number and consonant)

Use the counting principle to find the number of possible outcomes.

1. choosing lunch from 4 sandwiches, 3 drinks, and 5 deserts

2. choosing an outfit from 2 jackets, 3 pairs of pants, and 4 shirts

3. choosing a pizza with tomato, pesto, or cheese sauce and a topping of mushroom, meat, or eggplant

4. tossing a coin five times

5. spinning a spinner marked 1, 2, 3, 4 three times

6. spinning a spinner marked A, B, C, D, E two times

7. choosing a dessert from vanilla, chocolate, or strawberry frozen yogurt and a topping from fruit, nuts, granola, or sprinkles

8. choosing a letter from A, B, C, D and a number from 1, 2, 3, 4

Use the counting principle to find the probability.

9. A coin is tossed five times and a number cube is tossed. Find *P*(all tails and a multiple of 3).

10. A coin is tossed two times and a number cube is tossed. Find *P*(all heads and odd number).

11. A letter is chosen from A, B, C, D, E and a number is chosen from 1, 2, 3, 4. Find *P*(the letter B and a number less than 5).

A game is played by spinning the spinner and then choosing a card without looking. After each turn the card is replaced. Find the probability of each event.

1. *P*(1, then B)   2. *P*(2, then E)

3. *P*(1, then C)   4. *P*(1, then A)

5. *P*(3, then E)   6. *P*(3, then B)

7. *P*(2, then D)   8. *P*(2, then B)

9. *P*(3, then C)   10. *P*(2, then A)

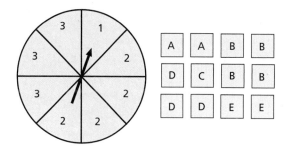

Ten cards are labeled from A to J and placed in a box. A card is taken from the box and is not replaced. Then a card is taken from the box again. Find the probability of each event.

11. *P*(B, then B)

12. *P*(A, then vowel)

13. *P*(C, then consonant)

14. *P*(E, then J)

1. A sample space has twelve equally likely outcomes. The payoffs for six of the outcomes is $4. The payoffs for six of the outcomes is $7. What is the expected value of the sample space?

2. A sample space has five equally likely outcomes. The payoffs for the outcomes are 1, 2, 3, 4 and 5. What is the expected value of the sample space?

3. A sample space has ten equally likely outcomes. The payoffs for four of the outcomes is 3. The expected value of the sample space is 12. What is the payoff for the tenth outcome?

4. Martin and Monika flip a coin 3 times. If one or two coins land tails up, then Martin scores 2 point. Otherwise, Monika scores 3 points. What is the expected value of the game?

5. Using a standard deck of cards, what is the expected value of drawing a king of hearts?

6. A school club sells calendars to raise money. The Calendar Company states that the probability of making $4000 is 0.75, and the probability of losing $2000 is 0.25. The Calendar Company claims that on average, schools make over $3000. Is the claim accurate?

Suppose a charity raffles off 10,000 tickets for a $1000 sound system. Each ticket costs $1.

7. What is the expected value for the purchase of one ticket?

8. Is $1 a fair price to pay for one ticket? Explain.

Chapter 11

Extra Practice 11–1 • Optical Illusions • pages 478–481

1. Write a statement about $\overline{AB}$ and $\overline{CD}$ that is suggested by the two figures.

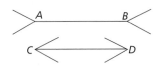

2. Write a statement about a circle that is suggested by four sets of curves.

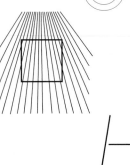

3. Write a statement about the size of circles P and Q that is suggested by the drawing.

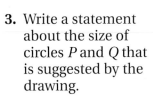

4. Write a statement about the perfect square that is suggested by the drawing.

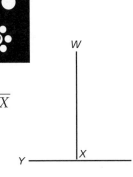

5. Write a statement about $\overline{WX}$ and $\overline{YZ}$ that is suggested by the picture.

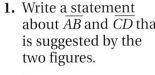

6. Write a statement about the length of the two figures suggested by the picture.

1. Use inductive reasoning to find the ones digit of any even-numbered power of 3.

2. Use the inductive reasoning to find the ones digit of any odd-numbered power of 3.

Use the given examples to complete each conjecture.

3. $15 - 9 = 6$ $23 - 13 = 10$ $47 - 19 = 28$ $31 - 19 = 12$

   The difference between the two odd numbers is an __?__ number.

4. $66 - 28 = 38$ $44 - 16 = 28$ $72 - 58 = 14$ $60 - 34 = 26$

   The difference between two even numbers is an __?__ number.

5. $19 + 32 = 51$ $47 + 16 = 63$ $67 + 22 = 89$ $25 + 66 = 91$

   The sum of an odd number and an even number is an __?__ number.

6. $19 + 35 = 54$ $27 + 23 = 50$ $71 + 13 = 84$ $63 + 29 = 92$

   The sum of two odd numbers is an __?__ number.

Examine each sequence of numbers. Give a pattern for the sequence and give the next two numbers.

7. 1, 2, 6, 24, 120, 720, . . .   8. 7, 4, 8, 5, 10, 7, 13, . . .   9. 1, 3, 6, 10, 15, . . .

10. Write any 3-digit number using 3 different digits. Reverse the digits and subtract the lesser number from the greater one. Add the first and last digits. What is their sum? Make a conjecture about the number that results from this process.

1. Write two conditional statements that can be made from these two sentences:

   The number is a prime. The number has exactly 2 factors.

2. Is the conditional statement true or false? If false, give a counterexample.

   a. If two numbers are even, then their quotient is even.

   b. If a number is divisible by 8, then it is divisible by 4.

Is the argument valid or invalid? Use a picture to help you decide.

3. If you have a textbook, then you will do your homework.
   You have your textbook.
   Therefore, you will do your homework.

4. If you are going shopping, then you walk to the store.
   You walk to the store.
   Therefore, you are going shopping.

5. If you forget your tickets, you cannot get on the plane.
   You did not get on the plane.
   Therefore, you forgot your tickets.

6. If the figure is a square, then it has four equal sides.
   The figure is a square.
   Therefore, the figure has four equal sides.

Use the Venn diagram to answer the following questions.

1. How many students wore a jacket to school?

2. How many students wore a jacket and a hat to school?

3. How many students wore a sweatshirt but not a jacket?

4. How many students wore a sweatshirt, jacket, and a hat to school?

The results of a poll focused on 150 mathematics, science, and U.S. History teachers are shown.

5. Make a Venn diagram for the survey.

6. How many teachers teach only mathematics?

7. How many teachers teach mathematics or science?

8. How many teachers teach mathematics and science but not U.S. History?

9. How many teachers teach only U.S. History?

10. How many teachers teach science and history but not mathematics?

Number of teachers	Subject taught
30	Mathematics and science
18	Science and U.S. history
12	Mathematics and U.S. history
10	All three subjects
68	Science
65	U.S. history

What conclusions can be made from the following statements?

1. Peter likes only chocolate ice cream.

   Peter is eating ice cream.

2. I have karate either on Wednesdays or Fridays.

   The karate studio is closed on Fridays.

3. Either Matt or Mike ate lunch at home today.

   Mike ate lunch in the cafeteria today.

Make a table to solve each problem.

4. Three brothers are studying math this semester. Their math books are titled: Algebra 1, Algebra 2, and Grade 5 Math. Tim passed Algebra 1 last year. Kelly is older than Paul. Who studies from which book?

5. My three sisters each have one son, but I have forgotten each son's name. I remember that their names are Bill, Omar and Trey. At Trey's birthday party I heard Franny say that she takes her son and Bill to baseball practice on Tuesday. Betty said Omar came over to play with her son last week. The names of Alice's children begin with vowels. Who is mother to whom?

1. How many different amounts of money can be made from four 1-dollar bills and three 5-dollar bills?

2. A 100-page book is being numbered by hand. How many times will the digit 3 be written?

3. If today is Wednesday, what day of the week will it be 200 days from today?

4. A train leaves Canon and travels 45 miles east toward Salt Rock. At that point it is $\frac{1}{4}$ of the way to its destination. How many miles is a round trip from Canon to Salt Rock and back?

5. A pair of in-line skates and a carrying case cost $148. The skates cost $19 more than the bag. How much does the bag cost?

6. If 3 lemons weigh as much as an apple and an apple weighs as much as 12 figs, how many figs weigh as much as a lemon?

For Exercises 7–9, suppose you have a 6 in. cube.

7. How many cuts would it take to cut the cube into 2 in. cubes?

8. How many cubes would result?

9. How many 4 in. cubes could you make using the 2 in. cubes?

Preparing for Standardized Tests

Becoming a Better Test-Taker

At some time in your life, you will probably have to take a standardized test. Sometimes this test may determine if you go on to the next grade level or course, or even if you will graduate from high school. This section of your textbook is dedicated to making you a better test-taker.

TYPES OF TEST QUESTIONS   In the following pages, you will see examples of four types of questions commonly seen on standardized tests. A description of each type is shown in the table below.

Type of Question	Description	See Pages
multiple choice	Four or five possible answer choices are given from which you choose the best answer.	772–775
gridded response	You solve the problem. Then you enter the answer in a special grid and shade in the corresponding circles.	776–779
short response	You solve the problem, showing your work and/or explaining your reasoning.	780–783
extended response	You solve a multi-part problem, showing your work and/or explaining your reasoning.	784–788

PRACTICE   After being introduced to each type of question, you can practice that type of question. Each set of practice questions is divided into five sections that represent the concepts most commonly assessed on standardized tests.

- Number and Operations
- Algebra
- Geometry
- Measurement
- Data Analysis and Probability

USING A CALCULATOR   On some tests, you are permitted to use a calculator. You should check with your teacher to determine if calculator use is permitted on the test you will be taking, and if so, what type of calculator can be used.

TEST-TAKING TIPS   In addition to the Test-Taking Tips like the one shown on the right, here are some additional thoughts that might help you.

- Get a good night's rest before the test. Cramming the night before does not improve your results.
- Budget your time when taking a test. Don't dwell on problems that you cannot solve. Just make sure to leave that question blank on your answer sheet.
- Watch for key words like NOT and EXCEPT. Also look for order words like LEAST, GREATEST, FIRST, and LAST.

> **Test-Taking Tip**
> If you are allowed to use a calculator, make sure you are familiar with how it works so that you won't waste time trying to figure out the calculator when taking the test.

Multiple-Choice Questions

Multiple-choice questions are the most common type of question on standardized tests. These questions are sometimes called *selected-response questions.* You are asked to choose the best answer from four or five possible answers.

To record a multiple-choice answer, you may be asked to shade in a bubble that is a circle or an oval or to just write the letter of your choice. Always make sure that your shading is dark enough and completely covers the bubble.

To make sure you have the correct solution, you must check to make sure that your answer satisfies the conditions of the original problem.

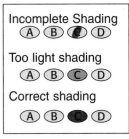

Example 1

Nathan wants to save $500 for a season ski pass. He has $200 and can save $25 per week. About how many months will he need to save in order to have enough money?

(A) 20 mo

(B) 15 mo

(C) 12 mo

(D) 8 mo

(E) 3 mo

Notice that the problem gives the amount that Nathan can save each *week,* but asks for the time it will take to save the money in *months.* Since Nathan can save $25 per week and there are about 4 weeks in a month, he can save about $4 \cdot \$25$ or $100 each month.

Let x represent the number of months. So, the amount he will have after x months is $200 + 100x$. Since Nathan wants to save $500, set the expression equal to 500. Write the equation $500 = 200 + 100x$.

Next, test each value given in the answer choices.

$500 = 200 + 100x$
$500 \stackrel{?}{=} 200 + 100(20)$ Replace x with 20.
$500 \neq 2200$

$500 = 200 + 100x$
$500 \stackrel{?}{=} 200 + 100(15)$ Replace x with 15.
$500 \neq 1700$

$500 = 200 + 100x$
$500 \stackrel{?}{=} 200 + 100(12)$ Replace x with 12.
$500 \neq 1400$

$500 = 200 + 100x$
$500 \stackrel{?}{=} 200 + 100(8)$ Replace x with 8.
$500 \neq 1000$

$500 = 200 + 100x$
$500 = 200 + 100(3)$ Replace x with 3.
$500 = 500$ ✓

The answer is E.

Many multiple-choice questions do not include a diagram. Drawing a diagram for the situation can help you to answer the question.

Example 2

Isabelle and Belinda take a hiking trip. They want to get to Otter's pond but cannot walk directly to it. They start from the beginning of the trail and follow the trail for 4 mi to the west. Then they turn south and walk 6 mi. How far is Otter's Pond from the start of the trail? Round to the nearest tenth of a mile.

Ⓐ **1.4 mi**   Ⓑ **4.5 mi**   Ⓒ **6.3 mi**   Ⓓ **7.2 mi**

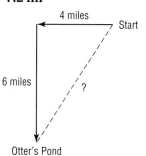

Draw a diagram of the hiking trip. Isabelle and Belinda have walked in a path that creates a right triangle.

Looking at the diagram, you can eliminate 1.4 mi since it is too small.

The legs of the right triangle are 4 and 6. Use the Pythagorean Theorem to find the hypotenuse.

$$c^2 = a^2 + b^2 \quad \text{Pythagorean Theorem}$$
$$c^2 = 6^2 + 4^2 \quad \text{Replace } a \text{ with 6 and } b \text{ with 4.}$$
$$c^2 = 36 + 16 \quad \text{Evaluate } 6^2 + 4^2.$$
$$c^2 = 52 \quad \text{Add 36 and 16.}$$
$$\sqrt{c^2} = \sqrt{52} \quad \text{Take the square root of each side.}$$
$$c \approx 7.2 \quad \text{Round to the nearest tenth.}$$

The answer is D, 7.2 miles.

Multiple-choice questions may require you to convert measurements to solve. Pay careful attention to each unit of measure in the question and the answer choices.

Example 3

Malik is planning to draw a large map of his neighborhood for a school project. He wants the scale for the map to be 1 in. = 8800 ft. His house is 2.5 mi from school. How far on the map will his house be from the school?

Ⓐ **0.5 in.**   Ⓑ **0.8 in.**   Ⓒ **1.0 in.**   Ⓓ **1.5 in.**   Ⓔ **1.8 in.**

The actual distance from Malik's house to the school is given in miles. You need to convert from miles to feet to solve the problem. Since 1 mi is equal to 5280 ft, 2.5 mi = 2.5(5280) or 13,200 ft.

Now write a proportion to find the distance.

$$\frac{1 \text{ in.}}{8800 \text{ ft}} = \frac{x}{13{,}200 \text{ ft}}$$
$$1 \cdot 13{,}200 = x \cdot 8800$$
$$\frac{13{,}200}{8800} = x$$
$$1.5 = x$$

On the map, Malik's house will be 1.5 in. from the school. D is correct.

Multiple-Choice Practice

Choose the best answer.

Number and Operations

1. In 2001, the population of China was approximately 1,273,000,000. Write the population in scientific notation.

   (A) 12.73×10^1 (B) 1273.0×10^6
   (C) 1.27×10^9 (D) 1.273×10^9

2. Tyler uses 4 gal of stain to cover 120 ft² of fence. He still has 520 ft² left to cover. Which proportion could he use to calculate how many more gallons of stain he will need?

   (A) $\dfrac{120}{520} = \dfrac{x}{4}$ (B) $\dfrac{4}{120} = \dfrac{x}{520}$

   (C) $\dfrac{120}{x} = \dfrac{520}{4}$ (D) $\dfrac{4}{520} = \dfrac{120}{x}$

3. If 1.5 c of nuts are in a bag of trail mix that serves 6 people, how many cups of nuts will be needed for a trail mix that will serve 9 people?

   (A) 1 c   (B) 1.5 c
   (C) 2.25 c   (D) 2.5 c

4. Hannah has a roll of ribbon for wrapping presents that is 10 yd long. If it takes $\dfrac{2}{3}$ yd of ribbon to wrap each present, how many presents can she wrap with the 10 yards?

   (A) 8
   (B) 10
   (C) 15
   (D) 20

Algebra

5. A rental store charges a down-payment of $100 and $75/mo to rent a television. Ms. Blackwell paid $550 to rent the television. How many months did she rent the television?

   (A) 4 mo
   (B) 6 mo
   (C) 8 mo
   (D) 10 mo

6. Six friends go to a movie and each buys a large container of popcorn. If a movie ticket costs $8.75 and a large popcorn costs $2.25, which expression can be used to find the total cost for all six people?

   (A) $2.25(8.75 + 5)$ (B) $2.25 + 8.75(6)$
   (C) $6(8.75 + 2.25)$ (D) $6(2.25) + 8.75$

7. Midtown Printing Company charges $50 to design a flyer and $0.25 per flyer for printing. If y is the total cost in dollars and x is the number of fliers, which equation describes the relationship between x and y?

   (A) $y = 50 - 0.25x$ (B) $y = 50x + 0.25$
   (C) $y = 50 + 0.25x$ (D) $y = 0.25x - 50$

8. The simple interest formula, $I = Prt$, gives the interest I earned for an amount of money invested P at a given rate r for t years. If Nicholas invests $2100 at an annual interest rate of 7.5%, how long will it take him to earn $3000? Round to the nearest year.

   (A) 3 yr
   (B) 19 yr
   (C) 20 yr
   (D) 52 yr

Geometry

9. Alyssa wants to redecorate her room. She makes a scale diagram of her room on paper. She then cuts out scale pictures of her bed, dresser, and desk. If she slides her dresser along the wall, what type of transformation is this?

   (A) reflection
   (B) rotation
   (C) dilation
   (D) translation

Test-Taking Tip

Question 8
Some multiple-choice questions have you use a formula to solve a problem. You can check your solution by replacing the variables with the given values and your answer. The answer choice that results in a true statement is the correct answer.

10. To get to work from her house, Amanda walks 6 blocks west then turns and walks 11 blocks south. If she could walk directly home from work, how many blocks would she need to walk? Round to the nearest block.

   Ⓐ 9 blocks

   Ⓑ 12 blocks

   Ⓒ 13 blocks

   Ⓓ 15 blocks

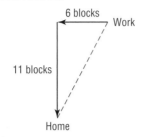

11. Andrés plans to lay sod in his yard. How many square feet of sod will he need?

   Ⓐ 5100 ft^2

   Ⓑ 8925 ft^2

   Ⓒ 9000 ft^2

   Ⓓ 9650 ft^2

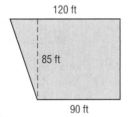

Measurement

12. The world's largest ball of Sisal Twine is located in Cawker City, Kansas. It consists of about 7,009,942 ft of twine. How many miles of twine is this?

   Ⓐ 586,080 mi   Ⓑ 1328 mi

   Ⓒ 1402 mi   Ⓓ 37,012,493,760 mi

13. The world's largest pizza was made on October 11, 1987 by Lorenzo Amato and Louis Piancone. The pizza measured 140 ft across. If a regular size pizza at a local restaurant measures 12 in. across, how many times more area does the largest pizza cover than the regular pizza?

   Ⓐ about 12 times   Ⓑ about 136 times

   Ⓒ 140 times   Ⓓ 19,600 times

14. A cookie recipe requires 6 c of chocolate chips. The recipe serves 20 people. How many cups of chocolate chips would be needed to serve 35 people?

   Ⓐ 10.5 c   Ⓑ 11 c

   Ⓒ 12 c   Ⓓ 15 c

15. Shopper's Mart sells two sizes of Corn Crunch cereal. The 16-oz box costs $4.95. The 12-oz box costs $3.55. What is true about these two cereals?

   Ⓐ The 16-oz box is a better buy.

   Ⓑ The 12-oz box is a better buy.

   Ⓒ They are the same cost per ounce.

   Ⓓ None of these statements are true.

Test-Taking Tip

Question 15
Always read every answer choice, particularly in questions that ask what is true about a given situation.

Data Analysis and Probability

16. The graph shows the number of hours that students in Mr. Cardona's math class watch television and the number of hours that they exercise in the same week. Based on the trend in the scatter plot, what number of hours of exercise would you expect a student to get that watches 15 h of television each week?

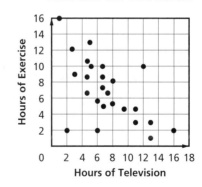

   Ⓐ 2 h   Ⓑ 6 h

   Ⓒ 10 h   Ⓓ 15 h

17. Mika received the following scores on the first four social studies tests: 89, 75, 82, and 77. By how many points will his average increase if he earns a 100 on the next test.

   Ⓐ 4   Ⓑ 20

   Ⓒ 25   Ⓓ 85

Gridded-Response Questions

Gridded-response questions are another type of question on standardized tests. These questions are sometimes called *student-produced response* or *grid in.*

For gridded response, you must mark your answer on a grid printed on an answer sheet. The grid contains a row of four or five boxes at the top, two rows of ovals or circles with decimal and fraction symbols, and four or five columns of ovals, numbered 0–9. At the right is an example of a grid from an answer sheet.

Example 1

Solve the equation $10 - 4a = -38$ for a.

What value do you need to find?

You need to find the value of a.

$$10 - 4a = -38 \qquad \text{Original equation}$$

$$10 - 4a - 10 = -38 - 10 \qquad \text{Undo addition.}$$

$$-4a = 248 \qquad \text{Simplify.}$$

$$\frac{-4a}{-4} = \frac{-48}{-4} \qquad \text{Undo multiplication.}$$

$$a = 12 \qquad \text{Simplify.}$$

How do you fill in the answer grid?
- Print your answer in the answer boxes.
- Print only one digit or symbol in each answer box.
- Do not write any digits or symbols outside the answer boxes.
- You may print your answer with the first digit in the left answer box, or with the last digit in the right answer box. You may leave blank any boxes you do not need on the right or the left side of your answer.
- Fill in only one bubble for every answer box that you have written in. Be sure not to fill in a bubble under a blank answer box.

Many gridded-response questions result in an answer that is a fraction or a decimal. These values can also be filled in on the grid.

Example 2

A recipe for orange chicken calls for 1 c of orange juice and serves 6 people. If Emily needs to serve 15 people, how many cups of orange juice will she need?

What value do you need to find?

You need to find the number of cups of orange juice Emily will need for 15 servings.

Write and solve a proportion for the problem. Let s represent the number of cups.

$$\frac{1\text{ c}}{6\text{ servings}} = \frac{s\text{ c}}{15\text{ servings}}$$

$\dfrac{1}{6} = \dfrac{s}{15}$ — Write the proportion.

$15 = 6s$ — Find the cross products.

$\dfrac{15}{6} = \dfrac{6s}{6}$ — Divide each side by 6.

$\dfrac{5}{2} = s$ — Simplify.

How do you fill in the answer grid?

You can either grid the fraction $\dfrac{5}{2}$, or rewrite it as 2.5 and grid the decimal. Be sure to write the decimal point or fraction bar in the answer box. The following are acceptable answer responses that represent $\dfrac{5}{2}$ and 2.5.

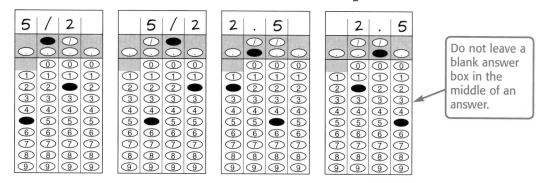

Do not leave a blank answer box in the middle of an answer.

Some problems may result in an answer that is a mixed number. Before filling in the grid, change the mixed number to an equivalent improper fraction or decimal. For example, if the answer is $1\dfrac{1}{2}$, do not enter 11/2, as this will be interpreted as $\dfrac{11}{2}$. Instead, enter 3/2 or 1.5.

Example 3

The Corner Candy Store sells 16 chocolates in a gift box. For Mother's Day, the store offers the chocolates in boxes of 20. What is the percent of increase?

Write the ratio for percent of increase.

$\text{percent of increase} = \dfrac{\text{amount of increase}}{\text{original amount}} \cdot 100$

Remember percent of increase has the original amount as the denominator.

$= \dfrac{20 - 16}{16} \cdot 100$ — Substitution

$= \dfrac{4}{16} \cdot 100$ — Subtraction

$= 0.25 \cdot 100$ — Rewrite as a decimal.

$= 25\%$

Since the question asks for the percent, be sure to grid 25, not 0.25.

Gridded-Response Practice

Solve each problem. Then copy and complete a grid like the one shown on page 592.

Number and Operations

1. The Downtown Department Store has a 75% markup on all clothing items. A certain sweater cost the store $20. What will be the selling price of the sweater in dollars?

2. The table shows the number of billionaires in the following countries in 2000.

Country	Number of Billionaires
USA	70
Germany	18
Japan	12
China	8
France/Mexico/Saudi Arabia	7

   What percent of the billionaires were in the USA? Round to the nearest percent.

3. Pepperoni is the most popular pizza topping. Each year approximately 251,770,000 lb of pepperoni are eaten. If this number were written in scientific notation, what would be the power of 10?

4. Dylan wants to buy an ice cream cone with three different flavors of ice cream. He has 18 flavors from which to choose. How many different ways can he have his cone?

5. Ignacio has $40 to spend at the mall. He spends half of it on a CD. He then spends $6 for lunch. Later he decides to go to a movie that costs $5.75. How much money does he have left in dollars?

6. There are 264 tennis players and 31 coaches at a sports camp. What is the ratio of tennis players to coaches as a decimal rounded to the nearest tenth?

Algebra

7. Find the x-intercept of the graph of the equation $2x + y = 5$.

8. Solve $4x - 5 = 2x + 3$ for x.

9. Ayana has $125 dollars to spend on CDs and DVDs. CDs cost $20. The equation $y = 125 - 20c$ represents the amount she has left to buy DVDs. If she buys 3 CDs, how much money in dollars will she have to buy DVDs?

10. For her birthday, Allison and her five friends went out for pizza. They ate an entire pizza that was cut into 16 pieces. If two of her friends ate 4 pieces, one of her friends ate 3 pieces and two of her friends ate 1 piece, how many pieces did Allison eat?

11. A parking garage has two different pay parking options. You can pay $11 for the day or $2 for the first hour and $0.75 for each additional half hour. How many hours would you need to stay for both rates to be the same?

Geometry

12. The bridge over the Hoover Dam in Lake Mead, Nevada, stretches 1324 ft. A model is built that is 60 ft long. What is the ratio of the length of the model to the length of the actual bridge? Round to the nearest hundredth.

13. Use the figure to find the value of x.

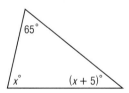

14. Quadrilateral $ABCD$ is translated 2 units to the right and 3 units down to get $A'B'C'D'$. What is the y-coordinate of A'?

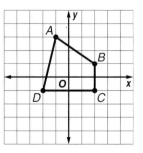

15. Toshiro swam across a river that was 20 ft wide. When he got across the river, the current had pushed him 40 ft farther down stream than when he had started. How far did he travel? Round to the nearest tenth of a foot.

16. A building casts a shadow that is 210 ft long. If the angle from the end of the shadow to the top of the building is 40°, what is the measure of the third angle?

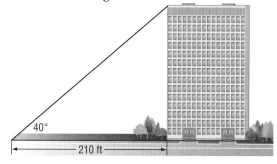

Measurement

17. The Student Council held a car wash for a fund-raiser. On Saturday, approximately 8 containers of car wash were used to wash 52 cars. Assuming the rate stayed the same, how many cars did they wash if they used 11 containers on Sunday?

18. Ethan wants to know how much water his sister's swimming pool holds. If the pool is 1 ft high and 6 ft across, what is the volume in cubic feet? Use 3.14 for π and round to the nearest tenth of a cubic foot.

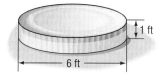

19. Marisa needs to make tablecloths for a wedding party. Each table is a rectangle measuring 8 ft by 3 ft. She wants the tablecloth to extend 1 ft on each of the four sides. In square feet, what will be the area of each tablecloth?

20. Darnell's commute takes 1.5 h. He drove at 20 mi/h for 0.75 h, and the rest of the time he drove at 50 mi/h. In miles, how far is his work from his house?

21. A juice company wants to make cylinder-shaped juice cans that hold approximately 196 in.³ of juice. The base has to have a diameter of 5 in. What will be the height of the can in inches? Use 3.14 for π and round to the nearest whole number.

Data Analysis and Probability

22. The table shows the average number of vacation days per year for people in selected countries. What is the mean number of vacation days per year for these countries? Round to the nearest whole day.

Average Number of Vacation Days per Year for 2000	
Country	Number of Days
Korea	25
Japan	25
United States	13
Brazil	34
Italy	42
France	37

Source: *The World Almanac*

23. Kaya and her family went on a vacation for spring break. The table shows the distance traveled for each hour. What is the rate of change in miles per hour between hours 3 and 4?

Time (hours)	Distance (miles)
1	60
2	132
3	208
4	273
5	328

24. If a standard six-sided die is rolled, what is the probability of rolling a multiple of 2?

Test-Taking Tip

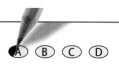

Question 24
Fractions do not have to be written in simplest form. Any equivalent fraction that fits the grid is correct.

Short-Response Questions

Short-response questions require you to provide a solution to the problem, as well as any method, explanation, and/or justification you used to arrive at the solution. These are sometimes called *constructed-response, open-response, open-ended, free-response,* or *student-produced questions.*

The following is a sample **rubric,** or scoring guide, for scoring short-response questions.

Credit	Score	Criteria
Full	2	Full credit: The answer is correct and a full explanation is provided that shows each step in arriving at the final answer.
Partial	1	Partial credit: There are two different ways to receive partial credit. • The answer is correct, but the explanation provided is incomplete or incorrect. • The answer is incorrect, but the explanation and method of solving the problem is correct.
None	0	No credit: Either an answer is not provided or the answer does not make sense.

On some standardized tests, no credit is given for a correct answer if your work is not shown.

Example

Nicole received grades of 79, 92, 68, 90, 72, and 92 on her history tests. What measure of central tendency would give her the highest grade for the term?

FULL CREDIT SOLUTION

First find the mean, median, and mode of her test grades.

The mean is the average of her test grades.

$$\frac{79 + 92 + 68 + 90 + 72 + 92}{6} = \frac{493}{6} \text{ or } 82.1\overline{6}$$

The steps, calculations and reasoning are clearly stated.

The mode is the number repeated most often in a set of data. Nicole received a 92 twice, so the mode is 92.

The median is the middle number in a set of data. Rearrange the test scores in order from least to greatest.

$$68, 72, 79, 90, 92, 92$$

The median is the average of 79 and 90 or $\frac{79 + 90}{2}$, or 84.5.

The solution of the problem is clearly stated.

Comparing the three values, the mode would give her the highest grade for the term.

PARTIAL CREDIT SOLUTION

In this sample solution, the answer is correct; however, there is no explanation or justification for the calculations.

$$\frac{79 + 92 + 68 + 90 + 72 + 92}{6} = 82$$

The mode is the highest number.

> There is no explanation of how the student will determine the answer

PARTIAL CREDIT SOLUTION

In this sample solution, the answer is incorrect because the student added Nicole's scores incorrectly. However, after the error, the calculations and reasoning are correct.

First find the mean of the test scores.

$$79 + 92 + 68 + 90 + 72 + 92 = 593$$

Divide by 6 to find that the mean is 98.8.

Next, find the median.

$$68, 72, 79, 90, 92, 92$$

$79 + 90 = 169$. The median is $\frac{169}{2}$ or 84.5.

The mode is 92 because she scored 92 the most number of times.

The mean of the test scores gives Nicole the highest score.

> The answer is incorrect, but the reasoning is correct.

NO CREDIT SOLUTION

In this sample solution, the answer is incorrect, and there is no explanation or justification for the calculations.

$79 + 92 + 68 + 90 + 72 = 401$

$68 + 72 + 79 + 90 + 92 + 92 = 493$

$493 + 401 = 894$

$894 \div 11 = 81.3$

Nicole's average score is 81.3.

> This solution shows no understanding of the problem.

Short-Response Practice

Solve each problem. Show all your work.

Number and Operations

1. Crispy Crunch cereal has a new box that says it is 25% larger than the original size. If the original box had 16 oz, how many ounces does the new box have?

2. Austin has $\frac{1}{3}$ gal of paint to paint his go cart. He knows that he will need $3\frac{1}{2}$ gal to paint his entire cart. How many more gallons does he need?

3. In 2001, the United States took in 6.484×10^9 from Canadian tourists. Express this value in standard form.

4. Prairie High School assigns every student a student identification code. The codes consist of one letter and five digits. What is the greatest number of students that can attend Prairie High School before codes will need to be reused?

5. Jacob works at a computer factory making $8.50 per hour. The company has not been selling as many computers lately so they have decreased each employee's wage by 6%. What will Jacob's new hourly wage be?

Algebra

6. Francisca ran for 1.25 h at an average rate of 5 miles per hour. What distance did Francisca run? (Use the formula $d = rt$, where d represents distance, r represents rate, and t represents time.)

7. Simplify the expression $-2(y + 4) - 3$.

8. Tariq plans to go to Raging Waters Water Park. The park charges a $12 admission and rents inner tubes for $1.50/h. If he has $15, for how many hours can he rent an inner tube?

9. Solve and graph the inequality $6 - 2m < 13$.

10. Roller Way Amusement park closes the park for a day and allows only schools to visit. The schools must bring students in buses that carry 30 students. If 32 buses are at the park, write an equation to represent the total number of students S that are at the park that day.

Geometry

11. Triangle RST is translated 3 units up and 1 unit to the left. Find the coordinates of translated $\triangle R'S'T'$.

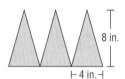

12. The Barnes family is installing a backsplash in their kitchen. If the total area of the triangles is 240 in.^{2}, how many triangles are in the backsplash?

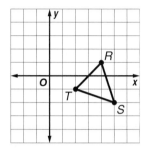

8 in.

⊢ 4 in. ⊣

13. Lawana has a window planter in the shape of a cylinder cut in half. She needs to fill it with dirt before she can plant flowers. What is the volume of the planter? Use 3.14 for π and round to the nearest tenth of a cubic foot.

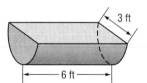

3 ft

6 ft

Test-Taking Tip

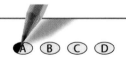

Question 13
After finding the solution, always go back and read the problem again to make sure your solution answers what the problem is asking.

14. $\angle JLK$ and $\angle KLM$ are complementary. If $m\angle KLM = 3x - 1$ and $m\angle JLK = x + 7$, find the measure of each angle.

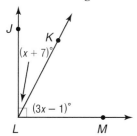

15. A cylindrical grain bin holds approximately 28,260 ft³ of grain. If the bin has a diameter of 50 ft, what is the bin's height? Use 3.14 for π and round to the nearest tenth of a foot.

$V = 28{,}260\ \text{ft}^3$

50 ft

Measurement

16. One kilometer is equal to about 0.62 mi. Tito is running a 10-km race. How many miles is this?

17. Jordan wants to make cone-shaped candles as shown in the diagram. What is the volume of one candle? Use 3.14 for π and round to the nearest tenth of a cubic inch.

8 in.

6 in.

18. Abigail is planning a cookout at her house. She wants to put trim around the edges of round tables. If the diameter of each table is 8 ft, how much trim will she need for each table? Use 3.14 for π and round to the nearest tenth of a cubic foot.

19. At Oakwood Lawncare Service, a lawn crew can do 5 jobs in 3 days. At this rate, how many days would it take the crew to do 20 jobs?

20. In a recent bicycle race, Kevin rode his bike at a pace of 16 mi/h. How many feet per minute is this?

21. Mateo wants to fill the cylindrical container shown with water. How many pints of water will he need? (Hint: One pint of liquid is equivalent to 28.875 in.³.) Use 3.14 for π and round to the nearest pint.

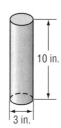

10 in.

3 in.

Data Analysis and Probability

22. Theo needs a four digit Personal Identification Number (PIN) for his checking account. If he can choose any digit from 0 to 9 for each of the digits, how many different PIN numbers are possible?

23. Matthew is choosing some CDs to take on a long car drive. He has 10 jazz CDs, 3 classical CDs, and 9 soundtracks. If he chooses two CDs without replacement, what is the probability that the first CD is a soundtrack and the second CD is a jazz CD?

24. Nikki is playing a game in which you spin the spinner below and then roll a six-sided die labeled 1 through 6. Each section of the spinner is equal in size. What is the probability that the spinner lands on blue and the die lands on an even number?

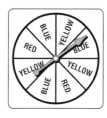

25. Two cards are drawn from a hand of eight cards numbered 1 to 8. The first card is not replaced after it is drawn. What is the probability that a 2 and a 5 are drawn?

Extended-Response Questions

Extended-response questions are often called *open-ended* or *constructed-response questions*. Most extended-response questions have multiple parts. You must answer all parts to receive full credit.

Extended-response questions are similar to short-response questions in that you must show all of your work in solving the problem and a rubric is used to determine whether you receive full, partial, or no credit. The following is a sample rubric for scoring extended-response questions.

Credit	Score	Criteria
Full	4	Full credit: A correct solution is given that is supported by well-developed, accurate explanations.
Partial	3, 2, 1	Partial credit: A generally correct solution is given that may contain minor flaws in reasoning or computation or an incomplete solution. The more correct the solution, the greater the score.
None	0	No credit: An incorrect solution is given indicating no mathematical understanding of the concept, or no solution is given.

> On some standardized tests, no credit is given for a correct answer if your work is not shown.

Make sure that when the problem says to *Show your work,* you show every aspect of your solution including figures, sketches of graphing calculator screens, or reasoning behind computations.

Example

Northern Sofa Store's delivery charges depend upon the distance furniture is delivered. The graph shows the charge for deliveries according to the distance.

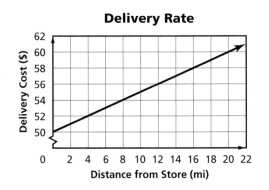

Delivery Rate

a. Write an equation to show the relationship between delivery cost y and distance from the store x.

b. Name the y-intercept and slope of the line that models the data. Explain what each means in this situation.

c. Suppose Mr. Hawkins wants a sofa delivered. His house is 21 mi from the store. What will be the cost of delivery?

FULL CREDIT SOLUTION

Part a A complete solution for writing the equation shows all the computations needed and the reasoning behind those computations.

> To write the equation, first I need to find the slope. I will use the two points marked on the graph, (0, 50) and (10, 55).

$$m = \frac{y_2 - y_1}{x_2 - x_1} \quad \text{slope formula}$$

$$= \frac{55 - 50}{10 - 0} \quad (x_i, y_i) = (0, 50) \text{ and } (x_2, y_2) = (10, 55)$$

$$= \frac{5}{10} \text{ or } 0.5$$

The slope of the line is 0.5 and the y-intercept is 50.

$$y = mx + b \qquad \text{Slope-intercept form}$$
$$y = 0.5x + 50 \quad m = 0.5 \text{ and } b = 50.$$

So an equation that fits the data is $y = 0.5x + 50$.

> The calculations and reasoning are clearly stated. The solution of the problem is also clearly stated.

Part b In this sample answer, the student demonstrates clear understanding of the y-intercept and slope of the graph.

> I can see from the graph that the y-intercept is 50 which means $50. This is the initial charge for any delivery and then additional money is charged for each mile. The slope of the line that I found in Part a is 0.5 which means $0.50. This is the additional charge per mile to deliver the furniture.

Part c In this sample answer, the student knows how to use the equation to find the delivery cost for a given distance.

> I will substitute 21 for x into the equation since Mr. Hawkins lives 21 mi from the store.
> $$y = 0.5x + 50$$
> $$y = 0.5(21) + 50$$
> $$y = 60.50$$
> The cost of delivery will be $60.50.

PARTIAL CREDIT SOLUTION

Part a This sample answer includes no explanations for the calculations performed. However, partial credit is given for correct calculations and a correct answer.

> $$m = \frac{y_2 - y_1}{x_2 - x_1} = \frac{55 - 50}{10 - 0} = \frac{1}{2}$$
>
> $$y = \frac{1}{2}x + 50$$

The equation is correct because $\frac{1}{2}$ is the same as 0.5.

> More credit would have been given if an explanation had been given.

Part b This part receives full credit because the student demonstrates understanding of the y intercept and slope.

> The y-intercept 50 is the charge for just making any delivery. It means $50. The slope $\frac{1}{2}$ is the charge per mile which is 50 cents.

Part c Partial credit is given for Part c because the student makes a calculation error.

> To get the delivery cost, substitute 21 mi into the equation.
>
> $y = \frac{1}{2}x + 50$
>
> $y = \frac{1}{2}(21) + 50$
>
> $y = 42 + 50 = 92$
>
> The cost is $92.

This sample answer might have received a score of 2, depending on the judgment of the scorer. Had the student gotten Part c correct, the score would probably have been a 3.

NO CREDIT SOLUTION

Part a The student demonstrates no understanding of how to write an equation for a line.

> If I use the points $(0, 50)$ and $(10, 55)$, an equation is $10y = 10x + 55$.

Part b The student does not understand the meaning of the *y*-intercept or the slope.

> The y-intercept is 0 because that is when the truck leaves the store. The slope is 10 because that is the distance from the store in miles when the truck makes its first stop.

Part c The student does not understand how to read the graph to find the cost or how to use an equation to find the cost.

> $21, because it is 21 mi from the store.

Extended-Response Practice

Solve each problem. Show all your work.

Number and Operations

1. In a recent survey, Funtime Amusement Park found that 6 out of 8 of their customers had been to the park before. In one week, 4500 people attended the park.

   a. What percent of customers have been to the park before?

   b. How many of the park goers in that week had been there before?

   c. If 3000 people were at the park the next week, how many would you expect had *not* been there before?

2. The following table shows the number of people of each age living in the U.S. in the year 2000.

Age	Number of People
Under 15	60,253,375
15 to 24	39,183,891
25 to 34	39,891,724
35 to 44	44,148,527
45 to 64	61,952,636
65 and over	34,991,753

   Source: U.S. Census Bureau

   a. To the nearest million, how many people were under age 25?

   b. What percent of people were under 15? Round to the nearest percent.

   c. What is the total population of the United States? Write in scientific notation.

Algebra

3. Victoria needs to study for a math exam. The exam is in 18 days. She has decided to begin right away by studying 15 min the first night and increasing her study time by 5 min each day.

   a. Write an expression for the total number of minutes that Victoria will study T for a given time d days from today.

   b. If she has 18 days to study, how many minutes will she study on the last day before the exam?

   c. Victoria begins to study on a Monday, 18 days before the exam. On what day will she study exactly one hour?

4. Jasmine wants to get her portrait taken for her senior pictures. She finds that three different portrait studios charge a sitting fee and charge a separate fee for each ordered picture. The table below shows their prices.

Studio	Sitting Fee	Cost per Portrait
Famous Photos	$50	$10
Picture Perfect	$80	$8
Timeless Portraits	$45	$15

   a. For each studio, write an equation that represents the total cost. In each of the three equations, use C to represent the total cost and p to represent the number of pictures.

   b. If Jasmine wants to order 30 portraits, which studio would be the least expensive?

   c. How many portraits will she need to order for Famous Photos and Picture Perfect to cost the same?

Geometry

5. A manufacturer ships its product in boxes that are 3 ft $\times$ 2 ft $\times$ 2 ft. The company needs to store some products in a warehouse space that is 32 ft $\times$ 8 ft $\times$ 10 ft.

   a. What is the volume of the storage space?

   b. What is the volume of each box?

   c. What is the greatest number of boxes the company can store in this space? (All the boxes must be stored in the same position.)

   d. How much storage space in *not* filled with boxes?

Test-Taking Tip

Question 5

Many standardized tests include any necessary formulas in the test booklet. It helps to be familiar with formulas such as the area of rectangular prisms, but use any formulas that are given to you.

---

6. Triangle *JKM* is shown.

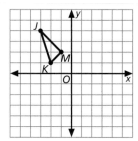

   a. What are the coordinates of the vertices of △*JKM*?

   b. Copy △*JKM* onto a sheet of grid paper. Label the vertices.

   c. Graph the image of △*JKM* after a translation 2 units left and 3 unit down. Label the translated image △*J'K'M'*.

   d. Graph the image of △*J'K'M'* after a reflection in the *y*-axis. Label the reflection △*J''K''M''*.

Measurement

7. The diagram shows a pattern for a garden that Marie plans to plant. Use 3.14 for π.

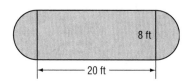

8 ft

20 ft

   a. What is the area of the garden in square feet?

   b. What is the area of the garden in square yards?

   c. Marie wants to put a stone border around the outside of the garden. What is the perimeter of the garden?

8. The table shows the speeds of several animals.

Animal	Speed (mi/h)
Cheetah	70
Zebra	40
Reindeer	32
Cat (domestic)	30
Wild Turkey	15

Source: *World Book*

   a. What is the rate of the cat in feet per minute?

   b. How many times faster is the cheetah than the wild turkey? Round to the nearest tenth.

Data Analysis and Probability

9. The table shows the life expectancy in years of people in different countries. The figures are for the year 2000.

Country	Life Expectancy (yr)
Afghanistan	45.9
Australia	79.8
Brazil	62.9
Canada	79.4
France	78.8
Haiti	49.2
Japan	80.7
Madagascar	55.0
Mexico	71.5
United States	77.1

Source: U.S. Census Bureau

   a. What is the median life expectancy?

   b. What is the mean life expectancy?

   c. In 2000, Cambodians had a life expectancy of 56.5. If Cambodia was added to the table, how would the mean be affected?

10. The table shows the results of a survey about the average time that individual students spend studying on weekday evenings.

Grade	Time (min)	Grade	Time (min)
2	20	6	60
2	15	6	45
2	20	6	55
4	30	6	60
4	20	8	70
4	25	8	80
4	40	8	75
4	30	8	60

   a. Make a scatter plot of the data.

   b. What are the coordinates of the point that represents the longest time spent on homework?

   c. Does a relationship exist between grade level and time spent studying? If so, write a sentence to describe the relationship. If not, explain why not.

Technology Reference Guide

Graphing Calculator Overview

This section summarizes some of the graphing calculator skills you might use in your mathematics classes using the TI-83 Plus or TI-84 Plus.

General Information

- Any yellow commands written above the calculator keys are accessed with the 2nd key, which is also yellow. Similarly, any green characters or commands above the keys are accessed with the ALPHA key, which is also green. In this text, commands that are accessed by the 2nd and ALPHA keys are shown in brackets. For example, 2nd [QUIT] means to press the 2nd key followed by the key below the yellow [QUIT] command.
- 2nd [ENTRY] copies the previous calculation so it can be edited or reused.
- 2nd [ANS] copies the previous answer so it can be used in another calculation.
- 2nd [QUIT] will return you to the home (or text) screen.
- 2nd [A-LOCK] allows you to use the green characters above the keys without pressing ALPHA before typing each letter.
- Negative numbers are entered using the (−) key, not the minus sign, −.
- The variable x can be entered using the X,T,θ,n key, rather than using ALPHA [X].
- 2nd [OFF] turns the calculator off.

Key Skills

Use this section as a reference for further instruction. For additional features, consult the TI-83 Plus or TI-84 Plus user's manual.

ENTERING AND GRAPHING EQUATIONS

Press Y=. Use the X,T,θ,n key to enter *any* variable for your equation. To see a graph of the equation, press GRAPH.

SETTING YOUR VIEWING WINDOW

Press WINDOW. Use the arrow or ENTER keys to move the cursor and edit the window settings. Xmin and Xmax represent the minimum and maximum values along the x-axis. Similarly, Ymin and Ymax represent the minimum and maximum values along the y-axis. Xscl and Yscl refer to the spacing between tick marks placed on the x- and y-axes. Suppose Xscl = 1. Then the numbers along the x-axis progress by 1 unit. Set Xres to 1.

THE STANDARD VIEWING WINDOW

A good window to start with to graph an equation is the **standard viewing window.** It appears in the WINDOW screen as follows.

To easily set the values for the standard viewing window, press ZOOM 6.

ZOOM FEATURES

To easily access a viewing window that shows only integer coordinates, press ZOOM 8 ENTER.

To easily access a viewing window for statistical graphs of data you have entered, press ZOOM 9.

USING THE TRACE FEATURE

To trace a graph, press TRACE. A flashing cursor appears on a point of your graph. At the bottom of the screen, x- and y-coordinates for the point are shown. At the top left of the screen, the equation of the graph is shown. Use the left and right arrow keys to move the cursor along the graph. Notice how the coordinates change as the cursor moves from one point to the next. If more than one equation is graphed, use the up and down arrow keys to move from one graph to another.

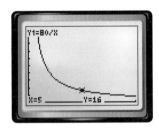

SETTING OR MAKING A TABLE

Press 2nd [TBLSET]. Use the arrow or ENTER keys to move the cursor and edit the table settings. Indpnt represents the x-variable in your equation. Set Indpnt to *Ask* so that you may enter any value for x into your table. Depend represents the y-variable in your equation. Set Depend to *Auto* so that the calculator will find y for any value of x.

USING THE TABLE

Before using the table, you must enter at least one equation in the Y= screen. Then press 2nd [TABLE]. Enter any value for x as shown at the bottom of the screen. The function entered as Y_1 will be evaluated at this value for x. In the two columns labeled X and Y_1, you will see the values for x that you entered and the resulting y-values.

PROGRAMMING ON THE TI–83 PLUS

When you press PRGM, you see three menus: EXEC, EDIT, and NEW. EXEC allows you to execute a stored program by selecting the name of the program from the menu. EDIT allows you to edit or change an existing program. NEW allows you to create a new program. For additional programming features, consult the TI–83 Plus or TI–84 Plus user's manual.

ENTERING INEQUALITIES

Press 2nd [TEST]. From this menu, you can enter the $=$, $\neq$, $>$, $\geq$, $<$, and $\leq$ symbols.

ENTERING AND DELETING LISTS

Press STAT ENTER. Under L_1, enter your list of numerical data. To delete the data in the list, use your arrow keys to highlight L_1. Press CLEAR ENTER. Remember to clear all lists before entering a new set of data.

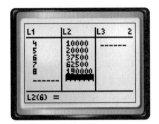

PLOTTING STATISTICAL DATA IN LISTS

Press Y=. If appropriate, clear equations. Use the arrow keys until Plot1 is highlighted. Plot1 represents a Stat Plot, which enables you to graph the numerical data in the lists. Press ENTER to turn the Stat Plot on and off. You may need to display different types of statistical graphs. To see the details of a Stat Plot, press 2nd [STAT PLOT] ENTER. A screen like the one below appears.

At the top of the screen, you can choose from one of three plots to store settings. The second line allows you to turn a Stat Plot on and off. Then you may select the type of plot: scatter plot, line plot, histogram, two types of box-and-whisker plots, or a normal probability plot. Next, choose which lists of data you would like to display along the x- and y-axes. Finally, choose the symbol that will represent each data point.

Cabri Jr. Overview

Cabri Junior for the TI-83 Plus and TI-84 Plus is a geometry application that is designed to reproduce the look and feel of a computer on a handheld device.

General Information

Starting Cabri Jr. To start Cabri Jr., press APPS and choose Cabri Jr. Press any key to continue. If you have not run the program on your calculator before, the F1 menu will be displayed. To leave the menu and obtain a blank screen, press CLEAR. If you have run the program before, the last screen that was in the program before it was turned off will appear. See Quitting Cabri Jr. for instructions on clearing this screen to obtain a blank screen.

In Cabri Jr., the four arrow keys (◄, ►, ▼, ▲), along with ENTER, operate as a mouse would on a computer. The arrows simulate moving a mouse, and ENTER simulates a left click on a mouse. For example, when you are to select an item, use the arrow keys to point to the selected item and then press ENTER. You will know you are accurately pointing to the selected item when the item, such as a point or line, is blinking.

Quitting Cabri Jr. To quit Cabri Jr., press 2nd [QUIT], or [OFF] to completely shut off the calculator. Leaving the calculator unattended for approximately 4 minutes will trigger the automatic power down. After the calculator has been turned off, pressing ON will result in the calculator turning on, but not Cabri Jr. You will need to press APPS and choose Cabri Jr. Cabri Jr. will then restart with the most current figure in its most recent state.

As Cabri Jr. resembles a computer, it also has dropdown menus that simulate the menus in many computer programs. There are five menus, F1 through F5.

Navigating Menus To navigate each menu, press the appropriate key for F1 through F5. The arrow keys will then allow you to navigate within each menu. The ▲ and ▼ keys allow you to move within the menu items. The ► and ◄ keys allow you to access a submenu of an item. If an item has an arrow to the right, this indicates there is a submenu. Although not displayed, the menu items are numbered. You can also select a menu item by pressing the number that corresponds to each item. For example, to select the fourth menu item in the list, press [4]. If you press a number greater than the number of items in the list, the last item will be selected. If you press [0], you will leave the menu without selecting an item. This is the same as pressing CLEAR.

Key Skills

Use this section as a reference for further instruction. For additional features, consult the TI-84 Plus user's manual.

[F1] MANAGING FIGURES

The menu below provides the basic operations when working in Cabri Jr. These are commands normally found in menus within computer applications.

[F2] CREATING OBJECTS

This menu provides the basic tools for creating geometric figures. You can create points in three different ways, a line and line segment by selecting two points, a circle by defining the center and radius, a triangle by finding three vertices, and a quadrilateral by finding four vertices.

[F3] CONSTRUCTING OBJECTS

This menu provides the tools to construct new objects from existing objects. You can construct perpendicular and parallel lines, a perpendicular bisector of a segment, an angle bisector, a midpoint of a line segment, a circle using the center and a point on the circle, and a locus.

[F4] TRANSFORMING OBJECTS

This menu provides the tools to transform geometric figures. Using figures that are already created, you can access this menu to create figures that are symmetrical to other figures, reflect figures over a line of reflection, translate figures using a line segment or two points that define the translation, rotate figures by defining the center of rotation and angle of rotation, and dilate figures using the center of the dilation and a scale factor.

[F5] COMPUTING OBJECTS

This menu provides the tools for displaying, labeling, measuring, and computing. You can make an object visible or invisible, label points on figures, alter the way objects are displayed, measure length, area, and angle measures, display coordinates of points and equations of lines, make calculations, and delete objects from the screen.

While a graphing calculator cannot do everything, it can make some tasks easier. To prepare for whatever lies ahead, you should try to learn as much as you can about the technology. The future will definitely involve technology and the people who are comfortable with it will be successful. Using a graphing calculator is a good start toward becoming familiar with technology.

Glossary/Glosario

A mathematics multilingual glossary is available at **www.math.glencoe.com/multilingual_glossary.** The glossary is available in the following languages.

Arabic   English   Korean   Tagalog
Bengali   Haitian Creole   Russian   Urdu
Cantonese   Hmong   Spanish   Vietnamese

English

Español

■ A ■

absolute value   (p. 105) The distance a number is from zero on a number line, represented by $|n|$.

valor absoluto   (p. 105) Valor que tiene una cifra por su figura, por ejemplo, en el número 567 el valor absoluto de $|es|$.

acute angle   (p. 550) An angle measuring between 0° and 90°.

acute triangle   (p. 553) A triangle having three acute angles.

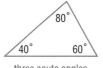
three acute angles
tres ángulos agudos

ángulo agudo   (p. 550) Angulo que mide menos de 90°.

triángulo acutángulo   (p. 553) Triángulo que sus tres ángulos son agudos.

addition property of opposites   (p. 104) When opposites are added, the sum is always zero.

propiedad de la adición en los números opuestos (p. 104) Si dos números opuestos se suman, el resultado es igual a cero.

adjacent angles   (p. 352) Angles that have a common vertex and a common side, but have no interior points in common.

ángulos adyacentes   (p. 352) Ángulos que tienen el mismo vértice y un lado común pero no comparten ningún punto interior.

alternate exterior angles   (p. 355) In the figure, transversal t intersects lines ℓ and m. $\angle 5$ and $\angle 3$, and $\angle 6$ and $\angle 4$ are alternate exterior angles.

alternate interior angles   (p. 355) In the figure, transversal t intersects lines ℓ and m. $\angle 1$ and $\angle 7$, and $\angle 2$ and $\angle 8$ are alternate interior angles.

ángulos alternos-externos   (p. 355) En la figura, la transversal t interseca las rectas ℓ y m. $\angle 5$ y $\angle 3$, y $\angle 6$ y $\angle 4$ son ángulos alternos internos.

ángulos alternos-internos   (p. 355) En la figura, la transversal t intersecta las rectos ℓ y m. $\angle 1$ y $\angle 7$, y $\angle 2$ y $\angle 8$ son ángulos alternos internos.

angle   (p. 550) A plane figure formed by two rays having a common endpoint.

ángulo   (p. 550) Figura plana formada por dos líneas rectas que parten del mismo vértice.

angle bisector   (p. 356) A ray that separates an angle into two congruent adjacent angles.

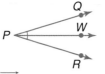
$\overrightarrow{PW}$ is the bisector of $\angle P$.
$\overrightarrow{PW}$ es la bisectriz del $\angle P$.

bisectriz   (p. 356) Línea recta que divide un ángulo en dos ángulos iguales.

area   (p. 66) The amount of surface enclosed by a geometric figure. Area is measured in square units.

área   (p. 66) Superficie que cubre una figura plana. El área se mide en unidades métricas al cuadrado.

English

Español

associative property (p. 119) If three or more numbers are added or multiplied, the numbers can be regrouped without changing the result. For example, $4 + (6 + 5) = (4 + 6) + 5$; $4 \times (6 \times 5) = (4 \times 6) \times 5$.

average (p. 10) See *mean*.

axes (p. 308) The perpendicular lines used for reference in a coordinate plane.

propiedad asociativa (p. 119) La suma o la multiplicación de tres o más números se puede reagrupar sin que se altere el resultado. Por ejemplo, $4 + (6 + 5) = (4 + 6) + 5$; $4 \times (6 \times 5) = (4 \times 6) \times 5$.

promedio (p. 10) Véase *media*.

ejes (p. 308) Las dos líneas perpendiculares que se usan como referencia en el plano de coordenadas o plano cartesiano.

■ B ■

back-to-back stem-and-leaf plots (p. 19) A stem-and-leaf plot with leaves on both sides of the stem.

bar graph (p. 28) A graphic representation of data using horizontal or vertical bars.

base (p. 132) The factor taken to a power in an expression in exponential form. For example, in 2^3 the base is 2.

binomial (p. 394) A polynomial with two terms.

box-and-whisker plot (p. 38) A graphical representation that identifies trends and summarizes information by showing the distribution of data by dividing it into four equal parts, of which the middle two are represented by a box and the outer two by whiskers.

diagramas de tallo y hoja uno tras de otro (p. 19) Diagrama de tallo y hoja con hojas en ambos lados del tallo.

gráfica de barra (p. 28) Representación gráfica de una serie de datos valiéndose de barras horizontales o verticales.

base (p. 132) Factor que es afectado por el exponente en una potencia. Por ejemplo, en 2^3 la base es 2.

binomio (p. 394) Expresión algebraica formada por dos términos.

diagrama de caja y pelo (p. 38) Representación gráfica que identifica tendencias y síntesis informativas mostrando la distribución de datos dividida en cuatro partes iguales, de las cuales las dos interiores se representan con una caja y las dos exteriores con pelo.

■ C ■

circle (p. 80) The set of all points in a plane the same distance from a given point called the center.

P is the center of the circle.
P es el centro del círculo.

círculo (p. 80) Plano curvo y cerrado limitado por la *circunferencia*.

circle graph (p. 20) A graphic representation in which data are expressed as parts of a whole in the form of a circle sliced into sectors. Also known as a *pie chart*.

circumference (p. 80) The distance around a circle.

closed set (p. 119) If an operation is performed on two numbers in a given set and the result is also a member of the set, then the set is closed with respect to that operation (closure property).

gráfica dentro del círculo (p. 20) Representación gráfica en la cual los datos se muestran como porciones de un todo representado por un círculo. Se le conoce también como *diagrama de sectores*.

circunferencia (p. 80) Línea curva cerrada cuyos puntos que la conforman se encuentran a la misma distancia de otro punto llamado centro. También se le define como la periferia del círculo.

conjunto cerrado (p. 119) Si una operación se lleva a cabo con dos números en un conjunto dado y el resultado es también miembro de ese conjunto, entonces el conjunto está cerrado con respecto a esa operación (propiedad concluyente).

English

clusters   (p. 16) Isolated groups of data values.

cluster sampling   (p. 6) Members of the population are chosen at random from a particular part of the population and then polled in clusters, not individually.

coefficient   (p. 394) A number by which a variable or group of variables is multiplied. For example, in $7ab^2$ the coefficient is 7.

collinear points   (p. 157) Points that lie on the same line.

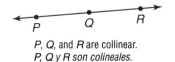

P, Q, and R are collinear.
P, Q y R son colineales.

commission   (p. 284) The amount of money earned by salespeople who work on commission that is a percent of their total sales. The percent is the *commission rate*. *Graduated commission* is a commission that increases as sales increase.

commission rate   (p. 284) See *commission*.

common factors   (p. 414) The factors that are the same for a given set of numbers are the common factors. For example, a common factor of 12 and 18 is 6.

commutative property   (p. 119) If two numbers are added or multiplied, the operations can be done in any order. For example, $4 \times 5 = 5 \times 4; 5 + 4 = 4 + 5$.

complementary angles   (p. 352) Two angles whose sum measures 90°.

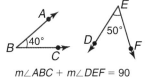

$m\angle ABC + m\angle DEF = 90$

conditional statement   (p. 488) One made from two simple sentences. It has an *if* part and a *then* part. The *if* part is called the *hypothesis* and the *then* part is called the *conclusion*.

cone   (p. 167) A three-dimensional figure that has a curved surface, one circular base and one vertex. The line segment from the vertex perpendicular to the base is the *altitude*. If the endpoint opposite the vertex is the center of the base, the figure is a *right cone*.

vertex
vértice

base
base

conjecture   (p. 482) A conclusion reached by the process of inductive reasoning.

Español

conglomerados   (p. 16) Grupos aislados de valores de datos.

muestra por conglomerado   (p. 6) De una parte de la población se escogen algunos miembros al azar y luego se les registra como conglomerado, no como individuos.

coeficiente   (p. 394) Número que multiplica a una variable o a un grupo de variables. Por ejemplo, en $7ab^2$ el coeficiente es 7.

puntos colineales   (p. 157) Dos o más puntos que forman parte de la misma línea.

comisión   (p. 284) Porcentaje que recibe un vendedor del total de su venta. Dicho porcentaje recibe el nombre de *tasa de comisión*.

tasa de comisión   (p. 284) Véase *comisión*.

factores comunes   (p. 414) Factores que son los mismos para un conjunto dado de números. Por ejemplo, un factor común de 12 y 18 es 6.

propiedad conmutativa   (p. 119) La suma y la multiplicación de dos números se pueden hacer en cualquier orden. Por ejemplo, $4 \times 5 = 5 \times 4; 5 + 4 = 4 + 5$.

ángulos complementarios   (p. 352) Dos ángulos cuyas medidas al sumarse dan como resultado 90°.

declaración condicional   (p. 488) Declaración compuesta por dos oraciones sencillas. Tiene dos partes: *si* y *entonces*. A la primera la precede *si*, y se llama *premisa* o *antecedente*; a la segunda la precede *entonces*, y se llama *conclusión* o *consecuente*.

cono   (p. 167) Figura tridimensional con una base circular y un vértice. La línea perpendicular a la base que inicia en el vértice viene siendo la *altura*. Si el punto extremo opuesto al vértice es el centro de la base, entonces será un *cono regular*.

conjetura   (p. 482) Conclusión a la que se llega a partir de un razonamiento inductivo.

English

constant (p. 394) A monomial that does not contain a variable.

convenience sampling (p. 6) The population is chosen only because it is easily available.

coordinate for a point (p. 308) A number associated with a point on a number line or an ordered pair of numbers associated with a point on a grid.

coordinate plane (p. 308) A plane formed by two perpendicular number lines that is used to identify the location of points in a way similar to using a map to identify locations of places.

coplanar points (p. 157) Points that lie in the same plane.

corresponding angles (p. 353) Angles that are in the same position relative to the transversal and the lines being intersected.

counterexample (p. 483) A case where the conjecture does not hold true.

counting principle (p. 450) A method used to find the number of possible outcomes of an experiment by multiplying the number of outcomes at each stage of the experiment.

cross-products (p. 84) For $\frac{a}{b}$ and $\frac{c}{d}$, the cross products are ad and bc. In a proportion, cross products are equal.

cube (p. 166) A rectangular prism with edges of equal length.

Customary (English) units (p. 53) Units used in the customary system of measurement. Frequently used customary units include the following: length—inch (in.), feet (ft), yard (yd), and mile (mi); capacity—cup (c), pint (pt), quart (qt), and gallon (gal); weight—ounce (oz), pound (lb), and ton (T); and temperature—degrees Fahrenheit.

cylinder (p. 167) A three-dimensional figure that has a curved surface and two identical, parallel, circular bases.

Español

constante (p. 394) Monomio que no contiene variables.

muestra por conveniencia (p. 6) Población escogida sólo porque está disponible.

coordenada para un punto (p. 308) Número asociado con un punto en la recta numérica, o un par números ordenados asociados con un punto en la cuadrícula.

plano de coordenadas o plano cartesiano (p. 308) Dos rectas numéricas perpendiculares que forman una cuadrícula. Se usa para localizar puntos de la misma manera que se usa un mapa para localizar algún lugar.

puntos coplanares (p. 157) Puntos que forman parte del mismo plano.

ángulos correspondientes (p. 353) Ángulos que están en la misma posición con relación a la secante y a las líneas paralelas.

contraejemplo (p. 483) Ejemplo donde la conjetura no es válida.

principio de conteo (p. 450) Método usado para encontrar el número de resultados posibles de un experimento multiplicando el número de resultados en cada etapa del experimento.

productos cruzados (p. 84) Para a/b y c/d, los productos cruzados son ad y bc. En una proporción los productos cruzados son iguales.

cubo (p. 166) Prisma regular con sus seis caras iguales.

unidades inglesas (p. 53) Unidades de medida que se usan en el sistema inglés. Algunas medidas que se usan con frecuencia son: longitud-pulgada (plg.), pies (pies), yarda (yd) y milla (mi); capacidad-taza (t), pinta (pt), cuarto de galón (ct) y galón (gal); peso-onza (oz), libra (lb) y tonelada (T); temperatura-grados Fahrenheit.

cilindro (p. 167) Figura tridimensional que tiene dos bases circulares las cuales son paralelas e idénticas.

English   Español

D

data   (p. 6) Pieces of information that can be gathered through interviews, records of events, or questionnaires. The word data is the plural form of the Latin word *datum.*

deductive reasoning   (p. 489) A process of reasoning in which a conclusion is drawn from a conditional statement and additional information.

degree   (p. 157, p. 550) A common unit of measurement for angles or temperature.

dependent events   (p. 457) Events such that the outcome of the one affects the outcome of another.

datos   (p. 6) Información estadística que puede conseguirse por medio de entrevistas, cuestionarios o grabaciones de eventos. En inglés, la palabra *data* es el plural del vocablo latín *datum.*

razonamiento deductivo   (p. 489) Razonamiento donde la conclusión se deduce de una declaración condicional y de información adicional.

grado   (p. 157, p. 550) Unidad de medida para ángulos o temperatura.

eventos dependientes   (p. 457) Dos eventos en los que el resultado del primero afecta el resultado del segundo y viceversa.

diagonal   (p. 362) A line segment that joins two vertices of a polygon and is not a side.

$\overline{SQ}$ is a diagonal.
$\overline{SQ}$ *es una diagonal.*

diagonal   (p. 362) Línea recta que conecta dos vértices no adyacentes de un polígono.

diameter   (p. 80) The fixed distance of a line segment that passes through the center of a circle and has endpoints on the circle. The diameter is twice the radius.

discount   (p. 270) The amount that the regular price is reduced.

distributive property   (p. 119) If one factor in a product is a sum, multiplying each addend by the other factor before adding does not change the product.

domain   (p. 314) All of the first values of the ordered pairs of a relation.

diámetro   (p. 80) Línea recta que pasa por el centro del círculo y que sus puntos extremos se localizan en la círcunferencia. El diámetro está formado por dos radios.

descuento   (p. 270) Diferencia entre el precio normal de un artículo y su precio de oferta.

propiedad distributiva   (p. 119) Si un factor en un producto es una suma, entonces al multiplicar cada sumando por el otro factor antes de hacer la suma el producto no cambia.

dominio   (p. 314) Todos los primeros valores de los pares ordenados en una relación.

E

endpoint   (p. 156) A point at the end of a segment or ray.

equation   (p. 208) A statement that two numbers or expressions are equal.

equiangular   (p. 160) Having angles of the same measure.

punto extremo   (p. 156) Punto al final de una línea recta.

ecuación   (p. 208) Planteamiento matemático donde dos números o expresiones algebraicas son iguales.

equiangular   (p. 160) Figura que tiene todos sus ángulos iguales.

equilateral triangle   (p. 161) A triangle with all three sides having the same length and all angles the same measure.

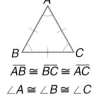

$\overline{AB} \cong \overline{BC} \cong \overline{AC}$
$\angle A \cong \angle B \cong \angle C$

triángulo equilátero   (p. 161) Triángulo que tiene los tres lados y los tres ángulos iguales.

English

equivalent ratios (p. 75) Two ratios that represent the same comparison. For example, 2:5, 4:10, 6:15 are equivalent ratios.

event (p. 436) An outcome or a combination of outcomes.

expected value (p. 464) The amount you can expect to win or lose in situations in which the winners are determined randomly.

experimental probability (p. 440) The probability of an event based on the results of an experiment.

exponent (p. 132) A number showing how many times the base is used as a factor. For example, in 2^3 the exponent is 3.

exponential form (p. 132) A number written with a base and an exponent. For example, $2 \times 2 \times 2 \times 2 = 2^4$.

Español

razones equivalentes (p. 75) Dos razones que representan la misma comparación. Por ejemplo, 2:5, 4:10, 6:15 son razones equivalentes.

evento (p. 436) Resultado o la combinación de varios resultados.

valor esperado (p. 464) Número que se espera que sea el ganador o perdedor en circunstancias donde el ganador se determina al azar.

probabilidad experimental (p. 440) La probabilidad de un evento basádose en los resultados de un experimento.

exponente (p. 132) Número que muestra las veces que la base se usa como factor. Por ejemplo, en 2^3 el exponente es 3.

forma exponencial (p. 132) Número expuesto con una base y un exponente. Por ejemplo, $2 \times 2 \times 2 \times 2 = 2^4$.

■ F ■

factor (p. 391) Any number multiplied by another number to produce a product.

factoring (p. 414) The reversing of the distributive property to find what factors were multiplied to obtain the product.

favorable outcome (p. 436) A particular outcome that you are calculating the likelihood of its occurrence. Also called the *desired outcome*.

formula (p. 232) An equation stating a relationship between two or more variables.

frequency table (p. 24) A table that shows how often an item appears in a set of data. A tally mark is used to record each response. The total number of marks for a given response is the frequency of that response.

function (p. 314) A relation in which each value of the domain is paired with one and only one value of the range.

function rule (p. 315) The description of a function.

factor (p. 391) Cualquier número multiplicado por otro número que da como resultado un producto.

factorizar (p. 414) Encontrar los factores que al multiplicarse nos den el número dado. Es lo opuesto de la propiedad distributiva.

resultado favorable (p. 436) Resultado particular cuya posibilidad de que se dé se calcula. También se le llama *resultado esperado*.

fórmula (p. 232) Ecuación que determina las relaciones entre dos o más variables.

tabla de frecuencia (p. 24) Una tabla de frecuencia muestra las veces que un número o un evento se manifiesta en un conjunto de datos.

función (p. 314) Relación en la cual cada valor del dominio se aparea solamente con un valor del alcance.

regla de la función (p. 315) Descripción de una función.

■ G ■

graph of the equation (p. 318) The set of all points whose coordinates are solutions of an equation.

graph of the function (p. 318) A graph of an equation that represents the function.

greatest common factor (GCF) (p. 414) The greatest numerical factor and variable or variables of greatest degree.

gráfica de la ecuación (p. 318) Conjunto de puntos cuyas coordenadas son las soluciones de una ecuación.

gráfica de la función (p. 318) Gráfica de una ecuación que representa la función.

máximo factor común (MFC) (p. 414) El número entero mayor que es factor de dos o más números.

English
Español

horizon line (p. 178) A line in a perspective drawing on which the vanishing point lies.

línea horizontal (p. 178) Línea recta en la cual el punto de fuga está tendido.

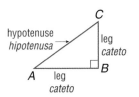

hypotenuse (p. 334) The side opposite the right angle in a right triangle.

hipotenusa (p. 334) Lado opuesto al ángulo recto en un triángulo rectángulo.

hypothesis (p. 488) The first part of a conditional statement containing information that leads to a conclusion.

antecedente (p. 488) Primera parte de una declaración condicional que contiene información de la que se deduce la conclusión.

identity property of addition (p. 105) The sum of any number and 0 is that number.

propiedad de identidad en la adición (p. 105) Todo número al que se le suma 0 el resultado será el mismo número.

image (p. 370) The new figure of a transformation.

imagen (p. 370) Nueva figura que resulta de una transformación.

income taxes (p. 274) Taxes paid based on income.

impuestos (p. 274) Contribución que se paga en base a los ingresos.

independent events (p. 456) Events such that the outcome of one does not depend on the outcome of another.

eventos independientes (p. 456) Dos eventos en los que el resultado del primero no depende del resultado del segundo y viceversa.

inductive reasoning (p. 482) The process of reaching a conclusion based on a set of specific examples.

razonamiento inductivo (p. 482) Proceso lógico donde una conclusión se desprende de un conjunto de ejemplos específicos.

inequality (p. 240) A mathematical sentence that contains one of the symbols $<$, $>$, $\leq$ or $\geq$.

desigualdad (p. 240) Declaración matemática que usa uno de los símbolos $<$, $>$, $\leq$ o $\geq$.

integers (p. 118) The set of whole numbers and their opposites.

enteros (p. 118) Conjunto de los números enteros y sus opuestos.

intercept (p. 319) The point where a line crosses an axis. The x-intercept is the x-coordinate of the point $(x, 0)$ where the graph crosses the x-axis. The y-intercept is the y-coordinate of the point $(0, y)$ where the graph crosses the y-axis.

intersección (p. 319) Punto donde una línea se cruza con otra. La intersección x se da en el punto $(x, 0)$ de la coordenada x. La intersección y se da en el punto $(0, y)$ de la coordenada y.

interest (p. 280) Money paid to an individual or institution for the privilege of using their money.

interés (p. 280) Cantidad de dinero que se paga a una persona o institución por el uso de su dinero.

intersecting lines (p. 353) Lines that have exactly one point in common.

líneas que se intersectan (p. 353) Líneas que tienen un punto en común.

inverse operations (p. 104) Operations that undo each other such as addition and subtraction or multiplication and division.

operaciones inversas (p. 104) Operaciones opuestas tales como la suma y la resta o la multiplicación y la división.

English

irrational number (p. 118) Numbers such as π that are non-terminating non-repeating decimals.

isometric drawing (p. 175) A three-dimensional, "corner" view of an object in which three sides of the object are shown and the unseen sides can be indicated with dashed lines. The dimensions of isometric drawings are proportional to the object's actual dimensions, and the object's parallel edges are parallel line segments.

isosceles triangle (p. 553) A triangle having at least two sides the same length and at least two angles the same measure.

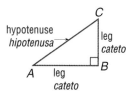

▪ L ▪

laws of exponents (p. 136) Rules that govern how to perform operations with numbers in exponential form. Examples are: To multiply numbers with the same base, add the exponents. To divide numbers with the same base, subtract the exponents.

legs of a triangle (p. 334) In a right triangle, the two sides that are not the hypotenuse.

like terms (p. 228, 395) See *terms*.

line (p. 156) A set of points that extends without end in two opposite directions. Two points determine a line.

linear equation (p. 318) An equation that represents a linear function.

linear function (p. 318) A function that is represented by a straight line.

line graph (p. 29) A graphic representation of data using points and line segments. A line graph shows trends, or changes, in data over a period of time.

line of best fit (p. 35) A line drawn near most of the points on a scatter plot. Also known as a *trend line*.

Español

número irracional (p. 118) Números con decimales como π que no terminan y que además no llegan a una cifra que se repita.

dibujo isométrico (p. 175) vista tridimensional de una esquina interior que muestra tres lados del objeto, y los que no se ven pueden ser indicados con líneas punteadas. Las dimensiones de los dibujos isométricos son proporcionales a las del objeto real, y los límites paralelos del objeto son rectas paralelas.

triángulo isósceles (p. 553) Triángulo que tiene dos lados iguales y uno desigual.

reglas de los exponentes (p. 136) Reglas para las operaciones con números en forma exponencial. Por ejemplo, en una multiplicación de números con la misma base, se suma los exponentes. En una dividisión de números con la misma base, se restan los exponentes.

catetos de un triángulo (p. 334) En un triángulo rectángulo, los dos lados que no son la hipotenusa.

hypotenuse / *hipotenusa*

leg / *cateto*

leg / *cateto*

fracciones comunes (p. 228, 395) Fracciones que tienen el mismo denominador. Por ejemplo, $\frac{5}{8}$ y $\frac{1}{8}$.

línea (p. 156) Conjunto infinito de puntos que se extiende en dos direcciones opuestas. Los dos puntos extremos determinan una línea.

ecuación lineal (p. 318) Ecuación que representa una función lineal.

función lineal (p. 318) Función que se representa con una línea recta.

gráfica de línea (p. 29) Representación gráfica de datos valiéndose de puntos y líneas.

línea de mejor ajuste (p. 35) La línea que puede dibujarse cerca de la mayoría de los puntos en un diagrama de dispersión que muestra una relación entre dos conjuntos de datos.

English   Español

line of symmetry   (p. 375) A line on which a figure can be folded, so that when one part is reflected over that line it matches the other part exactly.

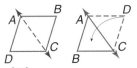

AC is a line of symmetry.
AC es un eje de simetría.

línea simétrica   (p. 375) Línea que divide a una figura en dos parte idénticas o iguales.

line segment   (p. 156) The set of points containing two endpoints and all points between them.

línea recta   (p. 156) Conjunto infinito puntos que apuntan hacia la misma dirección contenidos entre dos puntos extremos.

line symmetry   (p. 375) The property of a figure that can be divided into two matching parts by a line drawn through it.

simetría lineal   (p. 375) Propiedad que tiene cualquier figura que pueda ser dividida en dos partes idénticas o iguales al ser cortada por una línea simétrica.

■ M ■

mean   (p. 10) The sum of the data divided by the number of items of data. Also known as the *arithmetic average*.

media   (p. 10) La suma de un conjunto de números dividida entre la cantidad de números que forman el conjunto. También se conoce como *promedio*.

measures of central tendency   (p. 10) Statistical measures that locate centers of a set of data. These measures are the *mean*, the *median*, and the *mode*.

medidas de tendencia central   (p. 10) Medidas que se usan en la estadística para analizar datos. Estas medidas son la *media*, la *mediana* y la *moda*.

median   (p. 10) The middle value when the data are arranged in numerical order.

mediana   (p. 10) El valor que queda en el medio al ordenarse los datos de menor a mayor. Si quedaran dos valores, la mediana sería el promedio de ambos.

metric units   (p. 53) Units used in the metric system of measurement. Frequently used metric units include the following: length—meter (m), millimeter (mm), centimeter (cm), and kilometer (km); liquid capacity— liter (L) and milliliter (mL); mass—kilogram (kg) and gram (g); and temperature—degrees Celsius.

unidades métricas   (p. 53) Unidades que se usan en el sistema métrico decimal. Algunas unidades métricas que se usan frecuentemente son: longitud—metro (m), milímetro (mm), centímetro (cm) y kilómetro (Km); capacidad líquida-litro (l) y mililitro (ml); masa— kilogramo (Kg) y gramo (g); temperatura—grados centígrados.

midpoint   (p. 356) The point that separates a line segment into two line segments of equal length.

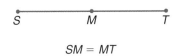

SM = MT

punto medio   (p. 356) Punto que divide a una línea recta en dos rectas de igual medida.

mode   (p. 10) The value that occurs most frequently in a set of data.

moda   (p. 10) Número o artículo que se da con más frecuencia en un conjunto de datos.

monomial   (p. 394) An expression that is a number, a variable, or the product of a number and one or more variables.

monomio   (p. 394) Expresión que puede ser un número, una variable o el producto de un número y de una o más variables.

English   Español

■ N ■

negative correlation (p. 35) A relationship between the sets of data on a scatter plot such that the slope of the line of best fit is down and to the right, or in other words, as the horizontal axis value increases, the vertical axis value decreases.

negative slope (p. 325) The slope of a line that slopes downward from left to right.

net (p. 167) A two-dimensional pattern that, when folded, forms a three-dimensional figure. Dotted lines indicate folds.

net pay (p. 274) The amount of money that a person is paid after taxes are subtracted. Also known as *take-home pay.*

correlación negativa (p. 35) Relación tal entre los conjuntos de datos en un diagrama disperso que la inclinación de la línea más apropiada es hacia abajo y a la derecha, o en otras palabras, cuando el valor del eje horizontal aumenta y el valor del eje vertical disminuye.

inclinación negativa (p. 325) Inclinación de una línea que desciende de izquierda a derecha.

red (p. 167) Patrón bimensional que, cuando se dobla, forma una figura tridimensional. Las líneas punteadas indican los dobleces.

pago neto (p. 274) Cantidad de dinero que se le paga a una persona después de que se le deducen los impuestos.

noncollinear points (p. 157) Points that do not lie on the same line.

A, B, and *C* are noncollinear.
A, B, y *C* son puntos no colineales.

puntos no colineales (p. 157) Puntos que no forman parte de la misma línea.

noncoplanar points (p. 157) Points that do not lie in the same plane.

nonlinear function (p. 338) A function represented by an equation whose graph is not a straight line.

puntos no coplanares (p. 157) Puntos que no forman parte del mismo plano.

función no lineal (p. 338) Función representada por una ecuación que su gráfica no es una línea recta.

■ O ■

obtuse angle (p.155) Any angle measuring between 90° and 180°.

obtuse triangle (p. 161) A triangle having one obtuse angle.

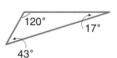

odds of an event (p. 437) A numerical measure of chance comparing the number of favorable outcomes to the number of unfavorable outcomes.

one-point perspective (p. 178) A perspective drawing that uses a single vanishing point to create depth.

open sentence (p. 208) A sentence that contains one or more variables. It can be true or false, depending upon what values are substituted for the variables.

ángulo obtuso (p. 155) Cualquier ángulo mayor de 90° y menor de 180°.

triángulo obtusángulo (p. 161) Triángulo que tiene un ángulo obtuso.

disparidades en un evento (p. 437) Medida numérica casual que se obtiene comparando la cantidad de resultados favorables con la cantidad de resultados desfavorables.

punto de perspectiva (p. 178) Punto de fuga que se utiliza para dar el sentido de la profundidad en un dibujo.

enunciado abierto (p. 208) Enunciado que contiene una o más variables. Puede ser verdadero o falso, dependiendo de los valores que sustituyan a las variables.

English

opposites   (p. 104) Two numbers the same distance from 0 but in opposite directions, for example, −27 and 27.

ordered pair   (p. 308) The coordinates of a point.

order of operations   (p. 114) Steps followed to simplify expressions of more than one kind of operation.

origin   (p. 308) The point of intersection of the *x*-axis and *y*-axis in a coordinate plane.

orthogonal drawing or orthographic projection (p. 179) A three-dimensional view of an object in which the top, front and side views of the object are shown without distorting the objects dimensions. These views appear as if your line of sight is perpendicular to the object's top, front and side.

outcomes   (p. 436) Possible results of an experiment.

outliers   (p. 16) Data values that are much greater than or much less than most of the other values.

Español

opuestos   (p. 104) Dos números que están a la misma distancia del cero pero en direcciones opuestas. Por ejemplo −27 y 27.

par ordenado   (p. 308) Las coordenadas de un punto.

orden de las operaciones   (p. 114) Pasos que se siguen para simplificar una expresión o a más de un tipo de operación.

origen   (p. 308) El punto de intersección del eje *x* y del eje *y* en un plano coordenado o cartesiano.

dibujo ortogonal o proyección ortogonal   (p. 179) Vista tridimensional de un objeto en el cual desde uno de los vértices superiores se muestran las aristas de una figura sin distorsionar las dimensiones de la misma. Esta perspectiva es posible si la visión de la línea es perpendicular a la parte superior de la figura.

resultados   (p. 436) Resultados posibles de un experimento.

datos extremos   (p. 16) Valores que son mucho mayores o mucho menores que la gran mayoría del conjunto de valores.

■ P ■

parallel lines   (p. 352) Lines lying on the same plane that do not intersect.

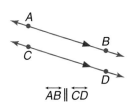

$$\overrightarrow{AB} \parallel \overrightarrow{CD}$$

líneas paralelas   (p. 352) Líneas que están en un mismo plano y no se intersectan.

parallelogram   (p. 154) A quadrilateral having both pairs of opposite sides parallel.

paralelogramo   (p. 154) Cuadrilátero que tiene ambos pares de lados opuestos paralelos.

percent   (p. 260) A ratio that compares a number to 100. Percent means "per one hundred."

por ciento   (p. 260) *Por cien.* Razón que equipara un número con 100.

percent of decrease   (p. 291) The percent of the amount of decrease from the original number.

por ciento a la baja   (p. 291) El porcentaje original dado a una cantidad, disminuye.

percent of increase   (p. 290) The percent of the amount of increase from the original number.

por ciento a la alta   (p. 290) El porcentaje original dado a una cantidad, aumenta.

perfect square   (p. 142) A number whose square root is an integer. For example, 4 is a perfect square.

cuadrado perfecto   (p. 142) Número cuya raíz cuadrada es un entero. Por ejemplo, 4 es un cuadrado perfecto.

perimeter   (p. 62) The distance around a plane figure.

perímetro   (p. 62) La distancia alrededor de una figura.

perpendicular lines   (p. 352) Intersecting lines that form right angles.

líneas perpendiculares   (p. 352) Líneas que se intersecan formando ángulos rectos.

line *m* ⊥ line *n*
recta *m* ⊥ recta *n*

English

perspective drawing (p. 178) A drawing made on a two-dimensional surface in such a way that three-dimensional objects appear true-to-life.

pictograph (p. 24) A graph that displays data with graphic symbols or pictures. The key identifies the number of data items represented by each symbol.

plane (p. 156) A flat surface that extends without end in all directions.

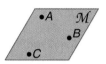

point (p. 156) A specific location in space.

polygon (p. 160) A two-dimensional, closed plane figure formed by joining three or more line segments at their endpoints. Each line segment joins exactly two others and is called a *side* of the polygon. Each point where two sides meet is a *vertex*. Polygons are classified by their numbers of sides.

polyhedron (p. 162) (plural: *polyhedra*). A three-dimensional, closed figure formed by joining three or more polygons at their sides. Each polygon of the polyhedron is called a *face* and joins multiple polygons along their sides. The line segment at which two faces meet is called an *edge*. The point where three or more edges meet is called the *vertex*.

polynomial (p. 394) The sum or difference of two or more monomials. Each monomial is called a *term* of the polynomial.

positive correlation (p. 35) A relationship between the sets of data on a scatter plot such that the slope of the line of best fit is up and to the right, or in other words, as the horizontal axis value increases, so does the vertical axis value.

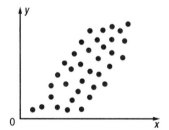

positive slope (p. 325) The slope of a line that slants upward from left to right.

power (p. 391) A number that can be expressed using an exponent. Read 2^4 as "2 to the fourth power."

power rule (p. 137) To raise an exponential number to a power, multiply the exponents. For example, $(3^2)^4 = 3^{2 \times 4} = 3^8$.

precision (p. 52) The exactness to which a measurement is made.

preimage (p. 370) The original figure of a transformation.

Español

dibujo en perspectiva (p. 178) Dibujo de una figura hecho en una superficie bimensional de tal manera que se vea tridimensional.

pictografía (p. 24) Gráfica que usa figuras o símbolos para representar datos. La clave identifica el número de artículos representados por cada símbolo.

plano (p. 156) Superficie llana que se extiende sin fin en todas direcciones.

punto (p. 156) Lugar específico en el espacio.

polígono (p. 160) Figura bidimensional, plana y cerrada que se forma al conectarse tres o más línea rectas en sus puntos extremos. Cada línea se conecta con otras dos y se le llama *lado del polígono*. Cada punto donde dos lados se unen es un *vértice*. Los polígonos se clasifican de acuerdo al número de sus lados.

poliedro (p. 162) Figura cerrada tridimensional formada por la unión de tres o más polígonos. Cada polígono recibe el nombre de *cara* y cada uno de sus lados es el punto de intersección con otro polígono. Esta línea de encuentro se llama *arista*. El punto de encuentro de tres o más aristas se llama *vértice*.

polinomio (p. 394) La suma o resta de dos o más monomios. Cada monomio es un *término* del polinomio.

correlación positiva (p. 35) Relación tal entre los conjuntos de datos en un diagrama disperso que la inclinación de la línea más apropiada es hacia arriba y a la derecha, o en otras palabras, cuando el valor del eje horizontal y vertical aumentan.

pendiente positiva (p. 325) La pendiente de una recta que se inclina de izquierda a derecha.

potencia (p. 391) Número que puede ser expresado usando un exponente. Léase 2^4 como "2 a la cuarta potencia".

regla de la potencia (p. 137) Para elevar un número exponencial a una potencia, se multiplican los exponentes. Por ejemplo, $(3^2)^4 = 3^{2 \times 4} = 3^8$.

precisión (p. 52) El grado de exactitud al que se toma una medida.

pre-imagen (p. 370) Figura original donde se da una transformación.

English

Español

principal   (p. 280) The amount of money that is earning interest or that you are borrowing.

capital   (p. 280) Cantidad de dinero que genera intereses, sea dinero invertido o prestado.

prism   (p. 166) A polyhedron with two identical faces, called *bases*. The bases are congruent polygons. The other faces are parallelograms.

prisma   (p. 166) Poliedro con dos caras idénticas llamadas *bases*. Las bases son polígonos congruentes. Las otras caras son paralelogramos.

probability   (p. 436) A ratio written as a percent, fraction or decimal of the number of favorable outcomes to the number of possible outcomes. Probability is a numerical measure of chance that is between 0 and 1.

probabilidad   (p. 436) Razón escrita como por ciento, fracción o decimal del número de resultados favorables al número total de resultados posibles. La probabilidad es una medida numérica posible que está entre 0 y 1.

property taxes   (p. 275) A tax paid by homeowners based on the value of their house and property. These taxes help pay for services such as schools, libraries and a police force.

impuestos sobre la renta   (p. 275) Impuestos pagados en base al valor de una casa o una propiedad. Estos impuestos se usan para dar servicios tales como el de la educación, el de bibliotecas o el de seguridad pública.

proportion   (p. 84) An equation stating that two ratios are equivalent.

proporción   (p. 84) Ecuación que establece que dos razones son equivalentes.

pyramid   (p. 167) A polyhedron with only one base. The other faces are triangles. A pyramid is named by the shape of its base.

pirámide   (p. 167) Poliedro con una sola base. Las otras caras son triángulos. Una pirámide recibe el nombre de acuerdo a la forma de su base.

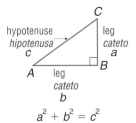

Pythagorean Theorem   (p. 334) A relationship between the legs of a right triangle and the hypotenuse in which the sum of the squares of the legs is equal to the square of the hypotenuse.

$$a^2 + b^2 = c^2$$

teorema de Pitágoras   (p. 334) En un triángulo rectángulo, la suma de los cuadrados de las longitudes de los catetos es igual al cuadrado de la longitud de la hipotenusa.

■ Q ■

quadrant   (p. 308) One of the four regions formed by the axes of the coordinate plane.

cuadrante   (p. 308) Una de las cuatro regiones formadas por los ejes del plano de coordenadas.

quadrilateral   (p. 154) A polygon having four sides.

cuadrilátero   (p. 154) Polígono que tiene cuatro lados.

quartiles   (p. 38) The numbers that separate a set of data into four equal parts.

cuarteros   (p. 38) Números que transforman un conjunto de datos en cuatro partes iguales

■ R ■

radical sign   (p. 142) The symbol $\sqrt{}$, used to indicate a nonnegative square root.

signo radical   (p. 142) El símbolo $\sqrt{}$, que se usa para indicar la raíz cuadrada no negativa.

Glossary/Glosario

English

Español

radius (p. 80) The fixed distance between the center of a circle and the circle itself.

radio (p. 80) Distancia que hay entre el centro del círculo y cualquier punto de la circunferencia.

random sampling (p. 6) Each member of the population has an equal chance of being selected.

muestra aleatoria (p. 6) Cada miembro de una población tiene la misma posibilidad de ser seleccionado.

range (p. 10) The difference between the greatest and least values in a set of data.

alcance (p. 10) La diferencia entre el número mayor y el menor en un conjunto de datos.

range (p. 314) All of the second values of the ordered pairs in a relation.

rango (p. 314) Todos los segundos valores de los pares ordenados en una relación.

rate (p. 280) The percent charged per year for the use of money over a given period of time.

tasa (p. 280) Razón que compara dos tipos diferentes de cantidades.

ratio (p. 74) A comparison of one number to another. A ration can be written three different ways: 2 to 5, 2:5, or $\frac{2}{5}$.

razón (p. 74) Comparación de un número con otro que se representa de las tres siguientes formas: 2 es a 5, 2:5, o $\frac{2}{5}$.

rational number (p. 118) Any number that can be expressed in the form $\frac{a}{b}$, where a is any integer and b is any integer except 0.

número racional (p. 118) Cualquier número que pueda ser expresado en forma $\frac{a}{b}$, donde a es cualquier entero y b es también cualquier entero excepto 0.

ray (p. 156) A part of a line having one endpoint and extending without end in one direction.

rayo o línea recta (p. 156) Parte de una línea que tiene un punto extremo y se extiende sin fin en una dirección.

real numbers (p. 118) The set of irrational and rational numbers together.

números reales (p. 118) Conjunto formado por los números racionales e irracionales.

reciprocals (p. 219) Two numbers are reciprocals when their product is 1.

recíprocos (p. 219) Dos números son recíprocos cuando su producto es 1.

rectangle (p. 154) A parallelogram having four right angles.

rectángulo (p. 154) Paralelogramo que tiene cuatro ángulo rectos.

reflection (p. 374) A transformation in which a figure is flipped, or reflected, over a line of reflection.

reflexión (p. 374) Transformación en la cual una figura se voltea o se refleja sobre una línea de reflejo.

regular polygon (p. 160) A polygon with all sides of equal length (equilateral) and all angles of equal measure (equiangular).

polígono regular (p. 160) Polígono que tiene todos sus lados iguales.

relation (p. 314) A set of paired data, or ordered pairs.

relación (p. 314) Conjunto de datos en pares o pares ordenados.

rhombus (p. 154) A parallelogram with all sides the same length.

rombo (p. 154) Paralelogramo que tiene sus cuatro lados iguales.

right angle (p. 155) An angle having a measure of 90°.

ángulo recto (p. 155) Angulo que mide 90°.

right triangle (p. 161) A triangle having a right angle.

triángulo rectángulo (p. 161) Triángulo que tiene un ángulo recto.

English

rotation   (p. 380) A transformation in which a figure is turned, or rotated, about a point.

rotational symmetry   (p. 381) The property of a figure that can be fitted exactly over its original position by rotating it. The *order of rotational symmetry* is the number of times a figure fits exactly over itself in the process of a complete turn.

Español

rotación   (p. 380) Tranformación donde una figura gira o rota sobre un punto.

simetrí rotacional   (p. 381) Propiedad de una figura que puede quedar exactamente en su posición original después de haber rotado. El orden de la simetría racional es el número de veces que una figura queda exactamente sobre sí misma en el proceso de una rotación completa.

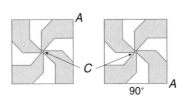

■ S ■

sale price   (p. 270) The regular price minus the discount.

sample   (p. 6) A part of a population.

sample space   (p. 446) The set of all possible outcomes in a probability experiment.

sampling   (p. 6) See *cluster sampling, convenience sampling, random sampling, and systematic sampling.*

scale drawing   (p. 85) A drawing that represents a real object. All lengths in the drawing are proportional to the actual lengths in the object. The ratio of the size of the drawing to the size of the actual object is called the *scale* of the drawing.

scalene triangle   (p. 161) A triangle with no sides the same length and no angles the same measure.

scatter plot   (p. 34) A graphic representation which displays two sets of data on the same set of axes.

scientific notation   (p. 133) A notation for writing a number as the product of a number between 1 and 10 and a power of 10.

sector   (p. 20) A slice of a circle graph that represents a percentage of the total number of data.

set   (p. 118) A group.

side   (p. 160) See *polygon.*

simple interest   (p. 280) Interest that is paid only on the principal.

simplify   (p. 48) To simplify an expression is to perform as many of the indicated operations as possible.

slope   (p. 324) The ratio of the change in y to the change in x, or the ratio of the rise to the run.

precio de oferta   (p. 270) Precio normal menos el descuento.

muestra   (p. 6) Parte de una población.

espacio de muestra   (p. 446) El conjunto de todos los resultados posibles en un experimento probable.

muestra   (p. 6) Véase *muestra por conglomerado, muestra por conveniencia, muestra aleatoria y muestra sistemática.*

dibujo a escala   (p. 85) Dibujo que representa un objeto real. Todas las longitudes en el dibujo son proporcionales a las longitudes reales del objeto. La razón del tamaño del dibujo al tamaño del objeto real se llama la *escala* del dibujo.

triángulo escaleno   (p. 161) Triángulo que tiene tanto sus lados como sus ángulos desiguales.

diagrama disperso   (p. 34) Representación gráfica que muestra dos conjuntos de datos en el mismo sistema de coordenadas.

notación científica   (p. 133) Notación para escribir un número como el producto de un número entre 1 y 10 y una potencia de 10.

sector   (p. 20) Parte de una gráfica dentro del círculo que representa un porcentaje del número total de datos.

conjunto   (p. 118) Grupo.

lado   (p. 160) Véase *polígono.*

interés simple   (p. 280) Interés que se paga en base a un capital.

simplificar   (p. 48) Simplificar una expresión es hacer todas las operaciones indicadas que sean posibles.

pendiente   (p. 324) Razón del cambio en y al cambio en x, o razón del cambio en la subida (cambio vertical) a la recorrida (cambio horizontal).

English

Español

slope-intercept form   (p. 328) A linear equation in the form $y = mx + b$, where m is the slope of the graph of the equation and b is the y-intercept.

forma de pendiente e intersección   (p. 328) Ecuación lineal en forma de $y = mx + b$ donde m es la pendiente de la gráfica de la ecuación y b es el interceptor en y.

sphere   (p. 167) A three-dimensional figure that is the set of all points in space that are the same distance from a given point, called the center of the sphere.

C is the center of the sphere.
C es el centro de la esfera.

esfera   (p. 167) Figura tridimensional formada por puntos que están a la misma distancia de un punto dado llamado el centro de la esfera.

square (geometric)   (p. 154) A parallelogram with four right angles and all sides the same length.

cuadrado (geométrico)   (p. 154) Paralelogramo con cuatro ángulos rectos y todos los lados iguales.

square (numeric)   (p. 142) The product of a number and itself.

cuadrado (numérico)   (p. 142) Producto de un número multiplicado por sí mismo.

square root   (p. 142) One of two equal factors of a number.

raíz cuadrada   (p. 142) Uno de los dos factores iguales de un número.

standard form   (p. 394) The arrangement of the terms of a polynomial in order from greatest to least powers of one of the variables.

forma estándar   (p. 394) La forma $Ax + By = C$ donde A y B son números reales y ambos no son iguales a cero.

stem-and-leaf plot   (p. 16) A graphic representation that organizes and displays data by designating digits from the data values as leaves and stems. Leaves are the last digits of the data values, and stems are the digits in front of the leaves.

diagrama de tallo y hoja   (p. 16) Representación gráfica donde se organizan y se muestran datos donde algunos dígitos se usan como tallos, y otros como hojas. Las hojas representan los últimos dígitos del conjunto de valores, y los tallos representan los dígitos que están frente a las hojas.

stems   (p. 16) See *stem-and-leaf plot*.

tallos   (p. 16) Véase *diagrama de tallo y hoja*.

straight angle   (p. 155) An angle having a measure of 180°.

ángulo plano   (p. 155) Angulo que mide 180°.

supplementary angles   (p. 352) Two angles whose sum measures 180°.

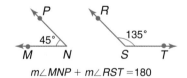

$m\angle MNP + m\angle RST = 180$

ángulos suplementarios   (p. 352) Dos ángulos cuya suma de sus medidas es igual a 180°.

surface area   (p. 194) The number of square units it would take to cover a surface.

superficie del área   (p. 194) Número de unidades al cuadrado que cubren una superficie.

survey   (p. 6) A means of collecting data for the analysis of some aspect of a group or area.

encuesta   (p. 6) Manera de recolectar datos para analizar algún aspecto de un determinado grupo o área.

systematic sampling   (p. 6) After a population is ordered in some way, its members are chosen according to a pattern.

muestra sistemática   (p. 6) Después de que una población se ordena de algún modo, los miembros de la población se eligen de acuerdo a un patrón.

English   Español

■ T ■

tax (p. 274) A charge, usually a percentage, that is imposed by an authority.

impuesto (p. 274) Cobro, por lo general un porcentaje, impuesto por las autoridades políticas.

terms (p. 84, 228, 394) The parts of an expression separated by addition or subtraction signs. In the expression $2a + 3b + 4a - 5b^2$, the terms $2a$ and $4a$ are *like terms* because they have identical variable parts. The terms $3b$, $4a$, and $5b^2$ are *unlike terms* because they have different variable parts.

términos (p. 84, 228, 394) Partes de una expresión algebraica separadas por el signo de la suma o de la resta. En la expresión $2a + 3b + 4a - 5b^2$, los términos $2a$ y $4a$ son *términos semejantes* porque sus variables son idénticas. Los términos $3b$, $4a$, y $5b^2$, son *términos diferentes* porque sus variables son diferentes.

tessellation (p. 381) A pattern in which identical copies of a figure fill a plane so that there are no gaps or overlaps. In a *regular tessellation*, each shape is a regular polygon, and all the shapes are congruent.

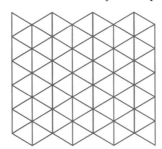

mosaico (p. 381) Patrón de una figura en donde copias idénticas caben sin dejar huecos y sin sobrepasar los bordes.

time (p. 280) Refers to the period of time during which the principal remains in a bank account. It also refers to the period of time that borrowed money has not been paid back.

plazo (p. 280) Período que un capital dura depositado en una cuenta bancaria. O también el período que un capital que se ha prestado dura sin pagarse.

transformation (p. 370) A movement of a figure by translation, reflection or rotation. The new figure is called the *image* of the original, and the original is called the *preimage* of the new.

transformación (p. 370) Movimiento de una figura ya sea por traslación, reflexión o rotación. La nueva figura recibe el nombre de *imagen* del original, y al original se le llama *pre-imagen*.

translation (p. 370) A transformation in which all the points of a figure are slid along a plane the same distance and the same direction producing a new figure that is exactly like the original.

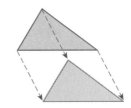

traslación (p. 370) Transformación en la cual todos los puntos de una figura se deslizan la misma distancia y hacia la misma dirección produciendo una nueva figura que es idéntica a la original.

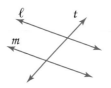

transversal (p. 353) A line that intersects two or more lines in a plane at different points.

transversal (p. 353) Línea que interseca dos o más líneas en puntos distintos de un plano.

Line *t* is a transversal.
La recta t es una transversal.

trapezoid (p. 154) A quadrilateral having one and only one pair of parallel sides.

trapezoide (p. 154) Cuadrilátero que tiene solamente un par de lados paralelos.

tree diagram (p. 446) A diagram that shows the total number of possible outcomes in a probability experiment.

diagrama de árbol (p. 446) Diagrama que muestra todos los resultados posibles en un experimento probable.

English / Español

English	**Español**

trend line   (p. 35) See *line of best fit.*

triangle   (p. 160) A polygon having three sides.

trinomial   (p. 394) A polynomial with three terms.

línea tendencial   (p. 35) Véase *línea más apropiada.*

triángulo   (p. 160) Polígono que tiene tres lados.

trinomio   (p. 394) Polinomio compuesto por tres términos.

■ U ■

undefined slope   (p. 325) The slope of a vertical line.

inclinación indefinida   (p. 325) Inclinación de una línea vertical.

■ V ■

vanishing point   (p. 178) A point at which parallel lines seem to intersect and at which a perspective drawing seems to disappear.

variable expressions   (p. 124) Expressions with at least one variable.

variable   (p. 124) An unknown number represented by a letter, such as *n, x,* or *y.*

Venn diagram   (p. 492) A diagram in which sets of data are represented by overlapping circles. These circles show relationships among the sets so that all possible combinations have a distinct area in the diagram.

vertex   (p. 160) (plural: *vertices*) The point at which two sides of a polygon meet or where three or more edges of a polyhedron meet.

punto de fuga   (p. 178) Punto en el cual las líneas paralelas parecen hacer intersección pero que en un dibujo a perspectiva tiende a desaparecer.

expresiones variables   (p. 124) Expresiones que contienen por lo menos una variable.

variables   (p. 124) Número desconocido que se representa con una letra, tal como *n, x,* o *y.*

diagrama de Venn   (p. 492) Diagrama en el cual un conjunto de datos se representan. En un diagrama de Venn cada cosa se representa por una región circular dentro de un rectángulo. El rectángulo se llama el *conjunto universal.*

vértice   (p. 160) Punto donde se intersecan dos lados de un polígono o tres o más aristas de un poliedro.

vertical angles   (p. 353) The opposite angles, formed by the intersection of two lines in a plane, that are not adjacent to each other and have the same measure.

∠1 and ∠3 are vertical angles.
∠2 and ∠4 are vertical angles.
∠1 *y* ∠3 *son ángulos opuestos por el vértice.*
∠2 *y* ∠4 *son ángulos opuestos por el vértice.*

ángulos opuestos por el vértice   (p. 353) Ángulos de la misma medida formados por dos líneas que se intersecan.

vertical line test   (p. 315) A test used to tell if a graphed relation is a function. If any vertical line goes through the plotted graph more than once, then the relation is not a function.

volume   (p. 184) The number of cubic units enclosed by three-dimensional object.

prueba de la línea vertical   (p. 315) Prueba que se usa para señalar si una relación graficada es una función. Si cualquier línea vertical pasa a través de la gráfica diagramada más de una vez, entonces la relación no es una función.

volumen   (p. 184) Número de unidades cúbicas que se necesita para llenar un objeto tridimensional.

■ X ■

x-**axis**   (p. 308) The horizontal number line in the coordinate plane.

eje *x* o de las abscisas   (p. 308) Recta numérica horizontal en un plano cartesiano.

English

x-coordinate (p. 314) The first number in an ordered pair.

x-intercept (p. 319) See *intercept*.

■ Y ■

y-axis (p. 308) The vertical number line in the coordinate plane.

y-coordinate (p. 314) The second number in an ordered pair.

y-intercept (p. 319) See *intercept*.

■ Z ■

zero slope (p. 325) The slope of a horizontal line.

Español

coordenada x (p. 314) Primer número en un par ordenado.

intersección en x (p. 319) Véase *intersección*.

eje *y* o de las ordenadas (p. 308) Recta numérica vertical en un plano cartesiano.

coordenada *y* (p. 314) Segundo número en un par ordenado.

intersección en *y* (p. 319) Véase *intersección*.

inclinación cero (p. 325) Grado de inclinación de una línea horizontal.

Glossary/Glosario

Selected Answers

Chapter 1: Data and Graphs

Lesson 1-1, pages 6–9

1. cluster sampling   3. Answers will vary.   5. personal business For 7 and 9, answers will vary.   11. c
For 13 and 15, advantages and disadvantages will vary.
13. cluster   15. random   17. Spenser   19. none of them
For 21 and 23, advantages and disadvantages will vary.
21. random   23. systematic   25. Answers will vary. Possible answer: poll every fifth customer at the food court.
27. Answers will vary.   29. orange juice   31. Answers will vary.   33. Answers will vary. Possible answer: Convenience—ask the first fifty people in line. Random—randomly draw student i.d. numbers from the school directory and poll them. Cluster—Poll only the freshmen.
For 35 and 37, answers will vary.   35. Most people who live near railroad tracks would like the tracks moved because trains make too much noise.   37. Answers will vary.
39. 8569   41. 33,090   43. 90   45. 270

Lesson 1-2, pages 10–13

1. 8.8   3. 7.0   5. The mean incorporates all of the data and is the average.   7. median   9. 9.7   11. 8.7, 9.2   13. The median—it is not affected by the 15.0 time.
15. 31.5 mi/h   17. 30 mi/h   19. mean   21. median
23. $69,620.83   25. none   27. The median because the mean is too much affected by the President's salary.
29. 191   31. about 5740   33. 388   35. 148   37. 1.5 h
39. 1.5 h   41. mean: 2.1 h, median: 1.5 h, mode: 1.5 h
43. 9.8 calories, 9.2 calories   45. The mean and the median both equal the middle number.   47. The median income did not change. The median value does not change just because minimum values change.   49. No; sample data: 14, 16   51. Answers will vary.
53. 1,034,500   1,035,000   1,030,000
55. 379,900   380,000   380,000

Review and Practice Your Skills, pages 14–15

1. systematic   3. cluster   5. The number is increasing.
7. Store A   9. 0.8   11. 8   13. 24   15. 3   17. 13
19. 150   21. 14   23. 13   25. 6   27. random
29. systematic   31. Older students join more clubs.
33. mean = 5   35. mean = 36.25
median = 5   median = 37.5
range = 7   range = 35
mode = 3, 5   mode = 40

Lesson 1-3, pages 16–19

1.
```
3 | 2 2 6 7 7 8
4 | 3 4 6 9
5 | 2 5 5 6 8 9
6 | 1 1 2 2 3
7 | 0 2 3 7
8 | 2 4
9 | 6
```
4|3 represents 43.

3. 32   5. 99   7. 34 and 99   9. high 40s, low 80s
11. 24   13. 5   15. 74.6, 75, 75, 48   17. 27   19. 8
21. 0   23. low 20s, mid 40s

25.
```
5 | 3
4 | 2
3 | 3 5
2 | 2 3 5
1 | 1 3 3 5 7 8
0 | 8
```
5|3 represents 5.3 mi/day.

39. not correct;

Age at inauguration		Age at death
	8	0 5
	7	3
8 7 7	5	

7|5 represents 57 years old.

27. 594 calories
29. Answers will vary.
31. 24   33. Kennedy, 46
35. Cleveland was also the 22nd president.
37. inauguration: 50's, death: mid 60's, high 70's/ low 80's

41. mean

Lesson 1-4, pages 20–21

1. 8   3. 28   5. 26.5%   7. 35.1%   9. about 3 times
11. 49%   13. 7 hours

15.
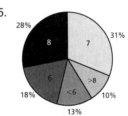

To estimate the size of a sector, set up the equation $\frac{x}{360} = \frac{\%}{100}$, where x is the angle measure of the sector and % is the whole number given in the table. Solve for x.

17.
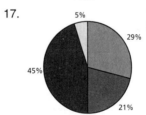

19. 5, 5.2, 5.20
21. 58, 58.0, 57.98

Review and Practice Your Skills, pages 22–23

1. 53   3. 41   5. 34
7.
```
6 | 0 2 4 5 6 6
7 | 0 2 4 4 6 6 8 9
8 | 0 1 4 6
```
6|0 represents a score of 60.

9.
```
1 | 2 3 4 6 6 7 8 8
2 | 2 3 4 5 5 6
3 | 1
```
1|2 represents 12,000 points.

11. 5%   13. 18%   15. 30%   17. 21%   19. 64%
21. 6%   23. systematic

25. mean: 242.14
median: 245
range: 25
mode: 245

27. mean: 2.54
median: 3
range: 4
mode: 3

29.

1	4 4 4 6 6 7 8 8
2	0 0 2 5
3	0 1 8
4	0 1

1|4 represents 14 years old.

31. 45%   **33.** 15%

Lesson 1-5, pages 24–27

1.

Number of Individuals Who Have Flown in Space

Austria	人
Canada	人 人 人 人 人
Cuba	人
France	人 人 人 人 人 人 人
Germany	人 人 人 人 人 人 人 人 人
India	人
Italy	人 人 人
Japan	人 人 人 人 人
Mexico	人
Saudi Arabia	人
U.K.	人
Vietnam	人

Key: 人 = 1 individual

3. There are too many individuals in relation to the other countries.
5. pita bread, 200 mg
7. $10\frac{3}{4}$ **9.** Answers will vary.

11.

Art Prices (dollars)

Price	Tally	Frequency
300					3
450					3
600	||||-|	6
750	||||-	5
1200				2
1350			1
1800				2
2700				2

13. $931.25   **15.** They both give data representations, but frequency tables use exact numbers, whereas pictographs use symbols as estimations.   **17.** 200 students   **19.** 150 students

21.

Weight of Dog Food Packages (pounds)

19.6	⋈
19.7	⋈ ⋈ ⋈ ⋈
19.8	⋈ ⋈ ⋈ ⋈ ⋈
19.9	⋈ ⋈ ⋈
20.0	⋈ ⋈ ⋈ ⋈ ⋈ ⋈ ⋈
20.1	⋈ ⋈ ⋈ ⋈ ⋈ ⋈
20.2	⋈ ⋈ ⋈ ⋈
20.3	
20.4	⋈

23. 23 bags
25. Yes; no; the median would only be the response with the highest frequency when it is the mode.
27. game 8

29.

Hockey Team Points

Player	Points
Kelly	⬤⬤⬤⬤⬤⬤⬤◖
Green	⬤⬤⬤⬤⬤⬤⬤⬤⬤◖
Tookey	⬤⬤⬤⬤⬤◖
Currie	⬤⬤⬤⬤◖
Smith	⬤⬤⬤⬤
Charron	⬤⬤⬤⬤⬤⬤

Key: ⬤ = 8 points

31. 50.5   **33.** none
35. Answers will vary.
37. 740.5   **39.** 3805

Lesson 1-6, pages 28–31

1. Monday, 40   **3.** Friday, Saturday   **5.** 4 million

7.

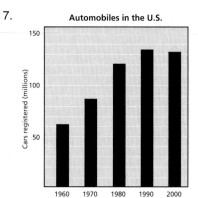

Automobiles in the U.S.

9. 11 times
11. 1971
13. Answers will vary. Possible answer: A line graph illustrates increases and decreases better than other graphs.
15. Black Sea, Arctic Ocean, Med. Sea, Hudson Bay   **17.** about $125   **19.** 1985
21. 79.8 yr

23. Answers will vary, but answers around 81 yr are consistent with the graph.   **25.** boys   **27.** Answers will vary.   **29.** 84.802   **31.** 129.21

Review and Practice Your Skills, pages 32–33

1.

Test Scores

Score	Frequency
75	2
80	3
85	7
90	4
95	3
100	1

3. 86.5

5.

Ages

Score	Frequency
13	3
14	6
15	7
16	11
17	5

7. 15.28   **9.** Clothing   **11.** $21,000   **13.** 17 h
15. Between 1982 and 1985 taxes fluctuated with relative stability between $2000 and $4000. Taxes then skyrocketed to $9000 from 1985 to 1988.   **17.** 1985 and 1986   **19.** 38%   **21.** cluster

23.

Test Scores

6	5 5
7	0 0 0 0 5 5 5
8	0 0 0 0 0 0 5 5
9	0 0 0 0 0

6|5 represents a score of 65.

Lesson 1-7, pages 34–37

1. $2500   **3.** neither   **5.** July   **7.** 100   **9.** Answers will vary.   **11.** about $1400   **13.** positive   **15.** Answers will vary.   **17.** Answers will vary.   **19.** yes   **21.** living conditions and medical technology   **23.** negative
25. women, the line is steeper   **27.** The line of best fit does not connect all the data points. A line graph is best used when there is only one vertical axis data value for every horizontal axis data value.   **29.** $40;

Years	Simple interest	Final balance
1	$5	$105
2	10	110
3	15	115
4	20	120
5	25	125
6	30	130
7	35	135
8	40	140

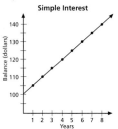

Simple Interest

31.

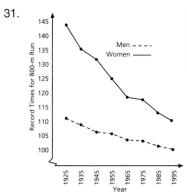

33. line graph
35. line graph or scatter plot
37. Diplodocus
39. 2

35.

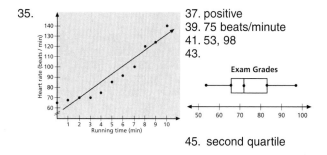

37. positive
39. 75 beats/minute
41. 53, 98
43.

45. second quartile

Lesson 1-8, pages 38–41

1. 50%   3. when there is a large gap in the middle of the data   5. 20, 26   7. 45, 30   9. 52, 50   11. $450, $200   13. $300   15. $235–$300   17. Yes. 139.0, 300.0, 103.0   19. produce—63.74, use—85.73; use   21. U.S. use—93.36   23. Russia, Japan, Germany, Canada, India   25. boys   27. median, range   29. Check students' work.

31.

33. The second quartile has a smaller range.
35. Answers will vary, but should include that 25% of the data is equal to either the first or third quartile.

37.

39. $0–$9
41. bar graph

Chapter 1 Review, pages 42–44

1. h   3. d   5. f   7. k   9. b   11. cluster   13. In this case, cluster sampling may be better. Random sampling from phone books may yield apartment dwellers, condo owners, etc.   15. The people being interviewed are those that do the shopping and determine what they will buy for their households.   17. 72   19. 45   21. mode   23. 5, 6, 7, 8, and 9   25. 5   27. 55–61, 62–66, 75–79, 91–96   29. 15   31. 21
33.

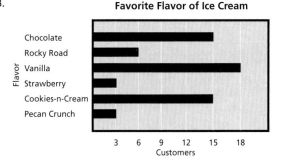

Chapter 2: Measurement

Lesson 2-1, pages 52–55

1. decimeter   3. foot   5. tape measure   7. cup
9. 47.5 cm   11. Answers will vary. A ruler is more accurate.   13. ounce   15. centimeter   17. teaspoon
19. ruler   21. thermometer   23. ruler or tape measure
25. yard   27. teaspoon   29. 1 gal   31. 6.3 mm
33. 2000 gal   35. A tailor should always use precise measurements to make sure clothes will fit.   37. 6 in.

For 39 and 41, answers will vary.   43. A gallon is too small of a unit to measure volume of such large bodies of water.   For 45 and 47, answers will vary.   For 49 and 51, answers will vary.

53.

2	0 2 2 5 6 9 9
3	1 2 4 6 6 7 8 9
4	1 1 3 4 4 7 8

2|6 represents 26.

55. nowhere   57. 0

Lesson 2-2, pages 56–59

1. 60   3. 52   5. 2000   7. 1240 yd   9. 8 yd 1 ft
11. 1 ft 3 in.   13. 2855   15. 120   17. 2.5   19. 12
21. 9; 4   23. 4; 8   25. 7   27. 15; 1   29. 2000
31. 1.254   33. 3.5   35. 6 in.   37. 7 gal 1 qt   39. 9 lb
41. 5 in.   43. 26   45. 175   47. 650   49. 3
51. Convert lb to oz and compare.   53. 9:33 P.M.
55. 4 gal 2 qt   57. 2 gal 1 pt   59. 7 ft   61. 4 fl oz

63.

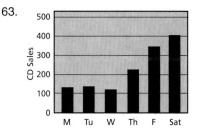

65. bar graph

Review and Practice Your Skills, pages 60–61

1. foot   3. pint   5. second   7. stopwatch
9. thermometer   11. measuring cup   13. gram
15. inch   17. 1 L   19. 4 yd   21. 2.3 mm   23. 7.5
25. 0.85   27. 0.4   29. 0.1   31. 16,000   33. 250
35. 126   37. 10,500   39. 3 lb 8 oz   41. 11 yd 1 ft
43. 67 c 4 fl oz   45. 31 in.   47. 12,600   49. 5.45
51. 1540   53. 1900   55. yard   57. milligram   59. 30 m
61. 4.75 kg   63. 3500   65. 85   67. 4500   69. 3 ft 4 in.
71. 4 qt   73. 125   75. 700

Lesson 2-3, pages 62–65

1. 29.6 m   3. 30 cm   5. 32 in.   7. 5.25 in.   9. 340 cm
11. Divide feet of wallpaper by 3. Convert any fractional amount to 1 yd.   13. 24 ft   15. 28 ft   17. 64 ft
19. 24 ft   21. 356 mm   23. 104 cm   25. 24 m
27. 160 cm   29. The perimeter of the lake is less than the length of the road since it is inside it.   31. $\approx$10 cm
33. 43.6 m   35. 200.8 m   37. 12 m   39. 12 m
41. Answers will vary.   43. mean = 4629 ft,   median = 5116 ft,   mode = 5600 ft   45. 5

Lesson 2-4, pages 66–69

1. 136 in.^{2} 3. 304 m^2 5. 468 in.^{2} 7. 324 in.^{2}
9. Check students' answers.   11. 1890 m^2 13. 78.44 km^2
15. 3136 mm^2 17. 572.2575 in.^{2} 19. 384 in.^{2}
21. 408 in.^{2} 23. 576 cm^2 25. 1280 m^2 27. 13 acres
29. No. Many perimeters yield an area of 200 mi^2.
31. 2106 ft^2, 195.63 m^2 33. 748 ft^2, 69.46 m^2
35. 45 ft^2, 4.16 m^2 37. B   39. 78 in.^{2} 41. Divide by 9. 1 square yard is the same as 3 ft $\times$ 3 ft, or 9 ft^2.
43. 15   45. 5500   47. 3, 1

Review and Practice Your Skills, pages 70–71

1. 16 ft   3. 18 ft   5. 24 ft.   7. 43 in.   9. 11 m
11. 1200 ft   13. 120 in.   15. 140 in.   17. 100 ft
19. 1.2 m^2 21. 144 in.^{2} 23. 150 cm^2 25. 21 yd^2
27. 225 cm^2 29. 66 mm^2 31. 27 in.^{2} 33. 45 yd^2
35. 200 mi   37. 1.3 L   39. 5100   41. 400   43. 300
45. 36 in.   47. 33 cm   49. 42 ft^2 51. 1.44 cm^2

Lesson 2-5, pages 72–73

1. P 3. A 5. 43.8 m   perimeter   7. 54 ft   9. 40 ft
11. 80 ft 6 in.   13. $96.66   15. $71.60   17. $144.10
19. 30 yd^2 21. 12 yd^2 23. $960   25. $384
27. $2880   29. 3120 ft^2 31. 140 ft   33. 8735 ft^2
35. systematic

Lesson 2-6, pages 74–77

1. 1:2, $\frac{1}{2}$ 3. 10:9, $\frac{10}{9}$ 5. $\frac{8}{3}$ 7. $\frac{5}{1}$ 9–11. Sample

answers given.   9. 2:3, 8:12, 12:18   11. $\frac{3}{5}, \frac{6}{10}, \frac{9}{15}$

13. no   15. $\frac{17}{20}$ = 0.85   17. $\frac{10}{3}$, 10 to 3   19. 1:3, 1 to 3

21. $\frac{7}{10}$, 7 to 10   23. 4:1, $\frac{4}{1}$ 25. $\frac{1}{5}$ 27. $\frac{16}{3}$ 29. $\frac{28}{11}$

31–37. Sample answers given.   31. $\frac{5}{6}, \frac{10}{12}, \frac{20}{24}$

33. $\frac{4}{5}, \frac{8}{10}, \frac{12}{15}$ 35. $\frac{14}{16}, \frac{21}{24}, \frac{28}{32}$ 37. $\frac{6}{20}, \frac{9}{30}, \frac{12}{40}$

39. no   41. no   43. yes   45. no   47. $\frac{1}{6}$ 49. $\frac{\text{Calories}}{\text{Weight}}$

while studying: 110 lb: $\frac{1.2}{110} \approx$ 0.0109   125 lb: $\frac{1.4}{125} \approx$

0.0112   150 lb: $\frac{1.7}{150} \approx$ 0.0113   175 lb: $\frac{1.9}{175} \approx$ 0.0109

200 lb: $\frac{2.2}{200} \approx$ 0.0110   51. 16   parts cornstarch

53. 3:8   55. midsize ratio is greater;   midsize = $\frac{2}{3}$

economy = $\frac{8}{15}$ 57. Answers will vary. Example: 9:10

59. 3:4   61. 6:17:12   63. A: 3 6 9 12 15   B: 4 8 12 16

20   C: 12 24 36 48 60   65. 0.005   67. 0.5   69. 0.6

71. 20 c   73. 3000 g

Review and Practice Your Skills, pages 78–79

1. A 3. P 5. P 7. 180 ft^2 9. 270 ft^2 11. $\frac{1}{8}$, 1 to 8

13. 10:3, 10 to 3   15. $\frac{12}{7}$, 12:7   17. $\frac{5}{2}$, 5 to 2   19. $\frac{5}{1}$

21. $\frac{4}{1}$ 23. $\frac{1}{4}$ 25–31. Sample answers given.

25. $\frac{5}{4}, \frac{15}{12}, \frac{20}{16}$ 27. $\frac{2}{5}, \frac{4}{10}, \frac{6}{15}$ 29. $\frac{4}{5}, \frac{8}{10}, \frac{12}{15}$

31. $\frac{8}{6}, \frac{12}{9}, \frac{16}{12}$ 33. no   35. yes   37. yes   39. no

41. $\frac{10}{24} = \frac{5}{12}$ 43. 8.2 cm   45. 13$\frac{1}{3}$ 47. 11 m

49. 3 yd^2 51. A 53. yes   55. yes

Lesson 2-7, pages 80–83

1. 254.3 cm^2 3. 572.3 in.^{2} 5. 87.9 cm   7. 50.2 ft^2
9. 78.5 cm^2 11. 30.8 m   13. 153.9 m^2 15. 5544 mm^2
17. 2464 ft^2 19. 263.8 mm   21. 175.8 ft   23. 283.4 cm^2
25. 143.1 ft^2 27. 59.7 cm   29. 42.4 ft   31. Yerkes
C = 320.4 cm   Lick C = 285.9 cm Paris C = 263.9 cm
33. 30 cm, 94.2 cm, 706.5 cm^2;   50 ft, 314 ft, 7850 ft;
1000 in., 3140 in., 775,000 in.^{2};   0.05 mm, 0.314 mm,
0.00786 mm^2 35. 238.6 m   37. square with 8-in. sides
39. 99.3 yd^2 41. 56.5 cm   43. 52.4 m   45. No; some may be better shooters.

Lesson 2-8, pages 84–87

1. yes   3. no   5. yes   7. 6   9. 12   11. 2 cm   13. 18
games   15. Answers will vary.   17. no   19. yes   21. yes
23. yes   25. 8   27. 1   29. 49   31. 4   33. 36   35. 30
37. 18 m   39. 6 in.   41. 11 cm   43. 12 parts   45. 15 mm
47. 11 in.   49. 1 ft 6 in.   51. h = 4 in., l = 9 in.
53. $\frac{9}{35} = \frac{27}{105}$ 55. $\frac{1}{2} = \frac{3}{6}, \frac{1}{2} = \frac{4}{8}, \frac{3}{6} = \frac{4}{8}$

$\frac{1}{3} = \frac{2}{6}, \frac{2}{6} = \frac{3}{9}, \frac{3}{4} = \frac{6}{8}, \frac{2}{3} = \frac{4}{6}, \frac{4}{6} = \frac{6}{9}, \frac{1}{4} = \frac{2}{8}$
57. Answers will vary. See below for sample.

	English rounded	Metric rounded
Height from base to torch	150 ft/ 18.75 in.	46.5 m/ 18.6 cm
Length of hand	16 ft/ 2 in.	5 m/ 2 cm
Index finger	8 ft/ 1 in.	2.5 m/ 1 cm
Distance across eye	2 ft/ 0.25 in.	1 m/ 0.4 cm
Length of nose	4 ft/ 0.5 in.	1.5 m/ 0.6 cm
Length of right arm	42 ft/ 5.25 in.	13 m/ 5.2 cm
Width of mouth	2 ft/ 0.25 in.	1 m/ 0.4 cm
Length of tablet	22 ft/ 2.75 in.	7 m/ 2.8 cm
Width of tablet	12 ft/ 1.5 in.	4 m/ 1.6 cm

59. As the amount of rain increases, the grass grows faster.

Review and Practice Your Skills, pages 88–89

1. 153.9 in.^{2} 3. 15.2 cm^2 5. 44.0 in.   7. 13.8 cm
9. 1256 cm^2 11. 31,400 in.^{2} 13. 0.8 m^2
15. 706.5 ft^2 17. 125.6 cm   19. 628 in.   21. 3.1 m
23. 94.2 ft   25. no   27. yes   29. yes   31. no   33. yes
35. 12   37. 3   39. 18   41. 32   43. 48   45. 25
47. 20 cm   49. 6.5 cm   51. 16 m   53. 5 km   55. 5 kL
57. 27 m   59. 450,000   61. 1, 2   63. 10   65. 84 cm
67. 26 in.   69. 62 m^2 71. 108 in.^{2} 73. A   75. P
77. yes   79. no   81. yes   83. 153.9 in.^{2} 85. 201.0 m^2
87. 44 in.   89. 50.2 m   91. 4   93. 24   95. 20
97. 160 km   99. 75 m

Lesson 2-9, pages 90–93

1. 86 cm^2 3. 39 cm^2 5. 18,700 yd^2 7. Answers will vary, probably between 18 in.^{2} and 40 in.^{2} 9. 20 ft^2
11. ≈18.5 yd^2 For 13 and 15, answers will vary, but should use square units.   17. 821.1 m^2 19. 249.0 m^2
21. 2 gal   23. 304 cm^2 25. 688 cm^2 27. Answers will vary.   29. 15.7 cm^2 31. Answers will vary, but should be near 65.   33. Enclose the figure in familiar shapes.
35. 85.6 mm; 457.96 mm^2 37. 18.8 ft; 28.3 ft^2

Chapter 2 Review, pages 94–96

1. b   3. c   5. l   7. f   9. i   11. liter   13. cup   15. 72
17. 800,000   19. 17 yd, 1 ft   21. 4   23. 30 cm   25. 20 m
27. 21.5 in.   29. 92 mm   31. 126 m^2 33. 38.25 mm^2
35. 648 in^2 or 4.5 ft^2 37. A   39. P   41. $\frac{4}{5}$, 4:5
43. 50.2 cm^2 45. 475.1 in.^{2} 47. 854.9 mm^2
49. 522.5 cm^2 51. 12.6 ft, 103.6 mm, 9.7 yd, 81.0 cm
53. 32   55. 1 in. = 4 ft   57. 239.4 cm^2 59. 187.5 m^2

Chapter 3: Real Numbers and Variable Expressions

Lesson 3-1, pages 104–107

1. 9   3. −5   5. 8   7. 3.7   9. −3   11. −13   13. 0
15. gained 8 yd   17. −4   19. $\frac{1}{4}$ 21. 2.0   23. −8
25. −18   27. $-\frac{13}{10} = -1\frac{3}{10}$ 29. 64   31. $\frac{3}{35}$ 33. $-2\frac{5}{6}$
35. 1   37. 5   39. −6   41. negative   43. positive
45. negative   47. 24,777 seats   264 − 69 − 92 + 21
49. $124   51. >   53. =   55. <   57. 90°F   59. 12
61. −11   63. Concord   65. 154   67. 77

Lesson 3-2, pages 108–111

1. −56   3. 36   5. 3   7. 3   9. 7   11. −12   13. 6;
13.
15. $2   17. −9   19. −21
21. 4   23. 72   25. −5
27. −6   29. −2   31. 7
33. 28   35. −64   37. −125
39. −40   41. =   43. <
45. >   47. 15° F higher
49. decreased $33
51. 5 · (−8) = −40   53. −2 · (−6) = 12   55. about 7 times deeper   57. 50   59. −5   61. 0 ÷ 1 = 0 Check: 0 · 1 = 0, 1 ÷ 0 = n   Check: n · 0 ≠ 1; So 1 ÷ 0 ≠ n and

this is true for any number n.   63. ≈ 19   65. ≈ 3900

Review and Practice Your Skills, pages 112–113

1. −1   3. −12   5. −4.7   7. −12   9. −23
11. −5   13. −6.5   15. 2.2   17. −10   19. 8.6   21. 9.4
23. 28   25. positive   27. negative   29. negative
31. positive   33. positive   35. molten lava; 1250°C
37. −12   39. −8   41. −6   43. 4   45. −48   47. 4
49. −25   51. 96   53. 4   55. −54   57. −180   59. −6
61. =   63. >   65. >   67. −9   69. 11   71. 0   73. 12
75. −20   77. 3.9   79. −8   81. 21   83. −8

Lesson 3-3, pages 114–117

1. 2   3. $-\frac{1}{4}$ 5. 2.3   7. 10   9. 18   11. 5   13. 18 mi
15. 35   17. $\frac{5}{8}$ 19. 0.9   21. 67   23. −50   25. 2500
27. 0.63   29. 29   31. 24   33. 9   35. 15   37. −3
39. 28   41. −18   43. $6.00   45. 6 · 4 − 25
47. 5^2 · (30 − 28) ÷ 20   49. +   51. +   53. ÷
55. false; $\frac{1}{7} \cdot \left(0 + \frac{1}{8}\right)$ 57. true   59. false; 0.4^2 ÷ (0 + 0.8)   61. 23 mi   63. $4.25   65–69. Sample answers given.   65. 8 − 3 − 6 + 5 = 4   67. (12 + 0) ÷ (4 · 3) = 1   69. 12 + 3 − 4 + 5 = 16
71.

73. 11   75. Answers will vary. Explanations should include that two wedges both contain 3 of the total 12, yet they are not of equal size. One player out of 12 should have a wedge equal to 28.8°. It is not that measure. The 5-10 player wedge should be equal to 150°. It is not.

Lesson 3-4, pages 118–121

1. integer, rational, real   3. whole, integer, rational, real
5. 7, commutative   7. 8, associative   9. Sample answers given. 720, distributive   11. closed   13. $5.59
15. integer, rational, real   17. rational, real   19. natural, whole, integer, rational, real   21. rational, real   23. 4, distributive   25. 6, associative   27. $\frac{3}{7}$, associative
29–35. Sample answers given.   29. 2.32, commutative, associative   31. 20, commutative, associative   33. 73, commutative, associative   35. 60, commutative, associative   37. not closed   39. integer, rational, real
41. whole, integer, rational, real   43. rational, real
45. irrational, real   47. 55 points   49. Answers will vary.   51. True; it can only be used with addition and multiplication.   53. Answers will vary.   55. C = 5.34 m, A = 2.27 m^2

Review and Practice Your Skills, pages 122–123

1. 30   3. 14   5. 17   7. 78   9. −400   11. 45
13. 17   15. 3   17. 32   19. 30   21. 7   23. −1598
25. 1   27. 13   29. 30   31. rational, real   33. rational, real   35. whole, integer, rational, real   37. rational, real

39. 29, commutative   41. 2, distributive   43. −12,
commutative   45. 11, distributive   47. 9.6; commutative,
associative   49. 10,040; distributive   51. 48; associative
53. $\frac{1}{9}$; commutative, associative   55. 1214; commutative,
associative   57. not closed   59. closed   61. −11
63. −21   65. −7   67. 30   69. 36   71. −66   73. 22
75. 36   77. 85   79. rational, real   81. irrational, real
83. 9, distributive   85. 3, distributive   87. 13, associative

Lesson 3-5, pages 124–127

1. $n + 7$ 3. $\frac{n}{9}$ or $n \div 9$ 5. 10   7. −10   9. $s + 27$
11. $p \div 5$ or $\frac{p}{5}$ 13. 6; (Each person gets 6 golf balls.)
15. $\frac{1}{2}n$, $\frac{n}{2}$, or $n \div 2$ 17. 6n 19. $\frac{n}{7}$ or $n \div 7$
21. 4n + 17   23. 0   25. −6   27. 5   29. 4   31. 6
33. 28   35. 10   37. −2   39. −12   41. −896   43. −6
45. 2   47. $\frac{d}{3}$ or $d \div 3$ 49. $e - 2$ 51. $c - 5$
53. $65 + d$ 55. $3d + 4$ 57. 5   59. 6   61. 25   63. x
65. 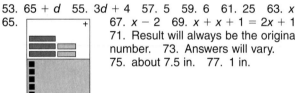   67. $x - 2$ 69. $x + x + 1 = 2x + 1$
71. Result will always be the original
number.   73. Answers will vary.
75. about 7.5 in.   77. 1 in.

Lesson 3-6, pages 128–129

1. C   3. 98,901, 87,912, 76,923, 65,934, 10,989 ·
5 = 54,945   5. $38   7. 512 tickets   9. Friday
11.   13. 10   15. 25   17. Answers will vary.
19. 29.4 ft²   21. 29.4 ft²

Review and Practice Your Skills, pages 130–131

1. $n - 5$ 3. 2n 5. $n + 3$ 7. 3$n + 2$ 9. 12
11. −21   13. 15   15. −3   17. −2   19. −2   21. 5
23. −7   25. −24   27. 15   29. −24   31. $\frac{m}{2} + 2$
33. 4$b + 10$ 35. $m - 5$ 37. a, 8.9, 10.1   39. c, 33, 13
41. c, 1.62, 4.86   43. PB-10806   45. −1   47. 6
49. 77   51. −356   53. 5, commutative   55. 13
57. −10   59. 6   61. a, 0.0287, 0.00287

Lesson 3-7, pages 132–135

1. 5³   3. $\frac{1}{4^4} = 4^{-4}$ 5. $\frac{1}{64}$ 7. 5629   9. 0.0000042
11. $9 \cdot 10^{-5}$ 13. first factor less than 1   second factor
not a power of 10   sum instead of product   15. 7⁴
17. $\frac{1}{3^3} = 3^{-3}$ 19. 8⁻⁶   21. 5⁵   23. $\frac{1}{4^4} = 4^{-4}$
25. −16   27. $\frac{1}{729}$ 29. $\frac{1}{1296}$ 31. $-\frac{1}{343}$ 33. 1
35. 216   37. 0.0000006876   39. 0.0071

41. 8,723,100,000   43. 0.0000091302   45. $1.2 \cdot 10^{-4}$
47. $2.5 \cdot 10^{-3}$ 49. $8.92 \cdot 10^2$ 51. $1.96 \cdot 10^{-11}$
53. $1.961048 \cdot 10^6$ 55. $7.5 \cdot 10^{-1}$ 57. No. 3² = 9 while
2³ = 8.   59. less than   61. 4096   63. equal   65. 17
67. $\frac{13}{36}$ 69. 7 h   71. 9   73. Answers will vary.
75. Poll every 10th customer alphabetically.

Lesson 3-8, pages 136–139

1. 10⁶   3. m^{28} 5. 9⁶   7. p^{10} 9. 3⁴⁸   11. x^{16}
13. 10⁵⁰³   15. 10³ times more   17. 10¹⁰   19. 6⁵
21. d^{11} 23. a^{15} 25. 5³   27. 6   29. x^{14} 31. n^3
33. 9⁶⁴   35. 11²⁰   37. x^{100} 39. p^9 41. 3⁸   43. 15³²
45. 6²⁵   47. 2⁹⁹   49. $b = 2$ 51. $f = 7$ 53. 10⁵ =
100,000 cm   55. 10² = 100 times longer   57. >
59. =   61. It is the number of unit cubes in a cube
of side 2.
63.

Review and Practice Your Skills, pages 140–141

1. 4⁶   3. 5⁸   5. 10⁵   7. 125   9. 1   11. 18   13. 1
15. 24,000   17. 8,900,000   19. $4.678 \cdot 10^5$ 21. 9.1 ·
10^{-2} 23. $6.3 \cdot 10^{-3}$ 25. $3.46 \cdot 10^7$ 27. $3.846 \cdot 10^3$
29. $1.05 \cdot 10^{-3}$ 31. $8 \cdot 10^{-5}$ m   33. 6⁵   35. 2⁷   37. c^6
39. 10⁸   41. 3⁵   43. 9²   45. 12   47. y^4 49. 13⁶
51. 2²⁴   53. x^6 55. q^{24} 57. 7³   59. 2⁷   61. 4⁵
63. 8   65. $a = 3$ 67. $h = 1$ 69. $d = 2$ 71. $x = 3$
73. −4   75. 10   77. 9   79. 22   81. 25   83. 61
85. 2, distributive   87. 1   89. 10   91. −14
93. c, 6.04, 6   95. $\frac{1}{36}$ 97. $\frac{1}{1,000,000}$ 99. 30,200
101. 602,130,000   103. $5.8 \cdot 10^{-3}$ 105. $7 \cdot 10^{-2}$
107. $6.75 \cdot 10^7$ 109. $4.43 \cdot 10^5$ 111. $2.5 \cdot 10^{-5}$
113. 8⁸   115. 10⁶   117. m^2 119. w^8 121. 2
123. 4⁴   125. n^3 127. h^5 129. 3⁸   131. 1⁹⁶ = 1
133. $q^0 = 1$ 135. x^{24} 137. $h = 3$ 139. $x = 4$

Lesson 3-9, page 142–145

1. 196   3. $\frac{25}{64}$ 5. $\frac{9}{13}$ 7. $\frac{12}{17}$ 9. 11   11. $\frac{1}{5}$
13. 4.690   15. 7.616   17. 4.8 cm   19. 441   21. 841
23. 2.89   25. 0.0064   27. $\frac{1}{25}$ 29. $\frac{64}{81}$ 31. 19
33. 24   35. 12   37. −10   39. −0.7   41. $-\frac{2}{9}$
43. $-\frac{4}{11}$ 45. $\frac{14}{23}$ 47. −100   49. −0.5   51. 22.956
53. 19.209   55. 21.954   57. 15.460   59. 9.487
61. 26.683   63. 3.9 sec or about 4 sec

65. Answers will vary. 67. 240 ft 69. 2.9 sec
71.

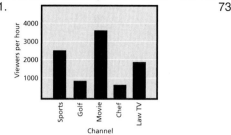

73. Movie 24

Chapter 3 Review, pages 146–148

1. b 3. k 5. f 7. l 9. g 11. −13 13. $\frac{1}{4}$ 15. 17
17. −5 19. 6 21. −3 23. 19 25. 24 27. 1 29. 30
31. rational, real 33. irrational, real 35. 24; commutative
37. 5; associative 39. $\frac{n-3}{8}$ 41. −6 43. −3 45. −45
47. −11 49. $12, $57 51. July 53. $\frac{1}{125}$
55. 0.0000478 57. x^6 59. d^{20} 61. 6^{72} 63. f^3 65. 36
67. 0.0001 69. $-\frac{4}{5}$ 71. $-\frac{2}{3}$ 73. 10.2 75. 62

Chapter 4: Two- and Three- Dimensional Geometry

Lesson 4-1, pages 156–159

1.

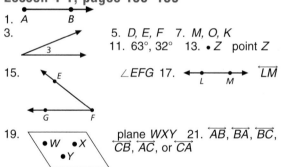

5. *D, E, F* 7. *M, O, K*
11. 63°, 32° 13. •*Z* point *Z*

15.

∠*EFG* 17. $\overleftrightarrow{LM}$

19.

plane *WXY* 21. $\overleftrightarrow{AB}$, $\overleftrightarrow{BA}$, $\overleftrightarrow{BC}$, $\overleftrightarrow{CB}$, $\overleftrightarrow{AC}$, or $\overleftrightarrow{CA}$

23 and 25. Answers will vary. 27. 25° 29. 140°
31. 115° 33. 30° 39. N and S, NE and SW, NW and
SE 41. Answers will vary. 43 and 45. Answers will vary.
47. −43 49. −14 51. 65-yd line

Lesson 4-2, pages 160–163

1 and 3. Answers will vary. 5. 6, 8, 12 7. 5, 6, 9
9. pentagon, no 11. quadrilateral, no 13. true
15. false 17. scalene, acute 19. scalene, right
21. 5, 6, 9 23. 6, 8, 12 25. Answers will vary.
27. octagon, yes 29. quadrilateral, no 31. 3.74 cm
33. 58° 35. 48° 37. 13.5 ft 39. 132.6 cm

Review and Practice Your Skills, pages 164–165

1.

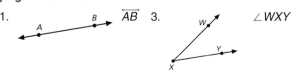

$\overrightarrow{AB}$ 3.

∠*WXY*

5.

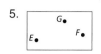

plane *EFG* 7.

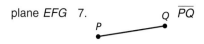

$\overline{PQ}$

9. *E, M, P* 11. *C, G, P* or *C, P, X* or *C, A, P* or *G, A, P*
13. 75° 15. 30° 17. 135° 19. 45° 29. hexagon, not
regular 31. pentagon, not regular 33. equiangular
35. scalene, right 37. 5, 5, 8 39. 7, 10, 15
41.

∠*DEF* 43.

$\overleftrightarrow{PQ}$

49. scalene, right 51. equilateral
53. 6, 8, 12 55. 4, 4, 6

Lesson 4-3, pages 166–169

1. triangular prism 3. square pyramid 5. cone
7. square pyramid 9. cylinder 11. pentagonal prism
13. rectangular pyramid 15. triangular pyramid
17. pentagonal prism 19. Answers will vary. 21. 5, 6,
9, 2 23. 7, 10, 15, 2 25. 5, 5, 8, 2 27. Sum of faces
and vertices minus the number of edges is always equal
to 2. 29. The building is shaped like a pentagonal prism.
31. 2*t* 33. Answers will vary. 35. *C* = 50.2 in.,
A = 201.0 in.² 37. Answers will vary.

Lesson 4-4, pages 170–171

3. hexagonal pyramid 5. rectangular prism
7.   9. For each pyramid,
                                he needs 1 square
                                and 4 identical
                                isosceles triangles.

13. 166, 206, 251
15. 22, 31, 32 17. $\frac{32}{972}$, $\frac{64}{2916}$, $\frac{128}{8748}$

Review and Practice Your Skills, pages 172–173

1. cone 3. hexagonal prism 5. triangular pyramid
7. pentagonal pyramid 9. triangular prism 11. cylinder
13. hexagonal prism 15. square pyramid on top of a
cube 17. Students' nets should have 4 trapezoids and
a rectangle on top; 4 larger trapezoids and a rectangle
for the bottom 19. *B, A, D* 21. 105° 23. 45°
25. isosceles 27. equilateral 29. square pyramid
31. cylinder 33. 7 faces, 10 vertices, 15 edges
35. 4 faces, 4 vertices, 6 edges 37. cone

Lesson 4-5, pages 174–177

1–3.

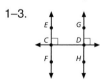

5.

7.

9. 10 11. Answers will vary. Parallel lines appear parallel.
Perpendicular lines don't necessarily appear
perpendicular. 13. $a \perp e, b \perp e, a \perp f, b \perp f$
15.

17. perpendicular

19.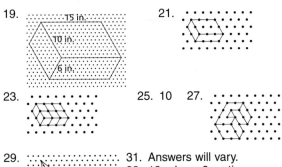
15 in.
10 in.
6 in.

21.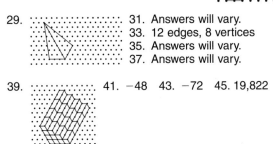

23.

25. 10 27.

29.

31. Answers will vary.
33. 12 edges, 8 vertices
35. Answers will vary.
37. Answers will vary.

39.

41. −48 43. −72 45. 19,822

Lesson 4-6, pages 178–181

1.

3.
A
y

5.

7. D

9.
A
y

11.
A
y

13.

15.

17. Top Front Side

19. Top Front Side

21. Top Front Side

23. Answers will vary. 25. Answers will vary.
27. Answers will vary. 29. 77 in.^{2} 31. 21 in.^{2}
33. −6 35. −10 37. 7

Review and Practice Your Skills, pages 182–183

1, 3, and 5. 5. They are perpendicular.
Sample answer:
D F C
A B

7.
14 ft
14 ft
14 ft

9.
2.3 cm 2.3 cm
2.3 cm

11.
11 yd
11 yd
11 yd

13. 6 15. 9

19.

21.

23. Top
Front
Side

25. Top
Front
Side

27. right, isosceles 29. scalene, acute 31. 5, 6, 9; 5, 5, 8
33. rectangular prism 35.
8 units
2 units
5 units

37. 11

Lesson 4-7, pages 184–187

1. 288 ft^3 3. 474.6 in.^{3} 5. 1650 cm^3 7. 2411.5 in.^{3}
9. 78.5 ft^3 11. 3128 ft^3 13. 1,009,936 m^3 15. 11 m^3
17. 0.1 cm^3 19. 1,601,613 mm^3 21. 70.3 ft^3
23. figure D 25. 3 ft, 226.1 ft^3 27. 13.8 in., 627.9 in.^{3}
29. No, multiplication is commutative. 31. 165.6 ft^3
33. 5 cm 35. 48.16 L 37. 5 : 1 39. 125 : 1 41. 64
43. 4096 45. 150 47. 12

Lesson 4-8, pages 188–191

1. 53.3 cm^3 3. 726 cm^3 5. 2,592,100 m^3 7. 4823.0 in.^{3}
9. 72 ft^3 11. 71.4 m^3 13. 500 cm^3 15. 320 ft^3
17. 72 in.^{3} 19. 2142.6 m^3 21. 6358.5 m^3
23. approximately 990,125 m^3 25. 1200 cm^3
27. 900 ft^2 29. 12 in. 31. Tanisha 33. 2.0 m
35. 9 in^2. 37. right iscoceles 39. acute equilateral
41. irregular octagon

Review and Practice Your Skills, pages 192–193

1. 180 in.^{3} 3. 10,890 ft^3 5. 40.5 cm^3 7. 100.5 yd^3
9. 785 cm^3 11. 471 ft^3 13. 840 ft^3 15. 62.8 cm^3

17. 448 ft³   19. $281\frac{2}{3}$ cm³   21. 177.8 m³   23. 300 in.³
25. 264 in.³   27. 506.6 cm³   29. 4.2 yd³   31. *M, O, P*
33. 100°   35. 150°   37. equilateral   39. obtuse
41. trapezoidal prism   43. pentagonal pyramid
45. 5, 6, 9   47. cylinder   49. cube   53. 10
55.    57.   59. 200 cm³
61. 84 ft³

Lesson 4-9, pages 194–197

1. 337 cm²   3. 288 m²   5. 9809 in.²   7. 27 cm²
9. 312 in.²   11. 3347 ft²   13. 2098 yd²   15. 864 cm²
17. 280 cm²   19. 432 yd²   21. 1800 in.²   23. 31 gal
25. 122 min   27. 39.25 in.²   29. 25,481 ft²
31. Commutative   33. Distributive   35. Associative
37. 1   39. 10   41. 254.34 in.³

Chapter 4 Review, pages 198–200

1. 1   3. k   5. a   7. d   9. g   11. •*A*
13. 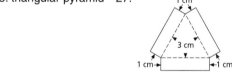   15.   17.

19. 5, 6, 9   21. 6, 8, 12   23. triangular prism
25. triangular pyramid   27.

29. *p* ∥ *q*   31. points B and C   33. 8
35. 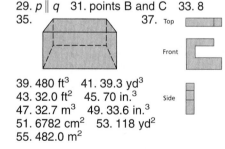   37.

39. 480 ft³   41. 39.3 yd³
43. 32.0 ft²   45. 70 in.³
47. 32.7 m³   49. 33.6 in.³
51. 6782 cm²   53. 118 yd²
55. 482.0 m²

Chapter 5: Equations and Inequalities

Lesson 5-1, pages 208–211

1. false   3. true   5. open   7. 4   9. 4   11. 5
13. 24 mi   15. Answers will vary.   17. false   19. false
21. open   23. 0   25. −3   27. 12   29. −1   31. −4.5
33. 2   35. 4   37. −1   39. 3   41. 7   43. −3   45. −6
47. 24   49. $43   51. No. 36 ÷ 3 = 12   53. *y* + 78 =
1973   55. 6 cm   57. Answers will vary.   59. 5, −5
61. 3, −3   63. 8   65. 1   67. 3–5

Lesson 5-2, pages 212–215

1. *y* = −11   3. *x* = 9   5. *y* = −10   7. *h* = 10   9. *a* = 34
11. 116°   13. 58 cards   15. *c* = 3   17. *a* = 8
19. *g* = 4.0   21. *y* = −12   23. *k* = −7   25. *n* = −1.8
27. *k* = 11   29. *d* = −7   31. *r* = 30   33. $y = -1\frac{1}{4}$
35. $t = 2\frac{3}{8}$ 37. *c* = −39.6   39. $2\frac{1}{4}$ h   41. 5′ 9″
43. Add 16 to 42.   45. *x* − 250 = 310; *x* = 560 ft
47. Answers will vary.   49. *n* + 5 = 12   51. *s* − 6 = 1
53. 2 − *k* = 3   55. *k* = 9   57. *p* = −4   59. *x* = −4.4
61. *x* = 2   63. 10.68   65. 11.18   67. 24

Review and Practice Your Skills, pages 216–217

1. false   3. false   5. false   7. true   9. false   11. −1
13. −18   15. −6   17. −3   19. 6   21. 9   23. −2
25. 4   27. −7   29. −2   31. −4   33. 5   35. −15
37. −8   39. −15   41. −2   43. −14   45. 4   47. −4
49. −10   51. 0   53. 1.2   55. 9.8   57. −1.3   59. $\frac{3}{8}$
61. $-1\frac{1}{10}$ 63. $2\frac{1}{2}$ 65. false   67. open   69. false
71. 6   73. −7   75. $-\frac{1}{3}$ 77. −10   79. −7   81. 4
83. −10   85. 10.3   87. 2.5   89. $-\frac{1}{2}$

Lesson 5-3, pages 218–221

1. *t* = 3   3. *x* = −4   5. *c* = −6   7. *a* = −15
9. *d* = 44   11. 6*g* = 4   13. $0.55   15. Both methods
isolate the variable. In the first equation, you divide both
sides by 6. In the second equation, you subtract 6 from
both sides.   17. *t* = −20   19. *w* = −1.25
21. $x = 4\frac{1}{2}$ 23. *c* = −4   25. $y = -\frac{8}{15}$ 27. *n* = −32
29. *g* = 4.16   31. *y* = 12   33. *m* = −4   35. $m = -1\frac{1}{3}$
37. (−) 42 ÷ (−) 12 ENTER   39. 0.495 ÷ 0.0165
ENTER
41. $\frac{-6}{y} = 6$ 43. $\frac{8}{r} = 3$ 45. 1045 = 16*s*; *s* ≈ 65.3 mi/h
47. 288 s   49. 750 = 12.5*m*; *m* = 60 min   51. walking
53. *n* = 2.5   55. *x* = −9   57. *x* = −4   59. no solution
61. angle   63. segment

Lesson 5-4, pages 222–225

1. *d* = 4   3. *p* = 1   5. *y* = 5   7. *a* = 60   9. *m* = 6
11. 5 hours   13. Answers will vary. Answers should
include reversing the order of operations.   15. $x = -\frac{3}{8}$
17. *n* = −3   19. *k* = 30   21. *x* = −3   23. *a* = 15
25. *m* = 52   27. *n* = 1.3   29. *f* = 26   31. *k* = 3
33. *p* = 7   35. *m* = −3.75   37. *x* = 77   39. 104
photographs   41. 24 = 3(5 + *x*)   43. *w* = 8, *P* = 64,
A = 192   45. 0.35 + 0.20(*x* − 1) = 1.05; *x* = 4.5 min
47. *w* = 160 ft, *l* = 360 ft   49. 14.2°C   51. 26.1°C
53. *x* = −2   55. 7*x* − 3 = 39; *x* = 6   57. Answers will
vary.   59. *SA* = 141.85 cm²; *V* = 122.72 cm³

Review and Practice Your Skills, pages 226–227

1. −16   3. $-\frac{1}{4}$ 5. $-\frac{1}{4}$ 7. −15   9. 27   11. 9   13. 0.4
15. −20   17. −80   19. −5   21. −19.2   23. $-\frac{1}{3}$

25. $\frac{1}{15}$ 27. -6 29. $-3w = 1$ 31. $\frac{1}{3} \div g = 2$

33. $\frac{-2}{d} = 8$ 35. -2 37. $\frac{4}{5}$ 39. $-1\frac{1}{2}$ 41. 3

43. $-1\frac{7}{8}$ 45. $5\frac{1}{2}$ 47. 1 49. 4 51. 12 53. -6

55. $-2\frac{1}{3}$ 57. 1 59. 2 61. 1 63. $3\frac{1}{2}$ 65. $-1\frac{2}{3}$

67. -11 69. 13 71. 4 73. 2 75. 5 77. -5

79. -15 81. -20 83. $\frac{5}{12}$ 85. 1 87. 8 89. $-\frac{6}{7}$

Lesson 5-5, pages 228–231

1. $15y$ 3. $9n + 2m$ 5. $-4f + 3$ 7. $2p + 5r + 4p = 6p$
$+ 5r$ 9. $x = -5$ 11. $m = 3$ 13. $y = 2$ 15. $2n + 45$
$= 180; n = 67.5$ 17. $5s + 5t$ 19. $-2x - 4$
21. $6a$ 23. $9n + 2x$ 25. $5x - 11$ 27. $p^2 + 3p - 3$
29. $n = 7$ 31. $r = 2.5$ 33. $n = -1.5$ 35. $x = 7$
37. $b = 8$ 39. $f = -3$ 41. $x = -3$ 43. $x = 5$
45. $m = 1.5$ 47. $n = -1$ 49. $7 = \frac{n}{-3 + 5}; n = 14$

51. $3x - 6 = 8 + (-4)x; x = 2$ 53. $75 + 2(x - 65) =$
129; $x = 92$ mi/h   55. $4x + 3x + 35 = 81.20; x = \$6.60$
57. $x = 15°$ 59. Answers will vary.   61. $25 + 0.50m$
63. Answers will vary.   65. Answers will vary.   67. -8
69. -7 71. -16 73. -48

Lesson 5-6, pages 232–235

1. $t = 1\frac{1}{3}$ h   3. $w = \frac{A}{l}$ 5. $N = 4H$ 7. $a = \frac{C}{9}$

9. Subtract 7 from both sides. Then divide both sides by 3.
11. 15.5 in.   13. $\$925$ 15. $s = p - 0.25p$ or $s = 0.75p$

17. $\$34$ 19. 152 chirps/min   21. $m = \frac{E}{c^2}$ 23. $s = \frac{P}{3}$

25. $l = \frac{P - 2w}{2}$ 27. $C = \frac{5}{9}(F - 32)$ 29. $M = 2O - R$

31. $a = \frac{v - r}{t}$ 33. $y = \frac{3z - ex}{-2}$ 35. 151.3 ft   37. No.

You can only drive 248 mi on one tank of gas.   39. yes
41. 4.9 in.   43. Answers will vary.   45. 63.1 ft; 195.6 ft^2

Review and Practice Your Skills, pages 236–237

1. $-d$ 3. $2m + 2n$ 5. $-5p + 4q$ 7. $b + 8$ 9. $4k$
$+ 3$ 11. -1 13. $-\frac{1}{2}$ 15. 1 17. $-2\frac{1}{2}$ 19. -2

21. $\frac{2}{5}$ 23. -1 25. -2 27. 8 29. $2\frac{1}{3}$ 31. $b = \frac{2A}{h}$

33. $w = \frac{V}{lh}$ 35. $w = \frac{A}{l}$ 37. 4 in.   39. $2\frac{2}{3}$ yd

41. $2\frac{1}{2}$ in.   43. 1.6 mm   45. 36 ft   47. 15 yd^2
49. 56.25 km   51. 6.75 m^2 53. -2 55. 9 57. -5
59. 14 61. -2 63. $-8\frac{1}{4}$ 65. $1\frac{1}{2}$ 67. $F = \frac{9}{5}C + 32$

Lesson 5-7, pages 238–239

1. $x + 0.15x$ 3. $\frac{1}{4}p$ 5. $0.5(x - 0.3x)$ 7. $\$0.22$

9. 5499 landfills   11. 1.65 in.   13. 19 years old
15. 78 ft, 156 ft   17. -17 19. 108 21. 0.0000116
23. positive

Lesson 5-8, pages 240–243

1.   3.

5.   7. $x > 300$ 9. $x \geq 7$ includes the number 7; $x > 7$ does not.

11.

13.

15.

17.

19.

21.

23.

25.

For 27, 29, 31, Sample answers given.   27. $18.5, 19, 19.5$
29. $-4, 0, 2$ 31. $-\frac{2}{5}, -1, -10$ 33. $p \leq 52,000$
35. $h \geq 2$ 37. $x \leq -3$ 39. $x = 2$ 41. $s \leq 20$
43. $x > 1,000,000$ 45. $x > 4,500,000$ 47. Answers
will vary.   49. $p \leq 720$ performances   51. $2d \geq 500$
53. 8 55. 5 57. True

Review and Practice Your Skills, pages 244–245

1. $0.4n$ 3. $\frac{1}{2}f$ 5. $x - 0.1x$ 7. $2x$ 9. $3x$

11.   13.

15.   17.

19.   21.

23.   25.

27.   29. $4, 3, 2$ 31. $5, 4, 3$ 33. $0,$
1, 2   35. $0, 1, 2$ 37. $1, 2, 3$

39. $-4, -5, -6$ 41. $n \geq 10$ 43. $n < -100$ 45. $n > 6$

47. 13 49. 5 51. 12 53. -5.3 55. -4 57. $-1\frac{1}{2}$

59. 27 61. -2 63. -14 65. $3\frac{1}{3}$ 67. -1 69. $1\frac{1}{4}$

71. 2 73. $m = \frac{y - b}{x}$ 75. $r = \frac{l}{pt}$ 77. 45 yd^2

79. $x - 0.35x$ 81. $\frac{x}{8}$

83.   85.

87.   89.

Lesson 5-9, pages 246–249

1. $x \geq 4$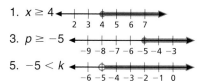
(number line: 2 3 4 5 6 7)

3. $p \geq -5$
(number line: −9 −8 −7 −6 −5 −4 −3)

5. $-5 < k$
(number line: −6 −5 −4 −3 −2 −1 0)

7. no more than $17   9. The equation has one solution, $x = -3$. The inequality has an infinite number of solutions less than -3.

11. $p \geq -4$
(number line: −5 −4 −3 −2 −1)

13. $n > 0$
(number line: −2 −1 0 1 2 3 4)

15. $6 < h$
(number line: 3 4 5 6 7 8 9)

17. $-3 \geq a$
(number line: −7 −6 −5 −4 −3 −2 −1)

19. $5 \leq c$
(number line: 3 4 5 6 7 8 9)

21. $k > -5$
(number line: −6 −5 −4 −3 −2 −1 0)

23. $y < 5$
(number line: 2 3 4 5 6 7 8)

25. $x < 2$
(number line: −2 −1 0 1 2 3 4)

27. $-3 < x$
(number line: −4 −3 −2 −1 0 1 2)

29. $1 > n$
(number line: −3 −2 −1 0 1 2 3)

31. $233   33. no more than 45 calculators
35. at least 92   37. Ryan, "no more than" means "less than or equal to"   39. Answers will vary.   41. $-7, -6, -5$ 43. $-4, -3$ 45. $x + 4 < 7; x < 3$ 47. 7 h and 9 h
49. $15 + 20.67x = 97.68; x = 4$ months   51. $d = -28$
53. $x = 4$ 55. $t = -32\frac{2}{9}$

Chapter 5 Review, pages 250–252

1. h   3. i   5. a   7. j   9. d   11. open   13. true   15. open
17. open   19. true   21. $\frac{1}{2}$ 23. 7   25. -9 27. 44   29. 3
31. -2 33. 27   35. -2.5 37. -25 39. $x - 18 = 29$;
$x = 47$ 41. -42 43. 12   45. -17 47. -72 49. $6h = -24$ 51. -1 53. 2   55. 4   57. 5   59. 6   61. 2   63. $7m$
65. $9d - 18$ 67. $5z$ 69. $-5y + 8x$ 71. $48 + 9w - 6r$
73. $m = \dfrac{y - b}{x}$ 75. $t = \dfrac{d}{r}$ 77. $h = \dfrac{3V}{\pi r^2}$ 79. $18
81. (number line: −4 −3 −2 −1 0 1 2)   83. (number line: −1 0 1 2 3 4 5)   85. $y < 2$
87. $x > -4$
89. $n \leq -4$;
(number line: −7 −6 −5 −4 −3 −2 −1 0 1)

Chapter 6: Equations and Percents

Lesson 6-1, pages 260–263

1. 69.6   3. 34%   5. 24   7. 27   9. $22.05   11. 11.83
$\boxed{\div}$ 215 $\boxed{\times}$ 100 $\boxed{\text{ENTER}}$ 13. Answers will vary.
For 15, 17, and 19, answers will vary.
21. 80%   23. 56.32   25. 35%   27. $33\frac{1}{3}$%
29. 24%   31. 70%   33. 4.5   35. 40%   37. 64.8
39. 88   41. $165,000   43. Sunday   45. *The Cosby Show* – 30,502,600 people, *Friends* – 16,141,500 people
47. Yes.   49. $7425.60
51.
(number line: −23 −21 −19)
53.
(number line: 0 1 2 3 4 5)

55. $5x \geq 250; x \geq 50$; 50 students or more

Lesson 6-2, pages 264–267

1. $36 = x \cdot 80$; 45%   3. $0.70x = 147$; 210
5. $0.75 \cdot 64 = x$; 48   7. $x = 0.0075 \cdot 80$; 0.6
9. 0.543 $\boxed{\times}$ 300 $\boxed{\text{ENTER}}$ 11. 80%

13. 14.3%   15. $142\frac{6}{7}$
17. 18   19. $66\frac{2}{3}$%   21. $274{,}166\frac{2}{3}$ 23. $33\frac{1}{3}$%
25. 960   27. 500%   29. 320   31. $66\frac{2}{3}$%   33. 40

35. 36   37. 60.4%   39. 59.4%   41. 59.2%   43. 89.1%
45. 7.1%   47. reducing   49. 8.3%   51. Answers will vary.   53. 200 lb   55. Answers will vary.   57. 5.8 cm

Review and Practice Your Skills, pages 268–269

1. 14   3. $8\frac{1}{3}$%   5. 1.2   7. 6400   9. 22.5   11. 233.6
13. $66\frac{2}{3}$%   15. 525   17. 63.6%   19. 5   21. 96
23. $6\frac{1}{4}$%   25. 35%   27. 3%   29. 112.5   31. 3.9
33. 200   35. 8.1   37. 20%   39. 420   41. 200%
43. 480   45. 40%   47. 50   49. 12.5   51. 80   53. 18.4
55. 37.8   57. 20%   59. 25   61. 472.7   63. 80%
65. 10.8   67. 1.5   69. 2%   71. 25%

Lesson 6-3, pages 270–273

1. $62.31, $353.07   3. $44.98, $44.97   5. $3.46, $111.72   7. 30%   9. 15%   11. $39.96   13. 33%
15. $15.05, $135.45   17. $127.65, $562.34   19. $6.00, $18.00   21. $66.44, $444.65   23. $69.25, $1315.75
25. 25%   27. 50%   29. 35%   31. $278.60   33. 12.3%
35. City View costs $825, Blue Sky costs $840
37. $390.99   39. $292.22   41. $1897.42   43. $169.99
45. Benjamin   47. 6, 12, 8   49. 4, 6, 4

Lesson 6-4, pages 274–277

1. $2.48, $47.48 3. $0.71, $11.70 5. $5300, $21,200
7. $1812.50 9. $5100 11. 6% 13. $111.25
15. $8.78, $143.78 17. $3.00, $52.99 19. $0.97,
$13.96 21. $14,960, $53,040 23. $4180, $17,820
25. $131.25, $618.75 27. $2070 29. $2850
31. $4420 33. 6% 35. 5% 37. 7.25% 39. $8257.00
41. $89.99 43. 15% 45. $116.57 47. The proportion
automatically includes division or multiplication by 100.
49. 512.50 m^2; 703.57 m^3 51. 120 cm^2; 54 cm^3
53. -8 55. -7

Review and Practice Your Skills, pages 278–279

1. $17.98, $71.92 3. $2.28, $43.22 5. $74.99, $50.00
7. 11.3% 9. 14% 11. 26.6% 13. $2.79, $35.59
15. $5.17, $74.16 17. $2.10, $32.05 19. $210;
$665/wk 21. $226.30; $503.70/wk 23. $13,020;
$33,480/yr 25. $5340 27. $7200 29. 8.5% 31. 7%
33. 8% 35. 46.4 37. 20% 39. 8.75 41. $14.99,
$34.96 43. 9.2% 45. $118, $1593 47. $94.92;
$583.08/wk 49. $212.50; $1037.50/wk

Lesson 6-5, pages 280–283

1. $160 3. $280.80 5. 7.5% 7. $130, $780
9. $675, $2175 11. Answers will vary. 13. $4.90,
$39.90 15. $1552.96, $6405.96 17. $358.28, $9315.28
19. $1936.66, $4137.41 21. $6,600, $14,600
23. $23.72, $93.47 25. 12% 27. $1,138.13,
$16,313.13 29. $16,250 31. 6.5% 33. $27,700
35. Answers will vary. 37. $4.5 \cdot 10^6$ 39. $6.0704 \cdot 10^8$
41. $6 \cdot 10^{-4}$ 43. $9 \cdot 10^{-5}$

Lesson 6-6, pages 284–287

1. $31.50 3. $546 5. 9% 7. 6.5% 9. Answers
will vary. 11. $53,375.00 13. $14.40 15. $640.00
17. $470.00 19. 9% 21. $595 23. $516
25. $23,125 27. $33.75 29. $435.07 31. Each
agent receives $3405. 33. $1375 35. $2295
37. $2069.50 39. 200% 41. 14.29% 43. 25%
45. 7:8

Review and Practice Your Skills, pages 288–289

1. $432 3. $1470 5. $261.25 7. 6% 9. 8%
11. $147.20, $607.20 13. $201, $871 15. $47.70,
$1637.70 17. $2.70 19. $22 21. $32 23. $1774
25. $4095.60 27. $4718 29. 4% 31. 7% 33. 12%
35. 7.5 37. 40 39. $17.67; $253.26 41. $5037.50

Lesson 6-7, pages 290–293

1. 5% 3. 87.5% 5. 40% 7. 10% 9. 20% 11. 10%
13. 500% 15. 10% 17. 25% 19. 18% 21. 25%
23. 10% 25. Answers will vary. 27. 7.9%, 9.0%
29. 0.8%, 2.2% 31. 2.6%, 1.7% 33. 1920–1930
35. Area = 12 square units 37. The 15% decrease is
applied to a larger number than the 15% increase.
39. false 41. true 43. 595.344 in.^{3}

Lesson 6-8, pages 294–295

1.

Method	Quarters	Dimes	Nickels
1	2	0	0
2	1	2	1
3	1	1	3
4	1	0	5
5	0	5	0
6	0	4	2
7	0	3	4
8	0	2	6
9	0	1	8
10	0	0	10

3. Answers will vary.

5.

	C	C/S	C/P	C/S/P/M	Works
Small	3	5	8	3	1
Medium	4	7	2	8	10
Large	4	5	8	12	8

7. large
9. 13%

11.

	Amount
Rent	$425.00
Car payment	$167.00
Car insurance	$72.00
Groceries	$81.07
Gas & electric	$34.50
Spending money	$100.00
Water	$11.72
Phone	$39.43
Credit card	$90.00

13. other bills 15. 4.59 ft^2

Chapter 6 Review, pages 296–298

1. b 3. h 5. e 7. g 9. i 11. $\frac{75}{100} = \frac{x}{64}$; 48
13. $\frac{x}{100} = \frac{5}{30}$; about 16.7 15. $50 17. $x = 0.07(80)$; 5.6
19. $1 = 40x$; 2.5% 21. $0.7x = 147$; 210 23. $225;
$674.98 25. 30% 27. $1000 29. $4302 31. $2.70;
$47.70 33. $669.60 35. $156,000 37. 14% 39. Even
though they have the same total cost, Plan (1) would yield
a $287.50 monthly payment where Plan (2) yields a
$479.17 monthly payment. 41. 2.5% 43. 12.5%
45. 10% 47. Ohio

Chapter 7: Functions and Graphs

Lesson 7-1, pages 306–307

1. B 3. D 5. A 7. Carla; Seishi 9. Seishi returning
home. 11. 15 min 13. Answers will vary. 15. Answers
will vary. 17. $375.00

Lesson 7-2, pages 308–311

1. (2, 2) 3. (0, 4) 5. (0, 0) 7. $D(4, -3)$
9 and 11. 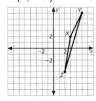 13.

15. The point is 2 units to the right of the origin and 3
units down. It is located in Quadrant IV. 17. $(-4, -4)$
19. $(-2, 0)$ 21. $(-3, 3)$ 23. Answers will vary.
25. $I(-4, -2)$, $B(-4, -4)$ 27. $J(2, -2)$, $K(5, -5)$,
$R(-3, 3)$

28–35.

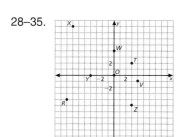

37.

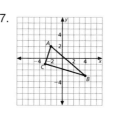

5.

x	y	(x, y)
−1	3	(−1, 3)
0	0	(0, 0)
2	−6	(2, −6)

39. (4, 4) **41.** *F*(7, 4)

43.

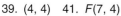
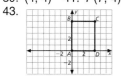

45.

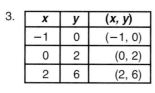

7. $\frac{4}{3}$; −4 **8.** −2; 8 **9.** $\frac{8}{3}$; −2 **11.**

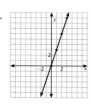

47. 15 **49.** Phil **51.** $8.25; $46.75 **53.** $53.75; $161.25

Review and Practice Your Skills, pages 312–313

1. time in hours **3.** 6 mi **5.** before **7.** 2 mi **9.** 4 mi
11. (−6, 4) **13.** (2, 4) **15.** (−5, 1) **17.** (2, −5) **19.** (−4, −3) **21.** (4, −3) **23.** *N*(−2, −4) **25.** III **27.** IV
29–40.

41. 9 mi **43.** 1 h **45.** (0, −1)
47. (3, 3) **49.** (−2, 3)
51. *C*(−2, 2)

52–59.

13. 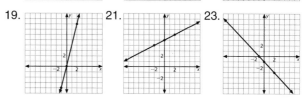 **15.** **17.**

19. **21.** **23.**

25. $\frac{3}{5}$; −3 **27.** $\frac{2}{3}$; 2 **29.** $\frac{1}{3}$; −1 **31.** *V* = {20, 10.2, 0.4, −9.4, −19.2, −29, −38.8} m/sec **33.** *y* = 0.25*x* + 1.1
35. a 1-min phone call **37.** Answers will vary.
39. *y* = *x* + 5

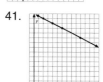

Lesson 7-3, pages 314–317

1. no **3.** yes **5.** yes **7.** {−1, 0, 4, 7} **9.** yes
11. *d*(3) = 135 The car travels 135 miles in 3 hours.
13. yes **15.** Hasina. Each element in the domain corresponds to only one element in the range.
17. yes **19.** no **21.** *D* = {0.1, 1.1, 1.2, 2.1}; *R* = {0.5, 3.2, 3.4, 3.6} **23.** *D* = {2, 3}; *R* = {−4, −2, 0, 2, 4}
25. yes **27.** no **29.** $401.85 **31.** $1232.50
33. *C*(*m*) = 15 + 18*m* **35.** *(*x*) = 3*x* + 2 **37.** 59
39. *A* = π*r*², 63.585 mm² **41.** *A* = $\frac{1}{2}$ *h*(*b*₁ + *b*₂), 2.59 m²

41.

43. Answers will vary.
45. *y* = −2*x* + 4.

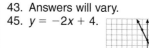

47. 17 **49.** −1

Review and Practice Your Skills, pages 322–323

1. yes **3.** yes **5.** no **7.** yes **9.** *D* = {−4, −2, 2, 4, 6}
R = {0, 1} **11.** *D* = {5, −2, −3, −2, 5} *R* = {−2, −1, 0, 1, 2}
13. *D* = {−10, −6, −2, 4, 8} *R* = {5, 3, 1, −2, −4}
15. *D* = {1, 2, 3} *R* = {1, 2} **17.** *D* = {−3, 2, −1, −2}
R = {2, 5, 0, −1, 3} **19.** yes **21.** no **23.** yes
25. no **27.** no

Lesson 7-4, pages 318–321

For 1, 3, Sample tables given.

1.

x	y	(x, y)
−1	−4	(−1, −4)
0	0	(0, 0)
2	8	(2, 8)

3.

x	y	(x, y)
−1	0	(−1, 0)
0	2	(0, 2)
2	6	(2, 6)

29.   31.   33.

5. $m = \frac{1}{3}$, $b = 3$; $y = \frac{1}{3}x + 3$ 7. 135

9. 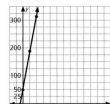   11. 3 years

35.   37.   39.

13. $m = 4$, $b = 1$ 15. $m = -2$, $b = -1$

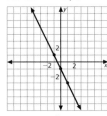   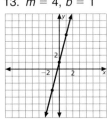

41. $-3, 3$ 43. $-5, -5$ 45. $-\frac{7}{2}, 7$ 47. 3, 15

49. $\frac{1}{4}, -1$ 51. $\frac{3}{2}, 3$ 53. $\frac{1}{2}$ h

17. $m = \frac{3}{2}$, $b = -5$ 19. $m = \frac{1}{4}$, $b = -4$

55–58.   59. $D = \{0.2, 0.1\}$
$R = \{-5, -3, 0, 3\}$
61. no

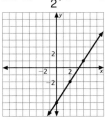   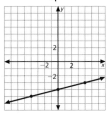

63. 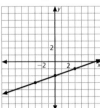   65.   67. $-5, -1$

21. $m = 0$, $b = 2$

Lesson 7-5, pages 324–327

1. $\frac{1}{4}$ 3. -2 5. $\frac{-5}{9}$ 7. negative   9. zero   11. $\frac{1}{15}$

13. positive   15. negative   17. zero   19. $\frac{5}{2}$ 21. -1

23. 0   25. 0   27. $\frac{3}{5}$ 29. $\frac{3}{2}$ 31. $-\frac{2}{3}$

33–36.   37.

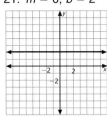

23. $m = -1$, $b = 1$; $y = -x + 1$
25. No. Her slope is $\frac{2}{3}$.   27. 5.00
29. $11.60   31. $y = 2x + 3$

33–34. 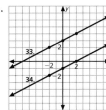   35. They are parallel.
37. $y = \frac{1}{2}x + 2$
39. 5   41. 13.56

39. $-\frac{1}{8}$ 41. Pablo wins. He has a greater slope, or rate.

43. $\frac{2}{3}, \frac{5}{3}, \frac{2}{3}, -\frac{1}{3}$; trapezoid   45. Answers will vary.

47. -12 49. 25   51. -8 53. 65

Lesson 7-6, pages 328–331

1. $m = \frac{1}{3}$; $b = -2$ 3. $m = \frac{3}{5}$; $b = -3$

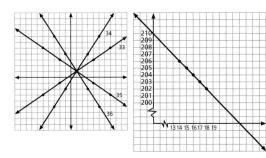

Review and Practice Your Skills, pages 332–333

1. $\frac{1}{2}$ 3. $\frac{2}{3}$ 5. -1

7. 2   9. $\frac{1}{2}$ 11. -1 13. 0   15. $\frac{7}{6}$ 17. undefined

19.   21.   23.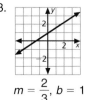

$m = 1$, $b = 3$ $m = -1$, $b = 2$ $m = \frac{2}{3}$, $b = 1$

25. **27.** **29.**

$m = \dfrac{-3}{2}$, $b = -2$ $m = \dfrac{-1}{2}$, $b = -1$ $m = \dfrac{-1}{3}$, $b = -1$

31. 0; 1.5; $y = 1.5$ **33.** -2; 2; $y = -2x + 2$
35. The person turned around and walked back towards the starting point.
36–38.   **39.** $D = \{3, 4, 5, 6, 7\}$
$R = \{-5, -2, -1\}$
41. yes   **43.** 0   **45.** $-\dfrac{2}{9}$
47. 1; -4

Lesson 7-7, pages 334–337

1. 2   **3.** 3   **5.** 5   **7.** 10   **9.** 13   **11.** 3.2   **13.** 5.4
15. 3.6   **17.** 10   **19.** 2.2   **21.** 5.7   **23.** 10.3   **25.** 2.2
27. 7.6   **29.** 3.2   **31.** 12.8   **33.** 9.2   **35.** 20 ft
37. $b = \sqrt{c^2 - a^2}$ **39.** 30 ft   **41.** 16.2; isosceles
43. $GH = \sqrt{18}$, $HK = 6$, $KG = \sqrt{18}$; yes   **45.** Answers
will vary.   **47.**   **49.** $101.87; $2060.87

Lesson 7-8, pages 338–341

1. yes   **3.** yes   **5.** yes   **7.** no   **9.** (2, 3), (3, 2),
(4, 3), (5.5, 3), (8, 4)   **11.** above   **13.** no   **15.** no
17. no   **19.** yes   **21.** yes   **23.** no   **25.** 79°F
27. above
29–30.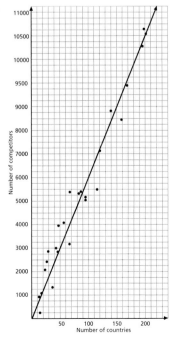

31. There is a linear direct correlation between the number of countries and number of competitors.
33. -9, -3, -2, 0, 4
35. 3

Chapter 7 Review, pages 342–344

1. c   **3.** g   **5.** j   **7.** k   **9.** l   **11.** B   **13.** 12 P.M.–1 P.M., 4 P.M.–
5 P.M.   **15.** 120 min   **17.** $(-3, -4)$ **19.** $(3, -4)$ **21.** yes
23. $D = \{-1, 0, 3, 8\}$, $R = \{4, 8, -2, 0\}$; yes   **25.** $\dfrac{7}{2}$, 7
27. insurance cost of $20   **29.** a: 0; b: negative; c: positive;
d: undefined   **31.** $m = 1$, b $= 2$, $y = x + 2$ **33.** $y = -\dfrac{5}{3}x$
35. about 3.16   **37.** yes   **39.** no   **41.** yes

Chapter 8: Relationships in Geometry

Lesson 8-1, pages 352–355

1. $\angle 1$, $\angle 3$, $\angle 7$ **3.** $\angle 1$ and $\angle 3$, $\angle 2$ and $\angle 4$, $\angle 5$ and $\angle 7$,
$\angle 6$ and $\angle 8$ **5.** equal, vertical   **7.** 35°   **9.** 125°   **11.** 30°
13. The creases are perpendicular. The angles created are all right angles.   **15.** 110°   **17.** 70°   **19.** 95°   **21.** 95°
23. If a transversal intersects a pair of lines at their intersection, the rules governing a transversal system do not follow.   **25.** $m\angle 1 = 22°$, $m\angle 2 = 68°$, $m\angle 3 = 68°$,
$m\angle 4 = 90°$ **27.** $m\angle 1 = 135°$, $m\angle 3 = 135°$, $m\angle 4 = 45°$
29. $m\angle QRN = 60°$; $m\angle PRN = 120°$ **31.** $\angle 6$ and $\angle 8$
33. Alternate interior angles are congruent. Alternate exterior angles are congruent.

35. $\overline{RT}$ 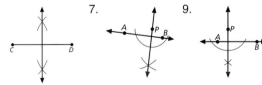   **37.** $\angle XYZ$

39. •v **41.** $\{-2, -1, 0, 1, 2\}$

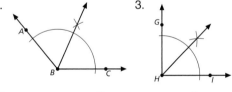

Lesson 8-2, pages 356–359

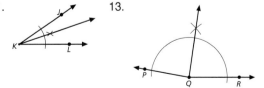

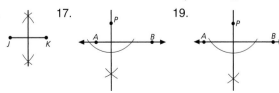

21. Constructions should show adjacent angles congruent to angles drawn at the beginning.

23.   25.

27. The perpendicular bisectors of each triangle intersect at the center of the circle within which each triangle is inscribed.

31. 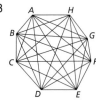   33. −3, 7   35. 7, −3

Review and Practice Your Skills, pages 360–361

1. 40° 3. 140° 5. 120° 7. 120° 9. 105° 11. 75°
13. 70° 15. 90° 17. 45° 19. 135° 21. 45°
29. 140° 31. 140° 33. 140° 35. 90° 37. 35°

Lesson 8-3, pages 362–365

1. 9, 720° 3. 14, 900° 5. 47° 7. 109° 9. 3, 540°, 5
11. 5, 900°, 14 13. 7, 1260°, 27 15. 120°

17.   19. A polygon with n sides has $n − 3$ diagonals and $n − 2$ nonoverlapping triangles. 21. 95° 23. 97 diagonals, 98 nonoverlapping triangles

25. 28 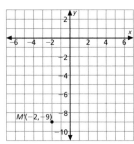   27. $m\angle R = 83°$, $m\angle S = 85°$, $m\angle Q = 131°$, $m\angle P = 61°$ 29. $m\angle L = 113°$, $m\angle O = 34°$, $m\angle P = 103°$, $m\angle W = 110°$
31. 720°, 9
33. 84

Lesson 8-4, pages 366–367

1.   3. Answers will vary, but are found by multiplying the number of seconds by 24.
5. isosceles triangle 7. isosceles triangle
9. 16 cm² 11. Observe students' work.
13. Answers will vary. 15. 17.49 units 17. 20.44 units; scalene 19. acute scalene triangle

Review and Practice Your Skills, pages 368–369

For 1–5, vertices may vary.
1. 540°   3. 720°

5. 540° 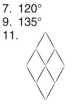   7. 120°
9. 135°
11.    13. hexagon 48 in.²
15. 110°
17. 70°

19. See Lesson 8–2 for examples.
21. 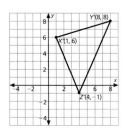   23.

Lesson 8-5, pages 370–373

1.   3.

5.   7.

For 9–13, see Exercises 1–4 for examples. 9. $Q'(−4, 3)$
11. $A'(−6, −6)$, $B'(−4, 4)$, $C'(−3, −4)$ 13. $D'(−8, 3)$, $E'(2, 3)$, $F'(0, −3)$, $G'(−6, −3)$
15.   17. 4 right, 3 down
19. 5 right, 2 down
21–23. Answers will vary.

25.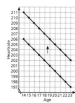

27. Answers will vary.

29.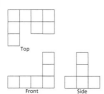

Lesson 8-6, pages 374–377

1. (4, 5) **3.** (2, 0) **5.** $G'(-1, -3)$, $H'(-3, -2)$, $I'(-5, -4)$;

7. 2 lines of symmetry

9. 1 line of symmetry

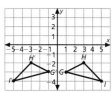

11. $(-2, -5)$
13. $(-7, -2)$ **15.**

17.

19. no **21.** Answers will vary. **23.** I, O, X
25. equilateral **27.** Pico is correct because a reflection across the *x*-axis results in an opposite *y*-value, and $0 \cdot -1 = 0$.

29. Answers will vary.

Equilateral triangle

31.

35. yes **37.** no
39. $-\dfrac{3}{4}$ **41.** yes

Review and Practice Your Skills, pages 378–379

1. $A'(0, -7)$ **3.** $C'(3, -3)$ **5.** $A'(-2, 0)$, $B'(-1, 3)$, $C'(2, -5)$ **7.** $X'(-2, 4)$, $Y'(-2, -3)$, $Z'(4, -2)$
9. $(-1, -3)$ **11.** $(5, 2)$ **13.** $(2, 0)$ **15.** $(2, 6)$
17. $(-4, -1)$ **19.** $(0, -2)$ **21.** $H'(-2, 3)$, $J'(-6, 1)$, $K'(-4, 6)$ **23.** $A'(4, -1)$, $B'(4, 4)$, $C'(-1, 4)$, $D'(-1, -1)$
25. $E'(-2, 4)$, $F'(-2, -3)$, $G'(-5, -3)$, $H'(-5, 4)$
27. no **29.** yes **31.** 105° **33.** 105° **35.** 540°
37. See Lesson 8-6 for examples. $T'(-5, -3)$, $R'(-3, 0)$, $S'(3, -7)$

Lesson 8-7, pages 380–383

1.   **3.** 3 **5.** 4 **7.** yes

9.   **11.** 1 **15.** yes;
13. 2

17.

Original frame   First rotation   Second rotation

19. Answers will vary. **21–23.** Answers will vary.
25.   **27.**   **29.** rectangular pyramid

Chapter 8 Review, pages 384–386

1. c **3.** j **5.** k **7.** h **9.** g **11.** $\angle DBA$ and $\angle ABE$; $\angle FBC$ and $\angle ABE$ **13.** $\angle DBA$ and $\angle CBF$; $\angle DBF$ and $\angle ABC$ **15.** 48°
17. 138° **19.** 30° **21.** 150° **23.** 30° **27.** 9 **29.** 1800° **31.** 6
33. $A'(0, -6)$, $B'(2, 2)$, $C'(1, -4)$ **35.** 5 units to the left and 7 units down **37.** (3, 2)
39.

41. 16 **43.** 4 **45.** 4
47. $R'(-5, 3)$, $S'(-1, 0)$, and $T'(-3, 4)$

Chapter 9: Polynomials

Lesson 9-1, pages 394–397

1. $2m^4 + m^3 + 4m^2 + 5m$ **3.** $5x^3 - 4x^2 - 3x + 2$
5. like **7.** like **9.** Although the variables are the same, the exponents for *x* differ.
11. $3c^2 - 7c$ **13.** $\dfrac{1}{2}n^4 + \dfrac{1}{2}n^3 + \dfrac{3}{4}n - \dfrac{3}{8}$
15. $-26^2 + 46 + 7$ **17.** $5n^5 + 2n^3 - 8n$ **19.** $-4m^3 + m^2 + 2m$ **21.** $2a^4 + 3a^3 + 2a^2 + 2a + 3$ **23.** $-t^3 + 4t^2 - 11t + 21$ **25.** $-14s^5 - s^3 + 9s^2 + 8s + 2$
27. unlike **29.** unlike **31.** unlike **33.** $3a - 3$
35. $-5a - 2$ **37.** $12r^2 + r$ **39.** $12s^2 - 2s$
41. $-9h^3 + 22h^2 - 2$ **43.** $\dfrac{5}{4}y^2 + \dfrac{5}{4}y + \dfrac{2}{3}$

45. Answers will vary but should include that like terms have the same variable to the same exponent.
47. $3q + 10d + 4n$; \$1.95 **49.** $120c + 800d + 9150e$
51. $5x$ mi/h **53.** The expression involves division by the variable *y*. **55.** $5xz^3 - 3yz^3$ **57.** $5xy + 2x^2 - x$
59. not possible **61.** 3

63.

65. length: $(2x + 40)$ mi width: $8x$ mi
67. 1 1 69. 160 71. 0.147

Lesson 9-2, pages 398–401

1. $5x^2 + 3x + 4$ 3. $3y^2 + 2y - 1$ 5. $9m - 1$ 7. $2y + 3$
9. $c^2 + 2c - 14$ 11. $6y - 3$ 13. $7m - 7$ 15. $5x^2$
$+ 6$ 17. $\frac{1}{3}m - n$ 19. $4y^2 - 5y - 1$ 21. $6x^2 + 3x - 3$
23. $\frac{5}{6}x$ 25. $4z - 7$ 27. $k + 3$ 29. $-3z + 2y$ 31. $5p^2$
$- 2p + 14$ 33. xy 35. $17m^3 - m^2 - 5m$ 37. $14d$
39. $7x$ 41. $2x - 5y - 90$ ft 43. $102d + 383r + 289f$
45. $97d + 401r + 326p$ 47. $5x + y$ 49. $4x + y$
51. $2p^2 - 6p + 11$ 53. $3x^2 - x - 2$ 55. $16abc$
57. $-3n^2 + 8n - 6$ 59. The resulting polynomial is
the opposite of the original polynomial. 61. 4 lb 9 oz
63. 18.33 65. $-\frac{6}{7}$

Review and Practice Your Skills, pages 402–403

1. $4t^2 + 2t - 8$ 3. $3h^2 - 5h + 8$
5. $-m^3 + 3m^2 - 2m + 4$ 7. $\frac{1}{3}g^4 + \frac{1}{2}g^2 - \frac{2}{3}g + \frac{1}{4}$
9. $d^5 + 0.2d^4 - 0.1d^3$ 11. like 13. unlike 15. like
17. unlike 19. $-g$ 21. $-4q$ 23. $6d - 6$ 25. $2y^2 -$
$2y$ 27. $2w^3 - 2w^2 + 2w$ 29. $0.3x^2 + 1.2x$ 31. $2d + 2$
33. $6m + 4$ 35. $-x - 7$ 37. $2x$ 39. $-4z^2 + 2z + 5$
41. $5p - 2$ 43. $-m^2 + m$ 45. $-4z + 5$ 47. $-4y^2 +$
$2y$ 49. $7g + 3h - 2$ 51. $-4p^2 - 2p + 7$ 53. like
55. unlike 57. like 59. $7n - 2$ 61. $9b$ 63. $2w^2 + 3w$
65. $-3y^3 + 7y^2 + y$ 67. $a^3 + 4a$ 69. $1.5h^2 + 0.5h +$
3.5 71. $-f - 2$ 73. $-4x^2 + 2x - 3$ 75. $6r - 3$
77. $-w^2 - 3w$ 79. $3e^3 - 2e^2 - 3e - 5$

Lesson 9-3, pages 404–407

1. $9xy$ 3. $-8gh$ 5. $-2y^3$ 7. $8y^3$ 9. $-m^6$ 11. $9x^4$
13. $6ab$ 15. $-12km$ 17. $-6kp$ 19. $7cd$ 21. $4a^2$
23. $-12y^5$ 25. $27m^6$ 27. $-10b^4$ 29. $6a^5$ 31. $-8p^8$
33. $243y^{20}$ 35. $-27x^6$ 37. $24x^5$ 39. $4a^4$ 41. $25y^{10}$
43. $12a^6b^3$ 45. $-3p^5r^5$ 47. $(2m)(3p) = 6mp$
49. Charts will vary. Examples will vary. Charts should
include the Product Rule, Power Rule, and Power of a
Product Rule. 51. $15x^y - 10x$ 53. 60 units^2
55. $\frac{1}{8}$ 61. $27a^2$ 63. $\frac{9}{2}p^3$ 65. $27b^3$ 67. 0.75 mi/min
69. 32 words/min

Lesson 9-4, pages 408–411

1. $4x^3 + 28x^2$ 3. $3g^4 - 21g^2$ 5. $12k^3 + 6k^2$ 7. $-3x^4$
$+ 3x^2$ 9. $-22c^3 + 10c^2$ 11. $\frac{1}{2}x^2 - \frac{1}{2}x$ 13. Using the
distributive property, multiply each term in the polynomial
by the monomial. 15. No. $x^2(3x) = 3x^3 \rightarrow 3x^3 + 2x^2$
17. $-8y^2 + 4y$ 19. $3x^2 + 5x$ 21. $-4y - 8$ 23. $k^3 -$
$4k^2 + k$ 25. $-12m^2 + 21m$ 27. $3x^3 + 12x^2 - 9x$

29. $2a^2 + 10a - 2$ 31. $-5y^3 + 15y^2 + 40y$ 33. $-2t^3$
$- 8t^2 + 24t$ 35. $4p^3 - 12p^2 - 8p$ 37. $-5a^3 - a^2 + 4a$
39. $-4a^4 - 12a^3 + 36a^2$ 41. $3y(y + 5)$, $3y^2 + 15y$
43. $x^3(x + 5)$, $x^4 + 5x^3$ 45. \$13.75 47. Both rectangles
have the same area, $y^2 + y$. The length and width of each
rectangle are different by 1 unit. 49. $8x^2 + 24x$
51. $-4p + 12$ 53. $x^3 + 3x^2y - 7xy^2$ 55. $16x^3y^2 -$
$12x^2y^3 + 28xy^4$ 57. $(25t + 24)$ sec 59. 92,814,000 mi
61. $4m(6m - 3)$, $24m^2 - 12m$ 63. 5.1 ft^2 65. 1.8 m^2
67. 1.5%

Review and Practice Your Skills, pages 412–413

1. $-8ab$ 3. $15st$ 5. $-3pq$ 7. $-3h^2$ 9. $-4y^4$
11. $-4r^5$ 13. $25b^2$ 15. $-10w^3$ 17. $-8b^3$
19. $2e^7$ 21. $-12a^9$ 23. $-75s^3$ 25. $20g^4$
27. $-32e^7$ 29. $36k^8$ 31. $18m^4n^3$ 33. $16c^5d^6$
35. $36p^{10}q^8$ 37. $16g^2 + 8g$ 39. $6z^3 + 12z$
41. $-6a^3 + 9a^2$ 43. $4k^2 - 8k$ 45. $18x^3 + 12x^2$
47. $10c^2 - 5c$ 49. $6r^3 - 3r^2 + 9r$ 51. $10g^3 - 6g^2 -$
$8g$ 53. $-4d^3 + 8d^2 + 12d$ 55. $8e^3 - 12e^2 + 12e$
57. $-12y^3 + 3y^2 + 18y$ 59. $-9j^3 - 6j^2 - 18j$
61. $-10x^3 + 6x^2 + 12x$ 63. $-8w^3 + 2w^2 + 6w$
65. $-40m^3 + 8m^2 + 28m$ 67. $12p^4 - 6p^3 + 4p^2$
69. $-20f^3 - 12f^2 - 8f$ 71. $6a^4 + 12a^3 + 18a^2$
73. $6x(x^2 + 4x + 8) = 6x^3 + 24x^2 + 48x$ 75. $-e^2 -$
$6e$ 77. $4a^2 + 4a + 3$ 79. $5x^3 - 2x^2$ 81. $b + 1$
83. $6x - 3$ 85. $2x^2 + 3x + 3$ 87. $-6ab$ 89. $18w^6$
91. $32m^5n^5$ 93. $3d^2 - 6d$ 95. $-8m^3 + 12m^2$
97. $15h^3 - 6h^2 + 12h$

Lesson 9-5, pages 414–417

1. $3y$ 3. $8x^2$ 5. $4(x + 3)$ 7. $y(yz - 9)$ 9. $(a + 2)$
$(a - 3)$ 11. $\frac{1}{4}(a + b + c + d)$ 13. $4y$ 15. $5ab$
17. $5s^2t$ 19. d 21. b 23. $6(1 - 2x)$ 25. $2x(x - 3)$
27. $9(1 - 2x)$ 29. $y(4x - y)$ 31. $4(n^2 - 4)$
33. $5mn(7 - 3mn)$ 35. $14m(2m - n)$
37. $3(2y^2 - 4y + 5)$ 39. $y(x + xy - y)$
41. $8ab(1 - 2a + 4b)$ 43. $3mn(9mn^2 - 6n - m)$
45. $(3 + k)(k - 1)$ 47. $(y + 7)(12 - y)$ 49. $z(z - 5)$
$(z - 1)$ 51. $-3a^2$ 53. $4y^2$ 55. $\frac{1}{3}xyz(1 - y - xz)$
57. $4(rs + st - tu + ru)$ 59. 12,277.4 yd^2
61. $r^2(\pi - 2)$ 63. 210, 290, 320 65. -20 67. 36

Lesson 9-6, pages 418–421

1. $-3x$ 3. -2 5. $9x^5y$ 7. b^8 and b^5 have the same
base. 9. $2y - 4$ 11. $a^2 + 2a + 6$ 13. $x^3 - 2x - 3$
15. $9h$ 17. $-6b$ 19. $3s$ 21. $4z$ 23. y 25. cd
27. ab^2 29. $4a^6$ 31. $-gh$ 33. $\frac{x^2}{z}$ 35. $a + 6$
37. $3a - 6b$ 39. $9x^2 - 3x + 1$ 41. $2x^2 - 3x + 1$
43. $3abc - 5c$ 45. length = $9a$ units 47. $2bc$
49. $7x^5y^4z^2$ 51. $4x^2 \div (-2x) = -2x$ 53. $8a^2 - 4b +$
$6ab^2$ 55. $2z - 3x + 4xy - 6zx^2y^2$ 57. $3xyz$
59. 163.1 in. 61. 42.7 in.

Review and Practice Your Skills, pages 422–423

1. $3d$ 3. $5c$ 5. $5m$ 7. $3g^2h$ 9. $3st$ 11. a
13. $3(3w + 1)$ 15. $5d(1 - 4d)$ 17. $5cd(2c - d)$

19. $x(-z + 1 - xz)$ 21. $ac(a - c + ac)$ 23. $3y(x + 4x^2 - 2y)$ 25. $(4 - n)(n + 2)$ 27. $(5 + 2z)(3 - z^2)$
29. $3c^2$ 31. $-2a^2b$ 33. $-2x$ 35. $5gh$ 37. $3a^2x$
39. $2ab^2$ 41. $3m^2 - 2n$ 43. $3t^2 - 4t$ 45. $3c - c^2$
47. $1 - 5x + 4x^2$ 49. $2m$ 51. $p^2 - 8pq + 10q^2$
53. $-8x^2 + 4x + 4$ 55. $a - 2b + 3$ 57. $-40f^5$
59. $10y^3 - 15y$ 61. $10m^4 + 6m^3 - 8m^2$ 63. $3(2xy + x - 3y)$ 65. $(2 + c)(c + 3)$ 67. $(3w^2 - 5)(2 - w)$

Lesson 9-7, pages 424–425

1. $x(x + 3) = x^2 + 3x$ 3. $-2y(y + 1) = -2y^2 - 2y$
5.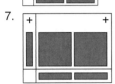
7. 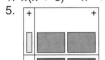   9. 6.48 ft²
11.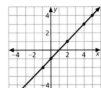
13. 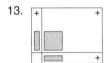   15.
17. $(x^2 + 10x)$m²   19. $(x^2 + x)$ ft²   21. $2y^2$ ft²
23. $(4x^2 + 4x)$ ft²   25. 4.1 m

Chapter 9 Review, pages 426–428

1. e   3. i   5. h   7. j   9. k   11. $8n$ 13. $-2x^3 + 12x^2 + 10x$ 15. $9r^2 + 8r$ 17. $-2p^2 - 2p$ 19. $8t^3 - t^2 - 9t$
21. $17n + 36$ 23. $6x^2 + 5x$ 25. $-m^2 + 2m + 8$ 27. $-n^3 + 7n^2$ 29. $3h + 9$ 31. -8 33. $7g^2 - 3g - 6$ 35. $4x^2 + x + 5$ 37. $2w + 3$ 39. $2u^2 - 21u - 11$ 41. $2y$ 43. $-z^2 - 6z + 5$ 45. x^6y^6 47. $12xy$ 49. $12tw$ 51. $-6x^3y$
53. $-12u^3v^3$ 55. $-3j^4k^3l$ 57. $-16a^2b^3$ 59. $25x^6y^2z^4$
61. $9w^2x^6y^4$ 63. $-5a^7b^7$ 65. $3x^3 + 6x^2$ 67. $24y^5 - 72y^4 + 56y^3$ 69. $-9b^2 + 6b$ 71. $-6y^3 + 6y^2$ 73. $7c^3 + 3c^2 - 4c$ 75. $8a^2 + 8a^3 - 24a$ 77. $12d^3 - 24d^2 - 45d$
79. $10m^4 - 14m^3 + 16m^2$ 81. $-2m^3 - \frac{1}{4}m^2 + \frac{7}{4}m$
83. $6^x(xy - 4)$ 85. $(y + 5)(y - 6)$ 87. $4(d^2 + 4)$
89. $15cd(1 + 2cd)$ 91. $12xy(3y - 4x)$ 93. $-3a^2b$
95. $2x^2 + 3x + 7$ 97. $3y^2$ 99. $25\frac{1}{4}$ ft

Chapter 10: Probability

Lesson 10-1, pages 436–439

1. $\frac{1}{2} = 50\%$ 3. $\frac{1}{3} = 33\frac{1}{3}\%$ 5. $\frac{5}{6} = 83\frac{1}{3}\%$ 7. 1 to 2
9. $\frac{2}{7}$ 11. 0   13. $P(\text{product even}) = \frac{3}{4}$, $P(\text{sum even}) = \frac{1}{2}$
15. $\frac{1}{3} = 33\frac{1}{3}\%$ 17. $\frac{1}{2} = 50\%$ 19. $\frac{1}{2} = 50\%$ 21. $\frac{5}{6} = 83\frac{1}{3}\%$ 23. $1 = 100\%$ 25. $\frac{1}{2}$ 27. $\frac{2}{1}$ 29. $\frac{1}{3}$ 31. $\frac{1}{3}$
33. 1   35. $\frac{1}{5}$ 37. $\frac{1}{6}$ 39. $\frac{1}{2}$ 41. $\frac{2}{3}$ 43. $\frac{1}{2}$

45. Sample answer: It is possible, but it does not seem likely.   47. $\frac{3}{4}$ 49. Answers will vary but should be close to 50%.   51. 100%   53. 7   55. $-\frac{3}{4}$
57. 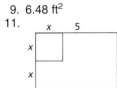   Tables will vary. Sample table:

x	y
0	−1
1	0
2	1

$y = x - 1$

Lesson 10-2, pages 440–443

1. $\frac{3}{20}$ 3. $\frac{1}{10}$ 5. $\frac{1}{5}$ 7. $\frac{11}{20}$ 9. $\frac{9}{20}$ 11. $\frac{7}{16}$ 13. $\frac{3}{4}$
15. Flip a coin 15 times. Add the probabilities of getting 12, 13, 14, and 15 heads.   17. $\frac{19}{30}$ 19. $\frac{5}{6}$ 21. $\frac{1}{5}$
23. $\frac{9}{10}$ 25. $\frac{9}{1000}$ 27. 1125   29. Answers will vary.
31. Yes; a larger sample provides a more accurate prediction.   33. 2   35. 90°   37. $\angle AXC$, $\angle CXD$, $\angle DXF$

Review and Practice Your Skills, pages 444–445

1. $\frac{1}{8}$, 12.5%   3. $\frac{1}{2}$, 50%   5. 0, 0%   7. 1 to 7   9. 1 to 1
11. $\frac{1}{6}$, 16.7%   13. $\frac{2}{6} = \frac{1}{3}$, 33.3%   15. $\frac{2}{3}$, 66.7%
17. 1 to 1   19. 1 to 5   21. $\frac{1}{6}$ 23. $\frac{1}{3}$ 25. $\frac{9}{50}$
27. $\frac{12}{25}$ 29. $\frac{1}{5}$ 31. $\frac{1}{5}$ 33. $\frac{3}{50}$ 35. $\frac{2}{5}$ 37. $\frac{1}{2}$, 50%
39. 1, 100%   41. $\frac{2}{3}$, 66.7%   43. 1 to 5   45. 1 to 5
47. $\frac{7}{50}$ 49. 0   51. $\frac{3}{50}$ 53. $\frac{21}{100}$

Lesson 10-3, pages 446–449

1. {H, T}   3. HHHH, HHHT, HHTH, HTHH, THHH, HHTT, HTHT, THHT, THTH, TTHH, HTTH, HTTT, THTT, TTHT, TTTH, TTTT (16 outcomes)
5. $\frac{1}{16}$ 7. 2 ways   9. {M, A, T, H, F, U, N}
11. 8 outcomes   13. 6 outcomes   15. $\frac{1}{18}$ 17. $\frac{2}{9}$
19. {A, B, C, D}   21. {A, B, C, D}   23. $\frac{1}{2}$ 25. Angles dividing sectors are all 90°. Areas of sectors are different.
27. $\frac{2}{5}$
29.
1	6	P	N		6	7	P	N
1	6	N	P		6	7	N	P
6	1	P	N		7	6	P	N
6	1	N	P		7	6	N	P
1	7	P	N		6	5	P	N
1	7	N	P		6	5	N	P
7	1	P	N		5	6	P	N
7	1	N	P		5	6	N	P
1	5	P	N		7	5	P	N
1	5	N	P		7	5	N	P
5	1	P	N		5	7	P	N
5	1	N	P		5	7	N	P

31. Answers will vary.

33. 205, 204, 203
    205, 203, 204
    204, 205, 203
    204, 203, 205
    203, 205, 204
    203, 204, 205

35. 6   37. 720   39. 1;
Explanations will vary.   41. $-7c$ $+ 8d - 15$ 43. $7j + 15 + 9jk$

45. $x = -4$ 47. $x = 5\frac{5}{6}$

Lesson 10-4, pages 450–453

1. 24   3. 216   5. $\frac{1}{216}$ 7. $\frac{1}{2^{10}}$ 9. $5 \cdot 3 \cdot 2 = 30$ ways

11. 12   13. 15   15. 8   17. 64   19. $\frac{1}{64}$ 21. $\frac{1}{7776}$

23. 17,576   25. 1,000,000   27. Answers will vary.

29. 50   31. 6   33. 2   35. 6   37. $\frac{1}{8}$ 39. $\frac{1}{8}$ 41. $\frac{1}{4}$

43. ←———•———•———→   45.
        N   P

47. Square, 4; rhombus, 4; equilateral triangle, 3; regular pentagon, 5

Review and Practice Your Skills, pages 454–455

1. {H, T}   3. {A, B, C, D}
5. 12   7. 16

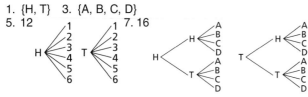

9. $\frac{1}{24}$ 11. $\frac{1}{12}$ 13. $\frac{1}{12}$ 15. 24   17. 26,000   19. 1296

21. $\frac{5}{52}$ 23. $\frac{16}{625}$ 25. $\frac{24}{36} = \frac{2}{3}$ 27. $\frac{3}{8}$ 29. $\frac{1}{6}$

31. $\frac{7}{20}$ 33. $6 \cdot 2 = 12$ 35. $2^8 = 256$

Lesson 10-5, pages 456–459

1. $\frac{8}{45}$ 3. $\frac{1}{9}$ 5. $\frac{1}{6}$ 7. Answers will vary.   9. $\frac{1}{150}$

11. $\frac{1}{30}$ 13. $\frac{3}{50}$ 15. $\frac{4}{75}$ 17. $\frac{1}{75}$ 19. $\frac{2}{75}$

21. independent   23. independent   25. 0   27. $\frac{2}{45}$

29. $\frac{1}{90}$ 31. Events become independent.   33. $\frac{24}{91}$

35. $\frac{5}{33}$ 37. $\frac{7}{22}$ 39. $\frac{1}{30}$ 41. $\frac{1}{6}$ 43. dependent

45. 0.096   47. $\frac{1}{6}$ 49. $\frac{1}{9}$ 51. pulling out a $20 bill and a $10 bill without replacement   53. pulling out a $20 bill, replacing it, then pulling out a $5 bill   55. 12 cm
57. 12.2 ft

Lesson 10-6, pages 460–461

1. Answers will vary.   3. 166,667 cans   5. wildlife shows
7. 25,000   9. 0.45   11. 0.16   13. 4500   15. 22,500

17. about $\frac{1}{6}$ of voters or 125,000   19. 112,500

21. Answers will vary.   23. Polling the school—it is a larger sample size.   25. $6x + 6y - 17$ 27. $-20g^3k - 80g^2k^2 + 120k^3$

Review and Practice Your Skills, pages 462–463

1. $\frac{1}{100}$ 3. 0   5. $\frac{1}{4}$ 7. $\frac{1}{10}$ 9. $\frac{1}{4}$ 11. $\frac{1}{10}$ 13. $\frac{1}{15}$

15. $\frac{1}{3}$ 17. $\frac{7}{15}$ 19. 10,000   21. 15,000   23. 25,000

25. 40,000   27. 5000   29. 0.36   31. 0.58   33. 0.36

35. 0.72   37. 11,000   39. 14,250   41. 25,250

43. 10,020   45. 7680   47. $\frac{7}{8}$ 49. $\frac{9}{50}$ 51. 0   53. 16

Lesson 10-7, pages 464–467

1. 2.5   3. $0.0004   5. 2   7. 3.5   9. 0.4 min   11. $0.60

13. $65,000   15. 5.6 jobs   17. $0.08   19. $1\frac{7}{8}$ points

21. no   23. Answers will vary but score will likely be greater than zero.

Chapter 10 Review, pages 468–470

1. f   3. k   5. c   7. l   9. h   11. $\frac{2}{5}$ 13. 1   15. $\frac{1}{8}$; $12\frac{1}{2}$%

17. $\frac{1}{2}$; 50%   19. $\frac{5}{8}$; $62\frac{1}{2}$%   21. $\frac{7}{8}$; 87.5%   23. $\frac{3}{20}$

25. $\frac{7}{20}$ 27. $\frac{3}{10}$ 29.   H ——— H ——— HH
                                                      T ——— HT
                                               T ——— H ——— TH
                                                      T ——— TT

31. 16   33. 16

35. 24   37. $\frac{3}{8}$

39. $\frac{1}{8}$ 41. $\frac{3}{20}$

43. $\frac{3}{50}$ 45. $\frac{1}{9}$

47. $\frac{1}{9}$ 49. 15,000   51. $\frac{2}{15}$ 53. The game is fair because the expected value is 0.

Chapter 11: Reasoning

Lesson 11-1, pages 478–481

1. $\overline{GH}$ is longer than $\overline{EF}$.   3. The sides of the figure with vertices A, B, and C curve toward the center of the circle at their midpoints.   5. Answers will vary.   7. 24 blocks
9. $\overline{BC}$ is longer than $\overline{AB}$; false.   11. The dots above the lower line are closer together than the dots above the upper line; false.   13. The picture can be viewed as a candlestick or as two people looking at each other.
15. No. Justifications will vary.   17. No. Justifications will vary.   19. Answers will vary.   21. The top level appears to be directly above the bottom level. However, the front of the top level is connected to the back of the bottom level in what appears to be a vertical column.   23. Predictions will vary. A longer strip with two twists is formed.
27. $A'(-2, -1)$, $B'(2, -4)$, $C'(4, -2)$

Lesson 11-2, pages 482–485

1. 6; Even-numbered powers of 4 end in the digit 6.   3. 2; The ones digit of powers of 2 follow the pattern 2, 4, 8, 6, 2, 4, 8, 6 . . .   5. No; counterexamples will vary.   7. No. She did not draw her conclusion based on the observation of several examples.   9. 4; the ones digit of even powers of 2 follow the pattern 4, 6, 4, 6 . . . 11. 9; The ones digit of odd powers of 9 is 9.   13. odd   15. even   17. If the numbers being multiplied at n-digit numbers, the first $n - 1$ digits of the product will be 5's, the next n digits will be 4's, and the final digit will be a 5.   19. No. The pattern is to add 7, 8, 9, 10, . . ., to find the next number. Nicki added 9 twice. The correct answer is 38, 49, 61.   21. True; Examples will vary.   23. yes, 90°   25. no

Review and Practice Your Skills, pages 486–487

1. Possible false conjecture: Line segment ST is bent.
3. Possible false conjecture: The inner black circle on the right looks larger than the black circle on the left.
5. Possible false conjecture: The diagram shows a white equilateral triangle.   7. 4   9. 6   11. 1   13. 0
15. 7   17. even   19. negative   21. Possible false conjecture: Line segment EF is shorter than line segment GH.   23. Possible false conjecture: The black and white circles are not the same size.   25. 6   27. odd

Lesson 11-3, pages 488–491

1. If there is no gasoline in the tank, then the car will not run. If the car will not run there is no gasoline in the tank.
3. true   5. true   7. invalid;

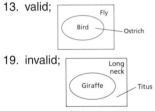

9. true   11. false; Counterexamples will vary.
13. valid;

15. If a number is even, it is divisible by 2.
17. If an animal is a whale, then it is a mammal.
19. invalid;

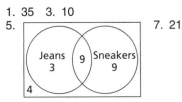

21. If a number is a whole number, then it is a rational number.

23. If an animal is a fish, then it is a sea creature. A shark is a fish. Therefore, a shark is a sea creature.
25. Answers will vary.   27. No; Sample answer: Let $a = 1$ and $b = 2$; then $1 * 2 = 1 + 2(2)$ or 5 but $2 * 1 = 2 + 2(1)$ or 4.   29. Conclusion: This dish is not wholesome. Argument: Puddings are nice. Nice things are not wholesome. This dish is a pudding. Therefore, this dish is not wholesome.   31. $8x$ 33. $24x^4$ 35. $\dfrac{1}{221}$

Lesson 11-4, pages 492–495

1. 35   3. 10
5.

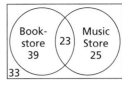

7. 21

9. 20   11. 6   13.

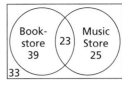

15. 25   17. 8
19. 12   21. 7
23. 8
25. 206
27. 22
29. Answers will vary.

31.

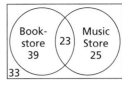

33. 38   35. 20   37. 144
39. Answers will vary, but should include that the picture on the top is not in the U.S. The headlights and taillights are on opposite sides of the streets.   41. 15

Review and Practice Your Skills, pages 496–497

1. false   3. true   5. false
7. invalid;   9. valid   11. If a number is even, then it is divisible by 2.   13. If a number is a power of 6, then it ends in the digit 6.
15. 23   17. 4   19. 9

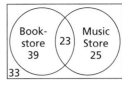

21.

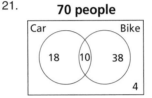

23. 10
25. Possible false conjecture: The diagonal lines are not parallel to each other.
27. valid

29.

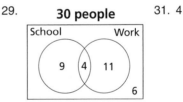

31. 4

Lesson 11-5, pages 498–501

1. Amy likes ice cream.   3. Brett will go golfing while on vacation.   5. Naomi earned 100% on the math test.
7. Terry is not married.   9. Tillie   11. Sue - Tom; DeeDee - Bill; Betty - Harry   13. Colette   15. Ely plays basketball, Amir plays football and Zeke plays baseball.
17. Queshia, Loraine or Hisa, Mila, Edna, Dana (greatest to least)   19. Answers will vary.   21. $\dfrac{1}{3}$ 23. $\dfrac{1}{18}$ 25. no

Lesson 11-6, pages 502–503

1. C   3. C   5. No. Although the ratio of their ages will approach 1, there will always be a 20 yr difference between their ages.   7. No. It is impossible to form a triangle with sides of 250 ft, 300 ft and 600 ft.
9. $4gh(4h - 7g)$ 11. $(t + 7)(x - 4)$ 13. $7m(2 - 7t^2)$
15. $-9y^4$ 17. $3d - 4c$ 19. $f^4 + 16f^5h^2 + 5f$

Review and Practice Your Skills, pages 504–505

1. Shantia will learn water skiing.   3. Julia is the oldest.
5. Kanya plays the clarinet.   7. Harriet likes biographies.

Fred prefers fiction. Wendy likes mysteries.   9. D   11. C
13. 1   15.

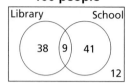

100 people

17. Ruth

Lesson 11-7, pages 506–509

1. 2:00 A.M.   3. 11 pieces   5. 35 triangles   7. two 16's
and four 17's   9. 20 posts   11. 6 cuts   13. 1 cube
15. 12 cubes   17. 0 cubes   19. Answers will vary.
21. No. Since he is already on the first floor, Miguel moves
seven stories to reach the eighth floor.   23. 14 squares
25. line   27. scatter

Chapter 11 Review, pages 510–512

1. d   3. h   5. f   7. k   9. c   11. The interior circles are
the same size; true   13. It could be a rabbit or a duck.
15. The result is always 6.   17. If it has been snowing,
then the streets are slippery. If the streets are slippery,
then it has been snowing.   19. valid   21. 183   23. 51
25. 0;

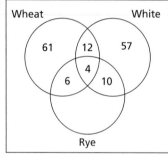

27. Rick; gardening, doctor; Carlos: painting, teacher;
Lee: fixing cars, lawyer   29. A
31.

Photo Credits

Index

Index

Index

Index

Index

■ Q ■

■ R ■

of cylinders, 194–197
of prisms, 194–197
of sides of pyramids, 191
Surveying, 3, 45, 97
Surveys, 6
Symbols on pictographs, 24
Symmetry, 170 line, *See* Line
    symmetry
  lines of, 375
  rotational, 381
Systematic sampling, 6